Student Solutions Manual

Daniel S. Miller

Niagara County Community College

Thinking Mathematically

 Edition

 Blitzer

PEARSON

Prentice Hall

Upper Saddle River, NJ 07458

Editor-in-Chief: Sally Yagan
Acquisitions Editor: Chuck Synovec
Supplement Editor: Joanne Wendelken
Executive Managing Editor: Kathleen Schiaparelli
Assistant Managing Editor: Karen Bosch Petrov
Production Editor: Jessica Barna
Supplement Cover Manager: Paul Gourhan
Supplement Cover Designer: Victoria Colotta
Manufacturing Buyer: Ilene Kahn
Manufacturing Manager: Alexis Heydt-Long

© 2008 Pearson Education, Inc.
Pearson Prentice Hall
Pearson Education, Inc.
Upper Saddle River, NJ 07458

Printed in the United States of America

10 9 8 7 6 5 4 3 2 1

ISBN 0-13-615539-1
      0-13-175220-0

Pearson Education Ltd., *London*
Pearson Education Australia Pty. Ltd., *Sydney*
Pearson Education Singapore, Pte. Ltd.
Pearson Education North Asia Ltd., *Hong Kong*
Pearson Education Canada, Inc., *Toronto*
Pearson Educación de Mexico, S.A. de C.V.
Pearson Education—Japan, *Tokyo*
Pearson Education Malaysia, Pte. Ltd.

Contents

Chapter 1
Problem Solving and Critical Thinking

Check Points 1.1

1. **a.** Add 6 each time.
$3 + 6 = 9$
$9 + 6 = 15$
$15 + 6 = 21$
$21 + 6 = 27$
$27 + 6 = 33$
$3, 9, 15, 21, 27, \underline{33}$

   b. Multiply by 5 each time.
$2 \times 5 = 10$
$10 \times 5 = 50$
$50 \times 5 = 250$
$250 \times 5 = 1250$
$2, 10, 50, 250, \underline{1250}$

   c. Cycle multiplying by 2, 3, 4.
$3 \times 2 = 6$
$6 \times 3 = 18$
$18 \times 4 = 72$
$72 \times 2 = 144$
$144 \times 3 = 432$
$432 \times 4 = 1728$
$1728 \times 2 = 3456$
$6, 18, 72, 144, 432, 1728, \underline{3456}$

   d. Cycle adding 8, adding 8, subtracting 14.
$1 + 8 = 9$
$9 + 8 = 17$
$17 - 14 = 3$
$3 + 8 = 11$
$11 + 8 = 19$
$19 - 14 = 5$
$5 + 8 = 13$
$13 + 8 = 21$
$21 - 14 = 7$
$9, 17, 3, 11, 19, 5, 13, 21, \underline{7}$

2. **a.** Starting with the third number, each number is the sum of the previous two numbers, $29 + 47 = 76$

   b. Starting with the second number, each number one less than twice the previous number, $2(129) - 1 = 257$

3. The shapes alternate between rectangle and triangle.
The number of little legs cycles from 1 to 2 to 3 and then back to 1.
Therefore the next figure will be a rectangle with 2 little legs.

4. **a.** Conjecture based on results: The original number is doubled.

Select a number.	4	10	0	3
Multiply the number by 4.	$4 \times 4 = 16$	$10 \times 4 = 40$	$0 \times 4 = 0$	$3 \times 4 = 12$
Add 6 to the product.	$16 + 6 = 22$	$40 + 6 = 46$	$0 + 6 = 6$	$12 + 6 = 18$
Divide this sum by 2.	$22 \div 2 = 11$	$46 \div 2 = 23$	$6 \div 2 = 3$	$18 \div 2 = 9$
Subtract 3 from the quotient.	$11 - 3 = 8$	$23 - 3 = 20$	$3 - 3 = 0$	$9 - 3 = 6$
Summary of results:	$4 \rightarrow 8$	$10 \rightarrow 20$	$0 \rightarrow 0$	$3 \rightarrow 6$

   b. Select a number: n
Multiply the number by 4: $4n$
Add 6 to the product: $4n + 6$
Divide this sum by 2: $\dfrac{4n + 6}{2} = \dfrac{4n}{2} + \dfrac{6}{2} = 2n + 3$
Subtract 3 from the quotient: $2n + 3 - 3 = 2n$

Exercise Set 1.1

1. Pattern: Add 4
$24 + 4 = 28$
$8, 12, 16, 20, 24, \underline{28}$

3. Pattern: Subtract 5
$17 - 5 = 12$
$37, 32, 27, 22, 17, \underline{12}$

5. Pattern: Multiply by 3
$243 \times 3 = 729$
3, 9, 27, 81, 243, <u>729</u>

7. Pattern: Multiply by 2
$16 \times 2 = 32$
1, 2, 4, 8, 16, <u>32</u>

9. Pattern: 1 alternates with numbers that are multiplied by 2
$16 \times 2 = 32$
1, 4, 1, 8, 1, 16, 1, <u>32</u>

11. Pattern: Subtract 2
$-4 - 2 = -6$
4, 2, 0, -2, -4, <u>-6</u>

13. Pattern: Add 4 to the denominator
$$\frac{1}{18+4} = \frac{1}{22}$$
$\frac{1}{2}, \frac{1}{6}, \frac{1}{10}, \frac{1}{14}, \frac{1}{18}, \frac{1}{\underline{22}}$

15. Pattern: Multiply the denominator by 3
$$\frac{1}{27 \times 3} = \frac{1}{81}$$
$1, \frac{1}{3}, \frac{1}{9}, \frac{1}{27}, \frac{1}{\underline{81}}$

17. Pattern: The second number is obtained by adding 4 to the first number. The third number is obtained by adding 5 to the second number. The number being added to the previous number increases by 1 each time. $33 + 9 = \underline{42}$

19. Pattern: The second number is obtained by adding 3 to the first number. The third number is obtained by adding 5 to the second number. The number being added to the previous number increases by 2 each time. $38 + 13 = \underline{51}$

21. Pattern: Starting with the third number, each number is the sum of the previous two numbers. $27 + 44 = \underline{71}$

23. Pattern: Cycle by adding 5, adding 5, then subtracting 7. $13 + 5 = \underline{18}$

25. Pattern: The second number is obtained by multiplying the first number by 2. The third number is obtained by subtracting 1 from the second number. Then multiply by 2 and then subtract 1, repeatedly. $34 - 1 = \underline{33}$

27. Pattern: Divide by -4
$$-1 \div (-4) = \frac{1}{4}$$
$64, -16, 4, -1, \frac{1}{\underline{4}}$

29. Pattern: The second value of each pair is 4 less than the first.
$3 - 4 = -1$
$(6,2), (0,-4), (7\frac{1}{2}, 3\frac{1}{2}), (2,-2), (3, \underline{-1})$

31. The figure cycles from square to triangle to circle and then repeats. So the next figure is

33. The pattern is to add one more letter to the previous figure and use the next consecutive letter in the alphabet. The next figure is shown at right.

d	d	d
d	d	

35. a. Conjecture based on results: The original number is doubled.

Select a number.	4	10	0	3
Multiply the number by 4.	$4 \times 4 = 16$	$10 \times 4 = 40$	$0 \times 4 = 0$	$3 \times 4 = 12$
Add 8 to the product.	$16 + 8 = 24$	$40 + 8 = 48$	$0 + 8 = 8$	$12 + 8 = 20$
Divide this sum by 2.	$24 \div 2 = 12$	$48 \div 2 = 24$	$8 \div 2 = 4$	$20 \div 2 = 10$
Subtract 4 from the quotient.	$12 - 4 = 8$	$24 - 4 = 20$	$4 - 4 = 0$	$10 - 4 = 6$
Summary of results:	4 → 8	10 → 20	0 → 0	3 → 6

b. $4n$

$4n+8$

$\dfrac{4n+8}{2}=\dfrac{4n}{2}+\dfrac{8}{2}=2n+4$

$2n+4-4=2n$

37. a. Conjecture based on results: The result is always 3.

Select a number.	4	10	0	3
Add 5 to the number.	$4+5=9$	$10+5=15$	$0+5=5$	$3+5=8$
Double the result.	$9\times2=18$	$15\times2=30$	$5\times2=10$	$8\times2=16$
Subtract 4.	$18-4=14$	$30-4=26$	$10-4=6$	$16-4=12$
Divide the result by 2.	$14\div2=7$	$26\div2=13$	$6\div2=3$	$12\div2=6$
Subtract the original number.	$7-4=3$	$13-10=3$	$3-0=3$	$6-3=3$
Summary of results:	$4\rightarrow3$	$10\rightarrow3$	$0\rightarrow3$	$3\rightarrow3$

b. $n+5$

$2(n+5)=2n+10$

$2n+10-4=2n+6$

$\dfrac{2n+6}{2}=\dfrac{2n}{2}+\dfrac{6}{2}=n+3$

$n+3-n=3$

39. Using inductive reasoning we predict $1+2+3+4+5+6=\dfrac{6\times7}{2}$.

Arithmetic verifies this result: $21=21$

41. Using inductive reasoning we predict $1+3+5+7+9+11=6\times6$.
Arithmetic verifies this result: $36=36$

43. Using inductive reasoning we predict $98765\times9+3=888,888$.
Arithmetic verifies this result:
$98765\times9+3=888,888$
$888,885+3=888,888$
$888,888=888,888$

45. The first multiplier increases by 33.
$132+33=165$
The second multiplier is 3367.
The product increases by 111,111.
$165\times3367=555,555$ is correct.

47. b; The resulting exponent is always the first exponent added to twice the second exponent.

49. c; The attendance started at approximately 1.6 and lost 0.06 per year.

51. deductive; The specific value was based on a general formula.

53. inductive; The general conclusion for all full-time four-year colleges was based on specific observations.

55. a. 1, 3, 6, 10, 15, and 21 are followed by
$21 + 7 = 28$
$28 + 8 = 36$
$36 + 9 = 45$
$45 + 10 = 55$
$55 + 11 = 66$
1, 3, 6, 10, 15, 21, 28, 36, 45, 55, and 66.

 b. $4 - 1 = 3$
$9 - 4 = 5$
$16 - 9 = 7$
$25 - 16 = 9$
The successive differences increase by 2.
$25 + 11 = 36$
$36 + 13 = 49$
$49 + 15 = 64$
$64 + 17 = 81$
$81 + 19 = 100$

 c. The successive differences are 4, 7, and 10. Since these differences are increasing by 3 each time. The next five numbers will be found by using differences of 13, 16, 19, 22, and 25.
$22 + 13 = 35$
$35 + 16 = 51$
$51 + 19 = 70$
$70 + 22 = 92$
$92 + 25 = 117$

 d. If a triangular number is multiplied by 8 and then 1 is added to the product, a <u>square</u> number is obtained.

61. The pattern suggests that the compatible expression is the square of the first number minus twice the product of the two numbers, plus the square of the second number.
$(11 - 7)^2 = 121 - 154 + 49$

63. Answers will vary. Possible answer: 5, 10, 15 or 5, 10, 20.

$5 \times 1 = 5$	$5 \times 2^0 = 5$
$5 \times 2 = 10$	$5 \times 2^1 = 10$
$5 \times 3 = 15$	$5 \times 2^2 = 20$

65. a. $6 \times 6 = 36$
$66 \times 66 = 4356$
$666 \times 666 = 443{,}556$
$6666 \times 6666 = 44{,}435{,}556$

 b. An additional digit of 6 is attached to the numbers being multiplied. An additional digit of 4 is attached to the left of the result and an additional digit of 5 is placed between the 3 and the 6.

 c. $66666 \times 66666 = 4{,}444{,}355{,}556$
$666{,}666 \times 666{,}666 = 444{,}443{,}555{,}556$

 d. Inductive reasoning; it uses an observed pattern and draws conclusions from that pattern.

Check Points 1.2

1. a. The digit to the right of the hundred millions digit is greater than 5. Thus, 295,734,134 rounded to the nearest hundred million is 300,000,000.

b. The digit to the right of the hundred thousands digit is less than 5. Thus, 295,734,134 rounded to the nearest hundred thousand is 295,700,000.

2. a. The digit to the right of the tenths digit is less than 5. Thus, 3.141593 rounded to the nearest tenth is 3.1.

b. The digit to the right of the ten-thousandths digit is greater than 5. Thus, 3.141593 rounded to the nearest ten-thousandth is 3.1416.

3. a. $\$2.40 + \$1.25 + \$4.60 + \$4.40 + \$1.40 + \$1.85 + 2.95 \approx \$2 + \$1 + \$5 + \$4 + \$1 + \$2 + 3 \approx \$18$

b. The bill of $21.85 is not reasonable. It is too high.

4. a. Round $52 per hour to $50 per hour and assume 40 hours per week.
$$\frac{40 \text{ hours}}{\text{week}} \times \frac{\$50}{\text{hour}} = \frac{\$2000}{\text{week}}$$
The architect's salary is $\approx \$2000$ per week.

b. Round 52 weeks per year to 50 weeks per year.
$$\frac{\$2000}{\text{week}} \times \frac{50 \text{ weeks}}{\text{year}} = \frac{\$100,000}{\text{year}}$$
The architect's salary is $\approx \$100,000$ per year.

5. a. $0.63 \times 178,586,613$

b. $0.6 \times 200,000,000 = 120,000,000$ non-owners

6. a. The yearly increase in life expectancy can be approximated by dividing the change in life expectancy by the change in time from 1952 to 2002. $\dfrac{79.9 - 71.1}{2002 - 1950} = \dfrac{8.8}{52} \approx 0.17$ yr for each subsequent birth year.

b.
$$\underset{\substack{\text{life expectancy} \\ \text{in 1950}}}{71.1} + \underset{\substack{\text{yearly} \\ \text{increase}}}{0.17} \underset{\substack{\text{number of years} \\ \text{from 1950 to 2050}}}{(2050 - 1950)} = 71.1 + 0.17(100)$$
$$= 71.1 + 17$$
$$= 88.1 \text{ yr}$$

7. a. about 118 democracies

b. The slowest rate of increase is indicated by the lowest slope of the graph. This occurs between 1981 and 1985.

c. There were 49 democracies in 1977.

8. a. The yearly increase can be approximated by dividing the change growth by the change in time from 1998 to 2004.
$$\frac{1.1 - 0.4}{2004 - 1998} = \frac{0.7}{6} \approx 0.12 \text{ million motorcycles per year}$$

b. Sales $M = \underset{\substack{\text{Sales in} \\ 1998}}{0.4} + \underset{\substack{\text{yearly} \\ \text{increase}}}{0.12} x$

    c.   2013 is 15 years after 1998. Thus, $0.4 + 0.12x = 0.4 + 0.12(15)$

$$= 0.4 + 1.8$$

$$= 2.2 \text{ million motorcycles}$$

Exercise Set 1.2

   1. **a.** 47,451,900

       b. 47,452,000

       c. 47,450,000

       d. 47,500,000

       e. 47,000,000

       f. 50,000,000

   3. 2.718

   5. 2.71828

   7. 2.718281828

   9. $350 + 600 = 950$
      Actual answer of 955 compares reasonably well

  11. $9 + 1 + 19 = 29$
      Actual answer of 29.23 compares quite well

  13. $32 - 11 = 21$
      Actual answer of 20.911 compares quite well

  15. $40 \times 6 = 240$
      Actual answer of 218.185 compares not so well

  17. $0.8 \times 400 = 320$
      Actual answer of 327.06 compares reasonably well

  19. $48 \div 3 = 16$
      Actual answer of 16.49 compares quite well

  21. 30% of 200,000 is 60,000
      Actual answer of 59,920.96 compares quite well

  23. $\$3.47 + \$5.89 + \$19.98 + \$2.03 + \$11.85 + \$0.23$

$$\approx \$3 + \$6 + \$20 + \$2 + \$12 + \$0$$

$$\approx \$43$$

  25. Round $19.50 to $20 per hour.
      40 hours per week
      (40 × $20) per week = $800/week
      Round 52 weeks to 50 weeks per year.
      50 weeks per year
      (50 × $800) per year = $40,000
      $19.50 per hour ≈ $40,000 per year

27. Round the $605 monthly payment to $600.
    3 years is 36 months.
    Round the 36 months to 40 months.
    $600 × 40 months = $24,000 total cost.
    $605 monthly payment for 3 years ≈ $24,000 total cost.

29. Round the raise of $310,000 to $300,000.
    Round the 294 professors to 300.
    $300,000 ÷ 300 professors = $1000 per professor.
    $310,000 raise ≈ $1000 per professor.

31. Round $61,500 to $60,000 per year.
    Round 52 weeks per year to 50 weeks per year.
    50 weeks × 40 hours per week = 2000 hours
    $60,000 ÷ 2000 hours = $30 per hour
    $61,500 per year ≈ $30 per hour

33. $80 \times 365 \times 24 = 700,800$ hr

35. $\dfrac{0.2 \times 100}{0.5} = \dfrac{20}{0.5} = 40$

    Actual answer of 42.03 compares quite reasonable.

37. The given information suggests $30 would be a good estimate per calculator.
    $30 \times 10 = \$300$ which is closest to choice b.

39. The given information suggests 65 mph would be a good rate estimate and 3.5 would be a good time estimate.
    $65 \times 3.5 = 227.5$ which is closest to choice c.

41. The given information suggests you can count 1 number per second.
    $\dfrac{10000}{60 \times 60} \approx 2.77$ or 3 hours

43. 50% of 200,000,000 is 100,000,000 American adults.

45. **a.** about 85 people per 100

    b. $(85 - 23) \times 87 \approx 5400$

47. **a.** $\dfrac{248 - 222}{2} = \$13$ per year

    b. $222 + 13(2010 - 2002) = \$326$

49. **a.** The maximum was reached in 1960. The amount that year was about 4100 cigarettes.

    b. The greatest rate of increase was from 1940 to 1950.

    c. The cigarette consumption was about 1500 in 1930.

51. a. $\dfrac{20,000-14,000}{2005-1998} \approx \$857$ per year

    b. $C = 14,000 + 857x$

    c. $C = 14,000 + 857(14) = \$25,998$ or $\$26,000$

67. a

69. b

71. $20 \times 16 \times 50 = 16,000$ hours .

$\dfrac{16,000}{24} \approx 667$ days

$\dfrac{667}{365} \approx 1.8$ yr

Check Points 1.3

1. The amount of money given to the cashier is unknown.

2. Step 1: Understand the problem.
Bottles: 128 ounces costs $5.39
Boxes: a 9-pack of 6.75 ounce boxes costs $3.15
We must determine whether bottles or boxes are the better value.
Step 2: Devise a plan.
Dividing the cost by the number of ounces will give us the cost per ounce. We will need to multiply 9 by 6.75 to determine the total number of ounces the boxes contain. The lower cost per ounce is the best value.
Step 3: Carry out the plan and solve the problem.
Unit price for the bottles: $\dfrac{\$5.39}{128 \text{ ounces}} \approx \$0.042$ per ounce
Unit price for the boxes: $\dfrac{\$3.15}{9 \times 6.75 \text{ ounces}} = \dfrac{\$3.15}{60.75 \text{ ounces}} \approx \$0.052$ per ounce
Bottles have a lower price per ounce and are the better value.
Step 4: Look back and check the answer.
This answer satisfies the conditions of the problem.

3. Step 1: Understand the problem.
We are given the cost of the computer, the amount of cash paid up front, and the amount paid each month. We must determine the number of months it will take to finish paying for the computer.
Step 2: Devise a plan.
Subtract the amount paid in cash from the cost of the computer. This results in the amount still to be paid. Because the monthly payments are $45, divide the amount still to be paid by 45. This will give the number of months required to pay for the computer.
Step 3: Carry out the plan and solve the problem.
The balance is $\$980 - \$350 = \$630$. Now divide the $630 balance by $45, the monthly payment.

$\$630 \div \dfrac{\$45}{\text{month}} = \$630 \times \dfrac{\text{month}}{\$45} = \dfrac{630 \text{ months}}{45} = 14$ months.

Step 4: Look back and check the answer.
This answer satisfies the conditions of the problem. 14 monthly payments at $45 each gives $14 \times \$45 = \$630$. Adding in the up front cash payment of $350 gives us $\$630 + \$350 = \$980$. $980 is the cost of the computer.

4. Step 1: Understand the problem.
   Step 2: Devise a plan.
   Make a list of all possible coin combinations. Begin with the coins of larger value and work toward the coins of smaller value.

   Step 3: Carry out the plan and solve the problem.

Quarters	Dimes	Nickels
1	0	1
0	3	0
0	2	2
0	1	4
0	0	6

   There are 5 combinations.
   Step 4: Look back and check the answer.
   Check to see that no combinations are omitted, and that those given total 30 cents. Also double-check the count.

5. Step 1: Understand the problem.
   We must determine the number of jeans/T-shirt combinations that we can make.
   For example, one such combination would be to wear the blue jeans with the beige shirt.
   Step 2: Devise a plan.
   Each pair of jeans could be matched with any of the three shirts. We will make a tree diagram to show all combinations.
   Step 3: Carry out the plan and solve the problem.

   There are 6 different outfits possible.
   Step 4: Look back and check the answer.
   Check to see that no combinations are omitted, and double-check the count.

6. Step 1: Understand the problem.
   There are many possible ways to visit each city once and then return home. We must find a route that costs less than $1460.
   Step 2: Devise a plan.
   From city A fly to the city with the cheapest available flight. Repeat this until all cities have been visited and then fly home. If this cost is above $1460 then use trial and error to find other alternative routes.
   Step 3: Carry out the plan and solve the problem.
   A to D costs $185, D to E costs $302, E to C costs $165, C to B costs $305, B back to A costs $500
   $185 + $302 + $165 + $305 + $500 = $1457
   The route A, D, E, C, B, A costs less than $1460
   Step 4: Look back and check the answer.
   This answer satisfies the conditions of the problem.

Trick Questions 1.3

1. The farmer has 12 sheep left since all but 12 sheep died.

2. All 12 months have [at least] 28 days.

3. The doctor and brother are brother and sister.

4. You should light the match first.

Exercise Set 1.3

1. The price of the computer is needed.

3. The number of words per page is needed.

5. Weekly salary is unnecessary information.
$212 - 200 = 12$ items sold in excess of 200
$12 \times \$15 = \$180$ extra is received.

7. How much the attendant was given is not necessary.
There were 5 hours of parking.
1st hour is $2.50
4 hours at $0.50/hr
$\$2.50 + (4 \times \$0.50) = \$2.50 + \$2.00$
$= \$4.50$
$4.50 was charged.

9. **a.** Step 1: Understand the problem.
Box #1: 15.3 ounces costs $3.37
Box #2: 24 ounces costs $4.59
We must determine whether Box #1 or Box #2 is the better value.
Step 2: Devise a plan.
Dividing the cost by the number of ounces will give us the cost per ounce. The lower cost per ounce is the best value.
Step 3: Carry out the plan and solve the problem.
Unit price for Box #1: $\dfrac{\$3.37}{15.3 \text{ ounces}} \approx \$0.22$ per ounce
Unit price for Box #2: $\dfrac{\$4.59}{24 \text{ ounces}} \approx \$0.19$ per ounce
The cereal that is 24 ounces for $4.59 is the better value.
Step 4: Look back and check the answer.
This answer satisfies the conditions of the problem.

    b. Unit price for Box #1: $0.22 per ounce
Unit price for Box #2: $\dfrac{\$4.59}{24 \text{ ounces}} \times \dfrac{16 \text{ ounces}}{\text{pound}} \approx \$3.06$ per pound

    c. No, explanations will vary.

11. Step 1: Comparing two yearly salaries
Step 2:
Convert the second person's wages to yearly salary.
Step 3:
The person that earns $3750/month earns
$12 \times \$3750 = \$45,000$/year. The person that earns $48,000/year gets $3000 more per year.
Step 4:
It appears to satisfy the conditions of the problem.

13. Step 1:
Find the difference between two methods of payment.
Step 2:
Compute total costs and compare two figures.
Step 3:
By spreading purchase out, the total comes to:
$100 + 14($50) = $100 + $700 = 800
$800 - $750 = 50 saved by paying all at once
Step 4:
It satisfies the conditions of problem.

15. Step 1:
Determine profit on goods sold.
Step 2:
Find total cost of buying product and comparing with gross sales.
Step 3:
Purchased: ($65 per dozen)(6 dozen) = 390
Sold: 6 dozen = 72 calculators
$\frac{72}{3} = 24$ groups of 3 at $20 per group.
$24 \times $20 = 480
$480 - $390 = 90 profit
Step 4:
It satisfies the conditions of the problem.

17. Step 1: Determine profit for ten-day period.
Step 2: Compare totals.
Step 3:
(200 slices)($1.50) = $300 for pizza
(85 sandwiches)($2.50) = $212.50 for sandwiches
For 10 day period:
Gross: $10($300) + 10($212.50) = $3000 + 2125.00
$= 5125.00
Expenses: 10($60) = $600
Profit: $5125.00 - $600 = 4525
Step 4:
It satisfies the conditions of the problem.

19. Step 1:
Compute total rental cost.
Step 2:
Add rental cost and mileage cost to get total cost.
Step 3:
Rental costs:
(2 weeks)($220 per week) = $440
Mileage: (500 miles)($0.25) = $125
Total: $440 + $125 = $565
Step 4:
It satisfies the conditions of problem.

21. Step 1:
A round trip was made; we need to determine how much was walked or ridden.
Step 2:
Add up the totals walked and ridden and compare.
Step 3:
It is 5 miles between the homes or a 10 mile round trip. The first 3 were covered with the bicycle, leaving 7 miles covered by walking.
7 miles − 3 miles = 4 miles more that was walked.
Step 4:
It satisfies the conditions of the problem.

23. Step 1:
Determine profit by comparing expenses with gross sales.
Step 2:
Calculate expenses and gross sales and compare.
Step 3:
Expense:
(25 calculators)($30) = $750
Gross Sales:
(22 calculators)($35.00) = $770
The storeowner receives $30 − $2 = $28 for each returned calculator.
(3 calculators)($28) = $84

Total Income:
$770 + $84 = 854
Profit = Income − Expenses
$= $854 - 750
$= 104
Step 4:
It satisfies the conditions of the problem.

25. The car depreciates at $\dfrac{23,000 - 2700}{7} = \$2900$ per year .

    $23,000 - 3(2900) = \$14,300$

27. Use a list.

2 Quarters	3 Dimes	5 Nickels
1	2	0
1	1	2
1	0	4
0	3	3
0	2	5

    There are 5 ways.

29. Make a list of all possible selections:
    Depp/Foxx, Depp/Stewart, Depp/Hilary,
    Foxx/Stewart, Foxx/Hilary,
    Stewart/Hilary
    There are 6 ways.

31. Use a list.

Pennies	Nickels	Dimes
21	0	0
16	1	0
11	2	0
11	0	1
6	3	0
6	1	1
1	4	0
1	2	1
1	0	2

    There are 9 ways.

33. Use a list.

1 pt	5 pt	10-pt	Total
3	0	0	3
2	1	0	7
1	2	0	11
2	0	1	12
0	3	0	15
1	1	1	16
0	2	1	20
1	0	2	21
0	1	2	25
0	0	3	30

There are 10 different totals.

35. The average expense is $\dfrac{42 + 10 + 26 + 32 + 30}{5} = \$28$

Thus, B owes $18 and C owes $2, A is owed $14, D is owed $4, and E is owed $2.
To resolve these discrepancies, B should give A $14 and give D $4, while C should give E $2.

37. Make a list of all possible orders:
TFFF, FTFF, FFTF, FFFT
The "True" could be written 1st, 2nd, 3rd, or 4th.
There are 4 ways.

39. The order the racers finished was; Andy, Darnell, Caleb, Beth, Ella.

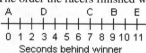

41. Home→Bank→Post Office→Dry Cleaners→Home will take 11.5 miles.

43. CO→WY→UT→AZ→NM→CO→UT

45. The problem states that the psychology major knocks on Jose's wall, and Jose's dorm is adjacent to Bob's dorm but not Tony's. Therefore Bob is the psychology major.

47. a.

5	22	18
28	15	2
12	8	25

b.

4	9	8
11	7	3
6	5	10

49.

9	6	7
0	1	4
3	2	5

51.

$$\begin{array}{r} 156 \\ 28\overline{)4368} \\ \underline{28} \\ 156 \\ \underline{140} \\ 168 \\ \underline{168} \\ 0 \end{array}$$

57. You should choose the dentist whose teeth show the effects of poor dental work because he took good care of the other dentist's teeth.

59. It is Friday. The first person is lying (as expected) because he told the truth on Thursday. The second is truthfully admitting that he lied the previous day.

61. Answers will vary.

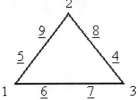

63. There is no missing dollar; in the end the customers paid a total of $27 of which $25 went to the restaurant and $2 was stolen by the waiter.

65. Answers will vary. One method is to start by multiplying 30 by each state's fraction of the population.

State A: $30 \times \dfrac{275}{1890} \approx 4.365$ or 4

State B: $30 \times \dfrac{383}{1890} \approx 6.079$ or 6

State C: $30 \times \dfrac{465}{1890} \approx 7.381$ or 7

State D: $30 \times \dfrac{767}{1890} \approx 12.175$ or 12

Notice that 4, 6, 7, and 12 add to 29, so there is 1 more representative to be allocated. We could give this extra representative to state C because it had the largest decimal part (0.381). This leads to an allocation of state A: 4, state B: 6, state C: 8, and state D: 12.

Chapter 1 Review Exercises

1. Deductive; the specific conclusion about *Carrie* was based on a general statement about all Stephen King books.

2. Inductive; the general conclusion for this next book was based on past specific observations.

3. Pattern: Add 5
$19 + 5 = 24$
4, 9, 14, 19, <u>24</u>

4. Pattern: Multiply by 2
$56 \times 2 = 112$
7, 14, 28, 56, <u>112</u>

5. Pattern: Numbers added increase by 1
$1 + 2 = 3$
$3 + 3 = 6$
$6 + 4 = 10$
$10 + 5 = 15$
$15 + 6 = 21$
1, 3, 6, 10, 15, <u>21</u>

6. Notice that $\dfrac{1}{2} = \dfrac{3}{6}$

Pattern: Add 1 to the denominator

$\dfrac{3}{7+1} = \dfrac{3}{8}$

$\dfrac{3}{4}, \dfrac{3}{5}, \dfrac{3}{6}, \dfrac{3}{7}, \dfrac{3}{\underline{8}}$ or

$\dfrac{3}{4}, \dfrac{3}{5}, \dfrac{1}{2}, \dfrac{3}{7}, \dfrac{3}{\underline{8}}$

7. Pattern: Divide by -2

$-5 \div (-2) = \dfrac{-5}{-2} = \dfrac{5}{2}$ or $2\dfrac{1}{2}$

$40, -20, 10, -5, \dfrac{5}{\underline{2}}$

8. Pattern: Subtract 60
    $-140 - 60 = -200$
    40, -20, -80, -140, $\underline{-200}$

9. Each number beginning with the third number is the sum of the previous two numbers. $16 + 26 = \underline{42}$

10. To get the second number, multiply the first number by 3. Then multiply the second number by 2 to get the third number. Then multiply by 3 and then by 2, repeatedly. $216 \times 2 = \underline{432}$

11. The pattern is alternating between square and circle while the line rotates 90° clockwise. The next figure is shown at right.

12. Using inductive reasoning we predict $2 + 4 + 8 + 16 + 32 = 64 - 2$.
    Arithmetic verifies this result:
    $2 + 4 + 8 + 16 + 32 = 64 - 2$
    $\qquad\qquad\qquad 62 = 62$

13. Using inductive reasoning we predict $444 \div 12 = 37$.
    Arithmetic verifies this result: $444 \div 12 = 37$
    $\qquad\qquad\qquad\qquad 37 = 37$

14. a. Conjecture based on results: The result is the original number.

Select a number.	4	10	0	3
Double the number.	$4 \times 2 = 8$	$10 \times 2 = 20$	$0 \times 2 = 0$	$3 \times 2 = 6$
Add 4 to the product.	$8 + 4 = 12$	$20 + 4 = 24$	$0 + 4 = 4$	$6 + 4 = 10$
Divide this sum by 2.	$12 \div 2 = 6$	$24 \div 2 = 12$	$4 \div 2 = 2$	$10 \div 2 = 5$
Subtract 2 from the quotient.	$6 - 2 = 4$	$12 - 2 = 10$	$2 - 2 = 0$	$5 - 2 = 3$
Summary of results:	4 → 4	10 → 10	0 → 0	3 → 3

   b. $2n$
       $2n + 4$
       $$\frac{2n + 4}{2} = \frac{2n}{2} + \frac{4}{2} = n + 2$$
       $n + 2 - 2 = n$

15. a. 4,115,600

    b. 4,116,000

    c. 4,100,000

    d. 4,000,000

16. a. 1.5

    b. 1.51

    c. 1.507

    d. 1.5065917

17. $2 + 4 + 10 = 16$
Actual answer: 15.71
quite reasonable

18. $9 \times 50 = 450$
Actual answer: 432.67
somewhat reasonable

19. $20 \div 4 = 5$
Actual answer: 4.79
quite reasonable

20. $0.60 \times 4000 = 2400$
Actual answer: 2397.0548
quite reasonable

21. $8.47 + $0.89 + $2.79 + $0.14 + $1.19 + $4.76
    $\approx$ $8 + $1 + $3 + $0 + $1 + $5
    $\approx$ $18

22. Round 78 hours to 80, round $6.85 to $7.00. $78 \times \$6.85 \approx 80 \times \$7.00 \approx \$560$

23. Round book price to $1.00 each.
Round chair price to $12.00 each.
Round plate price to $15.00.
$(21 \times \$0.85) + (2 \times \$11.95) + \$14.65$
    $\approx (21 \times \$1) + (2 \times \$12) + \$15$
    $\approx \$21 + \$24 + \$15$
    $\approx \$60$

24. 40% of 4,000,000 is 1,600,000 liberals.

25. The given information suggests $900 would be a good estimate for weekly salary.
$\$900 \times 10 \times 4 = \$36,000$ which is choice b.

26. $60 \times 60 \times 24 = 86,400$ which is closest to choice c.

27. a. called gay lesbian, or a homophobic name; women 13%; men 37%

b. 60% of 6,000,000 is 3,600,000 women

28. a. $\dfrac{45.8 - 39.8}{2004 - 2000} = 1.5$ million per year

b. $39.8 + 1.5(12) = 57.8$ million Americans

29. a. 1990; about 400,000 dropouts

b. 1985-1990

c. 1985

30. a. $\dfrac{3.0-1.7}{2004-1999} = 0.26$ billion per year

    b. $L = 1.7 + 0.26x$

    c. $L = 1.7 + 0.26(14) = \$5.43$ billion

31. The weight of the child is needed.

32. The unnecessary information is the customer giving the driver a \$20 bill.
For a 6 mile trip, the first mile is \$3.00, and the next 5 miles are \$0.50/half-mile or \$1.00/mile. The cost is
$$\$3.00 + (5 \times \$1.00) = \$3.00 + \$5.00$$
$$= \$8.00.$$

33. Total of $28 \times 2 = 56$ frankfurters would be needed. $\dfrac{56}{7} = 8$. Therefore, 8 pounds would be needed.

34. Rental for 3 weeks at \$175 per week is
$3 \times \$175 = \$525$. Mileage for 1200 miles at \$0.30 per mile is $1200 \times \$0.30 = \$360$. Total cost is $\$525 + \$360 = \$885$.

35.. Plan A is \$90 better.
Cost under Plan A: $100 + 0.80(1500) = \$1300$
Cost under Plan B: $40 + 0.90(1500) = \$1390$

36. The flight leaves Miami at 7:00 A.M. Pacific Standard Time. With a lay-over of 45 minutes, it arrives in San Francisco at 1:30 P.M. Pacific Standard Time, 6 hrs 30 min. – 45 min = 5 hours 45 minutes.

37. At steady decrease in value: $\dfrac{\$37,000 - \$2600}{8 \text{ years}} = \dfrac{\$34,400}{8 \text{ years}} = \$4300 / \text{ year}$
After 5 years: $\$4300 \times 5 = \$21,500$ decrease in value
Value of car: $\$37,000 - \$21,500 = \$15,500$

38. The machine will accept nickels, dimes, quarters.

nickels	dimes	quarters
7	0	0
5	1	0
3	2	0
2	0	1
1	3	0
0	1	1

There are 6 combinations.

Chapter 1 Test

1. deductive

2. inductive

3. $0 + 5 = 5$
   $5 + 5 = 10$
   $10 + 5 = 15$
   $15 + 5 = 20$
   $0, 5, 10, 15, \underline{20}$

4. $\dfrac{1}{6 \times 2} = \dfrac{1}{12}$

   $\dfrac{1}{12 \times 2} = \dfrac{1}{24}$

   $\dfrac{1}{24 \times 2} = \dfrac{1}{48}$

   $\dfrac{1}{48 \times 2} = \dfrac{1}{96}$

   $\dfrac{1}{6}, \dfrac{1}{12}, \dfrac{1}{24}, \dfrac{1}{48}, \underline{\dfrac{1}{96}}$

5. $3367 \times 15 = 50,505$

6. The outer figure is always a square. The inner figure appears to cycle from triangle to circle to square. The line segments at the bottom alternate from two to one. The next shape is shown at right.

7. **a.** Conjecture based on results: The original number is doubled.

Select a number.	4	10	3
Multiply the number by 4.	$4 \times 4 = 16$	$10 \times 4 = 40$	$3 \times 4 = 12$
Add 8 to the product.	$16 + 8 = 24$	$40 + 8 = 48$	$12 + 8 = 20$
Divide this sum by 2.	$24 \div 2 = 12$	$48 \div 2 = 24$	$20 \div 2 = 10$
Subtract 4 from the quotient.	$12 - 4 = 8$	$24 - 4 = 20$	$10 - 4 = 6$
Summary of results:	$4 \rightarrow 8$	$10 \rightarrow 20$	$3 \rightarrow 6$

   b. $4n$
   $4n + 8$
   $\dfrac{4n + 8}{2} = \dfrac{4n}{2} + \dfrac{8}{2} = 2n + 4$
   $2n + 4 - 4 = 2n$

8. 3,300,000

9. 706.38

10. Round $47.00 to $50.00.
    Round $311.00 to $310.00.
    Round $405.00 to $410.00.
    Round $681.79 to $680.00.
    Total needed for expenses:
    $47.00 + $311.00 + $405.00
    $\approx$ $50.00 + $310.00 + $410.00
    $\approx$ $770.00
    Additional money needed:
    $770.00 - $681.79 $\approx$ $770.00 - $680.00
    $\approx$ $90

11. Round \$485,000 to \$500,000.
    Round number of people to 20.
    $$\frac{\$485,000}{19 \text{ people}} \approx \frac{\$500,000}{20 \text{ people}}$$
    $$\approx \$25,000 \text{ per person}$$

12. $0.48992 \times 120 \approx 0.5 \times 120 \approx 60$

13. 70% of 40,000,000 is 28,000,000 cases.

14. $72,000 \div 30 = 2400$ which is choice a.

15. **a.** 2001; about 1220 discharges

    b. 1996

    c. $600 + 750 + 850 + 1000 + 1150 + 1050 + 1200 + 1200 + 900 + 800 = 9500$ discharges

16. **a.** $\dfrac{20.4 - 37.4}{2004 - 1970} = -0.5\%$ per year

    b. $S = 37.4 - 0.5x$

    c. $S = 37.4 - 0.5(40) = 17.4\%$

17. For 3 hours:
    Estes: \$9 per $\dfrac{1}{4}$ hour
    $3 \times 4 = 12$ quarter-hours $\rightarrow$ $12 \times \$9 = \$108$
    Ship and Shore: \$20 per $\dfrac{1}{2}$ hour
    $3 \times 2 = 6$ half-hours $\rightarrow$ $6 \times \$20 = \$120$
    Estes is a better deal by
    $\$120 - \$108 = \$12.00$.

18. 20 round trips mean 40 one-way trips at \$11/trip.
    (40 trips)(32 passengers)(\$11)
    = \$14,080 in one day

19. $\$960 - \$50 = \$910$ remaining to pay
    $$\frac{\$910}{\$35 \text{ per week}} = 26 \text{ weeks}$$

20. Belgium will have 160,000 more.
    Greece: $10,600,000 - 28,000(35) = 9,620,000$
    Belgium: $10,200,000 - 12,000(35) = 9,780,000$

Chapter 2
Set Theory

Check Points 2.1

1. Set L is the set of the first six lowercase letters in the English alphabet.

2. $M = \{$April, August$\}$

3. $O = \{1, 3, 5, 7, 9\}$

4. **a.** not the empty set; Many numbers meet the criteria to belong to this set.

   b. the empty set; No numbers meet the criteria, thus this set is empty

   c. not the empty set; "nothing" is not a set.

   d. not the empty set; This is a set that contains one element, that element is a set.

5. **a.** True; 8 is an element of the given set.

   b. True; r is not an element of the given set.

   c. False; {Monday} is a set and the set {Monday} is not an element of the given set.

6. **a.** $A = \{1, 2, 3\}$

   b. $B = \{15, 16, 17, \ldots\}$

   c. $O = \{1, 3, 5, \ldots\}$

7. **a.** $\{1, 2, 3, 4, \ldots, 199\}$

   b. $\{51, 52, 53, 54, \ldots, 200\}$

8. **a.** $n(A) = 5$; the set has 5 elements

   b. $n(B) = 1$; the set has only 1 element

   c. $n(C) = 8$; Though this set lists only five elements, the three dots indicate 12, 13, and 14 are also elements.

   d. $n(D) = 0$ because the set has no elements.

9. Yes, the sets are equivalent. Each language in the table is paired with a distinct number of 15–24-year olds.

10. **a.** True, {O, L, D} = {D, O, L} because the sets contain exactly the same elements.

    b. False, the two sets do not contain exactly the same elements.

Exercise Set 2.1

1. This is well defined and therefore it is a set.

3. This is a matter of opinion and not well defined, thus it is not a set.

5. This is well defined and therefore it is a set.

7. The set of known planets in our Solar System. Note to student: This exercise did not forget Pluto. In 2006, based on the requirement that a planet must dominate its own orbit, the International Astronomical Union removed Pluto from the list of planets.

9. The set of months that begin with J.

11. The set of natural numbers greater than 5.

13. The set of natural numbers between 6 and 20, inclusive.

15. {winter, spring, summer, fall}

17. {September, October, November, December}

19. {1, 2, 3}

21. {1, 3, 5, 7, 9, 11}

23. {1, 2, 3, 4, 5}

25. {6, 7, 8, 9, …}

27. {7, 8, 9, 10}

29. {10, 11, 12, 13, …, 79}

31. {2}

33. not the empty set

35. empty set

37. not the empty set
    Note that the number of women who served as U.S. president before 2000 is 0. Thus the number 0 is an element of the set.

39. empty set

41. empty set

43. not the empty set

45. not the empty set

47. True
    3 is a member of the set.

49. True
    12 is a member of the set.

51. False
    5 is *not* a member of the set.

53. True
    11 is *not* a member of the set.

55. False
    37 is a member of the set.

57. False
    4 is a member of the set.

59. True
    13 is *not* a member of the set.

61. False
    16 is a member of the set.

63. False
    The set {3} is *not* a member of the set.

65. True
    −1 is *not* a natural number.

67. $n(A) = 5$; There are 5 elements in the set.

69. $n(B) = 15$; There are 15 elements in the set.

71. $n(C) = 0$; There are *no* days of the week beginning with A.

73. $n(D) = 1$; There is 1 element in the set.

75. $n(A) = 4$; There is 4 elements in the set.

77. $n(B) = 5$; There is 5 elements in the set.

79. $n(C) = 0$; There are no elements in the set.

81. **a.** Not equivalent
        The number of elements is not the same.

    b. Not equal
        The two sets contain different elements.

83. **a.** Equivalent
        The number of elements is the same.

    b. Not equal
        The elements are not exactly the same.

85. **a.** Equivalent
        The number of elements is the same.

    b. Equal
        The elements are exactly the same.

87. **a.** Equivalent
        Number of elements is the same.

    b. Not equal
        The two sets contain different elements.

89. **a.** Equivalent
        Number of elements is the same.

    b. Equal
        The elements are exactly the same.

91. infinite

93. finite

95. finite

97. $\{x \mid x \in \mathbb{N} \text{ and } x \geq 61\}$

99. $\{x \mid x \in \mathbb{N} \text{ and } 61 \leq x \leq 89\}$

101. Answers will vary; an example is: $\{0, 1, 2, 3\}$ and $\{1, 2, 3, 4\}$.

103. Impossible. Equal sets have exactly the same elements. This would require that there also must be the same number of elements.

105. {New Zealand, Australia, United States}

107. {Australia, United States, United Kingdom, Switzerland, Ireland}

109. {United Kingdom, Switzerland, Ireland}

111. { }

113. {12, 19}

115. {20, 21}

117. There is not a one-to-one correspondence. These sets are not equivalent.

125. a

Check Points 2.2

1. a. $\not\subseteq$; because 6, 9, and 11 are not in set *B*.

    b. $\subseteq$; because all elements in set *A* are also in set *B*.

    c. $\subseteq$; because all elements in set *A* are also in set *B*.

2. a. Both $\subseteq$ and $\subset$ are correct.

    b. Both $\subseteq$ and $\subset$ are correct.

3. Yes, the empty set is a subset of any set.

4. a. 16 subsets, 15 proper subsets
There are 4 elements, which means there are 2^4 or 16 subsets. There are $2^4 - 1$ proper subsets or 15.

    b. 64 subsets, 63 proper subsets
There are 6 elements, which means there are 2^6 or 64 subsets. There are $2^6 - 1$ proper subsets or 63.

5. a. This set has 3 elements. Therefore there are $2^3 = 2 \times 2 \times 2 = 8$ subsets. This number means that there are 8 ways to make a selection, including the option to bring no books.

    b. { }
{*The Da Vinci Code*}
{*The Lord of the Rings*}
{*America* (*The Book*)}
{*The Da Vinci Code, The Lord of the Rings*}
{*The Da Vinci Code, America* (*The Book*)}
{*The Lord of the Rings, America* (*The Book*)}
{*The Da Vinci Code, The Lord of the Rings, America* (*The Book*)}

    c. 7 of the 8 subsets are proper subsets. {*The Da Vinci Code, The Lord of the Rings, America* (*The Book*)} is not a proper subset.

Exercise Set 2.2

1. $\subseteq$

3. $\nsubseteq$

5. $\nsubseteq$

7. $\nsubseteq$
   Subset cannot be larger than the set.

9. $\subseteq$

11. $\nsubseteq$

13. $\subseteq$ or $\subset$

15. $\subseteq$

17. $\subseteq$ or $\subset$

19. $\subseteq$

21. neither

23. True

25. False
    {Ralph} is a subset, not Ralph.

27. True

29. False
    The symbol "$\varnothing$" is not a member of the set.

31. True

33. False
    All elements of {1, 4} are members of {4, 1}

35. True

37. { } {Border Collie} {Poodle} {Border Collie, Poodle}

39. { } {t} {a} {b} {t, a} {t, b} {a, b} {t, a, b}

41. { } {0}

43. 16 subsets, 15 proper subsets
    There are 4 elements, which means there are 2^4 or 16 subsets. There are $2^4 - 1$ proper subsets or 15.

45. 64 subsets, 63 proper subsets
    There are 6 elements, which means there are 2^6 or 64 subsets. There are $2^6 - 1$ proper subsets or 63.

47. 128 subsets, 127 proper subsets
    There are 7 elements, which means there are 2^7 or 128 subsets. There are $2^7 - 1$ proper subsets or 127.

49. 8 subsets, 7 proper subsets
    There are 3 elements, which means there are 2^3 or 8 subsets. There are $2^3 - 1$ proper subsets or 7.

51. false; The set $\{1, 2, 3, ..., 1000\}$ has $2^{1000} - 1$ proper subsets.

53. true

55. false; $\varnothing \subseteq \{\varnothing, \{\varnothing\}\}$

57. true

59. true

61. true

63. false; The set of subsets of {a, e, i, o, u} contains 2^5 or 32 elements.

65. false; $D \subseteq T$

67. true

69. false; If $x \in W$, then $x \in D$.

71. true

73. true

75. $2^5 = 32$ option combinations

77. $2^6 = 64$ viewing combinations

79. $2^8 = 256$ city combinations

87. b

89. Number of proper subsets is $2^n - 1$, which means there are 128 total subsets (127 + 1).
    128 is 2^7, so there are 7 elements.

Check Points 2.3

1. **a.** $\{1, 5, 6, 7, 9\}$

   b. $\{1, 5, 6\}$

   c. $\{7, 9\}$

2. **a.** $\{a, b, c, d\}$

   b. $\{e\}$

   c. $\{e, f, g\}$

   d. $\{f, g\}$

3. $A' = \{b, c, e\}$; those are the elements in U but not in A.

4. **a.** $\{1, 3, 5, \underline{7}, \underline{10}\} \cap \{6, \underline{7}, \underline{10}, 11\} = \{7, 10\}$

   b. $\{1, 2, 3\} \cap \{4, 5, 6, 7\} = \varnothing$

   c. $\{1, 2, 3\} \cap \varnothing = \varnothing$

5. **a.** $\{1, 3, 5, 7, 10\} \cup \{6, 7, 10, 11\}$
      $= \{1, 3, 5, 6, 7, 10, 11\}$

   b. $\{1, 2, 3\} \cup \{4, 5, 6, 7\} = \{1, 2, 3, 4, 5, 6, 7\}$

   c. $\{1, 2, 3\} \cup \varnothing = \{1, 2, 3\}$

6. **a.** $A \cup B = \{b, c, e\}$
      $(A \cup B)' = \{a, d\}$

   b. $A' = \{a, d, e\}$
      $B' = \{a, d\}$
      $A' \cap B' = \{a, d\}$

7. **a.** $\{5\}$; region II

   b. $\{2, 3, 7, 11, 13, 17, 19\}$; the complement of region II

   c. $\{2, 3, 5, 7, 11, 13\}$; regions I, II, and III

   d. $\{17, 19\}$; the complement of regions I, II, and III

   e. $\{5, 7, 11, 13, 17, 19\}$; the complement of A united with B

   f. $\{2, 3\}$; A intersected with the complement of B

8. $n(A \cup B) = n(A) + n(B) - n(A \cap B)$
      $= 244 + 230 - 89$
      $= 385$

Exercise Set 2.3

1. U is the set of all composers.

3. U is the set of all brands of soft drinks.

5. $A' = \{c, d, e\}$

7. $C' = \{b, c, d, e, f\}$

9. $A' = \{6, 7, 8, \ldots, 20\}$

11. $C' = \{2, 4, 6, 8, \ldots, 20\}$

13. $A' = \{21, 22, 23, 24, \ldots\}$

15. $C' = \{1, 3, 5, 7, \ldots\}$

17. $A = \{1, 3, 5, 7\}$
      $B = \{1, 2, 3\}$
      $A \cap B = \{1, 3\}$

19. $A = \{1, 3, 5, 7\}$
      $B = \{1, 2, 3\}$
      $A \cup B = \{1, 2, 3, 5, 7\}$

21. $A = \{1, 3, 5, 7\}$
      $U = \{1, 2, 3, 4, 5, 6, 7\}$
      $A' = \{2, 4, 6\}$

23. $A' = \{2, 4, 6\}$
      $B' = \{4, 5, 6, 7\}$
      $A' \cap B' = \{4, 6\}$

25. $A = \{1, 3, 5, 7\}$
      $C' = \{1, 7\}$
      $A \cup C' = \{1, 3, 5, 7\}$

27. $A = \{1, 3, 5, 7\}$
      $C = \{2, 3, 4, 5, 6\}$
      $A \cap C = \{3, 5\}$
      $(A \cap C)' = \{1, 2, 4, 6, 7\}$

29. $A = \{1, 3, 5, 7\}$ $\qquad C = \{2, 3, 4, 5, 6\}$
      $A' = \{2, 4, 6\}$ $\qquad C' = \{1, 7\}$
      $A' \cup C' = \{1, 2, 4, 6, 7\}$

31. $A = \{1, 3, 5, 7\}$ $B = \{1, 2, 3\}$
      $(A \cup B) = \{1, 2, 3, 5, 7\}$
      $(A \cup B)' = \{4, 6\}$

33. $A = \{1, 3, 5, 7\}$
$A \cup \varnothing = \{1, 3, 5, 7\}$

35. $A \cap \varnothing = \varnothing$

37. $A \cup U = U$
$U = \{1, 2, 3, 4, 5, 6, 7\}$

39. $A \cap U = A$
$A = \{1, 3, 5, 7\}$

41. $A = \{a, g, h\}$
$B = \{b, g, h\}$
$A \cap B = \{g, h\}$

43. $A = \{a, g, h\}$
$B = \{b, g, h\}$
$A \cup B = \{a, b, g, h\}$

45. $A = \{a, g, h\}$
$U = \{a, b, c, d, e, f, g, h\}$
$A' = \{b, c, d, e, f\}$

47. $A' = \{b, c, d, e, f\}$
$B' = \{a, c, d, e, f\}$
$A' \cap B' = \{c, d, e, f\}$

49. $A = \{a, g, h\}$
$C' = \{a, g, h\}$
$A \cup C' = \{a, g, h\}$

51. $A = \{a, g, h\}$
$C = \{b, c, d, e, f\}$
$A \cap C = \varnothing$
$(A \cap C)' = \{a, b, c, d, e, f, g, h\}$

53. $A' = \{b, c, d, e, f\}$
$C' = \{a, g, h\}$
$A' \cup C' = \{a, b, c, d, e, f, g, h\}$

55. $A = \{a, g, h\}$
$B = \{b, g, h\}$
$A \cup B = \{a, b, g, h\}$
$(A \cup B)' = \{c, d, e, f\}$

57. $A \cup \varnothing = A$
$A = \{a, g, h\}$

59. $A \cap \varnothing = \varnothing$

61. $A = \{a, g, h\}$
$U = \{a, b, c, d, e, f, g, h\}$
$A \cup U = \{a, b, c, d, e, f, g, h\}$

63. $A = \{a, g, h\}$
$U = \{a, b, c, d, e, f, g, h\}$
$A \cap U = \{a, g, h\}$

65. $A = \{a, g, h\}$
$B = \{b, g, h\}$
$B' = \{a, c, d, e, f\}$
$A \cap B = \{g, h\}$
$(A \cap B) \cup B' = \{a, c, d, e, f, g, h\}$

67. $A = \{1, 3, 4, 7\}$

69. $U = \{1, 2, 3, 4, 5, 6, 7, 8, 9\}$

71. $A \cap B = \{3, 7\}$

73. $B' = \{1, 4, 8, 9\}$

75. $(A \cup B)' = \{8, 9\}$

77. $A = \{1, 3, 4, 7\}$
$B' = \{1, 4, 8, 9\}$
$A \cap B' = \{1, 4\}$

79. $B = \{\triangle, \text{two, four, six}\}$

81. $A \cup B = \{\triangle, \#, \$, \text{two, four, six}\}$

83. $n(A \cup B) = n(\{\triangle, \#, \$, \text{two, four, six}\}) = 6$

85. $n(A') = 5$

87. $(A \cap B)' = \{\#, \$, \text{two, four, six}, 10, 01\}$

89. $A' \cap B = \{\text{two, four, six}\}$

91. $n(U) - n(B) = 8 - 4 = 4$

93. $n(A \cup B) = n(A) + n(B) - n(A \cap B)$
$$= 17 \ + 20 \ - 6$$
$$= 31$$

95. $n(A \cup B) = n(A) + n(B) - n(A \cap B)$
$$= 17 \ + 17 \ - 7$$
$$= 27$$

97. $A = \{1, 3, 5, 7\}$
$B = \{2, 4, 6, 8\}$
$A \cup B = \{1, 2, 3, 4, 5, 6, 7, 8\}$

99. $U = \{1, 2, 3, 4, 5, 6, 7, 8\}$
$A = \{1, 3, 5, 7\}$
$A \cap U = \{1, 3, 5, 7\}$

101. $A = \{1, 3, 5, 7\}$
$C' = \{1, 6, 7, 8\}$
$A \cap C' = \{1, 7\}$

103. $U = \{1, 2, 3, 4, 5, 6, 7, 8\}$
$B = \{2, 4, 6, 8\}$
$C = \{2, 3, 4, 5\}$
$B \cap C = \{2, 4\}$
$(B \cap C)' = \{1, 3, 5, 6, 7, 8\}$

105. $A \cup (A \cup B)'$
$= \{23, 29, 31, 37, 41, 43, 53, 59, 61, 67, 71\}$

107. $n(U)\left[n(A \cup B) - n(A \cap B) \right] = 12\left[7 - 2 \right]$
$$= 12(5) = 60$$

109. {Ashley, Mike, Josh}

111. {Ashley, Mike, Josh, Emily, Hannah, Ethan}

113. {Ashley}

115. {Jacob}

117. Region I, The U.S. is in set A but not set B.

119. Region I, France is in set A but not set B.

121. Region IV, England is in neither set A nor set B.

123. Region I, 11 is in set A but not set B.

125. Region IV, 15 is in neither set A nor set B.

127. Region II, 454 is in set A and set B.

129. Region III, 9558 is in set B but not set A.

131. Region I, 9559 is in set A but not set B.

133. {spatial-temporal, sports equipment, toy cars and trucks} $\cap$ {dollhouses, spatial-temporal, sports equipment, toy cars and trucks}
= {spatial-temporal, sports equipment, toy cars and trucks}

135. {spatial-temporal, sports equipment, toy cars and trucks} $\cup$ {dollhouses, spatial-temporal, sports equipment, toy cars and trucks}
= {dollhouses, spatial-temporal, sports equipment, toy cars and trucks}

137. {toy cars and trucks } $\cap$ {dollhouses, domestic accessories, dolls, spatial-temporal, sports equipment}
= $\varnothing$

139. $n(A \cup B) = n(A) + n(B) - n(A \cap B)$
$$= 178 \ + 154 \ - 49$$
$$= 283 \text{ people}$$

153. d is true.

155.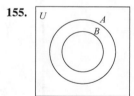

157.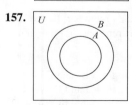

Check Points 2.4

1. **a.** $B \cap C = \{b, f\}$
        $A \cup (B \cap C) = \{a, b, c, d, f\}$

   b. $A \cup B = \{a, b, c, d, f\}$
        $A \cup C = \{a, b, c, d, f\}$
        $(A \cup B) \cap (A \cup C) = \{a, b, c, d, f\}$

   c. $C' = \{a, d, e\}$
        $B \cup C' = \{a, b, d, e, f\}$
        $A \cap (B \cup C') = \{a, b, d\}$

2. **a.** C is represented by regions IV, V, VI, and VII.
        Thus, $C = \{5, 6, 7, 8, 9\}$

   b. $B \cup C$ is represented by regions II, III, IV, V, VI, and VII.
        Thus, $B \cup C = \{1, 2, 5, 6, 7, 8, 9, 10, 12\}$

   c. $A \cap C$ is represented by regions IV and V.
        Thus, $A \cap C = \{5, 6, 7\}$

   d. B' is represented by regions I, IV, VII, and VIII.
        Thus, $B' = \{3, 4, 6, 8, 11\}$

   e. $A \cup B \cup C$ is represented by regions I, II, III, IV, V, VI, and VII.
        Thus,
        $A \cup B \cup C = \{1, 2, 3, 5, 6, 7, 8, 9, 10, 11, 12\}$

3.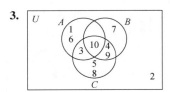

4. **a.** $A \cup B$ is represented by regions I, II, and III.
        Therefore $(A \cup B)'$ is represented by region IV.

   b. A' is represented by regions III and IV.
        B' is represented by regions I and IV.
        Therefore $A' \cap B'$ is represented by region IV.

   c. $(A \cup B)' = A' \cap B'$ because they both represent region IV.

5. **a.** $B \cup C$ is represented by regions II, III, IV, V, VI, and VII.
        Therefore $A \cap (B \cup C)$ is represented by regions II, IV, and V.

   b. $A \cap B$ is represented by regions II and V.
        $A \cap C$ is represented by regions IV and V.
        Therefore $(A \cap B) \cup (A \cap C)$ is represented by regions II, IV, and V.

   c. $A \cap (B \cup C) = (A \cap B) \cup (A \cap C)$ because they both represent region IV.

Exercises 2.4

1. $B \cap C = \{2, 3\}$
    $A \cup (B \cap C) = \{1, 2, 3, 5, 7\}$

3. $A \cup B = \{1, 2, 3, 5, 7\}$
    $A \cup C = \{1, 2, 3, 4, 5, 6, 7\}$
    $(A \cup B) \cap (A \cup C) = \{1, 2, 3, 5, 7\}$

5. $A' = \{2, 4, 6\}$ $C' = \{1, 7\}$
    $B \cup C' = \{1, 2, 3, 7\}$
    $A' \cap (B \cup C') = \{2\}$

7. $A' = \{2, 4, 6\}$ $C' = \{1, 7\}$
    $A' \cap B = \{2\}$
    $A' \cap C' = \emptyset$
    $(A' \cap B) \cup (A' \cap C') = \{2\}$

9. $A = \{1, 3, 5, 7\}$
    $B = \{1, 2, 3\}$
    $C = \{2, 3, 4, 5, 6\}$
    $A \cup B \cup C = \{1, 2, 3, 4, 5, 6, 7\}$
    $(A \cup B \cup C)' = \emptyset$

11. $A = \{1, 3, 5, 7\}$
     $B = \{1, 2, 3\}$
     $A \cup B = \{1, 2, 3, 5, 7\}$
     $(A \cup B)' = \{4, 6\}$
     $C = \{2, 3, 4, 5, 6\}$
     $(A \cup B)' \cap C = \{4, 6\}$

13. $B \cap C = \{b\}$
     $A \cup (B \cap C) = \{a, b, g, h\}$

15. $A \cup B = \{a, b, g, h\}$
$A \cup C = \{a, b, c, d, e, f, g, h\}$
$(A \cup B) \cap (A \cup C) = \{a, b, g, h\}$

17. $A' = \{b, c, d, e, f\}$
$C' = \{a, g, h\}$
$B \cup C' = \{a, b, g, h\}$
$A' \cap (B \cup C') = \{b\}$

19. $A' = \{b, c, d, e, f\}$
$A' \cap B = \{b\}$
$C' = \{a, g, h\}$
$A' \cap C' = \varnothing$
$(A' \cap B) \cup (A' \cap C') = \{b\}$

21. $A \cup B \cup C = \{a, b, c, d, e, f, g, h\}$
$(A \cup B \cup C)' = \varnothing$

23. $A \cup B = \{a, b, g, h\}$
$(A \cup B)' = \{c, d, e, f\}$
$(A \cup B)' \cap C = \{c, d, e, f\}$

25. II, III, V, VI

27. I, II, IV, V, VI, VII

29. II, V

31. I, IV, VII, VIII

33. $A = \{1, 2, 3, 4, 5, 6, 7, 8\}$

35. $A \cup B = \{1, 2, 3, 4, 5, 6, 7, 8, 9, 10, 11\}$

37. $A = \{1, 2, 3, 4, 5, 6, 7, 8\}$
$B = \{4, 5, 6, 9, 10, 11\}$
$A \cup B = \{1, 2, 3, 4, 5, 6, 7, 8, 9, 10, 11\}$
$(A \cup B)' = \{12, 13\}$

39. The set contains the elements in the two regions where the circles representing sets A and B overlap.
$A \cap B = \{4, 5, 6\}$

41. The set contains the element in the center region where the circles representing sets A, B, and C overlap.
$A \cap B \cap C = \{6\}$

43. $A \cap B \cap C = \{6\}$
$(A \cap B \cap C)' = \{1, 2, 3, 4, 5, 7, 8, 9, 10, 11, 12, 13\}$

45.
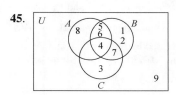

47.
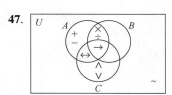

49. a. II

b. II

c. $A \cap B = B \cap A$

51. a. I, III, IV

b. IV

c. No, $(A \cap B)' \neq A' \cap B'$

53. Set A is represented by regions I and II.
Set A' is represented by regions III and IV.
Set B is represented by regions II and III.
Set B' is represented by regions I and IV.
$A' \cup B$ is represented by regions II, III, and IV.
$A \cap B'$ is represented by region I.
Thus, $A' \cup B$ and $A \cap B'$ are not equal for all sets A and B.

55. Set A is represented by regions I and II.
Set B is represented by regions II and III.
$(A \cup B)'$ is represented by region IV.
$(A \cap B)'$ is represented by regions I, III, and IV.
Thus, $(A \cup B)'$ and $(A \cap B)'$ are not equal for all sets A and B.

57. Set A is represented by regions I and II.
Set A' is represented by regions III and IV.
Set B is represented by regions II and III.
Set B' is represented by regions I and IV.

$\left(A' \cap B\right)'$ is represented by regions I, II, and IV.

$A \cup B'$ is represented by regions I, II, and IV.
Thus, $A' \cap B$ and $A \cup B'$ are equal for all sets A and B.

59. a. II, IV, V, VI, VII

    b. II, IV, V, VI, VII

    c. $(A \cap B) \cup C = (A \cup C) \cap (B \cup C)$

61. a. II, IV, V

    b. I, II, IV, V, VI

    c. No
The results in **a** and **b** show
$A \cap (B \cup C) \neq A \cup (B \cap C)$ because of the different regions represented.

63. The left expression is represented by regions II, IV, and V. The right expression is represented by regions II, IV, V, VI, and VII. Thus this statement is not true.

65. Both expressions are represented by regions II, III, IV, V, and VI. Thus this statement is true and is a theorem.

67. Both expressions are represented by region I. Thus this statement is true and is a theorem.

69. a. $A \cup \left(B' \cap C'\right) = \{c, e, f\}$
$\left(A \cup B'\right) \cap \left(A \cup C'\right) = \{c, e, f\}$

    b. $A \cup \left(B' \cap C'\right) = \{1, 3, 5, 7, 8\}$
$\left(A \cup B'\right) \cap \left(A \cup C'\right) = \{1, 3, 5, 7, 8\}$

    c. $A \cup \left(B' \cap C'\right) = \left(A \cup B'\right) \cap \left(A \cup C'\right)$

    d. $A \cup \left(B' \cap C'\right)$ and $\left(A \cup B'\right) \cap \left(A \cup C'\right)$ are both represented by regions I, II, IV, V, and VIII. Thus, the conjecture in part c is a theorem.

71. $\left(A \cap B'\right) \cap \left(A \cup B\right)$

73. $A' \cup B$

75. $\left(A \cap B\right) \cup C$

77. $A' \cap \left(B \cup C\right)$

79. {Ann, Jose, Al, Gavin, Amy, Ron, Grace}

81. {Jose}

83. {Lily, Emma}

85. {Lily, Emma, Ann, Jose, Lee, Maria, Fred, Ben, Sheila, Ellen, Gary}

87. {Lily, Emma, Al, Gavin, Amy, Lee, Maria}

89. {Al, Gavin, Amy}

91. The set of students who scored 90% or above on exam 1 and exam 3 but not on exam 2 is the empty set.

93. Region II

95. Region V

97. Region I

99. Region III

101. Region V

103. Region VII

105.

109. AB^+

111. No

Check Points 2.5

1. **a.** $55 + 20 = 75$

    b. $20 + 70 = 90$

    c. 20

    d. $55 + 20 + 70 = 145$

    e. 55

    f. 70

    g. 30

    h. $55 + 20 + 70 + 30 = 175$

2. Start by placing 700 in region II.
    Next place $1190 - 700$ or 490 in region III.
    Since half of those surveyed were women, place
    $1000 - 700$ or 300 in region I.
    Finally, place $2000 - 300 - 700 - 490$ or 510 in
    region IV.

    a. 490 men agreed with the statement and are
    represented by region III.

    b. 510 men disagreed with the statement and are
    represented by region IV.

3. Since 2 people collect all three items, begin by
    placing a 2 in region V.
    Since 29 people collect baseball cards and comic
    books, $29 - 2$ or 27 should be placed in region II.
    Since 5 people collect baseball cards and stamps,
    $5 - 2$ or 3 should be placed in region IV.
    Since 2 people collect comic books and stamps,
    $2 - 2$ or 0 should be placed in region VI.
    Since 108 people collect baseball cards,
    $108 - 27 - 3 - 2$ or 76 should be placed in region I.
    Since 92 people collect comic books,
    $92 - 27 - 2 - 0$ or 63 should be placed in region III.
    Since 62 people collect stamps, $62 - 3 - 2 - 0$ or 57
    should be placed in region VII.
    Since there were 250 people surveyed, place
    $250 - 76 - 27 - 63 - 3 - 2 - 0 - 57 = 22$ in region
    VIII.

    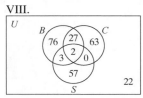

4. **a.** 63 as represented by region III.

    b. 3 as represented by region IV.

    c. 136 as represented by regions I, IV, and VII.

    d. 30 as represented by regions II, IV, and VI.

    e. 228 as represented by regions I through VII.

    f. 22 as represented by region VIII.

Exercise Set 2.5

1. 26

3. 17

5. 37

7. 7

9. Region I has $21 - 7$ or 14 elements.
    Region III has $29 - 7$ or 22 elements.
    Region IV has $48 - 14 - 7 - 22$ or 5 elements.

11. 17 as represented by regions II, III, V, and VI.

13. 6 as represented by regions I and II.

15. 28 as represented by regions I, II, IV, V, VI, and
    VII.

17. 9 as represented by regions IV and V.

19. 3 as represented by region VI.

21. 19 as represented by regions III, VI, and VII.

23. 21 as represented by regions I, III, and VII.

25. 34 as represented by regions I through VII.

27. Since $n(A \cap B) = 3$, there is 1 element in region II.
    Since $n(A \cap C) = 5$, there are 3 elements in region
    IV.
    Since $n(B \cap C) = 3$, there is 1 element in region
    VI.
    Since $n(A) = 11$, there are 5 elements in region I.
    Since $n(B) = 8$, there are 4 elements in region III.
    Since $n(C) = 14$, there are 8 elements in region VII.
    Since $n(U) = 30$, there are 6 elements in region
    VIII.

29. Since $n(A \cap B \cap C) = 7$, there are 7 elements in region V.
Since $n(A \cap B) = 17$, there are 10 elements in region II.
Since $n(A \cap C) = 11$, there are 4 elements in region IV.
Since $n(B \cap C) = 8$, there is 1 element in region VI.
Since $n(A) = 26$, there are 5 elements in region I.
Since $n(B) = 21$, there are 3 elements in region III.
Since $n(C) = 18$, there are 6 elements in region VII.
Since $n(U) = 38$, there are 2 elements in region VIII.

31. Since $n(A \cap B \cap C) = 2$, there are 2 elements in region V.
Since $n(A \cap B) = 6$, there are 4 elements in region II.
Since $n(A \cap C) = 9$, there are 7 elements in region IV.
Regions II, IV, and V contain a total of 13 elements, yet set A is stated to contain a total of only 10 elements. That is impossible.

33. $4 + 5 + 2 + 7 = 18$ respondents agreed with the statement.

35. $2 + 7 = 9$ women agreed with the statement.

37. 9 women who are not African American disagreed with the statement.

39. Parts b, c, and d are labeled.
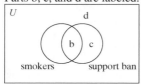

41. Parts b, c, and d are labeled.
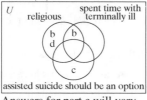
Answers for part e will vary.

43. Begin by placing 7 in the region that represents both newspapers and television.
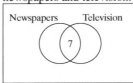

a. Since 29 students got news from newspapers, $29 - 7 = 22$ got news from only newspapers.

b. Since 43 students got news from television, $43 - 7 = 36$ got news from only television.
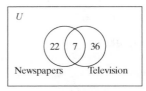

c. $22 + 7 + 36 = 65$ students who got news from newspapers or television.

d. Since 75 students were surveyed, $75 - 65 = 10$ students who did not get news from either.
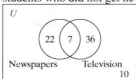

45. Construct a Venn diagram.
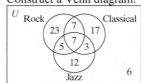

a. 23

b. 3

c. $17 + 3 + 12 = 32$

d. $23 + 17 + 12 = 52$

e. $7 + 3 + 5 + 7 = 22$

f. 6

47. Construct a Venn diagram.

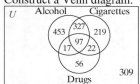

  a. 1500 (all eight regions)

  b. 1135 (the six regions of sets *A* and *C*)

  c. 56 (region VII)

  d. 327 (region II)

  e. 526 (regions I, IV, and VII)

  f. 366 (regions II, IV, and VI)

  g. 1191 (regions I through VII)

51. b

53. Under the conditions given concerning enrollment in math, chemistry, and psychology courses, the total number of students is 100, not 90.

Chapter 2 Review Exercises

1. the set of days of the week beginning with the letter T.

2. the set of natural numbers between 1 and 10, inclusive.

3. {m, i, s}

4. {8, 9, 10, 11, 12}

5. {1, 2, 3, …, 30}

6. not empty

7. empty set

8. $\in$
93 is an element of the set.

9. $\notin$
{d} is a subset, not a member; "d" would be a member.

10. 12
12 months in the year.

11. 15

12. $\neq$
The two sets do not contain exactly the same elements.

13. $\neq$
One set is infinite. The other is finite.

14. Equivalent
Same number of elements, but different elements.

15. Equal and equivalent
The two sets have exactly the same elements.

16. finite

17. infinite

18. $\subseteq$

19. $\nsubseteq$

20. $\subseteq$

21. $\subseteq$

22. both

23. False
Texas is not a member of the set.

24. False
4 is not a subset. {4} is a subset.

25. True

26. False
It is a subset but not a proper subset.

27. True

28. False
The set {six} has only one element.

29. True

30. $\varnothing$ {1} {5} {1, 5}
{1, 5} is not a proper subset.

31. There are 5 elements. This means there are $2^5 = 32$ subsets.
There are $2^5 - 1 = 31$ proper subsets.

32. {January, June, July}
There are 3 elements. This means there are $2^3 = 8$
subsets.
There are $2^3 - 1 = 7$ proper subsets.

33. $A \cap B = \{1, 2, 4\}$

34. $A \cup B' = \{1, 2, 3, 4, 6, 7, 8\}$

35. $A' \cap B = \{5\}$

36. $(A \cup B)' = \{6, 7, 8\}$

37. $A' \cap B' = \{6, 7, 8\}$

38. {4, 5, 6}

39. {2, 3, 6, 7}

40. {1, 4, 5, 6, 8, 9}

41. {4, 5}

42. {1, 2, 3, 6, 7, 8, 9}

43. {2, 3, 7}

44. {6}

45. {1, 2, 3, 4, 5, 6, 7, 8, 9}

46. $n(A \cup B) = n(A) + n(B) - n(A \cap B)$
$\qquad\qquad = 25 + 17 - 9$
$\qquad\qquad = 33$

47. $B \cap C = \{1, 5\}$
$A \cup (B \cap C) = \{1, 2, 3, 4, 5\}$

48. $A \cap C = \{1\}$
$(A \cap C)' = (2, 3, 4, 5, 6, 7, 8\}$
$(A \cap C)' \cup B = \{1, 2, 3, 4, 5, 6, 7, 8\}$

49. {c, d, e, f, k, p, r}

50. {f, p}

51. {c, d, f, k, p, r}

52. {c, d, e}

53. {a, b, c, d, e, g, h, p, r}

54. {f}

55.
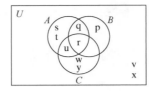

56. The shaded regions are the same for $(A \cup B)'$ and
$A' \cap B'$. Therefore $(A \cup B)' = A' \cap B'$

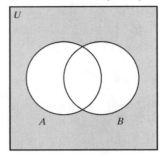

57. The statement is false because the shaded regions
are different.

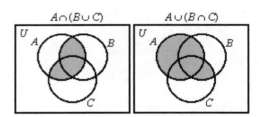

58. a. region I

    b. region III

    c. region VIII

    d. region II

    e. region V

59. a. Parts b and c are labeled.

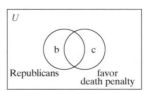

60. Begin by placing 400 in the region that represents both stocks and bonds.

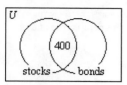

a. Since 650 respondents invested in stocks, $650 - 400 = 250$ invested in only stocks.

Furthermore, since 550 respondents invested in bonds, $550 - 400 = 150$ invested in only bonds. Place this data in the Venn diagram.

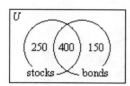

b. $250 + 400 + 150 = 800$ respondents invested in stocks or bonds.

c. Since 1000 people were surveyed, $1000 - 800 = 200$ respondents who did not invest in either.

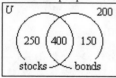

61. Construct a Venn diagram.

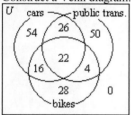

a. 50

b. 26

c. $54 + 26 + 50 = 130$

d. $26 + 16 + 4 = 46$

e. 0

Chapter 2 Test

1. $\{18, 19, 20, 21, 22, 23, 24\}$

2. False, $\{6\}$ is not an element of the set, but 6 is an element.

3. True, both sets have seven elements.

4. True

5. False, g is not an element in the larger set.

6. True

7. False, 14 is an element of the set.

8. False, Number of subsets: 2^N where N is the number of elements. There are 5 elements. $2^5 = 32$ subsets

9. False, $\varnothing$ is *not* a proper subset of itself.

10. $\varnothing$ $\{6\}$ $\{9\}$ $\{6, 9\}$
    $\{6, 9\}$ is not a proper subset.

11. $\{a, b, c, d, e, f\}$

12. $B \cap C = \{e\}$
    $(B \cap C)' = \{a, b, c, d, f, g\}$

13. $C' = \{b, c, d, f\}$
    $A \cap C' = \{b, c, d\}$

14. $A \cup B = \{a, b, c, d, e, f\}$
    $(A \cup B) \cap C = \{a, e\}$

15. $B' = \{a, b, g\}$
    $A \cup B' = \{a, b, c, d, g\}$
    $n(A \cup B') = 5$

16. $\{b, c, d, i, j, k\}$

17. $\{a\}$

18. $\{a, f, h\}$

19.

20. Both expressions are represented by regions III, VI, and VII. Thus this statement is true and is a theorem.

21. a. region V

    b. region VII

    c. region IV

    d. region I

    e. region VI

22. a.

    b. 263 (regions I, III, and VII)

    c. 25 (region VIII)

    d. 62 (regions II, IV, V, and VI)

    e. 0 (region VI)

    f. 147 (regions III, VI, and VII)

    g. 116 (region I)

Chapter 3
Logic

Check Points 3.1

1. a. Paris is not the capital of Spain.

   b. July is a month.

2. a. $\sim p$

   b. $\sim q$

3. Chicago O'Hare is not the world's busiest airport.

4. Some new tax dollars will not be used to improve education.
At least one new tax dollar will not be used to improve education.

Exercise Set 3.1

1. statement

3. statement

5. not a statement

7. statement

9. not a statement

11. statement

13. statement

15. It is not raining.

17. The Dallas Cowboys are the team with the most Super Bowl wins.

19. Chocolate in moderation is good for the heart.

21. $\sim p$

23. $\sim r$

25. Listening to classical music does not make infants smarter.

27. Sigmund Freud's father was 20 years older than his mother.

29. a. There are no whales that are not mammals.

   b. Some whales are not mammals.

31. a. At least one student is a business major.

   b. No students are business majors.

33. a. At least one thief is not a criminal.

   b. All thieves are criminals.

35. a. All Democratic presidents have not been impeached.

   b. Some Democratic presidents have been impeached.

37. Some Africans have Jewish ancestry.

39. Some rap is not hip-hop.

41. a. All birds are parrots.

   b. false; Some birds are not parrots.

43. a. No college students are business majors.

   b. false; Some college students are business majors.

45. a. All people like Sara Lee.

   b. Some people don't like Sara Lee.

47. a. No safe thing is exciting.

   b. Some safe things are exciting.

49. a. Some great actors are not Tom Hanks.

   b. All great actors are Tom Hanks.

   b. All generous philanthropists are Bill Gates.

51. b

53. true

55. false; Some college students in the United States are willing to marry without romantic love.

57. true

59. false; The sentence "5% of college students in Australia are willing to marry without romantic love" is a statement.

69. Answers will vary. Possible answer: Some mammals are not cats (true). Some cats are not mammals (false).

Check Points 3.2

1. a. $q \wedge p$ **b.** $\sim p \wedge q$

2. a. $p \vee q$ **b.** $q \vee \sim p$

3. a. $\sim p \rightarrow \sim q$ **b.** $q \rightarrow \sim p$

4. $\sim q \rightarrow p$

5. a. $q \leftrightarrow p$ **b.** $\sim p \leftrightarrow \sim q$

6. a. It is not true that he earns \$105,000 yearly and that he is often happy.
   b. He is not often happy and he earns \$105,000 yearly.

   c. It is not true that if he is often happy then he earns \$105,000 yearly.

7. a. If the plant is fertilized and the plant is watered, then the plant does not wilt.

   b. The plant is fertilized, and if the plant is watered then the plant does not wilt.

8. p: There is too much homework.; q: A teacher is boring.; r: I take the class.

   a. $(p \vee q) \rightarrow \sim r$ **b.** $p \vee (q \rightarrow \sim r)$

Exercise Set 3.2

1. $p \wedge q$

3. $q \wedge \sim p$

5. $\sim q \wedge p$

7. $p \vee q$; $\underset{p}{\underline{\text{I study}}}$ $\underset{\vee}{\underline{\text{or}}}$ $\underset{q}{\underline{\text{I pass the course.}}}$

9. $p \vee \sim q$; $\underset{p}{\underline{\text{I study}}}$ $\underset{\vee}{\underline{\text{or}}}$ $\underset{\sim q}{\underline{\text{I do not pass the course.}}}$

11. $p \rightarrow q$;
   $\underset{p}{\underline{\text{If this is an alligator,}}}$ $\underset{\rightarrow}{\underline{\text{then}}}$ $\underset{q}{\underline{\text{this is a reptile.}}}$

13. $\sim p \rightarrow \sim q$;
   $\underset{\sim p}{\underline{\text{If this is not an alligator,}}}$ $\underset{\rightarrow}{\underline{\text{then}}}$
   $\underset{\sim q}{\underline{\text{this is not a reptile.}}}$

15. $\sim q \rightarrow \sim p$

17. $p \rightarrow q$

19. $p \rightarrow \sim q$

21. $q \rightarrow \sim p$

23. $p \rightarrow \sim q$

25. $q \rightarrow \sim p$

27. $p \leftrightarrow q$;
   $\underset{p}{\underline{\text{The campus is closed}}}$ $\underset{\leftrightarrow}{\underline{\text{if and only if}}}$ $\underset{q}{\underline{\text{it is Sunday.}}}$

29. $\sim q \leftrightarrow \sim p$;
   $\underset{\sim q}{\underline{\text{It is not Sunday}}}$ $\underset{\leftrightarrow}{\underline{\text{if and only if}}}$
   $\underset{\sim p}{\underline{\text{the campus is not closed.}}}$

31. $q \leftrightarrow p$

33. The heater is not working and the house is cold.

35. The heater is working or the house is not cold.

37. If the heater is working then the house is not cold.

39. The heater is working if and only if the house is not cold.

41. It is July 4th and we are not having a barbeque.

43. It is not July 4th or we are having a barbeque.

45. If we are having a barbeque, then it is not July 4th.

47. It is not July 4th if and only if we are having a barbeque.

49. It is not true that Romeo loves Juliet and Juliet loves Romeo.

51. Romeo does not love Juliet and Juliet loves Romeo.

53. Neither Juliet loves Romeo nor Romeo loves Juliet.

55. Juliet does not love Romeo or Romeo loves Juliet.

57. Romeo does not love Juliet and Juliet does not love Romeo.

59. $(p \wedge q) \vee r$; $\left(\underset{p}{\text{The temperature outside is freezing}} \ \underset{\wedge}{\text{and}} \ \underset{q}{\text{the heater is working,}} \right) \underset{\vee}{\text{or}} \ \underset{r}{\text{the house is cold.}}$

61. $(p \vee \sim q) \to r$; $\left(\underset{}{\text{If}} \ \underset{p}{\text{the temperature outside is freezing}} \ \underset{\vee}{\text{or}} \ \underset{\sim q}{\text{the heater is not working,}} \right) \underset{\to}{\text{then}} \ \underset{r}{\text{the house is cold.}}$

63. $r \leftrightarrow (p \wedge \sim q)$; $\underset{r}{\text{The house is cold}} \ \underset{\leftrightarrow}{\text{if and only if}} \left(\underset{p}{\text{the temperature outside is freezing}} \ \underset{\wedge}{\text{and}} \ \underset{\sim q}{\text{the heater isn't working.}} \right)$

65. $(p \wedge \sim q) \to r$;

$\left(\underset{p}{\text{The temperature outside is freezing}} \ \underset{\wedge}{\text{and}} \ \underset{\sim q}{\text{the heater isn't working}} \right) \underset{\to}{\text{is a sufficient condition for}} \ \underset{r}{\text{the house being cold.}}$

67. If the temperature is above 85° and we have finished studying, then we go to the beach.

69. The temperature is above 85°, and if we finished studying then we go to the beach.

71. $\sim r \to (\sim p \vee \sim q)$; If we do not go to the beach, then the temperature is not above 85° or we have not finished studying.

73. If we do not go to the beach then we have not finished studying, or the temperature is above 85°.

75. $r \leftrightarrow (p \wedge q)$; We will go to the beach if and only if the temperature is above 85° and we have finished studying.

77. The temperature is above 85° if and only if we have finished studying, and we go to the beach.

79. If we do not go to the beach, then it is not true that both the temperature is above 85° and we have finished studying.

81. p: I like the teacher.; q: The course is interesting.; r: I miss class.; $(p \vee q) \to \sim r$

83. p: I like the teacher.; q: The course is interesting.; r: I miss class.; $p \vee (q \to \sim r)$

85. p: I like the teacher.; q: The course is interesting.; r: I miss class.; $r \leftrightarrow \sim (p \wedge q)$

87. p: I like the teacher.; q: The course is interesting.; r: I miss class.; $p \to (\sim r \leftrightarrow q)$

89. p: I like the teacher.; q: The course is interesting.; r: I miss class.; s: I spend extra time reading the book.; $(\sim p \wedge r) \to (\sim q \vee s)$

91. p: Being French is necessary for being a Parisian.; q: Being German is necessary for being a Berliner.; $\sim p \rightarrow \sim q$

93. p: You file an income tax report.; q: You file a complete statement of earnings.; r: You are a taxpayer.; s: You are an authorized tax preparer.; $(r \vee s) \rightarrow (p \wedge q)$

95. p: You are wealthy.; q: You are happy.; r: You live contentedly.; $\sim (p \rightarrow (q \wedge r))$

97. $[p \rightarrow (q \vee r)] \leftrightarrow (p \wedge r)$

99. $(p \rightarrow p) \leftrightarrow [(p \wedge p) \rightarrow \sim p]$

101. p: You can get rid of the family skeleton.
q: You may as well make it dance.
$\sim p \rightarrow q$

103. p: You know what you believe.
q: I can answer your questions.
$(p \rightarrow q) \wedge \sim q$

105. p: I am an intellectual.
q: I would be pessimistic about America.
$((p \rightarrow q) \wedge \sim p) \rightarrow \sim q$

115. $\left(\left(\underbrace{\text{Shooting unarmed civilians is morally justifiable}}_{p} \ \underbrace{\text{if and only if}}_{\leftrightarrow} \ \underbrace{\text{bombing unarmed civilians is morally justifiable,}}_{q} \right) \right.$

$\left. \underbrace{\text{and}}_{\wedge} \ \underbrace{\text{as the former is not morally justifiable,}}_{\sim p} \right) \rightarrow \left(\underbrace{\text{neither is the latter.}}_{\sim q} \right)$

Check Points 3.3

1. p: $3 + 5 = 8$ is true
q: $2 \times 7 = 20$ is false

a. $p \wedge q$
$T \wedge F$
F

b. $p \wedge \sim q$
$T \wedge \sim F$
$T \wedge T$
T

c. $\sim p \vee q$
$\sim T \vee F$
$F \vee F$
F

d. $\sim p \vee \sim q$
$\sim T \vee \sim F$
$F \vee T$
T

2. $\sim (p \vee q)$

p	q	$p \vee q$	$\sim (p \vee q)$
T	T	T	F
T	F	T	F
F	T	T	F
F	F	F	T

3. $\sim p \wedge \sim q$

p	q	$\sim p$	$\sim q$	$\sim p \wedge \sim q$
T	T	F	F	F
T	F	F	T	F
F	T	T	F	F
F	F	T	T	T

4. $(p\wedge\sim q)\vee\sim p$

p	q	$\sim p$	$\sim q$	$p\wedge\sim q$	$(p\wedge\sim q)\vee\sim p$
T	T	F	F	F	F
T	F	F	T	T	T
F	T	T	F	F	T
F	F	T	T	F	T

5. $p\wedge\sim p$ is false in all cases.

p	$\sim p$	$p\wedge\sim p$
T	F	F
F	T	F

6. p: I study hard.; q: I ace the final.; r: I fail the course.

a.

p	q	r	$q\vee r$	$p\wedge(q\vee r)$
T	T	T	T	T
T	T	F	T	T
T	F	T	T	T
T	F	F	F	F
F	T	T	T	F
F	T	F	T	F
F	F	T	T	F
F	F	F	F	F

b. false

7. $(p\vee q)\wedge\sim r$

$(T\vee F)\wedge\sim F$

$\quad T\wedge T$

$\qquad T$

Exercise Set 3.3

1. $\sim q$

$\sim F$

T

3. $p\wedge q$

$T\wedge F$

$\quad F$

5. $\sim p\wedge q$

$\sim T\wedge F$

$F\wedge F$

$\quad F$

7. $\sim p\wedge\sim q$

$\sim T\wedge\sim F$

$F\wedge T$

$\quad F$

9. $q\vee p$

$F\vee T$

$\quad T$

11. $p\vee\sim q$

$T\vee\sim F$

$T\vee T$

$\quad T$

13. $p\vee\sim p$

$T\vee\sim T$

$T\vee F$

$\quad T$

15. $\sim p\vee\sim q$

$\sim T\vee\sim F$

$F\vee T$

$\quad T$

17. $\sim p\wedge p$

p	$\sim p$	$\sim p\wedge p$
T	F	F
F	T	F

19. $\sim p\wedge q$

p	q	$\sim p$	$\sim p\wedge q$
T	T	F	F
T	F	F	F
F	T	T	T
F	F	T	F

21. $\sim(p\vee q)$

p	q	$p\vee q$	$\sim(p\vee q)$
T	T	T	F
T	F	T	F
F	T	T	F
F	F	F	T

23. $\sim p\wedge\sim q$

p	q	$\sim p$	$\sim q$	$\sim p\wedge\sim q$
T	T	F	F	F
T	F	F	T	F
F	T	T	F	F
F	F	T	T	T

25. $p \lor \sim q$

p	q	$\sim q$	$p \lor \sim q$
T	T	F	T
T	F	T	T
F	T	F	F
F	F	T	T

29. $(p \lor q) \land \sim p$

p	q	$\sim p$	$p \lor q$	$(p \lor q) \land \sim p$
T	T	F	T	F
T	F	F	T	F
F	T	T	T	T
F	F	T	F	F

27. $\sim(\sim p \lor q)$

p	q	$\sim p$	$\sim p \lor q$	$\sim(\sim p \lor q)$
T	T	F	T	F
T	F	F	F	T
F	T	T	T	F
F	F	T	T	F

31. $\sim p \lor (p \land \sim q)$

p	q	$\sim p$	$\sim q$	$p \land \sim q$	$\sim p \lor (p \land \sim q)$
T	T	F	F	F	F
T	F	F	T	T	T
F	T	T	F	F	T
F	F	T	T	F	T

33. $(p \lor q) \land (\sim p \lor \sim q)$

p	q	$\sim p$	$\sim q$	$p \lor q$	$\sim p \lor \sim q$	$(p \lor q) \land (\sim p \lor \sim q)$
T	T	F	F	T	F	F
T	F	F	T	T	T	T
F	T	T	F	T	T	T
F	F	T	T	F	T	F

35. $(p \land \sim q) \lor (p \land q)$

p	q	$\sim q$	$p \land \sim q$	$p \land q$	$(p \land \sim q) \lor (p \land q)$
T	T	F	F	T	T
T	F	T	T	F	T
F	T	F	F	F	F
F	F	T	F	F	F

37. $p \land (\sim q \lor r)$

p	q	r	$\sim q$	$\sim q \lor r$	$p \land (\sim q \lor r)$
T	T	T	F	T	T
T	T	F	F	F	F
T	F	T	T	T	T
T	F	F	T	T	T
F	T	T	F	T	F
F	T	F	F	F	F
F	F	T	T	T	F
F	F	F	T	T	F

39. $(r \wedge \sim p) \vee \sim q$

p	q	r	$\sim p$	$\sim q$	$r \wedge \sim p$	$(r \wedge \sim p) \vee \sim q$
T	T	T	F	F	F	F
T	T	F	F	F	F	F
T	F	T	F	T	F	T
T	F	F	F	T	F	T
F	T	T	T	F	T	T
F	T	F	T	F	F	F
F	F	T	T	T	T	T
F	F	F	T	T	F	T

41. $\sim (p \vee q) \wedge \sim r$

p	q	r	$p \vee q$	$\sim (p \vee q)$	$\sim r$	$\sim (p \vee q) \wedge \sim r$
T	T	T	T	F	F	F
T	T	F	T	F	T	F
T	F	T	T	F	F	F
T	F	F	T	F	T	F
F	T	T	T	F	F	F
F	T	F	T	F	T	F
F	F	T	F	T	F	F
F	F	F	F	T	T	T

43. a. p: You did the dishes.; q: You left the room a mess.; $\sim p \wedge q$

b. See truth table for Exercise 19.

c. The statement is true when p is false and q is true.

45. a. p: I bought a meal ticket.; q: I used it.; $\sim (p \wedge \sim q)$

b. $\sim(p \wedge \sim q)$

p	q	$\sim q$	$p \wedge \sim q$	$\sim(p \wedge \sim q)$
T	T	F	F	T
T	F	T	T	F
F	T	F	F	T
F	F	T	F	T

c. Answers will vary; an example is: The statement is true when p and q are true.

47. a. p: The student is intelligent.; q: The student is an overachiever.; $(p \vee q) \wedge \sim q$

b.

p	q	$\sim q$	$p \vee q$	$(p \vee q) \wedge \sim q$
T	T	F	T	F
T	F	T	T	T
F	T	F	T	F
F	F	T	F	F

c. The statement is true when p is true and q is false.

49. a. *p*: Married people are healthier than single people.; *q*: Married people are more economically stable than single people.; *r*: Children of married people do better on a variety of indicators.; $(p \wedge q) \wedge r$

   b.

p	*q*	*r*	$p \wedge q$	$(p \wedge q) \wedge r$
T	T	T	T	T
T	T	F	T	F
T	F	T	F	F
T	F	F	F	F
F	T	T	F	F
F	T	F	F	F
F	F	T	F	F
F	F	F	F	F

   c. The statement is true when *p*, *q*, and *r* are all true.

51. a. *p*: I go to office hours.; *q*: I ask questions.; *r*: My professor remembers me.; $(p \wedge q) \vee \sim r$

   b.

p	*q*	*r*	$\sim r$	$p \wedge q$	$(p \wedge q) \vee \sim r$
T	T	T	F	T	T
T	T	F	T	T	T
T	F	T	F	F	F
T	F	F	T	F	T
F	T	T	F	F	F
F	T	F	T	F	T
F	F	T	F	F	F
F	F	F	T	F	T

   c. Answers will vary; an example is: The statement is true when *p*, *q*, and *r* are all true.

53. $p \wedge (q \vee r)$
   $F \wedge (T \vee F)$
   $F \wedge T$
   F

55. $\sim p \vee (q \wedge \sim r)$
   $\sim F \vee (T \wedge \sim F)$
   $T \vee (T \wedge T)$
   $T \vee T$
   T

57. $\sim (p \wedge q) \vee r$
   $\sim (F \wedge T) \vee F$
   $\sim (F) \vee F$
   $T \vee F$
   T

59. $\sim (p \vee q) \wedge \sim (p \wedge r)$
   $\sim (F \vee T) \wedge \sim (F \wedge F)$
   $\sim (T) \wedge \sim (F)$
   $F \wedge T$
   F

61. $(\sim p \wedge q) \vee (\sim r \wedge p)$
   $(\sim F \wedge T) \vee (\sim F \wedge F)$
   $(T \wedge T) \vee (T \wedge F)$
   $T \vee F$
   T

63. $\sim [\sim (p \wedge \sim q) \vee \sim (\sim p \vee q)]$

p	*q*	$\sim [\sim (p \wedge \sim q) \vee \sim (\sim p \vee q)]$
T	T	F
T	F	F
F	T	F
F	F	F

65. $\left[(p \wedge \sim r) \vee (q \wedge \sim r)\right] \wedge \sim (\sim p \vee r)$

p	q	r	$\left[(p \wedge \sim r) \vee (q \wedge \sim r)\right] \wedge \sim (\sim p \vee r)$
T	T	T	F
T	T	F	T
T	F	T	F
T	F	F	T
F	T	T	F
F	T	F	F
F	F	T	F
F	F	F	F

67. p: You notice this notice.; q: You notice this notice is not worth noticing.; $(p \vee \sim p) \wedge q$

p	q	$\sim p$	$p \vee \sim p$	$(p \vee \sim p) \wedge q$
T	T	F	T	T
T	F	F	T	F
F	T	T	T	T
F	F	T	T	F

The statement is true when q is true.

69. p: $x \le 3$; q: $x \ge 7$; $\sim (p \vee q) \wedge (\sim p \wedge \sim q)$

p	q	$\sim p$	$\sim q$	$p \vee q$	$\sim (p \vee q)$	$\sim p \wedge \sim q$	$\sim (p \vee q) \wedge \sim (p \wedge \sim q)$
T	T	F	F	T	F	F	F
T	F	F	T	T	F	F	F
F	T	T	F	T	F	F	F
F	F	T	T	F	T	T	T

The statement is true when both p and q are false.

71. The chance of divorce peaks during the fourth year of marriage and the chance of divorce does not decrease after 4 years of marriage. This statement is false.

73. The chance of divorce does not peak during the fourth year of marriage and more than 2% of all divorces occur during the 25th year of marriage. This statement is false.

75. The chance of divorce peaks during the fourth year of marriage or the chance of divorce does not decrease after four years of marriage. This statement is true.

77. The chance of divorce does not peak during the fourth year of marriage or more than 2% of divorces occur during the 25th year of marriage. This statement is false.

79. The chance of divorce peaks during the fourth year of marriage and the chance of divorce decreases after four years of marriage, or more than 2% of divorces occur during the 25th year of marriage. This statement is true.

81. p: In 2004, 29.5% of college freshman were liberal.; q: In 2004, 25.1% of college freshman were conservative.; $\sim (p \wedge q)$; true

83. p: From 1984 through 2004, the percentage of liberal college freshman increased.; q: From 1984 through 2004, the percentage of conservative college freshman decreased.; r: From 1984 through 2004, the percentage of moderate college freshman increased.; $(p \vee q) \wedge \sim r$; true

85. a.   Hora Gershwin

b.   Bolera Mozart does not have a master's degree in music. Cha-Cha Bach does not either play three instruments or have five years experience playing with a symphony orchestra.

97. $p \veebar q$

p	q	$p \veebar q$
T	T	F
T	F	T
F	T	T
F	F	F

Check Points 3.4

1. $\sim p \rightarrow \sim q$

p	q	$\sim p$	$\sim q$	$\sim p \rightarrow \sim q$
T	T	F	F	T
T	F	F	T	T
F	T	T	F	F
F	F	T	T	T

The rightmost column shows that the statement is false when p is false and q is true; otherwise the statement is true.

2. $[(p \rightarrow q) \wedge \sim q] \rightarrow \sim p$ is a tautology because the final column is always true.

p	q	$\sim p$	$\sim q$	$p \rightarrow q$	$(p \rightarrow q) \wedge \sim q$	$[(p \rightarrow q) \wedge \sim q] \rightarrow \sim p$
T	T	F	F	T	F	T
T	F	F	T	F	F	T
F	T	T	F	T	F	T
F	F	T	T	T	T	T

3. a. p: You use Hair Grow.; q: You apply it daily.; r: You go bald.

p	q	r	$\sim r$	$p \wedge q$	$(p \wedge q) \rightarrow \sim r$
T	T	T	F	T	F
T	T	F	T	T	T
T	F	T	F	F	T
T	F	F	T	F	T
F	T	T	F	F	T
F	T	F	T	F	T
F	F	T	F	F	T
F	F	F	T	F	T

b.   No, the claim is not false under these conditions as shown in the third row.

4. $(p \vee q) \leftrightarrow (\sim p \to q)$ is a tautology.

p	q	$\sim p$	$p \vee q$	$\sim p \to q$	$(p \vee q) \leftrightarrow (\sim p \to q)$
T	T	F	T	T	T
T	F	F	T	T	T
F	T	T	T	T	T
F	F	T	F	F	T

5. $(p \wedge q) \to r$

$(T \wedge F) \to F$

$\quad F \to F$

$\qquad T$

Under these conditions, the claim is true.

Exercise Set 3.4

1. $p \to \sim q$

p	q	$\sim q$	$p \to \sim q$
T	T	F	F
T	F	T	T
F	T	F	T
F	F	T	T

3. $\sim(q \to p)$

p	q	$q \to p$	$\sim(q \to p)$
T	T	T	F
T	F	T	F
F	T	F	T
F	F	T	F

5. $(p \wedge q) \to (p \vee q)$

p	q	$p \wedge q$	$p \vee q$	$(p \wedge q) \to (p \vee q)$
T	T	T	T	T
T	F	F	T	T
F	T	F	T	T
F	F	F	F	T

7. $(p \to q) \wedge \sim q$

p	q	$p \to q$	$\sim q$	$(p \to q) \wedge \sim q$
T	T	T	F	F
T	F	F	T	F
F	T	T	F	F
F	F	T	T	T

9. $(p \vee q) \to r$

p	q	r	$p \vee q$	$(p \vee q) \to r$
T	T	T	T	T
T	T	F	T	F
T	F	T	T	T
T	F	F	T	F
F	T	T	T	T
F	T	F	T	F
F	F	T	F	T
F	F	F	F	T

11. $r \to (p \wedge q)$

p	q	r	$p \wedge q$	$r \to (p \wedge q)$
T	T	T	T	T
T	T	F	T	T
T	F	T	F	F
T	F	F	F	T
F	T	T	F	F
F	T	F	F	T
F	F	T	F	F
F	F	F	F	T

13. $\sim r \wedge (\sim q \rightarrow p)$

p	q	r	$\sim q$	$\sim r$	$\sim q \rightarrow p$	$\sim r \wedge (\sim q \rightarrow p)$
T	T	T	F	F	T	F
T	T	F	F	T	T	T
T	F	T	T	F	T	F
T	F	F	T	T	T	T
F	T	T	F	F	T	F
F	T	F	F	T	T	T
F	F	T	T	F	F	F
F	F	F	T	T	F	T

15. $\sim (p \wedge r) \rightarrow (\sim q \vee r)$

p	q	r	$\sim q$	$p \wedge r$	$\sim (p \wedge r)$	$\sim q \vee r$	$\sim (p \wedge r) \rightarrow (\sim q \vee r)$
T	T	T	F	T	F	T	T
T	T	F	F	F	T	F	F
T	F	T	T	T	F	T	T
T	F	F	T	F	T	T	T
F	T	T	F	F	T	T	T
F	T	F	F	F	T	F	F
F	F	T	T	F	T	T	T
F	F	F	T	F	T	T	T

17. $p \leftrightarrow \sim q$

p	q	$\sim q$	$p \leftrightarrow \sim q$
T	T	F	F
T	F	T	T
F	T	F	T
F	F	T	F

19. $\sim (p \leftrightarrow q)$

p	q	$p \leftrightarrow q$	$\sim (p \leftrightarrow q)$
T	T	T	F
T	F	F	T
F	T	F	T
F	F	T	F

21. $(p \leftrightarrow q) \rightarrow p$

p	q	$p \leftrightarrow q$	$(p \leftrightarrow q) \rightarrow p$
T	T	T	T
T	F	F	T
F	T	F	T
F	F	T	F

23. $(\sim p \leftrightarrow q) \to (\sim p \to q)$

p	q	$\sim p$	$\sim p \leftrightarrow q$	$\sim p \to q$	$(\sim p \leftrightarrow q) \to (\sim p \to q)$
T	T	F	F	T	T
T	F	F	T	T	T
F	T	T	T	T	T
F	F	T	F	F	T

25. $\left[(p \wedge q) \wedge (q \to p)\right] \leftrightarrow (p \wedge q)$

p	q	$p \wedge q$	$q \to p$	$(p \wedge q) \wedge (q \to p)$	$\left[(p \wedge q) \wedge (q \to p)\right] \leftrightarrow (p \wedge q)$
T	T	T	T	T	T
T	F	F	T	F	T
F	T	F	F	F	T
F	F	F	T	F	T

27. $(p \leftrightarrow q) \to \sim r$

p	q	r	$\sim r$	$p \leftrightarrow q$	$(p \leftrightarrow q) \to \sim r$
T	T	T	F	T	F
T	T	F	T	T	T
T	F	T	F	F	T
T	F	F	T	F	T
F	T	T	F	F	T
F	T	F	T	F	T
F	F	T	F	T	F
F	F	F	T	T	T

29. $(p \wedge r) \leftrightarrow \sim (q \vee r)$

p	q	r	$p \wedge r$	$q \vee r$	$\sim (q \vee r)$	$(p \wedge r) \leftrightarrow \sim (q \vee r)$
T	T	T	T	T	F	F
T	T	F	F	T	F	T
T	F	T	T	T	F	F
T	F	F	F	F	T	F
F	T	T	F	T	F	T
F	T	F	F	T	F	T
F	F	T	F	T	F	T
F	F	F	F	F	T	F

31. $\left[r \vee (\sim q \wedge p)\right] \leftrightarrow \sim p$

p	q	r	$\sim q$	$\sim q \wedge p$	$r \vee (\sim q \wedge p)$	$\sim p$	$\left[r \vee (\sim q \wedge p)\right] \leftrightarrow \sim p$
T	T	T	F	F	T	F	F
T	T	F	F	F	F	F	T
T	F	T	T	T	T	F	F
T	F	F	T	T	T	F	F
F	T	T	F	F	T	T	T
F	T	F	F	F	F	T	F
F	F	T	F	F	T	T	T
F	F	F	F	F	F	T	F

33. $[(p \to q) \land q] \to p$ is neither.

p	q	$p \to q$	$(p \to q) \land q$	$[(p \to q) \land q] \to p$
T	T	T	T	T
T	F	F	F	T
F	T	T	T	F
F	F	T	F	T

35. $\left[(p \to q) \land \sim q\right] \to \sim p$ is a tautology.

p	q	$\sim p$	$\sim q$	$p \to q$	$(p \to q) \land \sim q$	$\left[(p \to q) \land \sim q\right] \to \sim p$
T	T	F	F	T	F	T
T	F	F	T	F	F	T
F	T	T	F	T	F	T
F	F	T	T	T	T	T

37. $\left[(p \lor q) \land p\right] \to \sim q$ is neither.

p	q	$\sim q$	$p \lor q$	$(p \lor q) \land p$	$[(p \lor q) \land p] \to \sim q$
T	T	F	T	T	F
T	F	T	T	T	T
F	T	F	T	F	T
F	F	T	F	F	T

39. $(p \to q) \to (\sim p \lor q)$ is a tautology.

p	q	$\sim p$	$p \to q$	$\sim p \lor q$	$(p \to q) \to (\sim p \lor q)$
T	T	F	T	T	T
T	F	F	F	F	T
F	T	T	T	T	T
F	F	T	T	T	T

41. $(p \land q) \land (\sim p \lor \sim q)$ is a self-contradiction.

p	q	$\sim p$	$\sim q$	$p \land q$	$\sim p \lor \sim q$	$(p \land q) \land (\sim p \lor \sim q)$
T	T	F	F	T	F	F
T	F	F	T	F	T	F
F	T	T	F	F	T	F
F	F	T	T	F	T	F

43. $\sim(p \land q) \leftrightarrow (\sim p \land \sim q)$ is neither.

p	q	$\sim p$	$\sim q$	$p \land q$	$\sim(p \land q)$	$\sim p \land \sim q$	$\sim(p \land q) \leftrightarrow (\sim p \land \sim q)$
T	T	F	F	T	F	F	T
T	F	F	T	F	T	F	F
F	T	T	F	F	T	F	F
F	F	T	T	F	T	T	T

45. $(p \to q) \leftrightarrow (q \to p)$ is neither.

p	q	$p \to q$	$q \to p$	$(p \to q) \leftrightarrow (q \to p)$
T	T	T	T	T
T	F	F	T	F
F	T	T	F	F
F	F	T	T	T

47. $(p \to q) \leftrightarrow (\sim p \vee q)$ is a tautology.

p	q	$\sim p$	$p \to q$	$\sim p \vee q$	$(p \to q) \leftrightarrow (\sim p \vee q)$
T	T	F	T	T	T
T	F	F	F	F	T
F	T	T	T	T	T
F	F	T	T	T	T

49. $(p \leftrightarrow q) \leftrightarrow \left[(q \to p) \wedge (p \to q) \right]$ is a tautology.

p	q	$q \to p$	$p \to q$	$p \leftrightarrow q$	$(q \to p) \wedge (p \to q)$	$(p \leftrightarrow q) \leftrightarrow \left[(q \to p) \wedge (p \to q) \right]$
T	T	T	T	T	T	T
T	F	T	F	F	F	T
F	T	F	T	F	F	T
F	F	T	T	T	T	T

51. $(p \wedge q) \leftrightarrow (\sim p \vee r)$ is neither

p	q	r	$\sim p$	$p \wedge q$	$\sim p \vee r$	$(p \wedge q) \leftrightarrow (\sim p \vee r)$
T	T	T	F	T	T	T
T	T	F	F	T	F	F
T	F	T	F	F	T	F
T	F	F	F	F	F	T
F	T	T	T	F	T	F
F	T	F	T	F	T	F
F	F	T	T	F	T	F
F	F	F	T	F	T	F

53. $\left[(p \to q) \wedge (q \to r) \right] \to (p \to r)$ is a tautology.

p	q	r	$p \to q$	$q \to r$	$(p \to q) \wedge (q \to r)$	$p \to r$	$\left[(p \to q) \wedge (q \to r) \right] \to (p \to r)$
T	T	T	T	T	T	T	T
T	T	F	T	F	F	F	T
T	F	T	F	T	F	T	T
T	F	F	F	T	F	F	T
F	T	T	T	T	T	T	T
F	T	F	T	F	F	T	T
F	F	T	T	T	T	T	T
F	F	F	T	T	T	T	T

55. $\left[(q \to r) \land (r \to \sim p)\right] \leftrightarrow (q \land p)$ is neither.

p	q	r	$\left[(q \to r) \land (r \to \sim p)\right] \leftrightarrow (q \land p)$
T	T	T	F
T	T	F	F
T	F	T	T
T	F	F	F
F	T	T	F
F	T	F	T
F	F	T	F
F	F	F	F

57. a. p: You do homework right after class.; q: You fall behind.; $(p \to \sim q) \land (\sim p \to q)$

b.

p	q	$\sim p$	$\sim q$	$p \to \sim q$	$\sim p \to q$	$(p \to \sim q) \land (\sim p \to q)$
T	T	F	F	F	T	F
T	F	F	T	T	T	T
F	T	T	F	T	T	T
F	F	T	T	T	F	F

c. Answers will vary; an example is: The statement is true when p and q have opposite truth values.

59. a. p: You cut and paste from the Internet.; q: You cite the source.; r: You are charged with plagiarism.; $(p \land \sim q) \to r$

b.

p	q	r	$\sim q$	$p \land \sim q$	$(p \land \sim q) \to r$
T	T	T	F	F	T
T	T	F	F	F	T
T	F	T	T	T	T
T	F	F	T	T	F
F	T	T	F	F	T
F	T	F	F	F	T
F	F	T	T	F	T
F	F	F	T	F	T

c. Answers will vary; an example is: The statement is true when p, q, and r are all true.

61. a. p: You are comfortable in your room.; q: You are honest with your roommate.; r: You enjoy the college experience.; $(p \leftrightarrow q) \lor \sim r$

b.

p	q	r	$\sim r$	$p \leftrightarrow q$	$(p \leftrightarrow q) \lor \sim r$
T	T	T	F	T	T
T	T	F	T	T	T
T	F	T	F	F	F
T	F	F	T	F	T
F	T	T	F	F	F
F	T	F	T	F	T
F	F	T	F	T	T
F	F	F	T	T	T

c. Answers will vary; an example is: The statement is true when p, q, and r are all true.

63. a.   *p*: I enjoy the course.; *q*: I choose the class based on the professor.; *r*: I choose the class based on the course description.; $p \leftrightarrow (q \land \sim r)$

b.

p	*q*	*r*	$\sim r$	$q \land \sim r$	$p \leftrightarrow (q \land \sim r)$
T	T	T	F	F	F
T	T	F	T	T	T
T	F	T	F	F	F
T	F	F	T	F	F
F	T	T	F	F	T
F	T	F	T	T	F
F	F	T	F	F	T
F	F	F	T	F	T

c.   Answers will vary; an example is: The statement is true when *p*, *q*, and *r* are all false.

65. $\sim (p \to q)$
  $\sim (F \to T)$
  $\sim T$
  F

67. $\sim p \leftrightarrow q$
  $\sim F \leftrightarrow T$
  $T \leftrightarrow T$
  T

69. $q \to (p \land r)$
  $T \to (F \land F)$
  $T \to F$
  F

71. $(\sim p \land q) \leftrightarrow \sim r$
  $(\sim F \land T) \leftrightarrow \sim F$
  $(T \land T) \leftrightarrow T$
  $T \leftrightarrow T$
  T

73. $\sim [(p \to \sim r) \leftrightarrow (r \land \sim p)]$
  $\sim [(F \to \sim F) \leftrightarrow (F \land \sim F)]$
  $\sim [(F \to T) \leftrightarrow (F \land T)]$
  $\sim [T \leftrightarrow F]$
  $\sim F$
  T

75. $(p \to q) \leftrightarrow [(p \land q) \to \sim p]$

p	*q*	$(p \to q) \leftrightarrow [(p \land q) \to \sim p]$
T	T	F
T	F	F
F	T	T
F	F	T

77. $[p \to (\sim q \lor r)] \leftrightarrow (p \land r)$

p	*q*	*r*	$[p \to (\sim q \lor r)] \leftrightarrow (p \land r)$
T	T	T	T
T	T	F	T
T	F	T	T
T	F	F	F
F	T	T	F
F	T	F	F
F	F	T	F
F	F	F	F

79. p: You love a person.; q: You marry that person.; $(q \rightarrow p) \wedge (\sim p \rightarrow \sim q)$

p	q	$\sim p$	$\sim q$	$q \rightarrow p$	$\sim p \rightarrow \sim q$	$(q \rightarrow p) \wedge (\sim p \rightarrow \sim q)$
T	T	F	F	T	T	T
T	F	F	T	T	T	T
F	T	T	F	F	F	F
F	F	T	T	T	T	T

Answers will vary; an example is: The statement is true when both p and q are true.

81. p: You are happy.; q: You live contentedly.; r: You are wealthy.; $\sim [r \rightarrow (p \wedge q)]$

p	q	r	$p \wedge q$	$r \rightarrow (p \wedge q)$	$\sim [r \rightarrow (p \wedge q)]$
T	T	T	T	T	F
T	T	F	T	T	F
T	F	T	F	F	T
T	F	F	F	T	F
F	T	T	F	F	T
F	T	F	F	T	F
F	F	T	F	F	T
F	F	F	F	T	F

Answers will vary; an example is: The statement is true when p is false and both q and r are true.

83. p: Men averaged more hours than women for every year shown.; q: Men averaged more hours than women in 2002 through 2005.; r: Women averaged more hours than men in 2001.; $(\sim p \wedge q) \rightarrow r$; true

85. p: Men averaged 2.4 hours more than women in 2005.; q: Men averaged 1.2 hours more than women in 2004.; r: Men averaged fewer hours in 2005 than in 2002.; $(p \leftrightarrow q) \vee r$; true

87. The statement is of the form $(p \vee q) \leftrightarrow r$ with p false, q true, and r false.

$(p \vee q) \leftrightarrow r$

$(F \vee T) \leftrightarrow F$

$\quad T \leftrightarrow F$

$\qquad F$

Therefore the statement is false.

89. The statement is of the form $p \rightarrow (q \wedge r)$ with p true, q false, and r true.

$p \rightarrow (q \wedge r)$

$T \rightarrow (F \wedge T)$

$\quad T \rightarrow F$

$\qquad F$

Therefore the statement is false.

97. No, you cannot conclude you got an A. The person could still take you out to dinner if you received a different grade. That would be an example of an *if-then* statement with a false antecedent and a true consequent.

Check Points 3.5

1. a. $p \vee q$ and $\sim q \to p$ are equivalent.

p	q	$\sim q$	$p \vee q$	$\sim q \to p$
T	T	F	T	T
T	F	T	T	T
F	T	F	T	T
F	F	T	F	F

The statements are equivalent since their truth values are the same.

b. $\underline{ p } \overset{\vee}{} \underline{ q }$
I attend classes or I lose my scholarship.

...is equivalent to... $\underline{ \sim q } \overset{\to}{} \underline{ p }$
If I do not lose my scholarship, then I attend classes.

2. $\sim p$ and $\sim[\sim(\sim p)]$ are equivalent.

p	$\sim p$	$\sim(\sim p)$	$\sim[\sim(\sim p)]$
T	F	T	F
F	T	F	T

The statements are equivalent since their truth values are the same.

3. Given: If it's raining, then I need a jacket.
p: It's raining.
q: I need a jacket.
a: It's not raining or I need a jacket.
b: I need a jacket or it's not raining.
c: If I need a jacket, then it's raining.
d: If I do not need a jacket, then it's not raining.

The given is *not* equivalent to statement **(c)**

				Given	a	b	c	d
p	q	$\sim p$	$\sim q$	$p \to q$	$\sim p \vee q$	$q \vee \sim p$	$q \to p$	$\sim q \to \sim p$
T	T	F	F	T	T	T	T	T
T	F	F	T	F	F	F	T	F
F	T	T	F	T	T	T	F	T
F	F	T	T	T	T	T	T	T

4. a. If you're not driving too closely, then you can't read this.

b. If it's not time to do the laundry, then you have underwear.

c. If supervision during exams is required, then some students are not honest.

d. $q \to (p \vee r)$

5. Converse: If you don't see a Club Med, then you are in Iran.; Inverse: If you are not in Iran, then you see a Club Med.;
Contrapositive: If you see a Club Med, then you are not in Iran.

6. You do not have a fever and you have the flu.

7. Bart Simpson is not a cartoon character or Tony Soprano is not a cartoon character.

8. You do not leave by 5 P.M. and you arrive home on time.

9. **a.** Some horror movies are not scary or none are funny.

   b. Your workouts are not strenuous and you get stronger.

10. p: It is windy.
    q: We can swim.
    r: We can sail.
    The statement can be represented symbolically as $\sim p \rightarrow (q \wedge \sim r)$.
    Next write the contrapositive and simplify.
    $$\sim(q \wedge \sim r) \rightarrow \sim(\sim p)$$
    $$[\sim q \vee \sim(\sim r)] \rightarrow p$$
    $$(\sim q \vee r) \rightarrow p$$
    Thus, $\sim p \rightarrow (q \wedge \sim r) \equiv (\sim q \vee r) \rightarrow p$.
    The original statement is equivalent to "If we cannot swim or we can sail, then it is windy."

Exercise Set 3.5

1. **a.** $\sim p \rightarrow q$ and $p \vee q$ are equivalent.

p	q	$\sim p$	$\sim p \rightarrow q$	$p \vee q$
T	T	F	T	T
T	F	F	T	T
F	T	T	T	T
F	F	T	F	F

   b. The United States supports the development of solar-powered cars or it will suffer increasing atmospheric pollution.

3. not equivalent

5. equivalent

7. equivalent

9. not equivalent

11. not equivalent

13. equivalent

15. Given: I saw the original *King Kong* or the 2005 version.
p: I saw the original *King Kong*.
q: I saw the 2005 version.
a: If I did not see the original King Kong, I saw the 2005 version.
b: I saw both the original *King Kong* and the 2005 version.
c: If I saw the original *King Kong*, I did not see the 2005 version
d: If I saw the 2005 version, I did not see the original *King Kong*.

The given is equivalent to statement **(a)**

		Given	a	b	c	d
p	*q*	$p \vee q$	$\sim p \to q$	$p \wedge q$	$p \to \sim q$	$q \to \sim p$
T	T	T	T	T	F	F
T	F	T	T	F	T	T
F	T	T	T	F	T	T
F	F	F	F	F	T	T

17. Given: It is not true that Sondheim and Picasso are both musicians.
p: Sondheim is a musician.
q: Picasso is a musician.
a: Sondheim is not a musician or Picasso is not a musician.
b: If Sondheim is a musician, then Picasso is not a musician.
c: Sondheim is not a musician and Picasso is not a musician.
d: If Picasso is a musician, then Sondheim is not a musician.

The given is *not* equivalent to statement **(c)**

		Given	a	b	c	d
p	*q*	$\sim(p \wedge q)$	$\sim p \vee \sim q$	$p \to \sim q$	$\sim p \wedge \sim q$	$q \to \sim p$
T	T	F	F	F	F	F
T	F	T	T	T	F	T
F	T	T	T	T	F	T
F	F	T	T	T	T	T

19. Converse: If I am in Illinois, then I am in Chicago.
Inverse: If I am not in Chicago, then I am not in Illinois.
Contrapositive: If I am not in Illinois, I am not in Chicago.

21. Converse: If I cannot hear you, then the stereo is playing.
Inverse: If the stereo is not playing, then I can hear you.
Contrapositive: If I can hear you, then the stereo is not playing.

23. Converse: If you die, you don't laugh.
Inverse: If you laugh, you don't die.
Contrapositive: If you don't die, you laugh.

25. Converse: If all troops were withdrawn, then the president is telling the truth.
Inverse: If the president is not telling the truth, then some troops were not withdrawn.
Contrapositive: If some troops were not withdrawn, then the president was not telling the truth.

27. Converse: If some people suffer, then all institutions place profit above human need.
Inverse: If some institutions do not place profit above human need, then no people suffer.
Contrapositive: If no people suffer, then some institutions do not place profit above human need.

29. Converse: $\sim r \to \sim q$; Inverse: $q \to r$;
Contrapositive: $r \to q$

31. The negation of $p \to q$ is $p \wedge \sim q$: I am in Los Angeles and not in California.

33. The negation of $p \to q$ is $p \wedge \sim q$: It is purple and it is a carrot.

35. The negation of $p \to q$ is $p \wedge \sim q$: He doesn't, and I won't.

37. The negation of $p \to q$ is $p \wedge \sim q$: There is a blizzard, and some schools are not closed.

39. The negation of $\sim q \to \sim r$ is $\sim q \wedge r$

41. Australia is not an island or China is not an island.

43. My high school did not encourage creativity or did not encourage diversity.

45. Jewish scripture does not give a clear indication of a heaven and it does not give a clear indication of an afterlife.

47. The United States has eradicated neither poverty nor racism.

49. $\sim\left(\sim p \wedge q\right)$

$\sim\left(\sim p\right) \vee \sim q$

$p \vee \sim q$

51. p: You attend lecture.
q: You study.
r: You succeed.
The statement can be represented symbolically as $\left(p \wedge q\right) \to r$.
Next write the contrapositive and simplify.
$\sim r \to \sim\left(p \wedge q\right)$
$\sim r \to \left(\sim p \vee \sim q\right)$
Thus, $\left(p \wedge q\right) \to r \equiv \sim r \to \left(\sim p \vee \sim q\right)$.
The original statement is equivalent to "If you do not succeed, then you did not attend lecture or did not study."

53. p: He cooks.
q: His wife cooks.
r: His child cooks.
The statement can be represented symbolically as $\sim p \to \left(q \vee r\right)$.
Next write the contrapositive and simplify.
$\sim\left(q \vee r\right) \to \sim\left(\sim p\right)$
$\left(\sim q \wedge \sim r\right) \to p$
Thus, $\sim p \to \left(q \vee r\right) \equiv \left(\sim q \wedge \sim r\right) \to p$.
The original statement is equivalent to "If his wife does not cook and his child does not cook, then he does."

55. Write the contrapositive of $p \to \left(q \vee \sim r\right)$ and simplify.
$\sim\left(q \vee \sim r\right) \to \sim p$
$\left[\sim q \wedge \sim\left(\sim r\right)\right] \to \sim p$
$\left(\sim q \wedge r\right) \to \sim p$
Thus, $p \to \left(q \vee \sim r\right) \equiv \left(\sim q \wedge r\right) \to \sim p$.

57. I'm going to neither Seattle nor San Francisco.

59. I do not study and I pass.

61. I am going or he is not going.

63. A bill does not become law or it receives majority approval.

65. Write the negation of $p \vee \sim q$ and simplify.
$\sim\left(p \vee \sim q\right)$
$\sim p \wedge \sim\left(\sim q\right)$
$\sim p \wedge q$
Thus the negation of $p \vee \sim q$ is $\sim p \wedge q$.

67. Write the negation of $p \wedge (q \vee r)$ and simplify.

$$\sim\left[p \wedge (q \vee r)\right]$$

$$\sim p \vee \sim (q \vee r)$$

$$\sim p \vee (\sim q \wedge \sim r)$$

Thus the negation of $p \wedge (q \vee r)$ is $\sim p \vee (\sim q \wedge \sim r)$.

69. None are equivalent.

		a	b	c
p	q	$p \rightarrow \sim q$	$\sim p \vee q$	$\sim p \rightarrow q$
T	T	F	T	T
T	F	T	T	F
F	T	T	T	T
F	F	T	F	T

71. None are equivalent.

		a	b	c
p	q	$\sim (p \wedge \sim q)$	$\sim p \wedge q$	$p \vee \sim q$
T	T	T	F	T
T	F	F	F	T
F	T	T	T	F
F	F	T	F	T

73. None are equivalent.

			a	b	c
p	q	r	$p \rightarrow \sim (q \vee r)$	$(q \wedge r) \rightarrow \sim p$	$\sim p \rightarrow (q \wedge r)$
T	T	T	F	F	T
T	T	F	F	T	T
T	F	T	F	T	T
T	F	F	T	T	T
F	T	T	T	T	T
F	T	F	T	T	F
F	F	T	T	T	F
F	F	F	T	T	F

75. a and b are equivalent.

			a	b	c
p	q	r	$p \wedge (q \vee r)$	$p \wedge \sim (\sim q \wedge \sim r)$	$p \rightarrow (q \vee r)$
T	T	T	T	T	T
T	T	F	T	T	T
T	F	T	T	T	T
T	F	F	F	F	F
F	T	T	F	F	T
F	T	F	F	F	T
F	F	T	F	F	T
F	F	F	F	F	T

77. If there is no pain, there is no gain.; Converse: If there is no gain, then there is no pain.; Inverse: If there is pain, then there is gain.; Contrapositive: If there is gain, then there is pain.; Negation: There is no pain and there is gain.

79. If you follow Buddha's "Middle Way," then you are neither hedonistic nor ascetic.; Converse: If you are neither hedonistic nor ascetic, then you follow Buddha's "Middle Way."; Inverse: If you do not follow Buddha's "Middle Way," then you are either hedonistic or ascetic.; Contrapositive: If you are either hedonistic or ascetic, then you do not follow Buddha's "Middle Way."; Negation: You follow Buddha's "Middle Way" and you are either hedonistic or ascetic.

81. $p \wedge (\sim r \vee s)$ **82.** $p \wedge (r \wedge \sim s)$ **83.** $\sim p \vee (r \wedge s)$ **84.** $\sim p \wedge (\sim r \wedge \sim s)$

85. **a.**   false

b.   Smith does not comprise 1.006% or Johnson does not comprise 0.699%.

c.   true

87. **a.**   true

b.   Brown does not comprise 0.699% and Williams comprises 0.621%.

c.   false

89. **a.**   true

b.   Brown does not comprise 0.621% and Jones comprises 0.621%.

c.   false

91. **a.**   true

b.   Smith is not the top name or does not comprise 1.006%, and Johnson comprises 0.810%.

c.   false

103.   We will replace or repair the roof, *or* we will not sell the house.

Check Points 3.6

1.   The argument is valid. p: The U.S. must energetically support the development of solar-powered cars.
q: The U.S. must suffer increasing atmospheric pollution.

$p \vee q$

$\sim q$
 _____

$\therefore p$

p	q	$\sim q$	$p \vee q$	$(p \vee q) \wedge \sim q$	$\left[(p \vee q) \wedge \sim p\right] \rightarrow p$
T	T	F	T	F	T
T	F	T	T	T	T
F	T	F	T	F	T
F	F	T	F	F	T

2.   The argument is valid. p: I study for 5 hours. q: I fail.

$p \vee q$

$\sim p$
 _____

$\therefore q$

p	q	$\sim p$	$p \vee q$	$(p \vee q) \wedge \sim p$	$\left[(p \vee q) \wedge p\right] \rightarrow q$
T	T	F	T	F	T
T	F	F	T	F	T
F	T	T	T	T	T
F	F	T	F	F	T

3. The argument is invalid. **p:** You lower the fat in your diet. **q:** You lower your cholesterol. **r:** You reduce your risk of heart disease.

		p	q	r	$p \rightarrow q$	$q \rightarrow r$	$\sim p \rightarrow \sim r$	$[(p \rightarrow q) \wedge (q \rightarrow r)] \rightarrow (\sim p \rightarrow \sim r)$
$p \rightarrow q$		T	T	T	T	T	T	T
$q \rightarrow r$		T	T	F	T	F	T	T
$\therefore \sim p \rightarrow \sim r$		T	F	T	F	T	T	T
		T	F	F	F	T	T	T
		F	T	T	T	T	F	F
		F	T	F	T	F	T	T
		F	F	T	T	T	F	F
		F	F	F	T	T	T	T

4. a. $p \vee q$

   $\sim q$

   $\therefore p$

   This argument is valid by Disjunctive Reasoning.

b. $p \rightarrow q$

   q

   $\therefore p$

   This argument is invalid by Fallacy of the Converse.

c. $p \rightarrow q$

   $q \rightarrow r$

   $\therefore p \rightarrow r$

   This argument is valid by Transitive Reasoning.

5. The argument is valid. **p:** people are good. **q:** laws are needed to prevent wrongdoing. **r:** laws will succeed in preventing wrongdoing.

		p	q	r	$p \rightarrow \sim q$	$\sim p \rightarrow \sim r$	$\sim q \vee \sim r$	$[(p \rightarrow \sim q) \wedge (\sim p \rightarrow \sim r)] \rightarrow (\sim q \vee \sim r)$
$p \rightarrow \sim q$		T	T	T	F	T	F	T
$\sim p \rightarrow \sim r$		T	T	F	F	T	T	T
$\therefore \sim q \vee \sim r$		T	F	T	T	T	T	T
		T	F	F	T	T	T	T
		F	T	T	T	F	F	T
		F	T	F	T	T	T	T
		F	F	T	T	F	T	T
		F	F	F	T	T	T	T

6. Some schools across the country are not reexamining their educational objectives.

Exercise Set 3.6

1. This is an invalid argument.

p	q	$\sim p$	$\sim q$	$p \to q$	$(p \to q) \wedge \sim p$	$[(p \to q) \wedge \sim p] \to \sim q$
T	T	F	F	T	F	T
T	F	F	T	F	F	T
F	T	T	F	T	T	F
F	F	T	T	T	T	T

3. This is a valid argument.

p	q	$\sim p$	$\sim q$	$p \to \sim q$	$(p \to \sim q) \wedge q$	$[(p \to \sim q) \wedge q] \to \sim p$
T	T	F	F	F	F	T
T	F	F	T	T	F	T
F	T	T	F	T	T	T
F	F	T	T	T	F	T

5. This is a valid argument.

p	q	$\sim q$	$p \wedge \sim q$	$(p \wedge \sim q) \wedge p$	$[(p \wedge \sim q) \wedge p] \to \sim q$
T	T	F	F	F	T
T	F	T	T	T	T
F	T	F	F	F	T
F	F	T	F	F	T

7. This is an invalid argument.

p	q	$p \to q$	$q \to p$	$p \wedge q$	$[(p \to q) \wedge (q \to p)]$	$[(p \to q) \wedge (q \to p)] \to (p \wedge q)$
T	T	T	T	T	T	T
T	F	F	T	F	F	T
F	T	T	F	F	F	T
F	F	T	T	F	T	F

9. This is an invalid argument.

p	q	r	$p \to q$	$q \to r$	$r \to p$	$(p \to q) \wedge (q \to r)$	$[(p \to q) \wedge (q \to r)] \to (r \to p)$
T	T	T	T	T	T	T	T
T	T	F	T	F	T	F	T
T	F	T	F	T	T	F	T
T	F	F	F	T	T	F	T
F	T	T	T	T	F	T	F
F	T	F	T	F	T	F	T
F	F	T	T	T	F	T	F
F	F	F	T	T	T	T	T

11. This is a valid argument.

p	q	r	$p \to q$	$q \wedge r$	$p \vee r$	$(p \to q) \wedge (q \wedge r)$	$[(p \to q) \wedge (q \wedge r)] \to (p \vee r)$
T	T	T	T	T	T	T	T
T	T	F	T	F	T	F	T
T	F	T	F	F	T	F	T
T	F	F	F	F	T	F	T
F	T	T	T	T	T	T	T
F	T	F	T	F	F	F	T
F	F	T	T	F	T	F	T
F	F	F	T	F	F	F	T

13. This is a valid argument.

p	q	r	$\sim p$	$\sim r$	$p \leftrightarrow q$	$q \to r$	$\sim r \to \sim p$	$(p \leftrightarrow q) \wedge (q \to r)$	$[(p \leftrightarrow q) \wedge (q \to r)] \to (\sim r \to \sim p)$
T	T	T	F	F	T	T	T	T	T
T	T	F	F	T	T	F	F	F	T
T	F	T	F	F	F	T	T	F	T
T	F	F	F	T	F	T	F	F	T
F	T	T	T	F	F	T	T	F	T
F	T	F	T	T	F	F	T	F	T
F	F	T	T	F	T	T	T	T	T
F	F	F	T	T	T	T	T	T	T

15. This is a valid argument. p: It is cold. q: Motorcycle started.

$p \to \sim q$

q

$\therefore \sim p$

p	q	$\sim p$	$\sim q$	$p \to \sim q$	$(p \to \sim q) \wedge q$	$[(p \to \sim q) \wedge q \to \sim p]$
T	T	F	F	F	F	T
T	F	F	T	T	F	T
F	T	T	F	T	T	T
F	F	T	T	T	F	T

17. This an invalid argument. p: There is a dam. q: There is flooding.

$p \vee q$

q

$\therefore \sim p$

p	q	$\sim p$	$p \vee q$	$(p \vee q) \wedge q$	$[(p \vee q) \wedge q] \to \sim p$
T	T	F	T	T	F
T	F	F	T	F	T
F	T	T	T	T	T
F	F	T	F	F	T

19. p: We close the door.
q: There is less noise.

$p \to q$

q

$\therefore p$

Invalid, by fallacy of the converse.

21. $p \to q$

$\sim p \to q$

$\therefore q$

valid

23. p: We criminalize drugs.
q: We damage the future of young people.

$p \lor q$

$\underline{\sim q}$

$\therefore p$

Valid, by disjunctive reasoning.

25. This is an invalid argument. **p:** All people obey the law. **q:** No jails are needed.

$p \rightarrow q$

$\underline{\sim p}$

$\therefore \sim q$

p	q	$\sim p$	$\sim q$	$p \rightarrow q$	$(p \rightarrow q) \land \sim p$	$[(p \rightarrow q) \land \sim p] \rightarrow \sim q$
T	T	F	F	T	F	T
T	F	F	T	F	F	T
F	T	T	F	T	T	F
F	F	T	T	T	T	T

27. $p \rightarrow q$

$\underline{q \rightarrow r}$

$\therefore p \rightarrow r$

valid

29. $p \rightarrow q$

$\underline{q \rightarrow r}$

$\therefore r \rightarrow p$

invalid

31. This is a valid argument. **p:** Tim plays **q:** Janet plays **r:** Team wins

$(p \land q) \rightarrow r$

$\underline{p \land \sim r}$

$\therefore \sim q$

p	q	r	$\sim q$	$\sim r$	$p \land q$	$p \land \sim r$	$(p \land q) \rightarrow r$
T	T	T	F	F	T	F	T
T	T	F	F	T	T	T	F
T	F	T	T	F	F	F	T
T	F	F	T	T	F	T	T
F	T	T	F	F	F	F	T
F	T	F	F	T	F	F	T
F	F	T	T	F	F	F	T
F	F	F	T	T	F	F	T

$[(p \land q) \rightarrow r] \land (p \land \sim r)$	$[[(p \land q) \rightarrow r] \land (p \land \sim r)] \rightarrow \sim q$
F	T
F	T
F	T
T	T
F	T
F	T
F	T
F	T

33. This is a valid argument. *p*: It rains   *q*: It snows   *r*: I read

$(p \vee q) \rightarrow r$

$\sim r$

$\overline{\quad\quad\quad}$

$\therefore \sim(p \vee q)$

p	*q*	*r*	~*r*	*p* ∨ *q*	~(*p* ∨ *q*)	(*p* ∨ *q*) → *r*
T	T	T	F	T	F	T
T	T	F	T	T	F	F
T	F	T	F	T	F	T
T	F	F	T	T	F	F
F	T	T	F	T	F	T
F	T	F	T	T	F	F
F	F	T	F	F	T	T
F	F	F	T	F	T	T

$[(p \vee q) \rightarrow r] \wedge \sim r$	$\big[[(p \vee q) \rightarrow r] \wedge \sim r\big] \rightarrow \sim(p \vee q)$
F	T
F	T
F	T
F	T
F	T
F	T
F	T
T	T

35. This is an invalid argument.   *p*: It rains   *q*: It snows   *r*: I read

$(p \vee q) \rightarrow r$

r

$\overline{\quad\quad\quad}$

$\therefore p \vee q$

p	*q*	*r*	*p* ∨ *q*	(*p* ∨ *q*) → *r*	$[(p \vee q) \rightarrow r] \wedge r$	$\big[[(p \vee q) \rightarrow r] \wedge r\big] \rightarrow (p \vee q)$
T	T	T	T	T	T	T
T	T	F	T	F	F	T
T	F	T	T	T	T	T
T	F	F	T	F	F	T
F	T	T	T	T	T	T
F	T	F	T	F	F	T
F	F	T	F	T	T	F
F	F	F	F	T	F	T

37. This is an invalid argument.　*p*: It's hot.　*q*: It's humid.　*r*: I complain.

$(p \wedge q) \to r$

$\dfrac{\sim p \vee \sim q}{}$

$\therefore \sim r$

p	q	r	$\sim p$	$\sim q$	$\sim r$	$p \wedge q$	$\sim p \vee \sim q$	$(p \wedge q) \to r$
T	T	T	F	F	F	T	F	T
T	T	F	F	F	T	T	F	F
T	F	T	F	T	F	F	T	T
T	F	F	F	T	T	F	T	T
F	T	T	T	F	F	F	T	T
F	T	F	T	F	T	F	T	T
F	F	T	T	T	F	F	T	T
F	F	F	T	T	T	F	T	T

$[(p \wedge q) \to r] \wedge (\sim p \vee \sim q)$	$\big[[(p \wedge q) \to r] \wedge (\sim p \vee \sim q)\big] \to \sim r$
F	T
F	T
T	F
T	T
T	F
T	T
T	F
T	T

39. $p \to q$

$\dfrac{\sim p \to r}{}$

$\therefore q \vee r$

valid

41. $p \to q$

$q \to \sim r$

$\dfrac{r}{}$

$\therefore \sim p$

valid

43. *p*: A person is a chemist.

q: A person has a college degree.

$p \to q$

$\dfrac{\sim q}{}$

$\therefore \sim p$

My best friend is not a chemist. By contrapositive reasoning.

45. *p*: Writers improve.

q: "My Mother the Car" dropped from primetime.

$p \vee q$

$\dfrac{\sim p}{}$

$\therefore q$

"My Mother the Car" was dropped from primetime. By disjunctive reasoning.

47. *p*: All electricity off.

q: No lights work.

$p \to q$

$\dfrac{\sim q}{}$

$\therefore \sim p$

Some electricity is not off. By contrapositive reasoning.

49. *p*: I vacation in Paris.

q: I eat French pastries.

r: I gain weight.

$p \to q$

$\dfrac{q \to r}{}$

$\therefore p \to r$

If I vacation in Paris I gain weight. By transitive reasoning.

51. This is an invalid argument.

$p \rightarrow q$

$\dfrac{\sim p}{}$

$\therefore \sim q$

p	q	$\sim p$	$\sim q$	$p \rightarrow q$	$(p \rightarrow q) \wedge \sim p$	$[(p \rightarrow q) \wedge \sim p] \rightarrow \sim q$
T	T	F	F	T	F	T
T	F	F	T	F	F	T
F	T	T	F	T	T	F
F	F	T	T	T	T	T

53.
$p \vee q$

$\dfrac{\sim p}{}$

$\therefore q$

valid

55.
$p \rightarrow q$

$\dfrac{q}{}$

$\therefore p$

Invalid.

57. This is a valid argument.

$p \rightarrow q$

$\dfrac{\sim q}{}$

$\therefore \sim p$

p	q	$\sim p$	$\sim q$	$p \rightarrow q$	$(p \rightarrow q) \wedge \sim q$	$[(p \rightarrow q) \wedge \sim q] \rightarrow \sim p$
T	T	F	F	T	F	T
T	F	F	T	F	F	T
F	T	T	F	T	F	T
F	F	T	T	T	T	T

59.
$p \rightarrow q$

$\dfrac{q \rightarrow r}{}$

$\therefore p \rightarrow r$

valid

61. p: Poverty causes crime.
q: Crime sweeps American cities during the Great Depression.

$p \rightarrow q$

$\dfrac{\sim q}{}$

$\therefore \sim p$

Valid. By contrapositive reasoning.

63. h

65. i

67. c

69. a

71. j

73. d

83. p: You only spoke when spoken to, and I only speak when spoken to.
q: Nobody would ever say anything.

$p \rightarrow q$

$\dfrac{\sim q}{}$

$\therefore \sim p$

People sometimes speak without being spoken to.

p	q	$\sim p$	$\sim q$	$p \rightarrow q$	$(p \rightarrow q) \wedge \sim q$	$[(p \rightarrow q) \wedge \sim q] \rightarrow \sim p$
T	T	F	F	T	F	T
T	F	F	T	F	F	T
F	T	T	F	T	F	T
F	F	T	T	T	T	T

85. The doctor either destroys the base on which the placebo rests or jeopardizes a relationship built on trust.

Check Points 3.7

  1. The argument is valid.

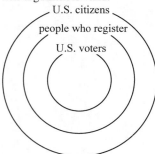

  2. The argument is invalid.

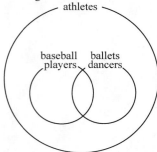

  3. The argument is valid.

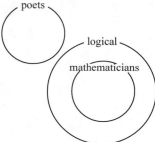

  4. The argument is invalid.

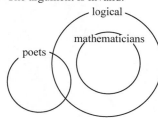

  5. The argument is invalid.

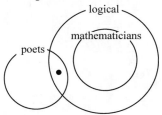

  6. The argument is invalid.
     The • is Euclid.

Exercise Set 3.7

  1. Valid.

  3. Invalid.

5. Valid.

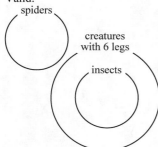

7. Invalid.

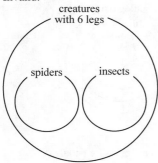

9. Invalid.

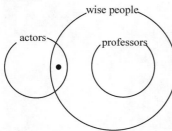

11. Valid.

13. Valid. The • is Savion Glover.

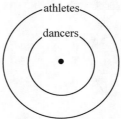

15. Invalid. The • is Savion Glover.

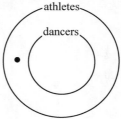

17. Invalid.

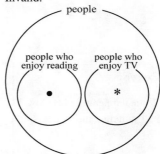

19. Valid.

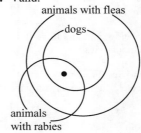

21. Valid.

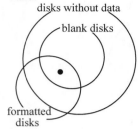

23. Valid. The • is 8.

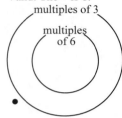

25. Invalid.

27. Invalid.

29. Valid.

31. Invalid.

33. Invalid.

35. Invalid.

37. Valid.

39. Valid.

43. b is true.

45. Some teachers are amusing people.

Chapter 3 Review Exercises

1. $(p \wedge q) \to r$; If the temperature is below 32° and we have finished studying, then we go to the movies.

2. $\sim r \to (\sim p \vee \sim q)$; If we do not go to the movies, then the temperature is not below 32° or we have not finished studying.

3. The temperature is below 32°, and if we finished studying, we go to the movies.

4. We will go to the movies if and only if the temperature is below 32° and we have finished studying.

5. It is not true that both the temperature is below 32° and we have finished studying.

6. We will not go to the movies if and only if the temperature is not below 32° or we have not finished studying.

7. $(p \wedge q) \vee r$

8. $(p \vee \sim q) \to r$

9. $q \to (p \leftrightarrow r)$

10. $r \leftrightarrow (p \wedge \sim q)$

11. $p \to r$

12. $q \to \sim r$

13. Some houses are not made with wood.

14. Some students major in business.

15. No crimes are motivated by passion.

16. All Democrats are registered voters.

17. Some new taxes will not be used for education.

18. neither

p	q	$\sim p$	$\sim p \wedge q$	$p \vee (\sim p \wedge q)$
T	T	F	F	T
T	F	F	F	T
F	T	T	T	T
F	F	T	F	F

19. neither

	p	q	~p	~q	~p∨~q
	T	T	F	F	F
	T	F	F	T	T
	F	T	T	F	T
	F	F	T	T	T

20. neither

p	q	~p	~p ∨ q	p → (~p ∨ q)
T	T	F	T	T
T	F	F	F	F
F	T	T	T	T
F	F	T	T	T

21. neither

p	q	~q	p ↔ ~q
T	T	F	F
T	F	T	T
F	T	F	T
F	F	T	F

22. tautology

p	q	~p	~q	p ∨ q	~(p ∨ q)	~p∧~q	~(p ∨ q) → (~p∧~q)
T	T	F	F	T	F	F	T
T	F	F	T	T	F	F	T
F	T	T	F	T	F	F	T
F	F	T	T	F	T	T	T

23. neither

p	q	r	~r	p ∨ q	(p ∨ q) → ~r
T	T	T	F	T	F
T	T	F	T	T	T
T	F	T	F	T	F
T	F	F	T	T	T
F	T	T	F	T	F
F	T	F	T	T	T
F	F	T	F	F	T
F	F	F	T	F	T

24. neither

p	q	r	p ∧ q	p ∧ r	(p ∧ q) ↔ (p ∧ r)
T	T	T	T	T	T
T	T	F	T	F	F
T	F	T	F	T	F
T	F	F	F	F	T
F	T	T	F	F	T
F	T	F	F	F	T
F	F	T	F	F	T
F	F	F	F	F	T

25. neither

p	q	r	$r \rightarrow p$	$q \vee (r \rightarrow p)$	$p \wedge [q \vee (r \rightarrow p)]$
T	T	T	T	T	T
T	T	F	T	T	T
T	F	T	T	T	T
T	F	F	T	T	T
F	T	T	F	T	F
F	T	F	T	T	F
F	F	T	F	F	F
F	F	F	T	T	F

26. a. p: I'm in class.; q: I'm studying.; $(p \vee q) \wedge \sim p$

b. $(p \vee q) \wedge \sim p$

p	q	$\sim p$	$p \vee q$	$(p \vee q) \wedge \sim p$
T	T	F	T	F
T	F	F	T	F
F	T	T	T	T
F	F	T	F	F

c. The statement is true when p is false and q is true.

27. a. p: You spit from a truck.; q: It's legal.; r: You spit from a car.; $(p \rightarrow q) \wedge (r \rightarrow \sim q)$

b.

p	q	r	$\sim q$	$p \rightarrow q$	$r \rightarrow \sim q$	$(p \rightarrow q) \wedge (r \rightarrow \sim q)$
T	T	T	F	T	F	F
T	T	F	F	T	T	T
T	F	T	T	F	T	F
T	F	F	T	F	T	F
F	T	T	F	T	F	F
F	T	F	F	T	T	T
F	F	T	T	T	T	T
F	F	F	T	T	T	T

c. The statement is true when p and q are both false.

28. $\sim (q \leftrightarrow r)$
$\sim (F \leftrightarrow F)$
$\sim T$
F

29. $(p \wedge q) \rightarrow (p \vee r)$
$(T \wedge F) \rightarrow (T \vee F)$
$F \rightarrow T$
T

30. $(\sim q \rightarrow p) \vee (r \wedge \sim p)$
$(\sim F \rightarrow T) \vee (F \wedge \sim T)$
$(T \rightarrow T) \vee (F \wedge F)$
$T \vee F$
T

31. $\sim [(\sim p \vee r) \rightarrow (q \wedge r)]$
$\sim [(\sim T \vee F) \rightarrow (F \wedge F)]$
$\sim [(F \vee F) \rightarrow F]$
$\sim [F \rightarrow F]$
$\sim T$
F

32. p: \$60.0 million was spent in 2002.; q: Spending decreased from 2002 to 2003.
$p \wedge \sim q$ is false.

33. p: More money was spent in 2003 than 2004.; q: \$67.9 million was spent in 2003.; r: \$53.6 million was spent in 2004.
$p \rightarrow (q \vee r)$ is true.

34. p: \$55.8 million was spent in 2001.; q: \$60.0 million was spent in 2002.; r: \$67.9 million was spent in 2004.
$(p \leftrightarrow q) \vee \sim r$ is true.

35. a. $\sim p \lor q \equiv p \rightarrow q$

p	q	$\sim p$	$\sim p \lor q$	$p \rightarrow q$
T	T	F	T	T
T	F	F	F	F
F	T	T	T	T
F	F	T	T	T

b. If the triangle is isosceles, then it has two equal sides.

36. c

37. not equivalent

p	q	$\sim(p \leftrightarrow q)$	$\sim p \lor \sim q$
T	T	F	F
T	F	T	T
F	T	T	T
F	F	F	T

38. equivalent

p	q	r	$\sim p \land (q \lor r)$	$(\sim p \land q) \lor (\sim p \land r)$
T	T	T	F	F
T	T	F	F	F
T	F	T	F	F
T	F	F	F	F
F	T	T	T	T
F	T	F	T	T
F	F	T	T	T
F	F	F	F	F

39. Converse: If I am in the South, then I am in Atlanta.
Inverse: If I am not in Atlanta, then I am not in the South.
Contrapositive: If I am not in the South, then I am not in Atlanta.

40. Converse: If today is not a holiday, then I am in class.
Inverse: If I am not in class, then today is a holiday.
Contrapositive: If today is a holiday, then I'm not in class.

41. Converse: If I pass all courses, then I worked hard.
Inverse: If I don't work hard, then I don't pass some courses.
Contrapositive: If I do not pass some course, then I did not work hard.

42. Converse: $\sim q \rightarrow \sim p$; Inverse: $p \rightarrow q$;
Contrapositive: $q \rightarrow p$

43. An argument is sound and it is not valid.

44. I do not work hard and I succeed.

45. $\sim r \land \sim p$

46. Chicago is not a city or Maine is not a city.

47. Ernest Hemingway was neither a musician nor an actor.

48. p: the number is positive.
q: the number is negative.
r: the number is zero.
The statement can be represented symbolically as $(\sim p \land \sim q) \rightarrow r$.
Next write the contrapositive and simplify.

$\sim(r) \rightarrow \sim(\sim p \land \sim q)$

$\sim r \rightarrow (p \lor q)$

Thus, $(\sim p \land \sim q) \rightarrow r \equiv \sim r \rightarrow (p \lor q)$.

The original statement is equivalent to "If a number is not zero, then the number is positive or negative."

49. I do not work hard and I succeed.

50. She is using her car or she is not taking a bus.

51. Write the negation of $\sim p \lor q$ and simplify.

$\sim(\sim p \lor q)$

$\sim(\sim p) \land \sim q$

$p \land \sim q$

52. a and **c** are equivalent.

		a	**b**	**c**
p	q	$p \rightarrow q$	$\sim p \rightarrow \sim q$	$\sim p \lor q$
T	T	T	T	T
T	F	F	T	F
F	T	T	F	T
F	F	T	T	T

53. a and **b** are equivalent.

		a	**b**	**c**
p	q	$\sim p \rightarrow q$	$\sim q \rightarrow p$	$\sim p \land \sim q$
T	T	T	T	F
T	F	T	T	F
F	T	T	T	F
F	F	F	F	T

54. a and **c** are equivalent.

		a	**b**	**c**
p	q	$p \lor \sim q$	$\sim q \rightarrow p$	$\sim(\sim p \land q)$
T	T	T	T	T
T	F	T	T	T
F	T	F	T	F
F	F	T	F	T

55. none

56. The argument is invalid.

p	q	$\sim q$	$p \rightarrow q$	$(p \rightarrow q) \wedge \sim q$	$[(p \rightarrow q) \wedge \sim q] \rightarrow p$
T	T	F	T	F	T
T	F	T	F	F	T
F	T	F	T	F	T
F	F	T	T	T	F

57. The argument is valid.

p	q	r	$p \wedge q$	$q \rightarrow r$	$p \rightarrow r$	$(p \wedge q) \wedge (q \rightarrow r)$	$[(p \wedge q) \wedge (q \rightarrow r)] \rightarrow (p \rightarrow r)$
T	T	T	T	T	T	T	T
T	T	F	T	F	F	F	T
T	F	T	F	T	T	F	T
T	F	F	F	T	F	F	T
F	T	T	F	T	T	F	T
F	T	F	F	F	T	F	T
F	F	T	F	T	T	F	T
F	F	F	F	T	T	F	T

58. The argument is invalid. **p:** Tony plays. **q:** Team wins.

$p \rightarrow q$

$\underline{q\qquad}$

$\therefore p$

p	q	$p \rightarrow q$	$(p \rightarrow q) \wedge q$	$[(p \rightarrow q) \wedge q] \rightarrow p$
T	T	T	T	T
T	F	F	F	T
F	T	T	T	F
F	F	T	F	T

59. The argument is invalid. **p:** Plant is fertilized. **q:** Plant turns yellow.

$p \vee q$

$\underline{q\qquad}$

$\therefore \sim p$

p	q	$\sim p$	$p \vee q$	$(p \vee q) \wedge q$	$[(p \vee q) \wedge q] \rightarrow \sim p$
T	T	F	T	T	F
T	F	F	T	F	T
F	T	T	T	T	T
F	F	T	F	F	T

60. The argument is valid. **p:** A majority of legislators vote for a bill. **q:** Bill does not become law.

$p \vee q$

$\underline{\sim p\qquad}$

$\therefore q$

p	q	$\sim p$	$p \vee q$	$(p \vee q) \wedge \sim p$	$[(p \vee q) \wedge \sim p] \rightarrow q$
T	T	F	T	F	T
T	F	F	T	F	T
F	T	T	T	T	T
F	F	T	F	F	T

61. The argument is valid. **p:** Good baseball player. **q:** Good hand–eye coordination.

$p \rightarrow q$

$\underline{\sim q\qquad}$

$\therefore \sim p$

p	q	$\sim p$	$\sim q$	$p \rightarrow q$	$(p \rightarrow q) \wedge \sim q$	$[(p \rightarrow q) \wedge \sim q] \rightarrow \sim p$
T	T	F	F	T	F	T
T	F	F	T	F	F	T
F	T	T	F	T	F	T
F	F	T	T	T	T	T

$$p \to \sim q$$
62. $\sim p \to q$
$$\therefore p \leftrightarrow \sim q$$
valid

$$p \to \sim q$$
63. $r \to q$
$$\therefore \sim r \to p$$
invalid

64. Invalid.

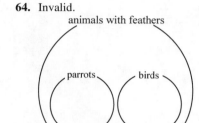

65. Valid.

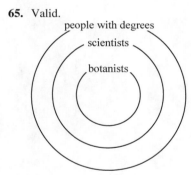

66. Valid.

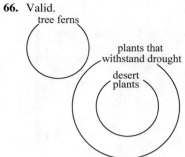

67. Invalid.

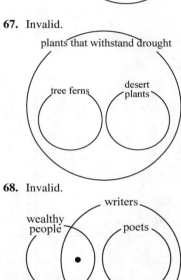

68. Invalid.

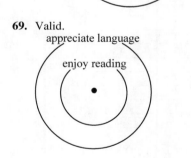

69. Valid.

appreciate language

enjoy reading

Chapter 3 Test

1. If I'm registered and I'm a citizen, then I vote.

2. I don't vote if and only if I'm not registered or I'm not a citizen.

3. I'm neither registered nor a citizen.

4. $(p \wedge q) \vee {\sim}r$

5. $({\sim}p \vee {\sim}q) \to {\sim}r$

6. $r \to q$

7. Some numbers are not divisible by 5.

8. No people wear glasses.

9. $p \wedge ({\sim}p \vee q)$

p	q	${\sim}p$	${\sim}p \vee q$	$p \wedge ({\sim}p \vee q)$
T	T	F	T	T
T	F	F	F	F
F	T	T	T	F
F	F	T	T	F

10. ${\sim}(p \wedge q) \leftrightarrow ({\sim}p \vee {\sim}q)$

p	q	${\sim}p$	${\sim}q$	$p \wedge q$	${\sim}(p \wedge q)$	$({\sim}p \vee {\sim}q)$	${\sim}(p \wedge q) \leftrightarrow ({\sim}p \vee {\sim}q)$
T	T	F	F	T	F	F	T
T	F	F	T	F	T	T	T
F	T	T	F	F	T	T	T
F	F	T	T	F	T	T	T

11. $p \leftrightarrow (q \vee r)$

p	q	r	$q \vee r$	$p \leftrightarrow (q \vee r)$
T	T	T	T	T
T	T	F	T	T
T	F	T	T	T
T	F	F	F	F
F	T	T	T	F
F	T	F	T	F
F	F	T	T	F
F	F	F	F	T

12. p: You break the law.; q: You change the law.; $(p \wedge q) \to {\sim} p$

p	q	${\sim}p$	$p \wedge q$	$(p \wedge q) \to {\sim} p$
T	T	F	T	F
T	F	F	F	T
F	T	T	F	T
F	F	T	F	T

Answers will vary; an example is: The statement is true when p is false.

13. $\sim(q \to r)$

$\sim(T \to F)$

$\sim F$

T

14. $(p \vee r) \leftrightarrow (\sim r \wedge p)$

$(F \vee F) \leftrightarrow (\sim F \wedge F)$

$F \leftrightarrow (T \wedge F)$

$F \leftrightarrow F$

T

15. **a.** p: There was a decrease in the percentage of students who considered "being very well off financially" essential.; q: There was a decrease in the percentage of students who considered "developing a meaningful philosophy of life" essential.; r: There was a decrease in the percentage of students who considered "raising a family" essential.; $\sim p \vee (q \wedge r)$

b. true

16. b

17. If it snows, then it is not August.

18. Converse: If I cannot concentrate, then the radio is playing.
Inverse: If the radio is not playing, then I can concentrate.

19. It is cold and we use the pool.

20. The test is not today and the party is not tonight.

21. The banana is not green or it is ready to eat.

22. a and b are equivalent.

		a	b	c
p	q	$\sim p \to q$	$p \vee q$	$p \to \sim q$
T	T	T	T	F
T	F	T	T	T
F	T	T	T	T
F	F	F	F	T

23. a and c are equivalent.

		a	b	c
p	q	$\sim(p \vee q)$	$\sim p \to \sim q$	$\sim p \wedge \sim q$
T	T	F	T	F
T	F	F	T	F
F	T	F	F	F
F	F	T	T	T

24. The argument is invalid. p: Parrot talks. q: It is intelligent.

$p \to q$

q

$\therefore p$

25. The argument is valid. p: I am sick. q: I am tired.

$p \vee q$

$\sim q$

$\therefore p$

26. The argument is invalid. p: I am going. q: You are going.

$p \leftrightarrow \sim q$

q

$\therefore p$

27. Invalid.

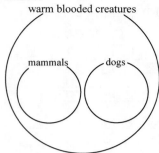

28. Valid.

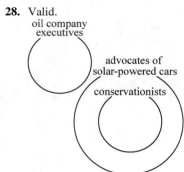

29. Invalid.

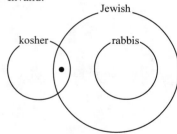

Chapter 4
Number Representation and Calculation

Check Points 4.1

1. **a.** $7^2 = 7 \times 7 = 49$

    b. $5^3 = 5 \times 5 \times 5 = 125$

    c. $1^4 = 1 \times 1 \times 1 \times 1 = 1$

    d. $10^5 = 10 \times 10 \times 10 \times 10 \times 10 = 100,000$

    e. $10^6 = 10 \times 10 \times 10 \times 10 \times 10 \times 10 = 1,000,000$

    f. $18^1 = 18$

2. **a.** $4026 = (4 \times 10^3) + (0 \times 10^2) + (2 \times 10^1) + (6 \times 1) = (4 \times 1000) + (0 \times 100) + (2 \times 10) + (6 \times 1)$

    b. $24,232 = (2 \times 10^4) + (4 \times 10^3) + (2 \times 10^2) + (3 \times 10^1) + (2 \times 1)$
    $$= (2 \times 10,000) + (4 \times 1000) + (2 \times 100) + (3 \times 10) + (2 \times 1)$$

3. **a.** $6000 + 70 + 3 = 6073$ **b.** $80,000 + 900 = 80,900$

4.
$$
\begin{array}{ccc}
\vee\vee\vee & << & <<<\vee \\
\downarrow & \downarrow & \downarrow
\end{array}
$$
$$(3 \times 60^2) + (20 \times 60) + (31 \times 1)$$
$$= (3 \times 3600) + (20 \times 60) + (31 \times 1)$$
$$= 10,800 + 1200 + 31$$
$$= 12,031$$

5.
$$
\begin{array}{rcrcr}
11 & \times & 7200 & = & 79,200 \\
3 & \times & 360 & = & 1080 \\
0 & \times & 20 & = & 0 \\
13 & \times & 1 & = & \underline{13} \\
& & & & 80,293
\end{array}
$$

Exercise Set 4.1

1. $5^2 = 5 \times 5 = 25$

3. $2^3 = 2 \times 2 \times 2 = 8$

5. $3^4 = 3 \times 3 \times 3 \times 3 = 81$

7. $10^5 = 10 \times 10 \times 10 \times 10 \times 10 = 100,000$

9. $36 = (3 \times 10^1) + (6 \times 1) = (3 \times 10) + (6 \times 1)$

11. $249 = (2\times10^2)+(4\times10^1)+(9\times1) = (2\times100)+(4\times10)+(9\times1)$

13. $703 = (7\times10^2)+(0\times10^1)+(3\times1) = (7\times100)+(0\times10)+(3\times1)$

15. $4856 = (4\times10^3)+(8\times10^2)+(5\times10^1)+(6\times1) = (4\times1000)+(8\times100)+(5\times10)+(6\times1)$

17. $3070 = (3\times10^3)+(0\times10^2)+(7\times10^1)+(0\times1) = (3\times1000)+(0\times100)+(7\times10)+(0\times1)$

19. $34,569 = (3\times10^4)+(4\times10^3)+(5\times10^2)+(6\times10^1)+(9\times1)$
$= (3\times10,000)+(4\times1000)+(5\times100)+(6\times10)+(9\times1)$

21. $230,007,004 = (2\times10^8)+(3\times10^7)+(0\times10^6)+(0\times10^5)+(0\times10^4)+(7\times10^3)+(0\times10^2)+(0\times10^1)+(4\times1)$
$= (2\times100,000,000)+(3\times10,000,000)+(0\times1,000,000)+(0\times100,000)$
$\quad +(0\times10,000)+(7\times1000)+(0\times100)+(0\times10)+(4\times1)$

23. $70 + 3 = 73$

25. $300 + 80 + 5 = 385$

27. $500,000 + 20,000 + 8000 + 700 + 40 + 3 = 528,743$

29. $7000 + 2 = 7002$

31. $600,000,000 + 2000 + 7 = 600,002,007$

33. $(10+10+1+1+1)\times1 = 23\times1 = 23$

35. $(10+10+1)\times60^1 + (1+1)\times1 = (21\times60)+2 = 1260+2 = 1262$

37. $(1+1+1)\times60^2 + (10+1+1)\times60^1 + (1+1+1)\times1 = (3\times3600)+(12\times60)+3 = 10,800+720+3 = 11,523$

39. $(10+1)\times60^3 + (10+1)\times60^2 + (10+1)\times60^1 + (10+1)\times1$
$= (11\times60^3)+(11\times60^2)+(11\times60^1)+(11\times1)$
$= (11\times216,000)+(11\times3600)+(11\times60)+11$
$= 2,376,000+39,600+660+11$
$= 2,416,271$

41. $14\times1=14$

43.
$$
\begin{array}{rcrcr}
19 & \times & 360 & = & 6840 \\
0 & \times & 20 & = & 0 \\
6 & \times & 1 & = & \underline{\quad 6} \\
& & & & 6846
\end{array}
$$

45.
$$
\begin{array}{rcrcr}
8 & \times & 360 & = & 2880 \\
8 & \times & 20 & = & 160 \\
8 & \times & 1 & = & \underline{\quad 8} \\
& & & & 3048
\end{array}
$$

47.
$$
\begin{array}{rcl}
2 & \times & 7200 = & 14,400 \\
0 & \times & 360 = & 0 \\
0 & \times & 20 = & 0 \\
11 & \times & 1 = & \underline{\quad 11} \\
& & & 14,411
\end{array}
$$

49.
$$
\begin{array}{rcl}
10 & \times & 7200 = & 72,000 \\
10 & \times & 360 = & 3,600 \\
0 & \times & 20 = & 0 \\
10 & \times & 1 = & \underline{\quad 10} \\
& & & 75,610
\end{array}
$$

51. $\left[(1)\times 60^2 + (10+10)\times 60^1 + (10+10+1)\times 1\right] + \left[(10+1)\times 60^2 + (10+10+10)\times 60^1 + (1+1+1+1)\times 1\right]$

$= \left[(1)\times 60^2 + (20)\times 60^1 + (21)\times 1\right] + \left[(11)\times 60^2 + (30)\times 60^1 + (4)\times 1\right]$

$= \left[3600 + 1200 + 21\right] + \left[39,600 + 1800 + 4\right]$

$= 4821 + 41,404$

$= 46,225$

$= (4\times 10^4) + (6\times 10^3) + (2\times 10^2) + (2\times 10^1) + (5\times 1)$

53. $\left[(1\times 360) + (6\times 20) + (6\times 1)\right] + \left[(5\times 360) + (0\times 20) + (13\times 1)\right]$

$= \left[360 + 120 + 6\right] + \left[1800 + 0 + 13\right]$

$= 486 + 1813$

$= 2299$

$= (2\times 10^3) + (2\times 10^2) + (9\times 10^1) + (9\times 1)$

55. 0.4759

57. 0.700203

59. 5000.03

61. 30,700.05809

63. 9734

65. 8097

67. $10^2 + 11^2 + 12^2 = 13^2 + 14^2$

$\ \ 100 + 121 + 144 = 169 + 196$

$365 = 365$

365 is the number of days in a non-leap year.

77. Change to Hindu-Arabic:

$$7 \times 360 = 2520$$
$$7 \times 20 = 140$$
$$7 \times 1 = \underline{7}$$
$$2667$$

Change to Babylonian:

$$= 2667$$
$$= 2640 + 27$$
$$= (44 \times 60) + 27$$
$$= (10 + 10 + 10 + 10 + 1 + 1 + 1 + 1) \times 60 + (10 + 10 + 1 + 1 + 1 + 1 + 1 + 1 + 1) \times 1$$

$$< < < < \vee \vee \vee \vee \qquad < < \vee \vee \vee \vee \vee \vee \vee$$

Check Points 4.2

1.
$$\begin{aligned} 3422_{\text{five}} &= (3 \times 5^3) + (4 \times 5^2) + (2 \times 5^1) + (2 \times 1) \\ &= (3 \times 5 \times 5 \times 5) + (4 \times 5 \times 5) + (2 \times 5) + (2 \times 1) \\ &= 375 + 100 + 10 + 2 \\ &= 487 \end{aligned}$$

2.
$$\begin{aligned} 110011_{\text{two}} &= (1 \times 2^5) + (1 \times 2^4) + (0 \times 2^3) + (0 \times 2^2) + (1 \times 2^1) + (1 \times 1) \\ &= (1 \times 32) + (1 \times 16) + (0 \times 8) + (0 \times 4) + (1 \times 2) + (1 \times 1) \\ &= 32 + 16 + 2 + 1 \\ &= 51 \end{aligned}$$

3.
$$\begin{aligned} \text{AD4}_{\text{sixteen}} &= (10 \times 16^2) + (13 \times 16^1) + (4 \times 1) \\ &= (10 \times 16 \times 16) + (13 \times 16) + (4 \times 1) \\ &= 2560 + 208 + 4 \\ &= 2772 \end{aligned}$$

4.
$$6_{\text{ten}} = (1 \times 5) + (1 \times 1) = 11_{\text{five}}$$

5. The place values in base 7 are $...7^4,\ 7^3,\ 7^2,\ 7^1,\ 1$ or $...2401, 343, 49, 7, 1$

$$\begin{array}{ccc}
1 & 0 & 3 \\
343\overline{)365} & 49\overline{)22} & 7\overline{)22} \\
\underline{343} & \underline{0} & \underline{21} \\
22 & 22 & 1
\end{array}$$

$$\begin{aligned} 365_{\text{ten}} &= (1 \times 343) + (0 \times 49) + (3 \times 7) + (1 \times 1) \\ &= (1 \times 7^3) + (0 \times 7^2) + (3 \times 7^1) + (1 \times 1) \\ &= 1031_{\text{seven}} \end{aligned}$$

6. The place values in base 5 are ...5^4, 5^3, 5^2, 5^1, 1 or ...3125, 625, 125, 25, 5, 1

$$625\overline{)2763} \qquad 125\overline{)263} \qquad 25\overline{)13} \qquad 5\overline{)13}$$

with quotients 4, 2, 0, 2 and:

$$\frac{2500}{263} \qquad \frac{250}{13} \qquad \frac{0}{13} \qquad \frac{10}{3}$$

$$2763_{ten} = (4\times625)+(2\times125)+(0\times25)+(2\times5)+(3\times1)$$
$$= (4\times5^4)+(2\times5^3)+(0\times5^2)+(2\times5^1)+(3\times1)$$
$$= 42023_{five}$$

Exercise Set 4.2

1. $(4\times5^1)+(3\times1)$
$$= 20+3$$
$$= 23$$

3. $(5\times8^1)+(2\times1)$
$$= 40+2$$
$$= 42$$

5. $(1\times4^2)+(3\times4^1)+(2\times1)$
$$= 16+12+2$$
$$= 30$$

7. $(1\times2^3)+(0\times2^2)+(1\times2^1)+(1\times1)$
$$= 8+0+2+1 = 11$$

9. $(2\times6^3)+(0\times6^2)+(3\times6^1)+(5\times1) = 432+0+18+5$
$$= 455$$

11. $(7\times8^4)+(0\times8^3)+(3\times8^2)+(5\times8^1)+(5\times1) = 28,672+0+192+40+5$
$$= 28,909$$

13. $(2\times16^3)+(0\times16^2)+(9\times16^1)+(6\times1) = 8192+0+144+6$
$$= 8342$$

15. $(1\times2^5)+(1\times2^4)+(0\times2^3)+(1\times2^2)+(0\times2^1)+(1\times1) = 32+16+0+4+0+1$
$$= 53$$

17. $(10\times16^3)+(12\times16^2)+(14\times16^1)+(5\times1) = 40,960+3072+224+5$
$$= 44,261$$

19. 12_{five} **25.** 31_{four}

21. 14_{seven} **27.** 101_{six}

23. 10_{two}

29. $25\overline{)87}^{\;3}$ $5\overline{)12}^{\;2}$

$\underline{75}$ $\underline{10}$

12 2

$87 = 322_{\text{five}}$

31. $64\overline{)108}^{\;1}$ $16\overline{)44}^{\;2}$ $4\overline{)12}^{\;3}$

$\underline{64}$ $\underline{32}$ $\underline{12}$

44 12 0

$108 = 1230_{\text{four}}$

33. $16\overline{)19}^{\;1}$ $8\overline{)3}^{\;0}$ $4\overline{)3}^{\;0}$ $2\overline{)3}^{\;1}$

$\underline{16}$ $\underline{0}$ $\underline{0}$ $\underline{2}$

3 3 3 1

$19 = 10011_{\text{two}}$

35. $81\overline{)138}^{\;1}$ $27\overline{)57}^{\;2}$ $9\overline{)3}^{\;0}$ $3\overline{)3}^{\;1}$

$\underline{81}$ $\underline{54}$ $\underline{0}$ $\underline{3}$

57 3 3 0

$138 = 12010_{\text{three}}$

37. $216\overline{)386}^{\;1}$ $36\overline{)170}^{\;4}$ $6\overline{)26}^{\;4}$

$\underline{216}$ $\underline{144}$ $\underline{24}$

170 26 2

$386 = 1442_{\text{six}}$

39. $343\overline{)1599}^{\;4}$ $49\overline{)227}^{\;4}$ $7\overline{)31}^{\;4}$

$\underline{1372}$ $\underline{196}$ $\underline{28}$

227 31 3

$1599 = 4443_{\text{seven}}$

41. $3052 = 3000 + 52$

$ = (50 \times 60^1) + (52 \times 1)$

$ = \;<<<<<\quad <<<<<\vee\vee$

43. $23,546 = 21,600 + 1920 + 26$

$ = (6 \times 60^2) + (32 \times 60^1) + (26 \times 1)$

$ = \;\vee\vee\vee\vee\vee\vee\quad <<<\vee\vee\quad <<\vee\vee\vee\vee\vee\vee$

45. 9307

$= 7200 + 1800 + 300 + 7$

$= (1 \times 7200) + (5 \times 360) + (15 \times 20) + (7 \times 1)$

•

‾‾‾‾‾

‗‗‗‗

• •

47. 28,704

$= 21,600 + 6840 + 260 + 4$

$= (3 \times 7200) + (19 \times 360) + (13 \times 20) + (4 \times 1)$

• • •

• • • •

‾‾‾‾‾

• • •

• • • •

49. $34_{\text{five}} = (3 \times 5^1) + (4 \times 1)$

$\phantom{34_{five}} = 15 + 4$

$\phantom{34_{five}} = 19_{\text{ten}}$

$19_{\text{ten}} = 14 + 5$

$\phantom{19_{ten}} = (2 \times 7^1) + (5 \times 1)$

$\phantom{19_{ten}} = 25_{\text{seven}}$

51. $110010011_{\text{two}} = 403_{\text{ten}}$

$403_{\text{ten}} = 384 + 16 + 3$

$\phantom{403_{ten}} = (6 \times 8^2) + (2 \times 8^1) + (3 \times 1)$

$\phantom{403_{ten}} = 623_{\text{eight}}$

53. Since A = 65, F = 70

$70 = 64 + 0 + 0 + 0 + 4 + 2 + 0$

$ = 1 \cdot 2^6 + 0 \cdot 2^5 + 0 \cdot 2^4 + 0 \cdot 2^3 + 1 \cdot 2^2 + 1 \cdot 2^1 + 0 \cdot 1$

$ = 1000110_{\text{two}}$

55. Since a = 97, m = 109

$109 = 64 + 32 + 0 + 8 + 4 + 0 + 1$

$ = 1 \cdot 2^6 + 1 \cdot 2^5 + 0 \cdot 2^4 + 1 \cdot 2^3 + 1 \cdot 2^2 + 0 \cdot 2^1 + 1 \cdot 1$

$ = 1101101_{\text{two}}$

57. 1010000_{two} 1000001_{two} 1001100_{two}

$64 + 16$ $64 + 1$ $64 + 8 + 4$

$\;80$ 65 76

$\;\text{P}$ A L

The word is PAL.

59.

M	o	m
77	111	109
$64+8+4+1$	$64+32+8+4+2+1$	$64+32+8+4+1$
1001101_{two}	1101111_{two}	1101101_{two}

The sequence is 100110111011111101101

65. Preceding: Following:

$$EC5_{sixteen} \qquad EC5_{sixteen}$$
$$\underline{-\ 1_{sixteen}} \qquad \underline{+\ 1_{sixteen}}$$
$$EC4_{sixteen} \qquad EC6_{sixteen}$$

Check Points 4.3

1.
$$\overset{1}{3}2_{five}$$
$$\underline{+44_{five}}$$
$$131_{five}$$
$$2+4=6=(1\times5^1)+(1\times1)=11_{five}$$
$$1+3+4=8=(1\times5^1)+(3\times1)=13_{five}$$

2.
$$\overset{1\ 1}{111}_{two}$$
$$\underline{+111}_{two}$$
$$1110_{two}$$
$$1+1=2=(1\times2^1)+(0\times1)=10_{two}$$
$$1+1+1=3=(1\times2^1)+(1\times1)=11_{two}$$

3.
$$\overset{3\ 6}{4\!1}_{five}$$
$$\underline{-23_{five}}$$
$$13_{five}$$

4.
$$\overset{4\ 8\ 3\ 11}{5144}_{seven}$$
$$\underline{-3236_{seven}}$$
$$1605_{seven}$$

5.
$$\overset{2}{4}5_{seven}$$
$$\underline{\times\ 3_{seven}}$$
$$201_{seven}$$

$$3\times5=15=(2\times7^1)+(1\times1)=21_{seven}$$
$$(3\times4)+2=14=(2\times7^1)+(0\times1)=20_{seven}$$

6.
$$2_{four}\overline{)112_{four}}$$
with quotient 23:
$$\begin{array}{r} 23 \\ 2_{four}\,\overline{)112_{four}} \\ \underline{10} \\ 12 \\ \underline{12} \\ 0 \end{array}$$
$$23_{four}$$

Exercise Set 4.3

Note: Numbers with no base specified are base 10.

1. $\overset{1}{2}3_{four}$
$+13_{four}$
$\overline{102_{four}}$

$3+3=6=(1\times4^1)+(2\times1)=12_{four}$
$1+2+1=4=(1\times4^1)+(0\times1)=10_{four}$

3. $\overset{1}{1}1_{two}$
$+11_{two}$
$\overline{110_{two}}$

$1+1+1=3=(1\times2^1)+(1\times1)=11_{two}$

5. $\overset{1\ 1}{3}42_{five}$
$+413_{five}$
$\overline{1310_{five}}$

$2+3=5=(1\times5^1)+(0\times1)=10_{five}$
$1+4+1=6=(1\times5^1)+(1\times1)=11_{five}$
$1+3+4=8=(1\times5^1)+(3\times1)=13_{five}$

7. $\overset{1\ 1}{6}45_{seven}$
$+324_{seven}$
$\overline{1302_{seven}}$

$5+4=9=(1\times7^1)+(2\times1)=12_{seven}$
$1+4+2=7=(1\times7^1)+(0\times1)=10_{seven}$
$1+6+3=10=(1\times7^1)+(3\times1)=13_{seven}$

9. $\overset{1\ 1\ 1}{6}784_{nine}$
$+7865_{nine}$
$\overline{15760_{nine}}$

$4+5=9=(1\times9^1)+(0\times1)=10_{nine}$
$1+8+6=15=(1\times9^1)+(6\times1)=16_{nine}$
$1+7+8=16=(1\times9^1)+(7\times1)=17_{nine}$
$1+6+7=14=(1\times9^1)+(5\times1)=15_{nine}$

11. $\overset{1\ 1}{1}4632_{seven}$
$+5604_{seven}$
$\overline{23536_{seven}}$

$6+6=12=(1\times7^1)+(5\times1)=15_{seven}$
$1+4+5=10=(1\times7^1)+(3\times1)=13_{seven}$

13. $\overset{2\ 6}{3}2_{four}$
-13_{four}
$\overline{13_{four}}$

15. $\overset{1\ 8}{2}3_{five}$
-14_{five}
$\overline{4_{five}}$

17. $\overset{6\ 13}{4}7\cancel{5}_{eight}$
-267_{eight}
$\overline{206_{eight}}$

19. $\overset{4\ 12\ \ 10}{5}6\cancel{3}_{seven}$
-164_{seven}
$\overline{366_{seven}}$

21. $\overset{0\ 1\ 2}{1}0\cancel{0}1_{two}$
-111_{two}
$\overline{10_{two}}$

23. $\overset{1\ 2\ 3}{1}2\cancel{0}\cancel{0}_{three}$
-1012_{three}
$\overline{111_{three}}$

25. $\overset{3}{2}5_{six}$
$\times4_{six}$
$\overline{152_{six}}$

$(2_{six}\times4_{six})+3_{six}=8_{ten}+3_{six}$
$=12_{six}+3_{six}$
$=15_{six}$

27.
$$11_{two}$$
$$\underline{\times \quad 1_{two}}$$
$$11_{two}$$

29.
$$\overset{3\ 2}{5\,4\,3}_{seven}$$
$$\underline{\times \qquad 5_{seven}}$$
$$4\,0\,1\,1_{seven}$$

$$3 \times 5 = 15 = (2 \times 7^1) + (1 \times 1) = 21_{seven}$$
$$(4 \times 5) + 2 = 22 = (3 \times 7^1) + (1 \times 1) = 31_{seven}$$
$$(5 \times 5) + 3 = 28 = (4 \times 7^1) + (0 \times 1) = 40_{seven}$$

31.
$$\overset{1\ 1}{6\,2\,3}_{eight}$$
$$\underline{\times \qquad 4_{eight}}$$
$$3\,1\,1\,4_{eight}$$

$$(3_{eight} \times 4_{eight}) = 12_{ten}$$
$$= (1 \times 8^1) + (4 \times 1)$$
$$= 14_{eight}$$
$$(2_{eight} \times 4_{eight}) + 1_{eight} = 8_{ten} + 1_{ten}$$
$$= 9_{ten}$$
$$= (1 \times 8^1) + (1 \times 1)$$
$$= 11_{eight}$$
$$(6_{eight} \times 4_{eight}) + 1_{eight} = 24_{ten} + 1_{ten}$$
$$= 25_{ten}$$
$$= (3 \times 8^1) + (1 \times 1)$$
$$= 31_{eight}$$

33.
$$21_{four}$$
$$\underline{\times 12_{four}}$$
$$102$$
$$\underline{210}$$
$$312_{four}$$

$$21_{four} \times 2_{four} = 9_{ten} \times 2_{ten}$$
$$= 18$$
$$= (1 \times 4^2) + (0 \times 4) + (2 \times 1)$$
$$= 102_{four}$$
$$21_{four} \times 1_{four} = 21_{four}$$

35.
$$2_{four}\overline{\smash{)}100_{four}}\ ^{\textstyle 20}$$
$$\underline{10}$$
$$00$$

$$20_{four}$$

37.
$$3_{five}\overline{\smash{)}224_{five}}\ ^{\textstyle 41}$$
$$\underline{22}$$
$$04$$
$$\underline{3}$$
$$1$$

$$41_{five} \text{ remainder of } 1$$

39.
$$\overset{1\ 1}{10110}_{two}$$
$$\underline{+\ 10100_{two}}$$
$$1000110_{two}$$

41.
$$\overset{1\ 1\ 1}{11111}_{two}$$
$$\underline{+\ 10110_{two}}$$
$$110101_{two}$$

$$110101_{two}$$
$$\underline{-\ \ \ \ 101_{two}}$$
$$110000_{two}$$

43.
$$1011_{two}$$
$$\underline{\times \quad 101_{two}}$$
$$1011_{two}$$
$$\underline{+\ 101100_{two}}$$
$$110111_{two}$$

45.
$$D3_{sixteen}$$
$$\underline{\times \quad 8A_{sixteen}}$$
$$83E_{sixteen}$$
$$\underline{+\ 6980_{sixteen}}$$
$$71BE_{sixteen}$$

47.
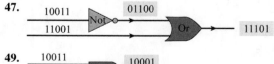

49.

51.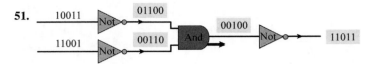

53. The circuit in Exercise 47 is a conditional gate.

57.
$$31_{\text{seven}} \overline{)2426_{\text{seven}}}^{\,56_{\text{seven}}}$$

$$\underline{215}$$
$$246$$
$$\underline{246}$$
$$0$$

Check Points 4.4

1. $100,000 + 100,000 + 100,000 + 100 + 100 + 10 + 10 + 1 + 1 = 300,222$

2. $2563 = 1000 + 1000 + 100 + 100 + 100 + 100 + 100 + 10 + 10 + 10 + 10 + 10 + 10 + 1 + 1 + 1$

〖〖⊚⊚⊚⊚⊚∩∩∩∩∩∩|||

3. $\text{MCCCLXI} = 1000 + 100 + 100 + 100 + 50 + 10 + 1 = 1361$

4. $\text{MCDXLVII} = \overset{M}{1000} + \overset{CD}{\overbrace{(500-100)}} + \overset{XL}{\overbrace{(50-10)}} + \overset{V}{5} + \overset{I}{1} + \overset{I}{1} = 1000 + 400 + 40 + 5 + 1 + 1 = 1447$

5. $399 = 100 + 100 + 100 + 90 + 9 = \overset{C}{100} + \overset{C}{100} + \overset{C}{100} + \overset{XC}{\overbrace{(100-10)}} + \overset{IX}{\overbrace{(10-1)}} = \text{CCCXCIX}$

6. $2693 = 2000 + 600 + 90 + 3$

二
千
六
百
九
十
三

7. $\omega\pi\varepsilon = 800 + 80 + 5 = 885$

Exercise Set 4.4

1. 322

3. 300,423

5. 132

7. $423 = (4 \times 100) + (2 \times 10) + (3 \times 1)$

⊚⊚⊚⊚∩∩|||

9. $1846 = (1\times1000)+(8\times100)+(4\times10)+(6\times1)$

   𓍢𓆼𓆼𓆼𓆼𓆼𓆼𓆼𓆼𓏲𓏲𓏲𓏲||||||

11. $23,547 = (2\times10,000)+(3\times1000)+(5\times100)+(4\times10)+(7\times1)$

   𓂭𓂭𓆼𓆼𓆼𓆼𓆼𓆼𓆼𓏲𓏲𓏲𓏲|||||||

13. XI = 11

15. XVI = 16

17. XL = 40

19. LIX = 59

21. CXLVI = 146

23. MDCXXI = 1621

25. MMDCLXXVII = 2677

27. $\overline{\text{IX}}$CDLXVI = 9466

29. 43 = XLIII

31. 129 = CXXIX

33. 1896 = MDCCCXCVI

35. 6892 = $\overline{\text{VI}}$DCCCXCII

37. $80 + 8 = 88$

$$\left.\begin{array}{r}8\\10\end{array}\right\}80$$
$$\left.8\right\}8$$

39. $500 + 20 + 7 = 527$

$$\left.\begin{array}{r}5\\100\end{array}\right\}500$$
$$\left.\begin{array}{r}2\\10\end{array}\right\}20$$
$$\left.7\right\}7$$

41. $2000+700+70+6=2776$

$$\left.\begin{array}{r}2\\1000\end{array}\right\}2000$$
$$\left.\begin{array}{r}7\\100\end{array}\right\}700$$
$$\left.\begin{array}{r}7\\10\end{array}\right\}70$$
$$\left.6\right\}6$$

43. 四
十
三

45. 五
百
八
十
三

47. 四
千
八
百
七
十

49. $\iota\beta = 12$

51. $\sigma\lambda\delta = 234$

53. $43 = \mu\gamma$

55. $483 = \upsilon\pi\gamma$

57. The value of this numeral is 2324.
Roman numeral: MMCCCXXIV
Chinese numeral:

二
千
三
百
二
十
四

59. The value of this numeral is 1741.
Egyptian numeral: 𓍢𓆼𓆼𓆼𓆼𓆼𓆼𓆼𓏲𓏲𓏲|
Chinese numeral:

一
千
七
百
四
十
一

61. The value of this numeral is 404.
$404 = 3104_{five}$

63. The value of this numeral is 192.
$192 = 1232_{five}$

65. ∩∩∩∩∩∩∩∩∩|||||||||

67. 1776 is the date the Declaration of Independence was signed.

69. Hindu-Arabic: 4,640,224
Roman numeral: $\overline{MMMMDCXLCCXXIV}$

77. Preceding: ⑨⑨⑨∩∩∩∩∩∩∩∩∩|||||||||

Following: ⑨⑨⑨

Chapter 4 Review Exercises

1. $11^2 = 11 \times 11 = 121$

2. $7^3 = 7 \times 7 \times 7 = 343$

3. $472 = (4 \times 10^2) + (7 \times 10^1) + (2 \times 1) = (4 \times 100) + (7 \times 10) + (2 \times 1)$

4. $8076 = (8 \times 10^3) + (0 \times 10^2) + (7 \times 10^1) + (6 \times 1) = (8 \times 1000) + (0 \times 100) + (7 \times 10) + (6 \times 1)$

5. $70,329 = (7 \times 10^4) + (0 \times 10^3) + (3 \times 10^2) + (2 \times 10^1) + (9 \times 1)$
$= (7 \times 10,000) + (0 \times 1000) + (3 \times 100) + (2 \times 10) + (9 \times 1)$

6. $706,953$

7. $740,000,306$

8. $<\!\vee \quad <\!\vee\vee\vee = (10+1) \times 60^1 + (10+1+1+1) \times 1$
$= (11 \times 60^1) + (13 \times 1)$
$= 660 + 13$
$= 673$

9. $\vee\vee \quad <\!< \quad <\!<\!<$
$= (1+1) \times 60^2 + (10+10) \times 60^1 + (10+10+10) \times 1$
$= (2 \times 60^2) + (20 \times 60) + (30 \times 1)$
$= (2 \times 3600) + 1200 + 30$
$= 7200 + 1230$
$= 8430$

10. $6 \times 360 = 2160$
$8 \times 20 = 160$
$11 \times 1 = \underline{\quad 11}$
2331

11. $9 \times 7200 = 64,800$
$2 \times 360 = 720$
$0 \times 20 = 0$
$16 \times 1 = \underline{\quad 16}$
$65,536$

12. Each position represents a particular value. The symbol in each position tells how many of that value are represented.

13. $34_{five} = (3 \times 5^1) + (4 \times 1)$
$= 15 + 4$
$= 19$

14. $110_{two} = (1 \times 2^2) + (1 \times 2^1) + (0 \times 1)$
$= 4 + 2 + 0$
$= 6$

15. $643_{seven} = (6 \times 7^2) + (4 \times 7^1) + (3 \times 1)$
$= 294 + 28 + 3$
$= 325$

16. $1084_{nine} = (1 \times 9^3) + (0 \times 9^2) + (8 \times 9^1) + (4 \times 1)$
$= 729 + 0 + 72 + 4$
$= 805$

17. $\text{FD3}_{\text{sixteen}} = (15 \times 16^2) + (13 \times 16^1) + (3 \times 1)$
$\quad\quad\quad\quad = 3840 + 208 + 3$
$\quad\quad\quad\quad = 4051$

18. $202202_{\text{three}} = (2 \times 3^5) + (0 \times 3^4) + (2 \times 3^3) + (2 \times 3^2) + (0 \times 3^1) + (2 \times 1)$
$\quad\quad\quad\quad\quad = 486 + 0 + 54 + 18 + 0 + 2$
$\quad\quad\quad\quad\quad = 560$

19. $89 = (3 \times 5^2) + (2 \times 5^1) + (4 \times 1)$
$\quad\quad = 324_{\text{five}}$

20. $21 = (1 \times 2^4) + (0 \times 2^3) + (1 \times 2^2) + (0 \times 2^1) + (1 \times 1)$
$\quad\quad = 10101_{\text{two}}$

21. $473 = (1 \times 3^5) + (2 \times 3^4) + (2 \times 3^3) + (1 \times 3^2) + (1 \times 3^1) + (2 \times 1)$
$\quad\quad\quad = 243 + 162 + 54 + 9 + 3 + 2$
$\quad\quad\quad = 122112_{\text{three}}$

22. $7093 = (2 \times 7^4) + (6 \times 7^3) + (4 \times 7^2) + (5 \times 7^1) + (2 \times 1)$
$\quad\quad\quad = 4802 + 2058 + 196 + 35 + 2$
$\quad\quad\quad = 26452_{\text{seven}}$

23. $9348 = (1 \times 6^5) + (1 \times 6^4) + (1 \times 6^3) + (1 \times 6^2) + (4 \times 6^1) + (0 \times 1)$
$\quad\quad\quad = 7776 + 1296 + 216 + 36 + 24$
$\quad\quad\quad = 111140_{\text{six}}$

24. $554 = (3 \times 12^2) + (A \times 12^1) + (2 \times 1)$
$\quad\quad\quad = 3A2_{\text{twelve}}$

25.
$$\begin{array}{r} \overset{1}{}46_{\text{seven}} \\ +\,53_{\text{seven}} \\ \hline 132_{\text{seven}} \end{array}$$

26.
$$\begin{array}{r} \overset{1\ \ 1}{}574_{\text{eight}} \\ +\,605_{\text{eight}} \\ \hline 1401_{\text{eight}} \end{array}$$

27.
$$\begin{array}{r} \overset{1\,1\,1\,1}{11011}_{\text{two}} \\ 10101_{\text{two}} \\ \hline 110000_{\text{two}} \end{array}$$

28.
$$\begin{array}{r} \overset{1}{}43C_{\text{sixteen}} \\ +\,694_{\text{sixteen}} \\ \hline AD0_{\text{sixteen}} \end{array}$$

29.
$$\begin{array}{r} \overset{2\ \ 10}{3\!\!\not4}_{\text{six}} \\ 25_{\text{six}} \\ \hline 5_{\text{six}} \end{array}$$

30.
$$\begin{array}{r} \overset{5\ \ 8\ \ 11}{624}_{\text{seven}} \\ -\,246_{\text{seven}} \\ \hline 345_{\text{seven}} \end{array}$$

31.
$$\begin{array}{r} \overset{0\ 1\ 2}{10\!\!\not0\!\!\not0 1}_{\text{two}} \\ -\,110_{\text{two}} \\ \hline 11_{\text{two}} \end{array}$$

32.
$$\begin{array}{r} \overset{3\ 6\ 1\ 6}{412\!\!\not1}_{\text{five}} \\ -\,1312_{\text{five}} \\ \hline 2304_{\text{five}} \end{array}$$

33.
$$\overset{1}{3}2_{\text{four}}$$
$$\underline{\times\ \ 3_{\text{four}}}$$
$$222_{\text{four}}$$

34.
$$\overset{2}{4}3_{\text{seven}}$$
$$\underline{\times\ \ 6_{\text{seven}}}$$
$$354_{\text{seven}}$$

35.
$$\overset{2\ 2}{1}23_{\text{five}}$$
$$\underline{\ \ 4_{\text{five}}}$$
$$1102_{\text{five}}$$

36.
$$2_{\text{four}}\overline{)\,332_{\text{four}}}\quad\frac{133}{}$$
$$\underline{2}$$
$$13$$
$$\underline{12}$$
$$12$$
$$\underline{12}$$
$$0$$
$$133_{\text{four}}$$

37.
$$4_{\text{five}}\overline{)\,103_{\text{five}}}\quad\frac{12}{}$$
$$\underline{4}$$
$$13$$
$$\underline{13}$$
$$0$$
$$12_{\text{five}}$$

38. 1246

39. 12,432

40. $2486 = (2 \times 1000) + (4 \times 100) + (8 \times 10) + (6 \times 1)$

41. $34{,}573 = (3 \times 10{,}000) + (4 \times 1000) + (5 \times 100) + (7 \times 10) + (3 \times 1)$

42. DDCCCBAAAA = 2314

43. 5492 = DDDDDCCCCBBBBBBBBBAA

44. Answers will vary.

45. CLXIII = 163

46. MXXXIV = 1034

47. MCMXC = 1990

48. 49 = XLIX

49. 2965 = MMCMLXV

50. If symbols increase in value from left to right, subtract the value of the symbol on the left from the symbol on the right.

51. $500 + 50 + 4 = 554$

$$\left.\begin{array}{r}5\\100\end{array}\right\}500$$

$$\left.\begin{array}{r}5\\10\end{array}\right\}50$$

$$\left.\begin{array}{r}\\4\end{array}\right\}4$$

52. $8000 + 200 + 50 + 3 = 8253$

$\left.\begin{array}{r}8 \\ 1000\end{array}\right\}8000$

$\left.\begin{array}{r}2 \\ 100\end{array}\right\}200$

$\left.\begin{array}{r}5 \\ 10\end{array}\right\}50$

$3\}3$

53. 二
百
七
十
四

54. 三
千
五
百
八
十
七

55. 365

56. 4520

57. G
Y
I
X
C

58. F
Z
H
Y
E
X
D

59. Answers will vary.

60. $\chi\nu\gamma = 653$

61. $\chi o\eta = 678$

62. $453 = \upsilon\nu\gamma$

63. $902 = \pi\beta$

64. UNG = 357

65. mhZRD = 37,894

66. rXJH = 80,618

67. 597 = WRG

68. $25,483 = $ lfVQC

Chapter 4 Test

1. $9 \times 9 \times 9 = 729$

2. $567 = (5 \times 10^2) + (6 \times 10^1) + (7 \times 1)$
$= (5 \times 100) + (6 \times 10) + (7 \times 1)$

3. $63,028 = (6 \times 10^4) + (3 \times 10^3) + (0 \times 10^2) + (2 \times 10^1) + (8 \times 1)$
$= (6 \times 10,000) + (3 \times 1000) + (0 \times 100) + (2 \times 10) + (8 \times 1)$

4. $7000 + 400 + 90 + 3 = 7493$

5. $400,000 + 200 + 6 = 400,206$

6. A number represents, "How many?" whereas a numeral is a symbol used to write a number

7. A symbol for zero is needed as a place holder when there are no values for a position.

8. $<< < \vee\vee \ \ <\vee = (10+10) \times 60^2 + (10+1+1) \times 60^1 + (10+1) \times 1$
$= (20 \times 60^2) + (12 \times 60) + (11 \times 1) = 72,000 + 720 + 11 = 72,731$

9. $4 \times 360 = 1440$
    $6 \times 20 = 120$
    $0 \times 1 = 0$
    $\overline{1560}$

10. $423_{\text{five}} = (4 \times 5^2) + (2 \times 5^1) + (3 \times 1) = 4 \times 25 + 10 + 3 = 100 + 10 + 3 = 113$

11. $267_{\text{nine}} = (2 \times 9^2) + (6 \times 9^1) + (7 \times 1) = 2 \times 81 + 54 + 7 = 162 + 54 + 7 = 223$

12. $110101_{\text{two}} = (1 \times 2^5) + (1 \times 2^4) + (0 \times 2^3) + (1 \times 2^2) + (0 \times 2^1) + (1 \times 1) = 32 + 16 + 0 + 4 + 0 + 1 = 53$

13. $77 = (2 \times 3^3) + (2 \times 3^2) + (1 \times 3^1) + (2 \times 1) = 2212_{\text{three}}$

14. $56 = (1 \times 2^5) + (1 \times 2^4) + (1 \times 2^3) + (0 \times 2^2) + (0 \times 2^1) + (0 \times 1) = 111000_{\text{two}}$

15. $1844 = (2 \times 5^4) + (4 \times 5^3) + (3 \times 5^2) + (3 \times 5^1) + (4 \times 1) = 1250 + 500 + 75 + 15 + 4 = 24334_{\text{five}}$

16. $\overset{1 1}{234_{\text{five}}}$
    $+423_{\text{five}}$
    $\overline{1212_{\text{five}}}$

17. $\overset{59}{5\cancel{6}2_{\text{seven}}}$
    -145_{seven}
    $\overline{414_{\text{seven}}}$

18. $\overset{2}{5}4_{\text{six}}$
    $\times 3_{\text{six}}$
    $\overline{250_{\text{six}}}$

19. $3_{\text{five}} \overline{)\,1213_{\text{five}}}$, quotient 221

    11
    $\overline{11}$
    11
    $\overline{03}$
    3
    $\overline{0}$

    221_{five}

20. 20,303

21. $32,634 = (3 \times 10,000) + (2 \times 1000) + (6 \times 100) + (3 \times 10) + (4 \times 1)$

    𝄇𝄇𝄇 ⚱⚱ ⚰ ͻͻͻͻͻͻͻ ∩∩∩ ||||

22. $\text{MCMXCIV} = \overbrace{1000}^{M} + \overbrace{(1000 - 100)}^{CM} + \overbrace{(100 - 10)}^{XC} + \overbrace{(5 - 1)}^{IV} = 1000 + 900 + 90 + 4 = 1994$

23. $459 = \overbrace{(500 - 100)}^{CD} + \overbrace{50}^{L} + \overbrace{(10 - 1)}^{IX} = \text{CDLIX}$

24. Answers will vary.

Chapter 5
Number Theory and the Real Number System

Check Points 5.1

1. The statement given in part (b) is true.

   a. False, 8 does not divide 48,324 because 8 does not divide 324.

   b. True, 6 divides 48,324 because both 2 and 3 divide 48,324. 2 divides 48,324 because the last digit is 4. 3 divides 48,324 because the sum of the digits, 21, is divisible by 3.

   c. False, 4 *does* divide 48,324 because the last two digits form 24 which is divisible by 4.

2.

   $$120 = 2^3 \cdot 3 \cdot 5$$

3. $225 = 3^2 \cdot 5^2$
   $825 = 3 \cdot 5^2 \cdot 11$
   Greatest Common Divisor: $3 \cdot 5^2 = 75$

4. $192 = 2^6 \cdot 3$
   $288 = 2^5 \cdot 3^2$
   Greatest Common Divisor: $2^5 \cdot 3 = 96$
   The largest number of people that can be placed in each singing group is 96.

5. $18 = 2 \cdot 3^2$
   $30 = 2 \cdot 3 \cdot 5$
   Least common multiple is: $90 = 2 \cdot 3^2 \cdot 5$

6. $40 = 2^3 \cdot 5$
   $60 = 2^2 \cdot 3 \cdot 5$
   Least common multiple is: $120 = 2^3 \cdot 3 \cdot 5$
   It will be 120 minutes, or 2 hours, until both movies begin again at the same time.
   The time will be 5:00 PM.

Exercise Set 5.1

1. 6944

   a. Yes. The last digit is four.

   b. No. The sum of the digits is 23, which is not divisible by 3.

   c. Yes. The last two digits form 44, which is divisible by 4.

   d. No. The number does not end in 0 or 5.

   e. No. The number is not divisible by both 2 and 3.

   f. Yes. The last three digits form 944, which is divisible by 8.

   g. No. The sum of the digits is 23, which is not divisible by 9.

   h. No. The number does not end in 0.

   i. No. The number is not divisible by both 3 and 4.

3. 21,408

   a. Yes. The last digit is eight.

   b. Yes. The sum of the digits is 15, which is divisible by 3.

   c. Yes. The last two digits form 08, which is divisible by 4.

   d. No. The number does not end in 0 or 5.

   e. Yes. The number is divisible by both 2 and 3.

   f. Yes. The last three digits form 408, which is divisible by 8.

   g. No. The sum of the digits is 15, which is not divisible by 9.

   h. No. The number does not end in 0.

   i. Yes. The number is divisible by both 3 and 4.

5. 26,428

    a. Yes. The last digit is 8.

    b. No. The sum of the digits is 22, which is not divisible by 3.

    c. Yes. The last 2 digits form 28, which is divisible by 4.

    d. No. The last digit is eight.

    e. No. The number is not divisible by both two and three.

    f. No. The last three digits form 428, which is not divisible by 8.

    g. No. The sum of the digits is 22, which is not divisible by 9.

    h. No. The number does not end in 0.

    i. No. The number is not divisible by 3 and 4.

7. 374,832

    a. Yes. The last digit is 2.

    b. Yes. The sum of the digits is 27, which is divisible by 3.

    c. Yes. The last two digits form 32, which is divisible by 4.

    d. No. The last digit is two.

    e. Yes. The number is divisible by 2 and 3.

    f. Yes. The last 3 digits form 832, which is divisible by 8.

    g. Yes. The sum of the digits is 27, which is divisible by 9.

    h. No. The last digit is 2.

    i. Yes. The number is divisible by both 3 and 4.

9. 6,126,120

    a. Yes. The last digit is 0.

    b. Yes. The sum of the digits is 18, which is divisible by 3.

    c. Yes. The last two digits form 20, which is divisible by 4.

    d. Yes. The last digit is 0.

    e. Yes. The number is divisible by both 2 and 3.

    f. Yes. The last 3 digits form 120, which is divisible by 8.

    g. Yes. The sum of the digits is 18, which is divisible by 9.

    h. Yes. The last digit is 0.

    i. Yes. The number is divisible by both 3 and 4.

11. True. $5958 \div 3 = 1986$
The sum of the digits is 27, which is divisible by 3.

13. True. $10,612 \div 4 = 2653$
The last two digits form 12, which is divisible by 4.

15. False

17. True. $104,538 \div 6 = 17,423$
The number is divisible by both 2 and 3.

19. True. $20,104 \div 8 = 2513$
The last three digits form 104, which is divisible by 8.

21. False

23. True. $517,872 \div 12 = 43,156$
The number is divisible by both 3 and 4.

25.

$$75 = 3 \cdot 5^2$$

27.

$$56 = 2^3 \cdot 7$$

29.

$$105 = 3 \cdot 5 \cdot 7$$

31.

$$500 = 2^2 \cdot 5^3$$

33.

$$663 = 3 \cdot 13 \cdot 17$$

35.

$$885 = 3 \cdot 5 \cdot 59$$

37.

$$1440 = 2^5 \cdot 3^2 \cdot 5$$

39.

$$1996 = 2^2 \cdot 499$$

41.

$$3675 = 3 \cdot 5^2 \cdot 7^2$$

43.

$$85,800 = 2^3 \cdot 3 \cdot 5^2 \cdot 11 \cdot 13$$

45. $42 = 2 \cdot 3 \cdot 7$

$56 = 2^3 \cdot 7$

Greatest Common Divisor: $2 \cdot 7 = 14$

47. $16 = 2^4$

$42 = 2 \cdot 3 \cdot 7$

Greatest Common Divisor: 2

49. $60 = 2^2 \cdot 3 \cdot 5$

$108 = 2^2 \cdot 3^3$

Greatest Common Divisor: $2^2 \cdot 3 = 12$

51. $72 = 2^3 \cdot 3^2$

$120 = 2^3 \cdot 3 \cdot 5$

Greatest Common Divisor: $2^3 \cdot 3 = 24$

53. $324 = 2 \cdot 3^2 \cdot 19$

$380 = 2^2 \cdot 5 \cdot 19$

Greatest Common Divisor: $2 \cdot 19 = 38$

55. $240 = 2^4 \cdot 3 \cdot 5$

$285 = 3 \cdot 5 \cdot 19$

Greatest Common Divisor: $3 \cdot 5 = 15$

57. $42 = 2 \cdot 3 \cdot 7$

$56 = 2^3 \cdot 7$

Least Common Multiple: $2^3 \cdot 3 \cdot 7 = 168$

59. $16 = 2^4$

$42 = 2 \cdot 3 \cdot 7$

Least Common Multiple: $2^4 \cdot 3 \cdot 7 = 336$

61. $60 = 2^2 \cdot 3 \cdot 5$

$108 = 2^2 \cdot 3^3$

Least Common Multiple: $2^2 \cdot 3^3 \cdot 5 = 540$

63. $72 = 2^3 \cdot 3^2$

$120 = 2^3 \cdot 3 \cdot 5$

Least Common Multiple: $2^3 \cdot 3^2 \cdot 5 = 360$

65. $342 = 2 \cdot 3^2 \cdot 19$

$380 = 2^2 \cdot 5 \cdot 19$

Least Common Multiple: $2^2 \cdot 3^2 \cdot 5 \cdot 19 = 3420$

67. $240 = 2^4 \cdot 3 \cdot 5$

$285 = 3 \cdot 5 \cdot 19$

Least Common Multiple

$= 2^4 \cdot 3 \cdot 5 \cdot 19$

$= 4560$

69. $d = 8$

$9 \vert 12,348$

71. $d = 6$

$8 \vert 76,523,456$

73. $d = 2,\ 6$

$4 \vert 963,232$ and $4 \vert 963,236$

75. 28 is a perfect number.

$28 = 1 + 2 + 4 + 7 + 14$

77. 20 is not a perfect number.

$20 \neq 1 + 2 + 4 + 5 + 10$

79. 41 is not an emirp because 14 is not prime.

81. 107 is an emirp because 701 is also prime.

83. 13 is not a Germain prime because $2(13) + 1 = 27$ is not prime.

85. 241 is not a Germain prime because

$2(241) + 1 = 483$ is not prime.

87. The GCD of 24 and 27 is 3.

The LCM of 24 and 27 is 216.

$3 \times 216 = 648$

$24 \times 27 = 648$

The product of the greatest common divisor and least common multiple of two numbers equals the product of the two numbers.

89. The numbers are the prime numbers less than 100.

91. $300 = 2^2 \cdot 3 \cdot 5^2$

$144 = 2^4 \cdot 3^2$

Greatest Common Divisor: $2^2 \cdot 3 = 12$

There would be 25 groups with 12 bottles of water each. There would be 12 groups with 12 cans of food each.

93. $310 = 2 \cdot 5 \cdot 31$

$460 = 2^2 \cdot 5 \cdot 23$

Greatest Common Divisor: $2 \cdot 5 = 10$

There would be 31 groups of 10 five-dollar bills.

There would be 46 groups of

10 ten-dollar bills.

95. $6 = 2 \cdot 3$

$10 = 2 \cdot 5$

Least Common Multiple is: $2 \cdot 3 \cdot 5 = 30$

It will be 30 more nights until both have the evening off, or July 1.

97. $15 = 3 \cdot 5$

$18 = 2 \cdot 3^2$

Least Common Multiple is: $2 \cdot 3^2 \cdot 5 = 90$

It takes 90 minutes.

111. a. GCD $= 2^{14} \cdot 3^{25} \cdot 5^{30}$

   b. LCM $= 2^{17} \cdot 3^{37} \cdot 5^{31}$

113. $85 + 15 = 100 = 2^2 \cdot 5^2$

$100 + 15 = 115 = 5 \cdot 23$

LCM $= 2^2 \cdot 5 \cdot 23 = 2300$

The films will begin at the same time

2300 min $\left(= 38\dfrac{1}{3} \text{hr} \right)$ after noon (today), or at 2:20

A.M. on the third day.

115. Yes, since 96 is divisible by 4, then 67,234,096 is divisible by 4.

117. Yes, since $4 + 8 + 2 + 0 + 1 + 6 + 5 + 1 = 27$ is divisible by 9, then 48,201,651 is divisible by 9.

Check Points 5.2

1.

2. **a.** $6 > -7$ because 6 is to the right of -2 on the number line.

   b. $-8 < -1$ because -8 is to the left of -1 on the number line.

   c. $-25 < -2$ because -25 is to the left of -2 on the number line.

   d. $-14 < 0$ because -14 is to the left of 0 on the number line.

3. **a.** $|-8| = 8$ because -8 is 8 units from 0.

   b. $|6| = 6$ because 6 is 6 units from 0.

4. **a.** $30 - (-7) = 30 + 7 = 37$

   b. $-14 - (-10) = -14 + 10 = -4$

   c. $-14 - 10 = -24$

5. $\overbrace{-412}^{2004\ \text{deficit}} - \overbrace{(-427)}^{2005\ \text{deficit}} = -412 + 427 = \$15$ billion

6. **a.** $(-5)^2 = (-5)(-5) = 25$

   b. $-5^2 = -(5 \cdot 5) = -25$

   c. $(-4)^3 = (-4)(-4)(-4) = -64$

   d. $(-3)^4 = (-3)(-3)(-3)(-3) = 81$

7. $7^2 - 48 \div 4^2 \cdot 5 + 2$
   $= 49 - 48 \div 16 \cdot 5 + 2$
   $= 49 - 3 \cdot 5 + 2$
   $= 49 - 15 + 2$
   $= 34 + 2$
   $= 36$

8. $(-8)^2 - (10 - 13)^2(-2)$
   $= (-8)^2 - (-3)^2(-2)$
   $= 64 - (9)(-2)$
   $= 64 - (-18)$
   $= 64 + (+18)$
   $= 82$

Exercise Set 5.2

1.

3.

5. $-2 < 7$ because -2 is to the left of 7 on the number line.

7. $-13 < -2$ because -13 is to the left of -2 on the number line.

9. $8 > -50$ because 8 is to the right of -50 on the number line.

11. $-100 < 0$ because -100 is to the left of 0 on the number line.

13. $|-14| = 14$ because -14 is 14 units from 0.

15. $|14| = 14$ because 14 is 14 units from 0.

17. $|-300,000| = 300,000$ because $-300,000$ is 300,000 units from 0.

19. $-7 + (-5) = -12$

21. $12 + (-8) = 4$

23. $6 + (-9) = -3$

25. $-9 + (+4) = -5$

27. $-9 + (-9) = -18$

29. $9 + (-9) = 0$

31. $13 - 8 = 5$

33. $8 - 15 = 8 + (-15) = -7$

35. $4 - (-10) = 4 + 10 = 14$

37. $-6-(-17)=-6+17=11$

39. $-12-(-3)=-12+3=-9$

41. $-11-17=-11+(-17)=-28$

43. $6(-9)=-54$

45. $(-7)(-3)=21$

47. $(-2)(6)=-12$

49. $(-13)(-1)=13$

51. $0(-5)=0$

53. $5^2=5\cdot 5=25$

55. $(-5)^2=(-5)\cdot(-5)=25$

57. $4^3=4\cdot 4\cdot 4=64$

59. $(-5)^3=(-5)(-5)(-5)=25(-5)=-125$

61. $(-5)^4=(-5)(-5)(-5)(-5)=625$

63. $-3^4=-\left[3\cdot 3\cdot 3\cdot 3\right]=-81$

65. $(-3)^4=(-3)(-3)(-3)(-3)=81$

67. $\dfrac{-12}{4}=-3$

69. $\dfrac{21}{-3}=-7$

71. $\dfrac{-90}{-3}=30$

73. $\dfrac{0}{-7}=0$

75. $\dfrac{-7}{0}$ is undefined

77. $(-480)\div 24=\dfrac{-480}{24}=-20$

79. $(465)\div(-15)=\dfrac{465}{-15}=-31$

81. $7+6\cdot 3=7+18=25$

83. $(-5)-6(-3)=-5+18=13$

85. $6-4(-3)-5=6-(-12)-5$
$=6+12-5$
$=18-5$
$=13$

87. $3-5(-4-2)=3-5(-6)$
$=3-(-30)$
$=3+30$
$=33$

89. $(2-6)(-3-5)=(-4)(-8)=32$

91. $3(-2)^2-4(-3)^2=3(4)-4(9)$
$=12-36$
$=-24$

93. $(2-6)^2-(3-7)^2=(-4)^2-(-4)^2$
$=16-16$
$=0$

95. $6(3-5)^3-2(1-3)^3=6(-2)^3-2(-2)^3$
$=6(-8)-2(-8)$
$=-48+16$
$=-32$

97. $8^2-16\div 2^2\cdot 4-3=64-16\div 4\cdot 4-3$
$=64-4\cdot 4-3$
$=64-16-3$
$=45$

99. $8-3\left[-2(2-5)-4(8-6)\right]$
$=8-3\left[-2(-3)-4(2)\right]$
$=8-3\left[6-8\right]$
$=8-3\left[-2\right]$
$=8+6$
$=14$

101. $-2^2+4\left[16\div(3-5)\right]$
$=-4+4\left[16\div(-2)\right]$
$=-4+4\left[-8\right]$
$=-4-32$
$=-36$

103. $4|10-(8-20)|$
$= 4|10-(-12)|$
$= 4|10+12|$
$= 4|22|$
$= 88$

105. $-10-(-2)^3 = -10-(-8) = -10+8 = -2$

107. $[2(7-10)]^2 = [2(-3)]^2 = [-6]^2 = 36$

109. $\overbrace{-416}^{\text{2000 deficit}} - \overbrace{(-805)}^{\text{2005 deficit}} = -416+805 = \$389 \text{ billion}$

111. $\overbrace{(-805)}^{\text{2005 deficit}} - 2\overbrace{(-390)}^{\text{2001 deficit}} = -805+780 = \$25 \text{ billion}$

113. The difference in elevation is
$19,321-(-436)$
$= 19,321+436$
$= 19,757 \text{ feet}$

115. $\$1853-\$2011 = -\$158 \text{ billion}$
There was a budget deficit in 2002.

117. $\overbrace{(1991-1863)}^{\text{2001 surplus}} - \overbrace{(2053-2479)}^{\text{2005 deficit}} = (128)-(-426)$
$= \$554 \text{ billion}$

119. $3°-(-4°) = 3°+4° = 7°\text{ F}$

121. $-24°-(-22°) = -24°+22° = -2°\text{ F}$

133. $(8-2)\cdot 3-4 = 14$

135. -36

Check Points 5.3

1. $72 = 2^3 \cdot 3^2$
$90 = 2 \cdot 5 \cdot 3^2$
Greatest Common Divisor is $2 \cdot 3^2$ or 18.
$\dfrac{72}{90} = \dfrac{72 \div 18}{90 \div 18} = \dfrac{4}{5}$

2. $2\dfrac{5}{8} = \dfrac{8 \cdot 2 + 5}{8} = \dfrac{16+5}{8} = \dfrac{21}{8}$

3. $\dfrac{5}{3} = 1\dfrac{2}{3}$

4. a. $\dfrac{3}{8} = 0.375$

$$
\begin{array}{r}
0.375 \\
8\overline{)3.000} \\
\underline{2\,4} \\
60 \\
\underline{56} \\
40 \\
\underline{40} \\
0
\end{array}
$$

   b. $\dfrac{5}{11} = 0.\overline{45}$

$$
\begin{array}{r}
0.4545\ldots \\
11\overline{)5.0000} \\
\underline{44} \\
60 \\
\underline{55} \\
50 \\
\underline{44} \\
60 \\
\underline{55} \\
5
\end{array}
$$

5. a. $0.9 = \dfrac{9}{10}$

   b. $0.86 = \dfrac{86}{100} = \dfrac{86 \div 2}{100 \div 2} = \dfrac{43}{50}$

   c. $0.053 = \dfrac{53}{1000}$

6. $n = 0.\overline{2}$

$n = 0.22222\ldots$

$10n = 2.22222\ldots$

$10n = 2.2222\ldots$

$\underline{-n = 0.2222\ldots}$

$9n = 2.0$

$n = \dfrac{2}{9}$

7. $n = 0.\overline{79}$

$n = 0.7979\ldots$

$100n = 79.7979\ldots$

$100n = 79.7979\ldots$

$\underline{-\ \ n = \ \ 0.7979\ldots}$

$99n = 79$

$n = \dfrac{79}{99}$

8. a. $\dfrac{4}{11} \cdot \dfrac{2}{3} = \dfrac{8}{33}$

b. $\left(-\dfrac{3}{7}\right)\left(-\dfrac{14}{4}\right) = \dfrac{42}{28} = \dfrac{42 \div 14}{28 \div 14} = \dfrac{3}{2}$ or $1\dfrac{1}{2}$

c. $\left(3\dfrac{2}{5}\right)\left(1\dfrac{1}{2}\right) = \dfrac{17}{5} \cdot \dfrac{3}{2} = \dfrac{51}{10}$ or $5\dfrac{1}{10}$

9. a. $\dfrac{9}{11} \div \dfrac{5}{4} = \dfrac{9}{11} \cdot \dfrac{4}{5} = \dfrac{36}{55}$

b. $-\dfrac{8}{15} \div \dfrac{2}{5} = -\dfrac{8}{15} \cdot \dfrac{5}{2} = -\dfrac{40}{30} = -\dfrac{4}{3}$ or $-1\dfrac{1}{3}$

c. $3\dfrac{3}{8} \div 2\dfrac{1}{4} = \dfrac{27}{8} \div \dfrac{9}{4} = \dfrac{27}{8} \cdot \dfrac{4}{9} = \dfrac{108}{72} = \dfrac{3}{2}$ or $1\dfrac{1}{2}$

10. a. $\dfrac{5}{12} + \dfrac{3}{12} = \dfrac{5+3}{12} = \dfrac{8}{12} = \dfrac{2}{3}$

b. $\dfrac{7}{4} - \dfrac{1}{4} = \dfrac{7-1}{4} = \dfrac{6}{4} = \dfrac{3}{2}$ or $1\dfrac{1}{2}$

c. $-3\dfrac{3}{8} - \left(-1\dfrac{1}{8}\right) = -\dfrac{27}{8} - \left(-\dfrac{9}{8}\right)$

$= -\dfrac{27}{8} + \dfrac{9}{8}$

$= \dfrac{-27+9}{8}$

$= \dfrac{-18}{8}$

$= -\dfrac{9}{4}$

or $-2\dfrac{1}{4}$

11. $\dfrac{1}{5} + \dfrac{3}{4} = \dfrac{1}{5} \cdot \dfrac{4}{4} + \dfrac{3}{4} \cdot \dfrac{5}{5} = \dfrac{4}{20} + \dfrac{15}{20} = \dfrac{19}{20}$

12. $\dfrac{3}{10} - \dfrac{7}{12} = \dfrac{3}{10} \cdot \dfrac{6}{6} - \dfrac{7}{12} \cdot \dfrac{5}{5} = \dfrac{18}{60} - \dfrac{35}{60} = -\dfrac{17}{60}$

13.   First, find the sum:

$\dfrac{1}{3} + \dfrac{1}{2} = \dfrac{1}{3} \cdot \dfrac{2}{2} + \dfrac{1}{2} \cdot \dfrac{3}{3} = \dfrac{2}{6} + \dfrac{3}{6} = \dfrac{5}{6}$

Next, divide by 2: $\dfrac{5}{6} \div \dfrac{2}{1} = \dfrac{5}{6} \cdot \dfrac{1}{2} = \dfrac{5}{12}$

14.   Amount of eggs needed

$= \dfrac{\text{desired serving size}}{\text{recipe serving size}} \times \text{eggs in recipe}$

$= \dfrac{7 \ \cancel{\text{dozen}}}{5 \ \cancel{\text{dozen}}} \times 2 \text{ eggs}$

$= \dfrac{14}{5} \text{ eggs}$

$= 2\dfrac{4}{5} \text{ eggs}$

$\approx 3 \text{ eggs}$

Exercise Set 5.3

1. $10 = 2 \cdot 5$

$15 = 3 \cdot 5$

Greatest Common Divisor is 5.

$\dfrac{10}{15} = \dfrac{10 \div 5}{15 \div 5} = \dfrac{2}{3}$

3. $15 = 3 \cdot 5$

$18 = 2 \cdot 3^2$

Greatest Common Divisor is 3.

$\dfrac{15}{18} = \dfrac{15 \div 3}{18 \div 3} = \dfrac{5}{6}$

5. $24 = 2^3 \cdot 3$

$42 = 2 \cdot 3 \cdot 7$

Greatest Common Divisor is $2 \cdot 3$ or 6.

$\dfrac{24}{42} = \dfrac{24 \div 6}{42 \div 6} = \dfrac{4}{7}$

7. $60 = 2^2 \cdot 3 \cdot 5$

$108 = 2^2 \cdot 3^3$

Greatest Common Divisor is $2^2 \cdot 3$ or 12.

$\dfrac{60}{108} = \dfrac{60 \div 12}{108 \div 12} = \dfrac{5}{9}$

9. $342 = 2 \cdot 3^2 \cdot 19$

$380 = 2^2 \cdot 5 \cdot 19$

Greatest Common Divisor is $2 \cdot 19$ or 38.

$\dfrac{342}{380} = \dfrac{342 \div 38}{380 \div 38} = \dfrac{9}{10}$

11. $308 = 2^2 \cdot 7 \cdot 11$

$418 = 2 \cdot 11 \cdot 19$

Greatest Common Divisor is $2 \cdot 11$ or 22.

$\dfrac{308}{418} = \dfrac{308 \div 22}{418 \div 22} = \dfrac{14}{19}$

13. $2\dfrac{3}{8} = \dfrac{8 \cdot 2 + 3}{8} = \dfrac{16 + 3}{8} = \dfrac{19}{8}$

15. $-7\dfrac{3}{5} = -\dfrac{5 \cdot 7 + 3}{5} = -\dfrac{35 + 3}{5} = -\dfrac{38}{5}$

17. $12\dfrac{7}{16} = \dfrac{16 \cdot 12 + 7}{16} = \dfrac{192 + 7}{16} = \dfrac{199}{16}$

19. $\dfrac{23}{5} = 4\dfrac{3}{5}$

21. $-\dfrac{76}{9} = -8\dfrac{4}{9}$

23. $\dfrac{711}{20} = 35\dfrac{11}{20}$

25. $\dfrac{3}{4} = 0.75$

$$
\begin{array}{r}
0.75 \\
4\overline{)3.00} \\
28 \\
\hline
20 \\
20 \\
\hline
0
\end{array}
$$

27. $\dfrac{7}{20} = 0.35$

$$
\begin{array}{r}
0.35 \\
20\overline{)7.00} \\
60 \\
\hline
100 \\
100 \\
\hline
0
\end{array}
$$

29. $\dfrac{7}{8} = 0.875$

$$
\begin{array}{r}
0.875 \\
8\overline{)7.000} \\
64 \\
\hline
60 \\
56 \\
\hline
40 \\
40 \\
\hline
0
\end{array}
$$

31. $\dfrac{9}{11} = 0.\overline{81}$

$$
\begin{array}{r}
0.8181\ldots \\
11\overline{)9.0000} \\
88 \\
\hline
20 \\
11 \\
\hline
90 \\
88 \\
\hline
20 \\
11 \\
\hline
9
\end{array}
$$

33. $\dfrac{22}{7} = 3.\overline{142857}$

$$
\begin{array}{r}
3.142857\ldots \\
7{\overline{\smash{\big)}\,22.000000}} \\
\underline{21} \\
10 \\
\underline{7} \\
30 \\
\underline{28} \\
20 \\
\underline{14} \\
60 \\
\underline{56} \\
40 \\
\underline{35} \\
50 \\
\underline{49} \\
10
\end{array}
$$

35. $\dfrac{2}{7} = 0.\overline{285714}$

$$
\begin{array}{r}
0.2857142\ldots \\
7{\overline{\smash{\big)}\,2.000000}} \\
\underline{14} \\
60 \\
\underline{56} \\
40 \\
\underline{35} \\
50 \\
\underline{49} \\
10 \\
\underline{7} \\
30 \\
\underline{28} \\
20 \\
\underline{14} \\
6
\end{array}
$$

37. $0.3 = \dfrac{3}{10}$

39. $0.4 = \dfrac{4}{10} = \dfrac{4 \div 2}{10 \div 2} = \dfrac{2}{5}$

41. $0.39 = \dfrac{39}{100}$

43. $0.82 = \dfrac{82}{100} = \dfrac{82 \div 2}{100 \div 2} = \dfrac{41}{50}$

45. $0.725 = \dfrac{725}{1000}$

$725 = 5^2 \cdot 29$

$1000 = 2^3 \cdot 5^3$

Greatest Common Divisor is 5^2 or 25.

$\dfrac{725}{1000} = \dfrac{725 \div 25}{1000 \div 25} = \dfrac{29}{40}$

47. $0.5399 = \dfrac{5399}{10,000}$

49. $n = 0.777\ldots$

$10n = 7.777\ldots$

$10n = 7.777\ldots$

$\underline{-n = 0.777\ldots}$

$9n = 7$

$n = \dfrac{7}{9}$

51. $n = 0.999\ldots$

$10n = 9.999\ldots$

$10n = 9.999\ldots$

$\underline{-n = 0.999\ldots}$

$9n = 9$

$n = 1$

53. $n = 0.3636\ldots$

$100n = 36.3636\ldots$

$100n = 36.3636\ldots$

$\underline{-n = 0.3636\ldots}$

$99n = 36$

$n = \dfrac{36}{99}$ or $\dfrac{4}{11}$

55. $n = 0.257257\ldots$

$1000n = 257.257257\ldots$

$1000n = 257.257257\ldots$

$\underline{-n = .257257\ldots}$

$999n = 257$

$n = \dfrac{257}{999}$

57. $\dfrac{3}{8} \cdot \dfrac{7}{11} = \dfrac{3 \cdot 7}{8 \cdot 11} = \dfrac{21}{88}$

59. $\left(-\dfrac{1}{10}\right)\left(\dfrac{7}{12}\right) = \dfrac{(-1)(7)}{10 \cdot 12} = \dfrac{-7}{120} = -\dfrac{7}{120}$

61. $\left(-\dfrac{2}{3}\right)\left(-\dfrac{9}{4}\right) = \dfrac{(-2)(-9)}{3 \cdot 4} = \dfrac{18}{12} = \dfrac{3}{2}$

63. $\left(3\dfrac{3}{4}\right)\left(1\dfrac{3}{5}\right) = \dfrac{15}{4} \cdot \dfrac{8}{5} = \dfrac{120}{20} = \dfrac{6}{1} = 6$

65. $\dfrac{5}{4} \div \dfrac{3}{8} = \dfrac{5}{4} \cdot \dfrac{8}{3} = \dfrac{5 \cdot 8}{4 \cdot 3} = \dfrac{40}{12} = \dfrac{10}{3}$

67. $-\dfrac{7}{8} \div \dfrac{15}{16} = -\dfrac{7}{8} \cdot \dfrac{16}{15}$
$= \dfrac{(-7)(16)}{8 \cdot 15}$
$= \dfrac{-112}{120}$
$= -\dfrac{14}{15}$

69. $6\dfrac{3}{5} \div 1\dfrac{1}{10} = \dfrac{33}{5} \div \dfrac{11}{10} = \dfrac{33}{5} \cdot \dfrac{10}{11} = \dfrac{330}{55} = \dfrac{6}{1} = 6$

71. $\dfrac{2}{11} + \dfrac{3}{11} = \dfrac{2+3}{11} = \dfrac{5}{11}$

73. $\dfrac{5}{6} - \dfrac{1}{6} = \dfrac{5-1}{6} = \dfrac{4}{6} = \dfrac{2}{3}$

75. $\dfrac{7}{12} - \left(-\dfrac{1}{12}\right) = \dfrac{7}{12} + \dfrac{1}{12} = \dfrac{7+1}{12} = \dfrac{8}{12} = \dfrac{2}{3}$

77. $\dfrac{1}{2} + \dfrac{1}{5} = \left(\dfrac{1}{2}\right)\left(\dfrac{5}{5}\right) + \left(\dfrac{1}{5}\right)\left(\dfrac{2}{2}\right)$
$= \dfrac{5}{10} + \dfrac{2}{10}$
$= \dfrac{5+2}{10}$
$= \dfrac{7}{10}$

79. $\dfrac{3}{4} + \dfrac{3}{20} = \left(\dfrac{3}{4}\right)\left(\dfrac{5}{5}\right) + \dfrac{3}{20}$
$= \dfrac{15}{20} + \dfrac{3}{20}$
$= \dfrac{15+3}{20}$
$= \dfrac{18}{20}$
$= \dfrac{9}{10}$

81. $\dfrac{5}{24} + \dfrac{7}{30} = \left(\dfrac{5}{24}\right)\left(\dfrac{5}{5}\right) + \left(\dfrac{7}{30}\right)\left(\dfrac{4}{4}\right)$
$= \dfrac{25}{120} + \dfrac{28}{120}$
$= \dfrac{25+28}{120}$
$= \dfrac{53}{120}$

83. $\dfrac{13}{18} - \dfrac{2}{9} = \dfrac{13}{18} - \dfrac{2}{9}\left(\dfrac{2}{2}\right)$
$= \dfrac{13}{18} - \dfrac{4}{18}$
$= \dfrac{13-4}{18}$
$= \dfrac{9}{18}$
$= \dfrac{1}{2}$

85. $\dfrac{4}{3} - \dfrac{3}{4} = \dfrac{4}{3}\left(\dfrac{4}{4}\right) - \dfrac{3}{4}\left(\dfrac{3}{3}\right)$
$= \dfrac{16}{12} - \dfrac{9}{12}$
$= \dfrac{16-9}{12}$
$= \dfrac{7}{12}$

87. $\dfrac{1}{15}-\dfrac{27}{50}$

$15 = 3 \cdot 5$

$50 = 2 \cdot 5^2$

Least Common Multiple is $2 \cdot 3 \cdot 5^2 = 6 \cdot 25 = 150$

$\dfrac{1}{15}\left(\dfrac{10}{10}\right)-\dfrac{27}{50}\left(\dfrac{3}{3}\right)=\dfrac{10}{150}-\dfrac{81}{150}$

$=\dfrac{10-81}{150}$

$=-\dfrac{71}{150}$

89. $3\dfrac{3}{4}-2\dfrac{1}{3}=3\dfrac{9}{12}-2\dfrac{4}{12}=1\dfrac{5}{12}$

91. $\left(\dfrac{1}{2}-\dfrac{1}{3}\right)\div\dfrac{5}{8}=\left[\left(\dfrac{1}{2}\right)\left(\dfrac{3}{3}\right)-\dfrac{1}{3}\left(\dfrac{2}{2}\right)\right]\div\dfrac{5}{8}$

$=\left(\dfrac{3}{6}-\dfrac{2}{6}\right)\div\dfrac{5}{8}$

$=\dfrac{1}{6}\div\dfrac{5}{8}$

$=\dfrac{1}{6}\cdot\dfrac{8}{5}$

$=\dfrac{1\cdot8}{6\cdot5}$

$=\dfrac{8}{30}$

$=\dfrac{4}{15}$

93. $\dfrac{1}{4}+\dfrac{1}{3}=\left(\dfrac{1}{4}\right)\left(\dfrac{3}{3}\right)+\left(\dfrac{1}{3}\right)\left(\dfrac{4}{4}\right)$

$=\dfrac{3}{12}+\dfrac{4}{12}$

$=\dfrac{3+4}{12}$

$=\dfrac{7}{12}$

$\dfrac{7}{12}\div2=\dfrac{7}{12}\cdot\dfrac{1}{2}=\dfrac{7}{24}$

95. $\dfrac{1}{2}+\dfrac{2}{3}=\left(\dfrac{1}{2}\right)\left(\dfrac{3}{3}\right)+\left(\dfrac{2}{3}\right)\left(\dfrac{2}{2}\right)$

$=\dfrac{3}{6}+\dfrac{4}{6}$

$=\dfrac{3+4}{6}$

$=\dfrac{7}{6}$

$\dfrac{7}{6}\div2=\dfrac{7}{6}\cdot\dfrac{1}{2}=\dfrac{7}{12}$

97. $-\dfrac{2}{3}+\left(-\dfrac{5}{6}\right)=\left(-\dfrac{2}{3}\right)\left(\dfrac{2}{2}\right)-\dfrac{5}{6}$

$=\dfrac{-4}{6}-\dfrac{5}{6}$

$=\dfrac{-4-5}{6}$

$=-\dfrac{9}{6}$

$-\dfrac{9}{6}\div2=-\dfrac{9}{6}\cdot\dfrac{1}{2}=-\dfrac{9}{12}=-\dfrac{3}{4}$

99. $\dfrac{13}{4}+\dfrac{13}{9}=\dfrac{13\cdot9}{4\cdot9}+\dfrac{13\cdot4}{9\cdot4}$

$=\dfrac{117}{36}+\dfrac{52}{36}$

$=\dfrac{117+52}{36}$

$=\dfrac{169}{36}$

$\dfrac{13}{4}\times\dfrac{13}{9}=\dfrac{13\cdot13}{4\cdot9}$

$=\dfrac{169}{36}$

Both are equal to $\dfrac{169}{36}$

101. $-\dfrac{9}{4}\left(\dfrac{1}{2}\right)+\dfrac{3}{4}\div\dfrac{5}{6}=-\dfrac{9}{8}+\dfrac{9}{10}$

$=-\dfrac{9}{40}$

103. $\dfrac{\dfrac{7}{9}-3}{\dfrac{5}{6}} \div \dfrac{3}{2}+\dfrac{3}{4} = \dfrac{\dfrac{-\dfrac{20}{9}}{\dfrac{5}{6}}} \div \dfrac{3}{2}+\dfrac{3}{4}$

$$= -\dfrac{8}{3} \times \dfrac{2}{3}+\dfrac{3}{4}$$

$$= -\dfrac{16}{9}+\dfrac{3}{4}$$

$$= -\dfrac{37}{36} \text{ or } -1\dfrac{1}{36}$$

105. $\dfrac{1}{4}-6(2+8) \div \left(-\dfrac{1}{3}\right)\left(-\dfrac{1}{9}\right)$

$$= \dfrac{1}{4}-6(10) \div 3$$

$$= \dfrac{1}{4}-60 \div 3$$

$$= \dfrac{1}{4}-20$$

$$= -19\dfrac{3}{4}$$

107. $\dfrac{5}{2^2 \cdot 3^2}-\dfrac{1}{2 \cdot 3^2} = \dfrac{5}{2^2 \cdot 3^2}-\dfrac{2}{2} \cdot \dfrac{1}{2 \cdot 3^2}$

$$= \dfrac{5}{2^2 \cdot 3^2}-\dfrac{2}{2^2 \cdot 3^2}$$

$$= \dfrac{3}{2^2 \cdot 3^2}$$

$$= \dfrac{1}{2^2 \cdot 3}$$

109. $\dfrac{1}{2^4 \cdot 5^3 \cdot 7}+\dfrac{1}{2 \cdot 5^4}-\dfrac{1}{2^3 \cdot 5^2}$

$$= \dfrac{5}{5} \cdot \dfrac{1}{2^4 \cdot 5^3 \cdot 7}+\dfrac{2^3 \cdot 7}{2^3 \cdot 7} \cdot \dfrac{1}{2 \cdot 5^4}-\dfrac{2 \cdot 5^2 \cdot 7}{2 \cdot 5^2 \cdot 7} \cdot \dfrac{1}{2^3 \cdot 5^2}$$

$$= \dfrac{5}{2^4 \cdot 5^4 \cdot 7}+\dfrac{56}{2^4 \cdot 5^4 \cdot 7}-\dfrac{350}{2^4 \cdot 5^4 \cdot 7}$$

$$= -\dfrac{289}{2^4 \cdot 5^4 \cdot 7}$$

111. $0.\overline{54} < 0.58\overline{3}$

$$\dfrac{6}{11} < \dfrac{7}{12}$$

113. $-0.8\overline{3} > -0.\overline{8}$

$$-\dfrac{5}{6} > -\dfrac{8}{9}$$

115. $\dfrac{86}{192} = \dfrac{43}{96}$

117. For each ingredient in the recipe, multiply the original quantity by $\dfrac{8}{16}$ or $\dfrac{1}{2}$.

$\dfrac{2}{3} \cdot \dfrac{1}{2} = \dfrac{1}{3}$ cup butter

$5 \cdot \dfrac{1}{2} = \dfrac{5}{2} = 2\dfrac{1}{2}$ ounces unsweetened chocolate

$1\dfrac{1}{2} \cdot \dfrac{1}{2} = \dfrac{3}{2} \cdot \dfrac{1}{2} = \dfrac{3}{4}$ cup sugar

$2 \cdot \dfrac{1}{2} = 1$ teaspoon vanilla

$2 \cdot \dfrac{1}{2} = 1$ egg

$1 \cdot \dfrac{1}{2} = \dfrac{1}{2}$ cup flour

119. For each ingredient in the recipe, multiply the original quantity by $\dfrac{20}{16}$ or $\dfrac{5}{4}$.

$\dfrac{2}{3} \cdot \dfrac{5}{4} = \dfrac{5}{6}$ cup butter

$5 \cdot \dfrac{5}{4} = \dfrac{25}{4} = 6\dfrac{1}{4}$ ounces unsweetened chocolate

$1\dfrac{1}{2} \cdot \dfrac{5}{4} = \dfrac{3}{2} \cdot \dfrac{5}{4} = \dfrac{15}{8} = 1\dfrac{7}{8}$ cups sugar

$2 \cdot \dfrac{5}{4} = \dfrac{5}{2} = 2\dfrac{1}{2}$ teaspoons vanilla

$2 \cdot \dfrac{5}{4} = \dfrac{5}{2} = 2\dfrac{1}{2}$ eggs

$1 \cdot \dfrac{5}{4} = \dfrac{5}{4} = 1\dfrac{1}{4}$ cups flour

121. Begin by dividing the 1 cup of butter by the quantity of butter needed for a 16-brownie batch:

$$1 \div \dfrac{2}{3} = 1 \times \dfrac{3}{2} = \dfrac{3}{2} = 1\dfrac{1}{2}$$

Thus, 1 cup of butter is enough for $1\dfrac{1}{2}$ batches.

Since each batch makes 16 brownies, $1\dfrac{1}{2}$ batches will make $16 \times 1\dfrac{1}{2} = 24$ brownies. Thus, 1 cup of butter is enough for 24 brownies.

123. $2\dfrac{2}{3} \cdot \dfrac{11}{8} = \dfrac{8}{3} \cdot \dfrac{11}{8} = \dfrac{88}{24} = \dfrac{11}{3}$ or $3\dfrac{2}{3}$ cups of water.

125. $3\frac{3}{10} - 2\frac{1}{2} = 3\frac{3}{10} - 2\frac{5}{10} = 2\frac{13}{10} - 2\frac{5}{10} = \frac{8}{10} = \frac{4}{5}$ yr

127. $2\frac{4}{5} - 1\frac{9}{10} = 2\frac{8}{10} - 1\frac{9}{10} = 1\frac{18}{10} - 1\frac{9}{10} = \frac{9}{10}$ yr

129. $1 - \frac{5}{12} - \frac{1}{4} = \frac{12}{12} - \frac{5}{12} - \frac{3}{12} = \frac{4}{12} = \frac{1}{3}$ ownership.

131. The total distance is their sum:

$\frac{3}{4} + \frac{2}{5} = \frac{15}{20} + \frac{8}{20} = \frac{23}{20}$ miles.

The difference is the amount farther:

$\frac{3}{4} - \frac{2}{5} = \frac{15}{20} - \frac{8}{20} = \frac{7}{20}$ mile.

133. $\frac{3}{5}$ of the total goes to relatives, so there is $\frac{2}{5}$ of the estate left. $\frac{1}{4}$ of that $\frac{2}{5}$ goes for AIDS research:

$\frac{1}{4} \cdot \frac{2}{5} = \frac{2}{20} = \frac{1}{10}$

147. 1st measure: $\frac{1}{4} + \frac{1}{4} + \frac{1}{8} + \frac{1}{8} = \frac{2}{8} + \frac{2}{8} + \frac{1}{8} + \frac{1}{8} = \frac{6}{8} = \frac{3}{4}$

2nd measure: $\frac{1}{4} + \frac{1}{4} + \frac{1}{4} = \frac{3}{4}$

3rd measure: $\frac{1}{4} + \frac{1}{8} + \frac{1}{8} + \frac{1}{8} + \frac{1}{8} = \frac{2}{8} + \frac{1}{8} + \frac{1}{8} + \frac{1}{8} + \frac{1}{8} = \frac{6}{8} = \frac{3}{4}$

4th measure: $\frac{1}{4} + \frac{1}{4} + \frac{1}{8} + \frac{1}{8} = \frac{2}{8} + \frac{2}{8} + \frac{1}{8} + \frac{1}{8} = \frac{6}{8} = \frac{3}{4}$

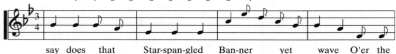

say　does　that　　Star-span-gled　Ban-ner　　yet　　wave　O'er　the

149. a. $\frac{197}{800} = 0.24625$

b. $\frac{4539}{3125} = 1.45248$

c. $\frac{7}{6250} = 0.00112$

Check Points 5.4

1. a. $\sqrt{12} = \sqrt{4 \cdot 3} = \sqrt{4} \cdot \sqrt{3} = 2\sqrt{3}$

b. $\sqrt{60} = \sqrt{4 \cdot 15} = \sqrt{4} \cdot \sqrt{15} = 2\sqrt{15}$

c. $\sqrt{55}$ cannot be simplified.

2. a. $\sqrt{3} \cdot \sqrt{10} = \sqrt{3 \cdot 10} = \sqrt{30}$

b. $\sqrt{10} \cdot \sqrt{10} = \sqrt{10 \cdot 10} = \sqrt{100} = 10$

c. $\sqrt{6} \cdot \sqrt{2} = \sqrt{6 \cdot 2} = \sqrt{12} = \sqrt{4} \cdot \sqrt{3} = 2\sqrt{3}$

3. a. $\frac{\sqrt{80}}{\sqrt{5}} = \sqrt{\frac{80}{5}} = \sqrt{16} = 4$

b. $\frac{\sqrt{48}}{\sqrt{6}} = \sqrt{\frac{48}{6}} = \sqrt{8} = \sqrt{4} \cdot \sqrt{2} = 2\sqrt{2}$

4. a. $8\sqrt{3} + 10\sqrt{3} = (8+10)\sqrt{3} = 18\sqrt{3}$

b. $4\sqrt{13} - 9\sqrt{13} = (4-9)\sqrt{13} = -5\sqrt{13}$

c. $7\sqrt{10} + 2\sqrt{10} - \sqrt{10} = (7+2-1)\sqrt{10} = 8\sqrt{10}$

5. a. $\sqrt{3} + \sqrt{12} = \sqrt{3} + \sqrt{4} \cdot \sqrt{3} = \sqrt{3} + 2\sqrt{3} = 3\sqrt{3}$

b. $4\sqrt{8} - 7\sqrt{18}$
$= 4\sqrt{4 \cdot 2} - 7\sqrt{9 \cdot 2}$
$= 4 \cdot 2\sqrt{2} - 7 \cdot 3\sqrt{2}$
$= 8\sqrt{2} - 21\sqrt{2}$
$= (8-21)\sqrt{2}$
$= -13\sqrt{2}$

6. a. $\dfrac{25}{\sqrt{10}} = \dfrac{25}{\sqrt{10}} \cdot \dfrac{\sqrt{10}}{\sqrt{10}} = \dfrac{25\sqrt{10}}{\sqrt{100}} = \dfrac{25\sqrt{10}}{10} = \dfrac{5\sqrt{10}}{2}$

b. $\sqrt{\dfrac{2}{7}} = \dfrac{\sqrt{2}}{\sqrt{7}} = \dfrac{\sqrt{2}}{\sqrt{7}} \cdot \dfrac{\sqrt{7}}{\sqrt{7}} = \dfrac{\sqrt{14}}{\sqrt{49}} = \dfrac{\sqrt{14}}{7}$

c. $\dfrac{5}{\sqrt{18}} = \dfrac{5}{\sqrt{18}} \cdot \dfrac{\sqrt{2}}{\sqrt{2}} = \dfrac{5\sqrt{2}}{\sqrt{36}} = \dfrac{5\sqrt{2}}{6}$

Exercise Set 5.4

1. $\sqrt{9} = 3$ because $3^2 = 9$.

3. $\sqrt{25} = 5$ because $5^2 = 25$.

5. $\sqrt{64} = 8$ because $8^2 = 64$.

7. $\sqrt{121} = 11$ because $11^2 = 121$.

9. $\sqrt{169} = 13$ because $13^2 = 169$.

11. a. $\sqrt{173} \approx 13.2$

b. $\sqrt{173} \approx 13.15$

c. $\sqrt{173} \approx 13.153$

13. a. $\sqrt{17,761} \approx 133.3$

b. $\sqrt{17,761} \approx 133.27$

c. $\sqrt{17,761} \approx 133.270$

15. a. $\sqrt{\pi} \approx 1.8$

b. $\sqrt{\pi} \approx 1.77$

c. $\sqrt{\pi} \approx 1.772$

17. $\sqrt{20} = \sqrt{4 \cdot 5} = \sqrt{4} \cdot \sqrt{5} = 2\sqrt{5}$

19. $\sqrt{80} = \sqrt{16 \cdot 5} = \sqrt{16} \cdot \sqrt{5} = 4\sqrt{5}$

21. $\sqrt{250} = \sqrt{25 \cdot 10} = \sqrt{25} \cdot \sqrt{10} = 5\sqrt{10}$

23. $7\sqrt{28} = 7\sqrt{4 \cdot 7}$
$= 7\sqrt{4} \cdot \sqrt{7}$
$= 7 \cdot 2 \cdot \sqrt{7}$
$= 14\sqrt{7}$

25. $\sqrt{7} \cdot \sqrt{6} = \sqrt{7 \cdot 6} = \sqrt{42}$

27. $\sqrt{6} \cdot \sqrt{6} = \sqrt{6 \cdot 6} = \sqrt{36} = 6$

29. $\sqrt{3} \cdot \sqrt{6} = \sqrt{3 \cdot 6}$
$= \sqrt{18}$
$= \sqrt{9 \cdot 2}$
$= \sqrt{9} \cdot \sqrt{2}$
$= 3\sqrt{2}$

31. $\sqrt{2} \cdot \sqrt{26} = \sqrt{2 \cdot 26}$
$= \sqrt{52}$
$= \sqrt{4 \cdot 13}$
$= \sqrt{4} \cdot \sqrt{13}$
$= 2\sqrt{13}$

33. $\dfrac{\sqrt{54}}{\sqrt{6}} = \sqrt{\dfrac{54}{6}} = \sqrt{9} = 3$

35. $\dfrac{\sqrt{90}}{\sqrt{2}} = \sqrt{\dfrac{90}{2}}$
$= \sqrt{45}$
$= \sqrt{9 \cdot 5}$
$= \sqrt{9} \cdot \sqrt{5}$
$= 3\sqrt{5}$

37. $\dfrac{-\sqrt{96}}{\sqrt{2}} = -\sqrt{\dfrac{96}{2}}$

$\qquad = -\sqrt{48}$

$\qquad = -\sqrt{16 \cdot 3}$

$\qquad = -\sqrt{16} \cdot \sqrt{3}$

$\qquad = -4\sqrt{3}$

39. $7\sqrt{3} + 6\sqrt{3} = (7+6)\sqrt{3} = 13\sqrt{3}$

41. $4\sqrt{13} - 6\sqrt{13} = (4-6)\sqrt{13} = -2\sqrt{13}$

43. $\sqrt{5} + \sqrt{5} = 1\sqrt{5} + 1\sqrt{5} = (1+1)\sqrt{5} = 2\sqrt{5}$

45. $4\sqrt{2} - 5\sqrt{2} + 8\sqrt{2} = (4-5+8)\sqrt{2} = 7\sqrt{2}$

47. $\sqrt{5} + \sqrt{20} = 1\sqrt{5} + \sqrt{4} \cdot \sqrt{5}$

$\qquad = 1\sqrt{5} + 2\sqrt{5}$

$\qquad = (1+2)\sqrt{5}$

$\qquad = 3\sqrt{5}$

49. $\sqrt{50} - \sqrt{18} = \sqrt{25} \cdot \sqrt{2} - \sqrt{9} \cdot \sqrt{2}$

$\qquad = 5\sqrt{2} - 3\sqrt{2}$

$\qquad = (5-3)\sqrt{2}$

$\qquad = 2\sqrt{2}$

51. $3\sqrt{18} + 5\sqrt{50} = 3\sqrt{9} \cdot \sqrt{2} + 5\sqrt{25} \cdot \sqrt{2}$

$\qquad = 3 \cdot 3 \cdot \sqrt{2} + 5 \cdot 5\sqrt{2}$

$\qquad = 9\sqrt{2} + 25\sqrt{2}$

$\qquad = (9+25)\sqrt{2}$

$\qquad = 34\sqrt{2}$

53. $\dfrac{1}{4}\sqrt{12} - \dfrac{1}{2}\sqrt{48} = \dfrac{1}{4}\sqrt{4} \cdot \sqrt{3} - \dfrac{1}{2}\sqrt{16} \cdot \sqrt{3}$

$\qquad = \dfrac{1}{4} \cdot 2 \cdot \sqrt{3} - \dfrac{1}{2} \cdot 4 \cdot \sqrt{3}$

$\qquad = \dfrac{1}{2}\sqrt{3} - \dfrac{4}{2}\sqrt{3}$

$\qquad = \left(\dfrac{1}{2} - \dfrac{4}{2}\right)\sqrt{3}$

$\qquad = -\dfrac{3}{2}\sqrt{3}$

55. $3\sqrt{75} + 2\sqrt{12} - 2\sqrt{48}$

$\qquad = 3 \cdot \sqrt{25} \cdot \sqrt{3} + 2 \cdot \sqrt{4} \cdot \sqrt{3} - 2 \cdot \sqrt{16} \cdot \sqrt{3}$

$\qquad = 3 \cdot 5 \cdot \sqrt{3} + 2 \cdot 2 \cdot \sqrt{3} - 2 \cdot 4 \cdot \sqrt{3}$

$\qquad = 15\sqrt{3} + 4\sqrt{3} - 8\sqrt{3}$

$\qquad = (15+4-8)\sqrt{3}$

$\qquad = 11\sqrt{3}$

57. $\dfrac{5}{\sqrt{3}} = \dfrac{5}{\sqrt{3}} \cdot \dfrac{\sqrt{3}}{\sqrt{3}} = \dfrac{5\sqrt{3}}{\sqrt{9}} = \dfrac{5\sqrt{3}}{3}$

59. $\dfrac{21}{\sqrt{7}} = \dfrac{21}{\sqrt{7}} \cdot \dfrac{\sqrt{7}}{\sqrt{7}} = \dfrac{21\sqrt{7}}{\sqrt{49}} = \dfrac{21\sqrt{7}}{7} = 3\sqrt{7}$

61. $\dfrac{12}{\sqrt{30}} = \dfrac{12\sqrt{30}}{\sqrt{30}\sqrt{30}}$

$\qquad = \dfrac{12\sqrt{30}}{\sqrt{900}}$

$\qquad = \dfrac{12\sqrt{30}}{30}$

$\qquad = \dfrac{2\sqrt{30}}{5}$

63. $\dfrac{15}{\sqrt{12}} = \dfrac{15}{\sqrt{4 \cdot 3}}$

$\qquad = \dfrac{15}{\sqrt{4}\sqrt{3}}$

$\qquad = \dfrac{15}{2\sqrt{3}}$

$\qquad = \dfrac{15\sqrt{3}}{2\sqrt{3}\sqrt{3}}$

$\qquad = \dfrac{15\sqrt{3}}{2\sqrt{9}}$

$\qquad = \dfrac{15\sqrt{3}}{2 \cdot 3}$

$\qquad = \dfrac{15\sqrt{3}}{6}$

$\qquad = \dfrac{5\sqrt{3}}{2}$

65. $\sqrt{\dfrac{2}{5}} = \dfrac{\sqrt{2}}{\sqrt{5}} = \dfrac{\sqrt{2}}{\sqrt{5}} \cdot \dfrac{\sqrt{5}}{\sqrt{5}} = \dfrac{\sqrt{10}}{\sqrt{25}} = \dfrac{\sqrt{10}}{5}$

67. $3\sqrt{8} - \sqrt{32} + 3\sqrt{72} - \sqrt{75}$

$\quad = 6\sqrt{2} - 4\sqrt{2} + 18\sqrt{2} - 5\sqrt{3}$

$\quad = 20\sqrt{2} - 5\sqrt{3}$

69. $3\sqrt{7} - 5\sqrt{14} \cdot \sqrt{2} = 3\sqrt{7} - 5\sqrt{28}$

$\quad\quad\quad\quad\quad\quad\quad = 3\sqrt{7} - 10\sqrt{7}$

$\quad\quad\quad\quad\quad\quad\quad = -7\sqrt{7}$

71. $\dfrac{\sqrt{32}}{5} + \dfrac{\sqrt{18}}{7} = \dfrac{4\sqrt{2}}{5} + \dfrac{3\sqrt{2}}{7}$

$\quad\quad\quad\quad\quad = \dfrac{28\sqrt{2}}{35} + \dfrac{15\sqrt{2}}{35}$

$\quad\quad\quad\quad\quad = \dfrac{43\sqrt{2}}{35}$

73. $\dfrac{\sqrt{2}}{\sqrt{3}} + \dfrac{\sqrt{3}}{\sqrt{2}}$

$\quad = \dfrac{\sqrt{2}}{\sqrt{3}} \cdot \dfrac{\sqrt{2}}{\sqrt{2}} + \dfrac{\sqrt{3}}{\sqrt{2}} \cdot \dfrac{\sqrt{3}}{\sqrt{3}}$

$\quad = \dfrac{2}{\sqrt{6}} + \dfrac{3}{\sqrt{6}}$

$\quad = \dfrac{5}{\sqrt{6}}$

$\quad = \dfrac{5}{\sqrt{6}} \cdot \dfrac{\sqrt{6}}{\sqrt{6}}$

$\quad = \dfrac{5\sqrt{6}}{6}$

75. $d(x) = \sqrt{\dfrac{3x}{2}}$

$\quad d(72) = \sqrt{\dfrac{3(72)}{2}}$

$\quad\quad\quad\quad = \sqrt{3(36)}$

$\quad\quad\quad\quad = \sqrt{3} \cdot \sqrt{36}$

$\quad\quad\quad\quad = 6\sqrt{3} \approx 10.4 \text{ miles}$

A passenger on the pool deck can see roughly 10.4 miles.

77. $v = \sqrt{20L}; L = 245$

$\quad v = \sqrt{20 \cdot 245} = \sqrt{4900} = 70$

The motorist was traveling 70 miles per hour, so he was speeding.

79. a. 41 in.

b. $h = 2.9\sqrt{x} + 20.1$

$\quad = 2.9\sqrt{50} + 20.1$

$\quad \approx 40.6 \text{ in.}$

The estimate from part (a) describes the median height obtained from the formula quite well.

81. a. At birth we have $x = 0$.

$\quad y = 2.9\sqrt{x} + 36$

$\quad\quad = 2.9\sqrt{0} + 36$

$\quad\quad = 2.9(0) + 36$

$\quad\quad = 36$

According to the model, the head circumference at birth is 36 cm.

b. At 9 months we have $x = 9$.

$\quad y = 2.9\sqrt{x} + 36$

$\quad\quad = 2.9\sqrt{9} + 36$

$\quad\quad = 2.9(3) + 36$

$\quad\quad = 44.7$

According to the model, the head circumference at 9 months is 44.7 cm.

c. At 14 months we have $x = 14$.

$\quad y = 2.9\sqrt{x} + 36$

$\quad\quad = 2.9\sqrt{14} + 36$

$\quad\quad \approx 46.9$

According to the model, the head circumference at 14 months is roughly 46.9 cm.

d. The model describes healthy children.

83. $R_f \sqrt{1 - \left(\dfrac{v}{c}\right)^2} = R_f \sqrt{1 - \left(\dfrac{0.8c}{c}\right)^2}$

$\quad = R_f \sqrt{1 - (0.8)^2}$

$\quad = R_f \sqrt{0.36}$

$\quad = 0.6 R_f$

If 100 weeks have passed for your friend on Earth, then you were gone for $0.6(100) = 60$ weeks.

93. $\sqrt{2} \approx 1.4$

$\quad \sqrt{2} < 1.5$

95. $\dfrac{-3.14}{2} = -1.5700$

$-\dfrac{\pi}{2} \approx -1.5708$

$\dfrac{-3.14}{2} > -\dfrac{\pi}{2}$

97. $-\sqrt{47} \approx -6.86$
Therefore $-\sqrt{47}$ is between -7 and -6.

99. Answers will vary.
Example: $\left(6+\sqrt{2}\right) - \left(1+\sqrt{2}\right) = 5$

Check Points 5.5

1. $\left\{-9,\ -1.3,\ 0,\ 0.\overline{3},\ \dfrac{\pi}{2},\ \sqrt{9},\ \sqrt{10}\right\}$

   a. Natural numbers: $\sqrt{9}$ because $\sqrt{9} = 3$

   b. Whole numbers: $0,\ \sqrt{9}$

   c. Integers: $-9,\ 0,\ \sqrt{9}$

   d. Rational numbers: $-9,\ -1.3,\ 0,\ 0.\overline{3},\ \sqrt{9}$

   e. Irrational numbers: $\dfrac{\pi}{2},\ \sqrt{10}$

   f. Real numbers: All numbers in this set.

2. a. Associative property of multiplication

   b. Commutative property of addition

   c. Distributive property of multiplication over addition

   d. Commutative property of multiplication

3. a. Yes, the natural numbers are closed with respect to multiplication.

   b. No, the integers are not closed with respect to division. Example: $3 \div 5 = 0.6$ which is not an integer.

Exercise Set 5.5

1. $\left\{-9,\ -\dfrac{4}{5},\ 0,\ 0.25,\ \sqrt{3},\ 9.2,\ \sqrt{100}\right\}$

   a. Natural numbers: $\sqrt{100}$ because $\sqrt{100} = 10$

   b. Whole numbers: $0,\ \sqrt{100}$

   c. Integers: $-9,\ 0,\ \sqrt{100}$

   d. Rational numbers: $-9,\ -\dfrac{4}{5},\ 0,\ 0.25,\ 9.2,\ \sqrt{100}$

   e. Irrational numbers: $\sqrt{3}$

   f. Real numbers: All numbers in this set.

3. $\left\{-11,\ -\dfrac{5}{6},\ 0,\ 0.75,\ \sqrt{5},\ \pi,\ \sqrt{64}\right\}$

   a. Natural numbers: $\sqrt{64}$ because $\sqrt{64} = 8$

   b. Whole numbers: 0 and $\sqrt{64}$

   c. Integers: $-11,\ 0,\ \sqrt{64}$

   d. Rational numbers: $-11,\ -\dfrac{5}{6},\ 0,\ 0.75,\ \sqrt{64}$

   e. Irrational numbers: $\sqrt{5},\ \pi$

   f. Real numbers: All numbers in this set.

5. 0 is the only whole number that is not a natural number.

7. Answers will vary. Possible answer: 0.5

9. Answers will vary. Possible answer: 7

11. Answers will vary. Possible answer: $\sqrt{3}$

13. $3 + (4+5) = 3 + (5+4)$

15. $9 \cdot (6+2) = 9 \cdot (2+6)$

17. $(4 \cdot 5) \cdot 3 = 4 \cdot (5 \cdot 3)$

19. $7 \cdot (4+5) = 7 \cdot 4 + 7 \cdot 5$

21. $5(6+\sqrt{2})=5\cdot6+5\cdot\sqrt{2}=30+5\sqrt{2}$

23. $\sqrt{7}(3+\sqrt{2})=\sqrt{7}\cdot3+\sqrt{7}\cdot\sqrt{2}=3\sqrt{7}+\sqrt{14}$

25. $\sqrt{3}(5+\sqrt{3})=\sqrt{3}\cdot5+\sqrt{3}\cdot\sqrt{3}=5\sqrt{3}+\sqrt{9}$
$$=5\sqrt{3}+3$$

27. $\sqrt{6}(\sqrt{2}+\sqrt{6})=\sqrt{6}\cdot\sqrt{2}+\sqrt{6}\cdot\sqrt{6}$
$$=\sqrt{12}+\sqrt{36}$$
$$=2\sqrt{3}+6$$

29. $6+(-4)=(-4)+6$
Commutative property of addition.

31. $6+(2+7)=(6+2)+7$
Associative property of addition.

33. $(2+3)+(4+5)=(4+5)+(2+3)$
Commutative property of addition.

35. $2(-8+6)=-16+12$
Distributive property of multiplication over addition.

37. $(2\sqrt{3})\cdot\sqrt{5}=2(\sqrt{3}\cdot\sqrt{5})$
Associative property of multiplication

39. Answers will vary.
Example: $1-2=-1$

41. Answers will vary.
Example: $\dfrac{-2}{8}=-\dfrac{1}{4}$

43. Answers will vary.
Example: $\sqrt{5}\sqrt{5}=\sqrt{25}=5$

45. true

47. false

49. $5(x+4)+3x$
$=(5x+20)+3x$ distributive property
$=(20+5x)+3x$ commutative property of addition
$=20+(5x+3x)$ associative property
$=20+(5+3)x$ distributive property
$=20+8x$
$=8x+20$ commutative property of addition

51. vampire

53. vampire

55. narcissistic; $3^3+7^3+1^3=371$

57. narcissistic; $9^4+4^4+7^4+4^4=9474$

59. a. distributive property

b. $\dfrac{D(A+1)}{24}=\dfrac{200(12+1)}{24}$
$$=\dfrac{200(13)}{24}$$
$$=\dfrac{2600}{24}$$
$$\approx108 \text{ mg}$$

$\dfrac{DA+D}{24}=\dfrac{200\cdot12+200}{24}$
$$=\dfrac{2400+200}{24}$$
$$=\dfrac{2600}{24}$$
$$\approx108 \text{ mg}$$

69. c is true

Check Points 5.6

1. a. $19^0=1$

b. $(3\pi)^0=1$

c. $(-14)^0=1$

d. $-14^0=-1$

2. a. $9^{-2}=\dfrac{1}{9^2}=\dfrac{1}{81}$

b. $6^{-3}=\dfrac{1}{6^3}=\dfrac{1}{216}$

c. $12^{-1}=\dfrac{1}{12}$

3. a. $7.4\times10^9=7,400,000,000$

b. $3.017\times10^{-6}=0.000003017$

4. a. $7,410,000,000=7.41\times10^9$

b. $0.000000092=9.2\times10^{-8}$

5. Because a billion is 10^9, this value can be expressed as 519×10^9.
   Convert 519 to scientific notation and simplify.
   $$519 \times 10^9 = (5.19 \times 10^2) \times 10^9$$
   $$= 5.19 \times (10^2 \times 10^9)$$
   $$= \$5.19 \times 10^{11}$$

6. $(1.3 \times 10^7) \times (4 \times 10^{-2}) = (1.3 \times 4) \times (10^7 \times 10^{-2})$
   $$= 5.2 \times 10^{7+(-2)}$$
   $$= 5.2 \times 10^5$$
   $$= 520,000$$

7. $\dfrac{6.9 \times 10^{-8}}{3 \times 10^{-2}} = \left(\dfrac{6.9}{3}\right) \times \left(\dfrac{10^{-8}}{10^{-2}}\right)$
   $$= 2.3 \times 10^{-8-(-2)}$$
   $$= 2.3 \times 10^{-6}$$
   $$= 0.0000023$$

8. a. $0.0036 \times 5,200,000$
   $$= 3.6 \times 10^{-3} \times 5.2 \times 10^6$$
   $$= (3.6 \times 5.2) \times (10^{-3} \times 10^6)$$
   $$= 18.72 \times 10^3$$
   $$= 1.872 \times 10 \times 10^3$$
   $$= 1.872 \times 10^4$$

   b. Based on part (a):
   $0.0036 \times 5,200,000$
   $$= 1.872 \times 10^4$$
   $$= 18,720$$

9. $\dfrac{13 \times 10^9}{5.1 \times 10^6} = \left(\dfrac{13}{5.1}\right) \times \left(\dfrac{10^9}{10^6}\right) \approx 2.5 \times 10^3 = \$2500$

Exercise Set 5.6

1. $2^2 \cdot 2^3 = 2^{2+3} = 2^5 = 32$

3. $4 \cdot 4^2 = 4^1 \cdot 4^2 = 4^{1+2} = 4^3 = 64$

5. $(2^2)^3 = 2^{2 \cdot 3} = 2^6 = 64$

7. $(1^4)^5 = 1^{4 \cdot 5} = 1^{20} = 1$

9. $\dfrac{4^7}{4^5} = 4^{7-5} = 4^2 = 16$

11. $\dfrac{2^8}{2^4} = 2^{8-4} = 2^4 = 16$

13. $3^0 = 1$

15. $(-3)^0 = 1$

17. $-3^0 = -1$

19. $2^{-2} = \dfrac{1}{2^2} = \dfrac{1}{4}$

21. $4^{-3} = \dfrac{1}{4^3} = \dfrac{1}{64}$

23. $2^{-5} = \dfrac{1}{2^5} = \dfrac{1}{32}$

25. $3^4 \cdot 3^{-2} = 3^{4+(-2)} = 3^2 = 9$

27. $3^{-3} \cdot 3 = 3^{-3} \cdot 3^1 = 3^{-3+1} = 3^{-2} = \dfrac{1}{3^2} = \dfrac{1}{9}$

29. $\dfrac{2^3}{2^7} = 2^{3-7} = 2^{-4} = \dfrac{1}{2^4} = \dfrac{1}{16}$

31. $2.7 \times 10^2 = 270$

33. $9.12 \times 10^5 = 912,000$

35. $8 \times 10^7 = 8.0 \times 10^7 = 80,000,000$

37. $1 \times 10^5 = 1.0 \times 10^5 = 100,000$

39. $7.9 \times 10^{-1} = 0.79$

41. $2.15 \times 10^{-2} = 0.0215$

43. $7.86 \times 10^{-4} = 0.000786$

45. $3.18 \times 10^{-6} = 0.00000318$

47. $370 = 3.7 \times 10^2$

49. $3600 = 3.6 \times 10^3$

51. $32,000 = 3.2 \times 10^4$

53. $220,000,000 = 2.2 \times 10^8$

55. $0.027 = 2.7 \times 10^{-2}$

57. $0.0037 = 3.7 \times 10^{-3}$

59. $0.00000293 = 2.93 \times 10^{-6}$

61. $820 \times 10^5 = (8.2 \times 10^2) \times 10^5 = 8.2 \times 10^7$

63. $0.41 \times 10^6 = (4.1 \times 10^{-1}) \times 10^6 = 4.1 \times 10^5$

65. $2100 \times 10^{-9} = (2.1 \times 10^3) \times 10^{-9} = 2.1 \times 10^{-6}$

67. $(2 \times 10^3)(3 \times 10^2) = (2 \times 3) \times (10^{3+2})$
$$= 6 \times 10^5$$
$$= 600,000$$

69. $(2 \times 10^9)(3 \times 10^{-5}) = (2 \times 3) \times (10^{9-5})$
$$= 6 \times 10^4$$
$$= 60,000$$

71. $(4.1 \times 10^2)(3 \times 10^{-4}) = (4.1 \times 3) \times (10^{2-4})$
$$= 12.3 \times 10^{-2}$$
$$= 1.23 \times 10 \times 10^{-2}$$
$$= 1.23 \times 10^{-1}$$
$$= 0.123$$

73. $\dfrac{12 \times 10^6}{4 \times 10^2} = \left(\dfrac{12}{4}\right) \times \left(\dfrac{10^6}{10^2}\right)$
$$= 3 \times 10^{6-2}$$
$$= 3 \times 10^4$$
$$= 30,000$$

75. $\dfrac{15 \times 10^4}{5 \times 10^{-2}} = \left(\dfrac{15}{5}\right) \times \left(\dfrac{10^4}{10^{-2}}\right)$
$$= 3 \times 10^{4-(-2)}$$
$$= 3 \times 10^6$$
$$= 3,000,000$$

77. $\dfrac{6 \times 10^3}{2 \times 10^5} = \left(\dfrac{6}{2}\right) \times \left(\dfrac{10^3}{10^5}\right)$
$$= 3 \times 10^{3-5}$$
$$= 3 \times 10^{-2}$$
$$= 0.03$$

79. $\dfrac{6.3 \times 10^{-6}}{3 \times 10^{-3}} = \left(\dfrac{6.3}{3}\right) \times \left(\dfrac{10^{-6}}{10^{-3}}\right)$
$$= 2.1 \times 10^{-6-(-3)}$$
$$= 2.1 \times 10^{-3}$$
$$= 0.0021$$

81. $(82,000,000)(3,000,000,000)$
$$= (8.2 \times 10^7)(3.0 \times 10^9)$$
$$= (8.2 \times 3.0) \times (10^{7+9})$$
$$= 24.6 \times 10^{16}$$
$$= 2.46 \times 10 \times 10^{16}$$
$$= 2.46 \times 10^{17}$$

83. $(0.0005)(6,000,000)$
$$= (5.0 \times 10^{-4})(6.0 \times 10^6)$$
$$= (5.0 \times 6.0)(10^{-4+6})$$
$$= 30 \times 10^2$$
$$= 3 \times 10 \times 10^2$$
$$= 3 \times 10^3$$

85. $\dfrac{9,500,000}{500} = \dfrac{9.5 \times 10^6}{5 \times 10^2}$
$$= \left(\dfrac{9.5}{5}\right) \times (10^{6-2})$$
$$= 1.9 \times 10^4$$

87. $\dfrac{0.00008}{200} = \dfrac{8 \times 10^{-5}}{2 \times 10^2}$
$$= \left(\dfrac{8}{2}\right) \times (10^{-5-2})$$
$$= 4 \times 10^{-7}$$

89. $\dfrac{480,000,000,000}{0.00012} = \dfrac{4.8 \times 10^{11}}{1.2 \times 10^{-4}}$
$$= \left(\dfrac{4.8}{1.2}\right) \times (10^{11-(-4)})$$
$$= 4 \times 10^{15}$$

91. $\dfrac{2^4}{2^5} + \dfrac{3^3}{3^5} = \dfrac{1}{2} + \dfrac{1}{3^2}$
$$= \dfrac{1}{2} + \dfrac{1}{9}$$
$$= \dfrac{11}{18}$$

93. $\dfrac{2^6}{2^4} - \dfrac{5^4}{5^6} = \dfrac{2^2}{1} - \dfrac{1}{5^2}$

$$= 4 - \dfrac{1}{25}$$

$$= \dfrac{99}{25}$$

$$= 3\dfrac{24}{25}$$

95. $\dfrac{\left(5\times10^3\right)\left(1.2\times10^{-4}\right)}{\left(2.4\times10^2\right)} = 2.5\times10^{-3}$

97. $\dfrac{\left(1.6\times10^4\right)\left(7.2\times10^{-3}\right)}{\left(3.6\times10^8\right)\left(4\times10^{-3}\right)} = 0.8\times10^{-4} = 8\times10^{-5}$

99. $46.5\times10^9 = (4.65\times10^1)\times10^9$

$$= 4.65\times(10^1\times10^9)$$

$$= \$4.65\times10^{10}$$

101. $(18.4-18.3)\times10^9 = 0.1\times10^9$

$$= (1\times10^{-1})\times10^9$$

$$= 1\times(10^{-1}\times10^9)$$

$$= \$1\times10^8$$

103. 20 billion $= 2\times10^{10}$

$\dfrac{2\times10^{10}}{3\times10^8} = \dfrac{2}{3}\times\dfrac{10^{10}}{10^8}$

$$\approx 0.6667\times10^{10-8}$$

$$= 0.6667\times10^2$$

$$= 6.7\times10^1 \approx 67$$

The average American consumes about 67 hotdogs each year.

105. 8 billion $= 8\times10^9$

$\dfrac{8\times10^9}{3.2\times10^7} = \dfrac{8}{3.2}\times\dfrac{10^9}{10^7}$

$$= 2.5\times10^{9-7}$$

$$= 2.5\times10^2 = 250$$

$2.5\times10^2 = 250$ chickens are raised for food each second in the U.S.

107. a. $\dfrac{519\times10^9}{48\times10^6} \approx 10.813\times10^3$

$$= \$1.08\times10^4$$

$$= \$10,813$$

b. $\dfrac{\$10,813}{12} \approx \$901$

109. Medicaid: $\dfrac{198\times10^9}{53.4\times10^6} \approx 3.708\times10^3$

$$= \$3708$$

Medicare: $\dfrac{294\times10^9}{42.3\times10^6} \approx 6.950\times10^3$

$$= \$6950$$

Medicare provides a greater per person benefit by $3242.

111. $20,000\left(5.3\times10^{-23}\right)$

$$= \left(2\times10^4\right)\left(5.3\times10^{-23}\right)$$

$$= (2\cdot5.3)\times\left(10^4\cdot10^{-23}\right)$$

$$= 10.6\times10^{4+(-23)}$$

$$= 10.6\times10^{-19}$$

$$= 1.06\times10^{-18}$$

The mass of 20,000 oxygen molecules is 1.06×10^{-18} grams.

113. $\dfrac{365 \text{ days}}{1 \text{ year}} \cdot \dfrac{24 \text{ hours}}{1 \text{ day}} = 8760$ hours/year

$$= 8.76\times10^3 \text{ hours/year}$$

$\dfrac{8.76\times10^3 \text{ hours}}{1 \text{ year}} \cdot \dfrac{60 \text{ minutes}}{1 \text{ hour}}$

$$= 525.6\times10^3 \text{ minutes/year}$$

$$= 5.256\times10^5 \text{ minutes/year}$$

$\dfrac{5.256\times10^5 \text{ minutes}}{1 \text{ year}} \cdot \dfrac{60 \text{ seconds}}{1 \text{ minute}}$

$$= 315.36\times10^5 \text{ seconds/year}$$

$$= 3.1536\times10^7 \text{ seconds/year}$$

There are 3.1536×10^7 seconds in a year.

125. d is true

127. $1-(2^{-1}+2^{-2}) = 1-\left(\dfrac{1}{2}+\dfrac{1}{4}\right)$

$$= \dfrac{4}{4} - \left(\dfrac{2}{4}+\dfrac{1}{4}\right)$$

$$= \dfrac{4}{4} - \dfrac{3}{4} = \dfrac{1}{4}$$

Check Points 5.7

1. $100, 100 + 20 = 120, 120 + 20 = 140, 140 + 20 = 160, 160 + 20 = 180, 180 + 20 = 200$
   $100, 120, 140, 160, 180,$ and 200

2. $8, 8 - 3 = 5, 5 - 3 = 2, 2 - 3 = -1, -1 - 3 = -4, -4 - 3 = -7$
   $8, 5, 2, -1, -4,$ and -7

3. $a_n = a_1 + (n-1)d$
   $a_9 = 6 + (9-1)(-5)$
   $= 6 + 8(-5)$
   $= 6 - 40$
   $= -34$

4. **a.** $a_n = a_1 + (n-1)d$
   $a_n = 12.0 + (n-1)1.16$
   $= 12.0 + 1.16n - 1.16$
   $= 1.16n + 10.84$

   b. $a_n = 1.16n + 10.84$
   $= 1.16(50) + 10.84$
   ≈ 68.8 million

5. $12, 12\left(-\dfrac{1}{2}\right) = -6, -6\left(-\dfrac{1}{2}\right) = 3, 3\left(-\dfrac{1}{2}\right) = -\dfrac{3}{2},$
   $-\dfrac{3}{2}\left(-\dfrac{1}{2}\right) = \dfrac{3}{4}, \dfrac{3}{4}\left(-\dfrac{1}{2}\right) = -\dfrac{3}{8}$
   $12, -6, 3, -\dfrac{3}{2}, \dfrac{3}{4}, -\dfrac{3}{8}$

6. $a_n = a_1 r^{n-1}$ with $a_1 = 5,$ $r = -3,$ and $n = 7$
   $a_7 = 5(-3)^{7-1} = 5(-3)^6 = 5(729) = 3645$

7. $a_n = a_1 r^{n-1}$ with $a_1 = 3$ and $r = \dfrac{6}{3} = 2$. Thus
   $a_n = 3(2)^{n-1}$
   $a_8 = 3(2)^{8-1} = 3(2)^7 = 3(128) = 384$

Exercise Set 5.7

1. $8, 8 + 2 = 10, 10 + 2 = 12, 12 + 2 = 14, 14 + 2 = 16, 16 + 2 = 18$
   $8, 10, 12, 14, 16,$ and 18

3. $200, 200 + 20 = 220, 220 + 20 = 240, 240 + 20 = 260, 260 + 20 = 280, 280 + 20 = 300$
   $200, 220, 240, 260, 280,$ and 300

5. $-7, -7 + 4 = -3, -3 + 4 = 1, 1 + 4 = 5, 5 + 4 = 9, 9 + 4 = 13$
   $-7, -3, 1, 5, 9,$ and 13

7. $-400, -400 + 300 = -100, -100 + 300 = 200, 200 + 300 = 500, 500 + 300 = 800, 800 + 300 = 1100$
   $-400, -100, 200, 500, 800,$ and 1100

9. $7, 7 - 3 = 4, 4 - 3 = 1, 1 - 3 = -2, -2 - 3 = -5, -5 - 3 = -8$
   $7, 4, 1, -2, -5,$ and -8

11. $200, 200 - 60 = 140, 140 - 60 = 80, 80 - 60 = 20, 20 - 60 = -40, -40 - 60 = -100$
    $200, 140, 80, 20, -40,$ and -100

13. $\dfrac{5}{2}, \dfrac{5}{2} + \dfrac{1}{2} = \dfrac{6}{2} = 3, \dfrac{6}{2} + \dfrac{1}{2} = \dfrac{7}{2}, \dfrac{7}{2} + \dfrac{1}{2} = \dfrac{8}{2} = 4, \dfrac{8}{2} + \dfrac{1}{2} = \dfrac{9}{2}, \dfrac{9}{2} + \dfrac{1}{2} = \dfrac{10}{2} = 5$
    $\dfrac{5}{2}, 3, \dfrac{7}{2}, 4, \dfrac{9}{2},$ and 5

15. $\dfrac{3}{2}$, $\dfrac{6}{4}+\dfrac{1}{4}=\dfrac{7}{4}$, $\dfrac{7}{4}+\dfrac{1}{4}=\dfrac{8}{4}=2$, $\dfrac{8}{4}+\dfrac{1}{4}=\dfrac{9}{4}$, $\dfrac{9}{4}+\dfrac{1}{4}=\dfrac{10}{4}=\dfrac{5}{2}$, $\dfrac{10}{4}+\dfrac{1}{4}=\dfrac{11}{4}$

$\dfrac{3}{2}$, $\dfrac{7}{4}$, 2, $\dfrac{9}{4}$, $\dfrac{5}{2}$, and $\dfrac{11}{4}$

17. 4.25, $4.25 + 0.3 = 4.55$, $4.55 + 0.3 = 4.85$, $4.85 + 0.3 = 5.15$, $5.15 + 0.3 = 5.45$, $5.45 + 0.3 = 5.75$
4.25, 4.55, 4.85, 5.15, 5.45, and 5.75

19. 4.5, $4.5 - 0.75 = 3.75$, $3.75 - 0.75 = 3$, $3 - 0.75 = 2.25$, $2.25 - 0.75 = 1.5$, $1.5 - 0.75 = 0.75$
4.5, 3.75, 3, 2.25, 1.5, and 0.75

21. $a_1 = 13$, $d = 4$
$a_6 = 13 + (6-1)(4)$
$\quad = 13 + 5(4)$
$\quad = 13 + 20$
$\quad = 33$

23. $a_1 = 7$, $d = 5$
$a_{50} = 7 + (50-1)(5)$
$\quad = 7 + 49(5)$
$\quad = 7 + 245$
$\quad = 252$

25. $a_1 = -5$, $d = 9$
$a_9 = -5 + (9-1)(9)$
$\quad = -5 + 8(9)$
$\quad = -5 + 72$
$\quad = 67$

27. $a_1 = -40$, $d = 5$
$a_{200} = -40 + (200-1)(5)$
$\quad = -40 + 199(5)$
$\quad = -40 + 995$
$\quad = 955$

29. $a_1 = -8$, $d = 10$
$a_{10} = -8 + (10-1)(10)$
$\quad = -8 + 9(10)$
$\quad = -8 + 90$
$\quad = 82$

31. $a_1 = 35$, $d = -3$
$a_{60} = 35 + (60-1)(-3)$
$\quad = 35 + 59(-3)$
$\quad = 35 + (-177)$
$\quad = -142$

33. $a_1 = 12$, $d = -5$
$a_{12} = 12 + (12-1)(-5)$
$\quad = 12 + 11(-5)$
$\quad = 12 + (-55)$
$\quad = -43$

35. $a_1 = -70$, $d = -2$
$a_{90} = -70 + (90-1)(-2)$
$\quad = -70 + 89(-2)$
$\quad = -70 + (-178)$
$\quad = -248$

37. $a_1 = 6$, $d = \dfrac{1}{2}$
$a_{12} = 6 + (12-1)\left(\dfrac{1}{2}\right)$
$\quad = 6 + 11\left(\dfrac{1}{2}\right)$
$\quad = \dfrac{12}{2} + \dfrac{11}{2}$
$\quad = \dfrac{23}{2}$

39. $a_1 = 14$, $d = -0.25$
$a_{50} = 14 + (50-1)(-0.25)$
$\quad = 14 + 49(-0.25)$
$\quad = 14 + (-12.25)$
$\quad = 1.75$

41. $a_n = a_1 + (n-1)d$ with $a_1 = 1$, $d = 4$
$a_n = 1 + (n-1)4$
$\quad = 1 + 4n - 4$
$\quad = 4n - 3$
Thus $a_{20} = 4(20) - 3 = 77$.

43. $a_n = a_1 + (n-1)d$ with $a_1 = 7$, $d = -4$

$a_n = 7 + (n-1)(-4)$

$= 7 - 4n + 4$

$= -4n + 11$

Thus $a_{20} = -4(20) + 11 = -69$.

45. $a_n = a_1 + (n-1)d$ with $a_1 = 9$, $d = 2$

$a_n = 9 + (n-1)2$

$= 9 + 2n - 2$

$= 2n + 7$

Thus $a_{20} = 2(20) + 7 = 47$.

47. $a_n = a_1 + (n-1)d$ with $a_1 = -20$, $d = -4$

$a_n = -20 + (n-1)(-4)$

$= -20 - 4n + 4$

$= -4n - 16$

Thus $a_{20} = -4(20) - 16 = -96$.

49. $a_1 = 4$, $r = 2$

$4, 4 \cdot 2 = 8, 8 \cdot 2 = 16, 16 \cdot 2 = 32, 32 \cdot 2 = 64,$

$64 \cdot 2 = 128$

$4, 8, 16, 32, 64, 128$

51. $a_1 = 1000$, $r = 1$

$1000, 1000 \cdot 1 = 1000, 1000 \cdot 1 = 1000, \ldots$

$1000, 1000, 1000, 1000, 1000, 1000$

53. $a_1 = 3$, $r = -2$

$3, 3(-2) = -6, -6(-2) = 12, 12(-2) = -24, -24(-2) =$

$48, 48(-2) = -96$

$3, -6, 12, -24, 48, -96$

55. $a_1 = 10$, $r = -4$

$10, 10(-4) = -40, -40(-4) = 160,$

$160(-4) = -640, -640(-4) = 2560,$

$2560(-4) = -10,240$

$10, -40, 160, -640, 2560,$ and $-10,240$

57. $a_1 = 2000$, $r = -1$

$2000, 2000(-1) = -2000,$

$-2000(-1) = 2000, \ldots$

$2000, -2000, 2000, -2000, 2000, -2000$

59. $a_1 = -2$, $r = -3$

$-2, -2(-3) = 6, 6(-3) = -18, -18(-3) = 54, 54(-3) =$

$-162, -162(-3) = 486$

$-2, 6, -18, 54, -162, 486$

61. $a_1 = -6$, $r = -5$

$-6, -6(-5) = 30, 30(-5) = -150,$

$-150(-5) = 750, 750(-5) = -3750,$

$-3750(-5) = 18,750$

$-6, 30, -150, 750, -3750, 18750$

63. $a_1 = \dfrac{1}{4}$, $r = 2$

$\dfrac{1}{4}, \dfrac{1}{4} \cdot 2 = \dfrac{1}{2}, \dfrac{1}{2} \cdot 2 = 1, 1 \cdot 2 = 2, 2 \cdot 2 = 4,$

$4 \cdot 2 = 8$

$\dfrac{1}{4}, \dfrac{1}{2}, 1, 2, 4, 8$

65. $a_1 = \dfrac{1}{4}$, $r = \dfrac{1}{2}$

$\dfrac{1}{4}, \dfrac{1}{4} \cdot \dfrac{1}{2} = \dfrac{1}{8}, \dfrac{1}{8} \cdot \dfrac{1}{2} = \dfrac{1}{16}, \dfrac{1}{16} \cdot \dfrac{1}{2} = \dfrac{1}{32},$

$\dfrac{1}{32} \cdot \dfrac{1}{2} = \dfrac{1}{64}, \dfrac{1}{64} \cdot \dfrac{1}{2} = \dfrac{1}{128}$

$\dfrac{1}{4}, \dfrac{1}{8}, \dfrac{1}{16}, \dfrac{1}{32}, \dfrac{1}{64}, \dfrac{1}{128}$

67. $a_1 = -\dfrac{1}{16}$, $r = -4$

$-\dfrac{1}{16}, -\dfrac{1}{16} \cdot (-4) = \dfrac{1}{4}, \dfrac{1}{4} \cdot (-4) = -1,$

$-1(-4) = 4, 4(-4) = -16, -16(-4) = 64$

$-\dfrac{1}{16}, \dfrac{1}{4}, -1, 4, -16, 64$

69. $a_1 = 2$, $r = 0.1$

$2, 2(0.1) = 0.2, 0.2(0.1) = 0.02,$

$0.02(0.1) = 0.002, 0.002(0.1) = 0.0002, 0.0002(0.1)$

$= 0.00002.$

$2, 0.2, 0.02, 0.002, 0.0002, 0.00002$

71. $a_1 = 4$, $r = 2$

$a_7 = 4(2)^{7-1}$

$= 4(2)^6$

$= 4(64)$

$= 256$

73. $a_1 = 2$, $r = 3$

$a_{20} = 2(3)^{20-1}$

$= 2(3)^{19}$

$= 2,324,522,934$

$\approx 2.32 \times 10^9$

75. $a_1 = 50, \ r = 1$

$a_{100} = 50(1)^{100-1}$

$= 50(1)^{99}$

$= 50$

77. $a_1 = 5, \ r = -2$

$a_7 = 5(-2)^{7-1}$

$= 5(-2)^6$

$= 320$

79. $a_1 = 2, \ r = -1$

$a_{30} = 2(-1)^{30-1}$

$= 2(-1)^{29}$

$= -2$

81. $a_1 = -2, \ r = -3$

$a_6 = -2(-3)^{6-1}$

$= -2(-3)^5$

$= 486$

83. $a_1 = 6, \ r = \dfrac{1}{2}$

$a_8 = 6\left(\dfrac{1}{2}\right)^{8-1}$

$= 6\left(\dfrac{1}{2}\right)^7$

$= \dfrac{6}{128}$

$= \dfrac{3}{64}$

85. $a_1 = 18, \ r = -\dfrac{1}{3}$

$a_6 = 18\left(-\dfrac{1}{3}\right)^{6-1}$

$= 18\left(-\dfrac{1}{3}\right)^5$

$= -\dfrac{18}{243}$

$= -\dfrac{2}{27}$

87. $a_1 = 1000, \ r = -\dfrac{1}{2}$

$a_{40} = 1000\left(-\dfrac{1}{2}\right)^{40-1}$

$= 1000\left(-\dfrac{1}{2}\right)^{39}$

$\approx -1.82 \times 10^{-9}$

89. $a_1 = 1,000,000, \ r = 0.1$

$a_8 = 1,000,000(0.1)^{8-1}$

$= 1,000,000(0.1)^7$

$= 0.1$

91. $a_n = a_1 r^{n-1}$ with $a_1 = 3$ and $r = \dfrac{12}{3} = 4$. Thus $a_n = 3(4)^{n-1}$

$a_7 = 3(4)^{7-1} = 3(4)^6 = 3(4096) = 12,288$

93. $a_n = a_1 r^{n-1}$ with $a_1 = 18$ and $r = \dfrac{6}{18} = \dfrac{1}{3}$. Thus $a_n = 18\left(\dfrac{1}{3}\right)^{n-1}$

$a_7 = 18\left(\dfrac{1}{3}\right)^{7-1} = 18\left(\dfrac{1}{3}\right)^6 = 18\left(\dfrac{1}{729}\right) = \dfrac{18}{729} = \dfrac{2}{81}$

95. $a_n = a_1 r^{n-1}$ with $a_1 = 1.5$ and $r = \dfrac{-3}{1.5} = -2$. Thus $a_n = 1.5(-2)^{n-1}$

$a_7 = 1.5(-2)^{7-1} = 1.5(-2)^6 = 1.5(64) = 96$

97. $a_n = a_1 r^{n-1}$ with $a_1 = 0.0004$ and $r = \dfrac{-0.004}{0.0004} = -10$. Thus $a_n = 0.0004(-10)^{n-1}$

$a_7 = 0.0004(-10)^{7-1} = 0.0004(-10)^6 = 0.0004(1,000,000) = 400$

99. The common difference of the arithmetic sequence is 4.
$2 + 4 = 6, 6 + 4 = 10, 10 + 4 = 14,$
$14 + 4 = 18, 18 + 4 = 22$
$2, 6, 10, 14, 18, 22, \ldots$

101. The common ratio of the geometric sequence is 3.
$5 \cdot 3 = 15, 15 \cdot 3 = 45, 45 \cdot 3 = 135, \quad 5, 15, 45, 135, 405, 1215, \ldots$
$135 \cdot 3 = 405, 405 \cdot 3 = 1215$

103. The common difference of the arithmetic sequence is 5.
$-7 + 5 = -2, -2 + 5 = 3, 3 + 5 = 8,$
$8 + 5 = 13, 13 + 5 = 18.$
$-7, -2, 3, 8, 13, 18, \ldots$

105. The common ratio of the geometric sequence is $\dfrac{1}{2}$.

$3 \cdot \dfrac{1}{2} = \dfrac{3}{2}, \dfrac{3}{2} \cdot \dfrac{1}{2} = \dfrac{3}{4}, \dfrac{3}{4} \cdot \dfrac{1}{2} = \dfrac{3}{8}, \dfrac{3}{8} \cdot \dfrac{1}{2} = \dfrac{3}{16}$

$\dfrac{3}{16} \cdot \dfrac{1}{2} = \dfrac{3}{32}$

$3, \dfrac{3}{2}, \dfrac{3}{4}, \dfrac{3}{8}, \dfrac{3}{16}, \dfrac{3}{32}, \ldots$

107. The common difference of the arithmetic sequence is $\dfrac{1}{2}$.

$\dfrac{1}{2} + \dfrac{1}{2} = 1, 1 + \dfrac{1}{2} = \dfrac{3}{2}, \dfrac{3}{2} + \dfrac{1}{2} = 2, 2 + \dfrac{1}{2} = \dfrac{5}{2},$

$\dfrac{5}{2} + \dfrac{1}{2} = 3$

$\dfrac{1}{2}, 1, \dfrac{3}{2}, 2, \dfrac{5}{2}, 3, \ldots$

109. The common ratio of the geometric sequence is -1.
$7(-1) = -7, -7(-1) = 7, 7(-1) = -7,$
$-7(-1) = 7, 7(-1) = -7$
$7, -7, 7, -7, 7, -7, \ldots$

111. The common difference of the arithmetic sequence is -14.
$7 - 14 = -7, -7 - 14 = -21, -21 - 14 = -35, -35 - 14 = -49, -49 - 14 = -63$
$7, -7, -21, -35, -49, -63, \ldots$

113. The common ratio of the geometric sequence is $\sqrt{5}$.

$\sqrt{5} \cdot \sqrt{5} = 5, 5 \cdot \sqrt{5} = 5\sqrt{5}, 5\sqrt{5} \cdot \sqrt{5} = 25,$

$25 \cdot \sqrt{5} = 25\sqrt{5}, 25\sqrt{5} \cdot \sqrt{5} = 125$

$\sqrt{5}, 5, 5\sqrt{5}, 25, 25\sqrt{5}, 125, \ldots$

115. arithmetic; use $S_n = \dfrac{n}{2}(a_1 + a_n)$

$S_{10} = \dfrac{10}{2}(4 + 58) = 310$

117. geometric; use $S_n = \dfrac{a_1(1 - r^n)}{1 - r}$

$S_{10} = \dfrac{2(1 - 3^{10})}{1 - 3} = 59,048$

119. geometric; use $S_n = \dfrac{a_1(1 - r^n)}{1 - r}$

$S_{10} = \dfrac{3\left(1 - (-2)^{10}\right)}{1 - (-2)} = -1023$

121. arithmetic; use $S_n = \dfrac{n}{2}(a_1 + a_n)$

$S_{10} = \dfrac{10}{2}(-10 + 26) = 80$

123. $1 + 2 + 3 + 4 + \cdots + 100$

$S_{100} = \dfrac{100}{2}(1 + 100) = 5050$

125. a. $a_n = 23.08 + 0.12n$

  b. $a_{40} = 23.08 + 0.12(40) = 27.88$ years old

127. Company A: $a_{10} = 24000 + (10 - 1)1600 = 38,400$
Company B: $b_{10} = 28000 + (10 - 1)1000 = 37,000$
Company A will pay $1400 more in year 10.

129. $a_1 = 1,\ r = 2$

$a_{15} = 1(2)^{15-1}$

$= 2^{14}$

$= 16,384$

On the 15th day you will put aside $16,384.

131. $a_7 = \$3,000,000(1.04)^{7-1}$

$\approx \$3,795,957$ salary in year 7.

133. a. $\dfrac{30.15}{29.76} \approx 1.013;$ $\dfrac{30.54}{30.15} \approx 1.013;$ $\dfrac{30.94}{30.54} \approx 1.013;$ $\dfrac{31.34}{30.94} \approx 1.013$

$\dfrac{31.75}{31.34} \approx 1.013;$ $\dfrac{32.16}{31.75} \approx 1.013;$ $\dfrac{32.58}{32.16} \approx 1.013;$ $r \approx 1.013$

b. $a_n = 29.76(1.013)^{n-1}$

c. $a_{11} \approx 29.76(1.013)^{11-1} \approx 33.86$ million in 2000.
The geometric sequence described the actual population very well.

143. d is true.

Chapter 5 Review Exercises

1. 238,632
2: Yes; The last digit is 2.
3: Yes; The sum of the digits is 24, which is divisible by 3.
4: Yes; The last two digits form 32, which is divisible by 4.
5: No; The last number does not end in 0 or 5.
6: Yes; The number is divisible by both 2 and 3.
8: Yes; The last three digits form 632, which is divisible by 8.
9: No; The sum of the digits is 24, which is not divisible by 9.
10: No; the last digit is not 0.
12: Yes; The number is divisible by both 3 and 4.
The number is divisible by 2, 3, 4, 6, 8, 12.

2. 421,153,470
2: Yes; The last digit is 0.
3: Yes; The sum of the digits is 27, which is divisible by 3.
4: No; The last two digits form 70, which is not divisible by 4.
5: Yes; The number ends in 0.
6: Yes; The number is divisible by both 2 and 3.
8: No; The last three digits form 470, which is not divisible by 8.
9: Yes; The sum of the digits is 27, which is divisible by 9.
10: Yes; The number ends in 0.
12: No; The number is not divisible by both 3 and 4.
The number is divisible by 2, 3, 5, 6, 9, 10.

3.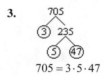
$705 = 3 \cdot 5 \cdot 47$

4.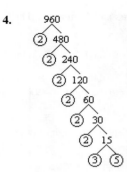
$960 = 2^6 \cdot 3 \cdot 5$

5. 6825
$6825 = 3 \cdot 5^2 \cdot 7 \cdot 13$

6. $30 = 2 \cdot 3 \cdot 5$
$48 = 2^4 \cdot 3$
Greatest Common Divisor = $2 \cdot 3 = 6$
Least Common Multiple = $2^4 \cdot 3 \cdot 5 = 240$

7. $36 = 2^2 \cdot 3^2$
$150 = 2 \cdot 3 \cdot 5^2$
Greatest Common Divisor = $2 \cdot 3 = 6$
Least Common Multiple = $2^2 \cdot 3^2 \cdot 5^2 = 900$

8. $216 = 2^3 \cdot 3^3$
    $254 = 2 \cdot 127$
    Greatest Common Divisor = 2
    Least Common Multiple = $2^3 \cdot 3^3 \cdot 127$
    $\qquad\qquad\qquad\qquad = 27,432$

9. $24 = 2^3 \cdot 3$
    $60 = 2^2 \cdot 3 \cdot 5$
    Greatest Common Divisor = $2^2 \cdot 3 = 12$
    There can be 12 people placed on each team.

10. $42 = 2 \cdot 3 \cdot 7$
    $56 = 2^3 \cdot 7$
    Least Common Multiple = $2^3 \cdot 3 \cdot 7 = 168$
    $168 \div 60 = 2.8$ or 2 hours and 48 minutes. They will begin again at 11:48 A.M.

11. $-93 < 17$ because -93 is to the left of 17 on the number line.

12. $-2 > -200$ because -2 is to the right of -200 on the number line.

13. $|-860| = 860$ because -860 is 860 units from 0 on the number line.

14. $|53| = 53$ because 53 is 53 units from 0 on the number line.

15. $|0| = 0$ because 0 is 0 units from 0 on the number line.

16. $8 + (-11) = -3$

17. $-6 + (-5) = -11$

18. $-7 - 8 = -7 + (-8) = -15$

19. $-7 - (-8) = -7 + 8 = 1$

20. $(-9)(-11) = 99$

21. $5(-3) = -15$

22. $\dfrac{-36}{-4} = 9$

23. $\dfrac{20}{-5} = -4$

24. $-40 \div 5 \cdot 2 = -8 \cdot 2 = -16$

25. $-6 + (-2) \cdot 5 = -6 + (-10) = -16$

26. $6 - 4(-3 + 2) = 6 - 4(-1) = 6 + 4 = 10$

27. $28 \div (2 - 4^2) = 28 \div (2 - 16)$
    $\qquad\qquad\quad = 28 \div (-14)$
    $\qquad\qquad\quad = -2$

28. $36 - 24 \div 4 \cdot 3 - 1 = 36 - 6 \cdot 3 - 1$
    $\qquad\qquad\qquad\qquad = 36 - 18 - 1$
    $\qquad\qquad\qquad\qquad = 18 - 1$
    $\qquad\qquad\qquad\qquad = 17$

29. $-57 - (-715) = -57 + 715 = \$658$ billion

30. $40 = 2^3 \cdot 5$
    $75 = 3 \cdot 5^2$
    Greatest Common Divisor is 5.
    $\dfrac{40}{75} = \dfrac{40 \div 5}{75 \div 5} = \dfrac{8}{15}$

31. $36 = 2^2 \cdot 3^2$
    $150 = 2 \cdot 3 \cdot 5^2$
    Greatest Common Divisor is $2 \cdot 3$ or 6.
    $\dfrac{36}{150} = \dfrac{36 \div 6}{150 \div 6} = \dfrac{6}{25}$

32. $165 = 3 \cdot 5 \cdot 11$
    $180 = 2^2 \cdot 3^2 \cdot 5$
    Greatest Common Divisor is $3 \cdot 5$ or 15.
    $\dfrac{165}{180} = \dfrac{165 \div 15}{180 \div 15} = \dfrac{11}{12}$

33. $5\dfrac{9}{11} = \dfrac{11 \cdot 5 + 9}{11} = \dfrac{64}{11}$

34. $-3\dfrac{2}{7} = -\dfrac{7 \cdot 3 + 2}{7} = -\dfrac{23}{7}$

35. $\dfrac{27}{5} = 5\dfrac{2}{5}$

36. $-\dfrac{17}{9} = -1\dfrac{8}{9}$

37. $\dfrac{4}{5} = 0.8$

$$5\overline{)4.0}$$
$$\underline{40}$$
$$0$$
$$0.8$$

38. $\dfrac{3}{7} = 0.\overline{428571}$

$$7\overline{)3.0000000}$$
$$0.4285714$$
$$\underline{28}$$
$$20$$
$$\underline{14}$$
$$60$$
$$\underline{56}$$
$$40$$
$$\underline{35}$$
$$50$$
$$\underline{49}$$
$$10$$
$$\underline{7}$$
$$30$$
$$\underline{28}$$
$$2$$

39. $\dfrac{5}{8} = 0.625$

$$8\overline{)5.000}$$
$$0.625$$
$$\underline{48}$$
$$20$$
$$\underline{16}$$
$$40$$
$$\underline{40}$$
$$0$$

40. $\dfrac{9}{16} = 0.5625$

$$16\overline{)9.0000}$$
$$0.5625$$
$$\underline{80}$$
$$100$$
$$\underline{96}$$
$$40$$
$$\underline{32}$$
$$80$$
$$\underline{80}$$
$$0$$

41. $0.6 = \dfrac{6}{10} = \dfrac{6 \div 2}{10 \div 2} = \dfrac{3}{5}$

42. $0.68 = \dfrac{68}{100}$

$68 = 2^2 \cdot 17$

$100 = 2^2 \cdot 5^2$

Greatest Common Divisor is 2^2 or 4.

$\dfrac{68 \div 4}{100 \div 4} = \dfrac{17}{25}$

43. $0.588 = \dfrac{588}{1000}$

$588 = 2^2 \cdot 3 \cdot 7^2$

$1000 = 2^3 \cdot 5^3$

Greatest Common Divisor is 2^2 or 4.

$\dfrac{588 \div 4}{1000 \div 4} = \dfrac{147}{250}$

44. $0.0084 = \dfrac{84}{10,000}$

$84 = 2^2 \cdot 3 \cdot 7$

$10,000 = 2^4 \cdot 5^4$

Greatest Common Divisor is 2^2 or 4.

$\dfrac{84 \div 4}{10,000 \div 4} = \dfrac{21}{2500}$

45. $n = 0.555\ldots$
$10n = 5.555\ldots$

$10n = 5.555\ldots$
$\underline{-\quad n = 0.555\ldots}$
$9n = 5$

$n = \dfrac{5}{9}$

46. $n = 0.3434\ldots$
$100n = 34.3434\ldots$

$100n = 34.3434\ldots$
$\underline{-\quad n = 0.3434\ldots}$
$99n = 34$

$n = \dfrac{34}{99}$

47.
$$n = 0.113113 \ldots$$
$$1000n = 113.113113 \ldots$$

$$1000n = 113.113113\ldots$$
$$- \quad n = \quad 0.113113\ldots$$
$$\overline{999n = 113}$$

$$n = \frac{113}{999}$$

48. $\dfrac{3}{5} \cdot \dfrac{7}{10} = \dfrac{3 \cdot 7}{5 \cdot 10} = \dfrac{21}{50}$

49. $\left(3\dfrac{1}{3}\right)\left(1\dfrac{3}{4}\right) = \dfrac{10}{3} \cdot \dfrac{7}{4} = \dfrac{70}{12} = \dfrac{35}{6}$ or $5\dfrac{5}{6}$

50. $\dfrac{4}{5} \div \dfrac{3}{10} = \dfrac{4}{5} \cdot \dfrac{10}{3} = \dfrac{4 \cdot 10}{5 \cdot 3} = \dfrac{40}{15} = \dfrac{8}{3}$

51. $-1\dfrac{2}{3} \div 6\dfrac{2}{3} = -\dfrac{5}{3} \div \dfrac{20}{3} = -\dfrac{5}{3} \cdot \dfrac{3}{20} = -\dfrac{15}{60} = -\dfrac{1}{4}$

52. $\dfrac{2}{9} + \dfrac{4}{9} = \dfrac{2+4}{9} = \dfrac{6}{9} = \dfrac{2}{3}$

53. $\dfrac{7}{9} + \dfrac{5}{12} = \dfrac{7}{9} \cdot \dfrac{4}{4} + \dfrac{5}{12} \cdot \dfrac{3}{3}$

$$= \dfrac{28}{36} + \dfrac{15}{36}$$

$$= \dfrac{28+15}{36}$$

$$= \dfrac{43}{36}$$

54. $\dfrac{3}{4} - \dfrac{2}{15} = \dfrac{3}{4} \cdot \dfrac{15}{15} - \dfrac{2}{15} \cdot \dfrac{4}{4}$

$$= \dfrac{45}{60} - \dfrac{8}{60}$$

$$= \dfrac{45-8}{60}$$

$$= \dfrac{37}{60}$$

55. $\dfrac{1}{3} + \dfrac{1}{2} \cdot \dfrac{4}{5} = \dfrac{1}{3} + \dfrac{1 \cdot 4}{2 \cdot 5}$

$$= \dfrac{1}{3} + \dfrac{4}{10}$$

$$= \dfrac{1}{3} + \dfrac{2}{5}$$

$$= \dfrac{1}{3} \cdot \dfrac{5}{5} + \dfrac{2}{5} \cdot \dfrac{3}{3}$$

$$= \dfrac{5}{15} + \dfrac{6}{15}$$

$$= \dfrac{11}{15}$$

56. $\dfrac{3}{8}\left(\dfrac{1}{2} + \dfrac{1}{3}\right) = \dfrac{3}{8}\left(\dfrac{1}{2} \cdot \dfrac{3}{3} + \dfrac{1}{3} \cdot \dfrac{2}{2}\right)$

$$= \dfrac{3}{8}\left(\dfrac{3}{6} + \dfrac{2}{6}\right)$$

$$= \dfrac{3}{8}\left(\dfrac{5}{6}\right)$$

$$= \dfrac{15}{48}$$

$$= \dfrac{5}{16}$$

57. $\dfrac{1}{7} + \dfrac{1}{8} = \dfrac{1}{7} \cdot \dfrac{8}{8} + \dfrac{1}{8} \cdot \dfrac{7}{7}$

$$= \dfrac{8}{56} + \dfrac{7}{56}$$

$$= \dfrac{15}{56}$$

$$\dfrac{15}{56} \div 2 = \dfrac{15}{56} \cdot \dfrac{1}{2} = \dfrac{15}{112}$$

58. $\dfrac{3}{4} + \dfrac{3}{5} = \dfrac{3}{4} \cdot \dfrac{5}{5} + \dfrac{3}{5} \cdot \dfrac{4}{4}$

$$= \dfrac{15}{20} + \dfrac{12}{20}$$

$$= \dfrac{27}{20}$$

$$\dfrac{27}{20} \div 2 = \dfrac{27}{20} \cdot \dfrac{1}{2} = \dfrac{27}{40}$$

59. $4\dfrac{1}{2} \cdot \dfrac{15}{6} = \dfrac{9}{2} \cdot \dfrac{15}{6} = \dfrac{135}{12} = \dfrac{45}{4}$ or $11\dfrac{1}{4}$ pounds.

60. $1-\left(\dfrac{1}{4}+\dfrac{1}{3}\right)=1-\left(\dfrac{1}{4}\cdot\dfrac{3}{3}+\dfrac{1}{3}\cdot\dfrac{4}{4}\right)$

$\qquad\qquad\quad =1-\left(\dfrac{3}{12}+\dfrac{4}{12}\right)$

$\qquad\qquad\quad =\dfrac{12}{12}-\dfrac{7}{12}$

$\qquad\qquad\quad =\dfrac{5}{12}$

At the end of the second day, $\dfrac{5}{12}$ of the tank is filled with gas.

61. $\sqrt{28}=\sqrt{4\cdot7}=\sqrt{4}\cdot\sqrt{7}=2\sqrt{7}$

62. $\sqrt{72}=\sqrt{36\cdot2}=\sqrt{36}\cdot\sqrt{2}=6\sqrt{2}$

63. $\sqrt{150}=\sqrt{25\cdot6}=\sqrt{25}\cdot\sqrt{6}=5\sqrt{6}$

64. $\sqrt{300}=\sqrt{100\cdot3}=\sqrt{100}\cdot\sqrt{3}=10\sqrt{3}$

65. $\sqrt{6}\cdot\sqrt{8}=\sqrt{6\cdot8}=\sqrt{48}=\sqrt{16}\cdot\sqrt{3}=4\sqrt{3}$

66. $\sqrt{10}\cdot\sqrt{5}=\sqrt{10\cdot5}$

$\qquad\qquad\; =\sqrt{50}$

$\qquad\qquad\; =\sqrt{25}\cdot\sqrt{2}$

$\qquad\qquad\; =5\sqrt{2}$

67. $\dfrac{\sqrt{24}}{\sqrt{2}}=\sqrt{\dfrac{24}{2}}=\sqrt{12}=\sqrt{4}\cdot\sqrt{3}=2\sqrt{3}$

68. $\dfrac{\sqrt{27}}{\sqrt{3}}=\sqrt{\dfrac{27}{3}}=\sqrt{9}=3$

69. $\sqrt{5}+4\sqrt{5}=1\sqrt{5}+4\sqrt{5}=(1+4)\sqrt{5}=5\sqrt{5}$

70. $7\sqrt{11}-13\sqrt{11}=(7-13)\sqrt{11}=-6\sqrt{11}$

71. $\sqrt{50}+\sqrt{8}=\sqrt{25}\cdot\sqrt{2}+\sqrt{4}\cdot\sqrt{2}$

$\qquad\qquad\quad =5\sqrt{2}+2\sqrt{2}$

$\qquad\qquad\quad =(5+2)\sqrt{2}$

$\qquad\qquad\quad =7\sqrt{2}$

72. $\sqrt{3}-6\sqrt{27}=\sqrt{3}-6\sqrt{9}\cdot\sqrt{3}$

$\qquad\qquad\quad =\sqrt{3}-6\cdot3\sqrt{3}$

$\qquad\qquad\quad =1\sqrt{3}-18\sqrt{3}$

$\qquad\qquad\quad =(1-18)\sqrt{3}$

$\qquad\qquad\quad =-17\sqrt{3}$

73. $2\sqrt{18}+3\sqrt{8}=2\sqrt{9}\cdot\sqrt{2}+3\sqrt{4}\cdot\sqrt{2}$

$\qquad\qquad\quad =2\cdot3\cdot\sqrt{2}+3\cdot2\cdot\sqrt{2}$

$\qquad\qquad\quad =6\sqrt{2}+6\sqrt{2}$

$\qquad\qquad\quad =(6+6)\sqrt{2}$

$\qquad\qquad\quad =12\sqrt{2}$

74. $\dfrac{30}{\sqrt{5}}=\dfrac{30}{\sqrt{5}}\cdot\dfrac{\sqrt{5}}{\sqrt{5}}=\dfrac{30\sqrt{5}}{\sqrt{25}}=\dfrac{30\sqrt{5}}{5}=6\sqrt{5}$

75. $\sqrt{\dfrac{2}{3}}=\dfrac{\sqrt{2}}{\sqrt{3}}=\dfrac{\sqrt{2}}{\sqrt{3}}\cdot\dfrac{\sqrt{3}}{\sqrt{3}}=\dfrac{\sqrt{6}}{\sqrt{9}}=\dfrac{\sqrt{6}}{3}$

76. $W=4\sqrt{2x}$

$\qquad =4\sqrt{2\cdot6}$

$\qquad =4\sqrt{12}$

$\qquad =8\sqrt{3}\approx13.9$ feet per second

77. $\left\{-17,\ -\dfrac{9}{13},\ 0,\ 0.75,\ \sqrt{2},\ \pi,\ \sqrt{81}\right\}$

   a. Natural numbers:
$\sqrt{81}$ because $\sqrt{81}=9$

   b. Whole numbers: $0,\ \sqrt{81}$

   c. Integers: $-17,\ 0,\ \sqrt{81}$

   d. Rational numbers:
$-17,\ -\dfrac{9}{13},\ 0,\ 0.75,\ \sqrt{81}$

   e. Irrational numbers: $\sqrt{2},\ \pi$

   f. Real numbers: All numbers in this set.

78. Answers will vary. Example: -3

79. Answers will vary. Example: $\dfrac{1}{2}$

80. Answers will vary: Example: $\sqrt{2}$

81. $3 + 17 = 17 + 3$
Commutative property of addition

82. $(6 \cdot 3) \cdot 9 = 6 \cdot (3 \cdot 9)$
Associative property of multiplication

83. $\sqrt{3}\left(\sqrt{5} + \sqrt{3}\right) = \sqrt{15} + 3$
Distributive property of multiplication over addition.

84. $(6 \cdot 9) \cdot 2 = 2 \cdot (6 \cdot 9)$
Commutative property of multiplication

85. $\sqrt{3}\left(\sqrt{5} + \sqrt{3}\right) = \left(\sqrt{5} + \sqrt{3}\right)\sqrt{3}$
Commutative property of multiplication

86. $(3 \cdot 7) + (4 \cdot 7) = (4 \cdot 7) + (3 \cdot 7)$
Commutative property of addition

87. Answers will vary. Example: $2 \div 6 = \dfrac{1}{3}$

88. Answers will vary. Example: $4 - 5 = -1$

89. $6 \cdot 6^2 = 6^1 \cdot 6^2 = 6^{1+2} = 6^3 = 216$

90. $2^3 \cdot 2^3 = 2^{3+3} = 2^6 = 64$

91. $(2^2)^2 = 2^{2 \cdot 2} = 2^4 = 16$

92. $(3^3)^2 = 3^{3 \cdot 2} = 3^6 = 729$

93. $\dfrac{5^6}{5^4} = 5^{6-4} = 5^2 = 25$

94. $7^0 = 1$

95. $(-7)^0 = 1$

96. $6^{-3} = \dfrac{1}{6^3} = \dfrac{1}{216}$

97. $2^{-4} = \dfrac{1}{2^4} = \dfrac{1}{16}$

98. $\dfrac{7^4}{7^6} = 7^{4-6} = 7^{-2} = \dfrac{1}{7^2} = \dfrac{1}{49}$

99. $3^5 \cdot 3^{-2} = 3^{5-2} = 3^3 = 27$

100. $4.6 \times 10^2 = 460$

101. $3.74 \times 10^4 = 37,400$

102. $2.55 \times 10^{-3} = 0.00255$

103. $7.45 \times 10^{-5} = 0.0000745$

104. $7520 = 7.52 \times 10^3$

105. $3,590,000 = 3.59 \times 10^6$

106. $0.00725 = 7.25 \times 10^{-3}$

107. $0.000000409 = 4.09 \times 10^{-7}$

108. $420 \times 10^{11} = \left(4.2 \times 10^2\right) \times 10^{11} = 4.2 \times 10^{13}$

109. $0.97 \times 10^{-4} = \left(9.7 \times 10^{-1}\right) \times 10^{-4} = 9.7 \times 10^{-5}$

110. $(3 \times 10^7)(1.3 \times 10^{-5}) = (3 \times 1.3) \times 10^{7-5}$
$$= 3.9 \times 10^2$$
$$= 390$$

111. $(5 \times 10^3)(2.3 \times 10^2) = (5 \times 2.3) \times 10^{3+2}$
$$= 11.5 \times 10^5$$
$$= 1.15 \times 10 \times 10^5$$
$$= 1.15 \times 10^6$$
$$= 1,150,000$$

112. $\dfrac{6.9 \times 10^3}{3 \times 10^5} = \left(\dfrac{6.9}{3}\right) \times 10^{3-5}$
$$= 2.3 \times 10^{-2}$$
$$= 0.023$$

113. $\dfrac{2.4 \times 10^{-4}}{6 \times 10^{-6}} = \left(\dfrac{2.4}{6}\right) \times 10^{-4-(-6)}$
$$= 0.4 \times 10^{-4+6}$$
$$= 0.4 \times 10^2$$
$$= 40$$

114. $(60,000)(540,000) = (6.0 \times 10^4)(5.4 \times 10^5)$
$$= (6.0 \times 5.4) \times 10^{4+5}$$
$$= 32.4 \times 10^9$$
$$= 3.24 \times 10 \times 10^9$$
$$= 3.24 \times 10^{10}$$

115. $(91,000)(0.0004) = (9.1 \times 10^4)(4 \times 10^{-4})$
$$= (9.1 \times 4) \times 10^{4-4}$$
$$= 36.4 \times 10^0$$
$$= 3.64 \times 10^1$$

116. $\dfrac{8,400,000}{4000} = \dfrac{8.4 \times 10^6}{4 \times 10^3}$
$$= \left(\dfrac{8.4}{4}\right) \times 10^{6-3}$$
$$= 2.1 \times 10^3$$

117. $\dfrac{0.000003}{0.00000006} = \dfrac{3 \times 10^{-6}}{6 \times 10^{-8}}$
$$= \left(\dfrac{3}{6}\right) \times 10^{-6-(-8)}$$
$$= 0.5 \times 10^2$$
$$= 5 \times 10^{-1} \times 10^2$$
$$= 5 \times 10^1$$

118. $\dfrac{10^9}{10^6} = 10^{9-6} = 10^3 = 1000$ years

119. $150 \times (3 \times 10^8) = 450 \times 10^8$
$$= 4.5 \times 10^2 \times 10^8$$
$$= \$4.5 \times 10^{10}$$

120. $2 \times (6.5 \times 10^9) = 13 \times 10^9$
$$= 1.3 \times 10^1 \times 10^9$$
$$= 1.3 \times 10^{10} \text{ people}$$

121. $a_1 = 7,\ d = 4$
7, 7 + 4 = 11, 11 + 4 = 15, 15 + 4 = 19, 19 + 4 = 23,
23 + 4 = 27
7, 11, 15, 19, 23, 27

122. $a_1 = -4,\ d = -5$
$-4, -4 - 5 = -9, -9 - 5 = -14, -14 - 5 = -19, -19 - 5$
$= -24, -24 - 5 = -29$
$-4, -9, -14, -19, -24, -29$

123. $a_1 = \dfrac{3}{2},\ d = -\dfrac{1}{2}$
$$\dfrac{3}{2}, \dfrac{3}{2} - \dfrac{1}{2} = \dfrac{2}{2} = 1, \dfrac{2}{2} - \dfrac{1}{2} = \dfrac{1}{2}, \dfrac{1}{2} - \dfrac{1}{2} = 0,$$
$$0 - \dfrac{1}{2} = -\dfrac{1}{2}, -\dfrac{1}{2} - \dfrac{1}{2} = -1$$
$$\dfrac{3}{2}, 1, \dfrac{1}{2}, 0, -\dfrac{1}{2}, -1$$

124. $a_1 = 5,\ d = 3$
$a_6 = 5 + (6 - 1)(3)$
$$= 5 + 5(3)$$
$$= 5 + 15$$
$$= 20$$

125. $a_1 = -8,\ d = -2$
$a_{12} = -8 + (12 - 1)(-2)$
$$= -8 + 11(-2)$$
$$= -8 + (-22)$$
$$= -30$$

126. $a_1 = 14,\ d = -4$
$a_{14} = 14 + (14 - 1)(-4)$
$$= 14 + 13(-4)$$
$$= 14 + (-52)$$
$$= -38$$

127. $a_n = a_1 + (n - 1)d$ with $a_1 = -7,\ d = 4$
$a_n = -7 + (n - 1)4$
$$= -7 + 4n - 4$$
$$= 4n - 11$$
Thus $a_{20} = 4(20) - 11 = 69$.

128. $a_n = a_1 + (n - 1)d$ with $a_1 = 200,\ d = -20$
$a_n = 200 + (n - 1)(-20)$
$$= 200 - 20n + 20$$
$$= -20n + 220$$
Thus $a_{20} = -20(20) + 220 = -180$.

129. $a_1 = 3,\ r = 2$
$3, 3 \cdot 2 = 6, 6 \cdot 2 = 12, 12 \cdot 2 = 24,$
$24 \cdot 2 = 48, 48 \cdot 2 = 96$
3, 6, 12, 24, 48, 96

130. $a_1 = \dfrac{1}{2}, \; r = \dfrac{1}{2}$

$$\dfrac{1}{2}, \; \dfrac{1}{2} \cdot \dfrac{1}{2} = \dfrac{1}{4}, \; \dfrac{1}{4} \cdot \dfrac{1}{2} = \dfrac{1}{8}, \; \dfrac{1}{8} \cdot \dfrac{1}{2} = \dfrac{1}{16},$$

$$\dfrac{1}{16} \cdot \dfrac{1}{2} = \dfrac{1}{32}, \; \dfrac{1}{32} \cdot \dfrac{1}{2} = \dfrac{1}{64}$$

$$\dfrac{1}{2}, \; \dfrac{1}{4}, \; \dfrac{1}{8}, \; \dfrac{1}{16}, \; \dfrac{1}{32}, \; \dfrac{1}{64}$$

131. $a_1 = 16, \; r = -\dfrac{1}{2}$

$$16, \; 16\left(-\dfrac{1}{2}\right) = -8, \; -8\left(-\dfrac{1}{2}\right) = 4, \; 4\left(-\dfrac{1}{2}\right) = -2,$$

$$-2\left(-\dfrac{1}{2}\right) = 1, \; 1\left(-\dfrac{1}{2}\right) = -\dfrac{1}{2}$$

$$16, \; -8, \; 4, \; -2, \; 1, \; -\dfrac{1}{2}$$

132. $a_1 = 2, \; r = 3$

$$a_4 = 2(3)^{4-1}$$
$$= 2(3)^3$$
$$= 2(27)$$
$$= 54$$

133. $a_1 = 16, \; r = \dfrac{1}{2}$

$$a_6 = 16\left(\dfrac{1}{2}\right)^{6-1}$$
$$= 16\left(\dfrac{1}{2}\right)^5$$
$$= \dfrac{16}{32}$$
$$= \dfrac{1}{2}$$

134. $a_1 = -3, \; r = 2$

$$a_5 = -3(2)^{5-1}$$
$$= -3(2)^4$$
$$= -3(16)$$
$$= -48$$

135. $a_n = a_1 r^{n-1}$ with $a_1 = 1$ and $r = \dfrac{2}{1} = 2$. Thus

$$a_n = 2^{n-1}$$
$$a_8 = 2^{8-1} = 2^7 = 128$$

136. $a_n = a_1 r^{n-1}$ with $a_1 = 100$ and $r = \dfrac{10}{100} = \dfrac{1}{10}$. Thus

$$a_n = 100\left(\dfrac{1}{10}\right)^{n-1}$$

$$a_8 = 100\left(\dfrac{1}{10}\right)^{8-1}$$

$$= 100\left(\dfrac{1}{10}\right)^7$$

$$= \dfrac{100}{10,000,000}$$

$$= \dfrac{1}{100,000}$$

137. The common difference in the arithmetic sequence is 5.
$4 + 5 = 9, \; 9 + 5 = 14, \; 14 + 5 = 19,$
$19 + 5 = 24, \; 24 + 5 = 29$
$4, 9, 14, 19, 24, 29, \ldots$

138. The common ratio in the geometric sequence is 3.
$2 \cdot 3 = 6, \; 6 \cdot 3 = 18, \; 18 \cdot 3 = 54, \; 54 \cdot 3 = 162, \; 162 \cdot 3$
$= 486$
$2, 6, 18, 54, 162, 486, \ldots$

139. The common ratio in the geometric sequence is $\dfrac{1}{4}$.

$$1 \cdot \dfrac{1}{4} = \dfrac{1}{4}, \; \dfrac{1}{4} \cdot \dfrac{1}{4} = \dfrac{1}{16}, \; \dfrac{1}{16} \cdot \dfrac{1}{4} = \dfrac{1}{64},$$

$$\dfrac{1}{64} \cdot \dfrac{1}{4} = \dfrac{1}{256}, \; \dfrac{1}{256} \cdot \dfrac{1}{4} = \dfrac{1}{1024}$$

$$1, \; \dfrac{1}{4}, \; \dfrac{1}{16}, \; \dfrac{1}{64}, \; \dfrac{1}{256}, \; \dfrac{1}{1024}, \ldots$$

140. The common difference in the arithmetic sequence is -7.
$0 - 7 = -7, \; -7 - 7 = -14, \; -14 - 7 = -21, \; -21 - 7 = -28,$
$-28 - 7 = -35$
$0, -7, -14, -21, -28, -35, \ldots$

141. a. $a_1 = 31.5$ and $d = -0.54$

$$a_n = a_1 + (n-1)d$$
$$= 31.5 + (n-1)(-0.54)$$
$$= 31.5 - 0.54n + 0.54$$
$$= -0.54n + 32.04$$

b. Vegetables:

$$a_n = -0.54n + 32.04$$
$$a_{18} = -0.54(18) + 32.04 = 22.32\%$$

142. a. Divide each value by the previous value:

$$\frac{5.9}{4.2} = 1.405$$

$$\frac{8.3}{5.9} = 1.407$$

$$\frac{11.6}{8.3} = 1.398$$

$$\frac{16.2}{11.6} = 1.397$$

$$\frac{22.7}{16.2} = 1.401$$

The population is increasing geometrically with $r \approx 1.4$.

b. $a_n = 4.2(1.4)^{n-1}$

c. 2080 is 9 decades after 1990 so $n = 9$.

$$a_n = 4.2(1.4)^{n-1}$$

$$a_8 = 4.2(1.4)^{9-1} \approx 62.0$$

In 2080, the model predicts the U.S. population, ages 85 and older, will be 62.0 million

Chapter 5 Test

1. 391,248

   2:   Yes; the last digit is 8.

   3:   Yes; the sum of the digits is 27, which is divisible by 3.

   4:   Yes; the last two digits form 48, which is divisible by 4.

   5:   No; the number does not end in 0 or 5.

   6:   Yes; the number is divisible by both 2 and 3.

   8:   Yes; the last three digits form 248, which is divisible by 8.

   9:   Yes; the sum of the digits is 27, which is divisible by 9.

   10: No; the number does not end in 0.

   12: Yes; the number is divisible by both 3 and 4.

   391, 248 is divisible by 2, 3, 4, 6, 8, 9, 12.

2.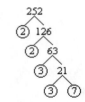

   $252 = 2^2 \cdot 3^2 \cdot 7$

3. $48 = 2^4 \cdot 3$

   $72 = 2^3 \cdot 3^2$

   Greatest Common Divisor $= 2^3 \cdot 3 = 24$

   Least Common Multiple $= 2^4 \cdot 3^2 = 144$

4. $-6 - (5 - 12) = -6 - (-7) = -6 + 7 = 1$

5. $(-3)(-4) \div (7 - 10) = (-3)(-4) \div (-3)$
   $$= 12 \div (-3)$$
   $$= -4$$

6. $(6-8)^2(5-7)^3 = (-2)^2(-2)^3$
   $$= 4(-8)$$
   $$= -32$$

7. $\dfrac{7}{12} = 0.58\overline{3}$

$$
\begin{array}{r}
0.5833\ldots \\
12{\overline{\smash{)}7.0000}} \\
\underline{60} \\
100 \\
\underline{96} \\
40 \\
\underline{36} \\
40 \\
\underline{36} \\
4
\end{array}
$$

8. $n = 0.6464\ldots$
   $100n = 64.6464\ldots$

   $100n = 64.6464\ldots$
   $\underline{-\quad n = \;0.6464\ldots}$
   $\quad 99n = 64$

   $n = \dfrac{64}{99}$

9. $\left(-\dfrac{3}{7}\right) \div \left(-2\dfrac{1}{7}\right) = \left(-\dfrac{3}{7}\right) \div \left(-\dfrac{15}{7}\right)$

$\qquad = \left(-\dfrac{3}{7}\right) \cdot \left(-\dfrac{7}{15}\right)$

$\qquad = \dfrac{(-3)(-7)}{7 \cdot 15}$

$\qquad = \dfrac{21}{105}$

$\qquad = \dfrac{1}{5}$

10. $\dfrac{19}{24} - \dfrac{7}{40} = \dfrac{19}{24} \cdot \dfrac{5}{5} - \dfrac{7}{40} \cdot \dfrac{3}{3}$

$\qquad = \dfrac{95}{120} - \dfrac{21}{120}$

$\qquad = \dfrac{95 - 21}{120}$

$\qquad = \dfrac{74}{120}$

$\qquad = \dfrac{37}{60}$

11. $\dfrac{1}{2} - 8\left(\dfrac{1}{4} + 1\right) = \dfrac{1}{2} - 8\left(\dfrac{5}{4}\right)$

$\qquad = \dfrac{1}{2} - 10$

$\qquad = \dfrac{1}{2} - \dfrac{20}{2}$

$\qquad = -\dfrac{19}{2}$

12. $\dfrac{1}{2} + \dfrac{2}{3} = \dfrac{1}{2} \cdot \dfrac{3}{3} + \dfrac{2}{3} \cdot \dfrac{2}{2}$

$\qquad = \dfrac{3}{6} + \dfrac{4}{6}$

$\qquad = \dfrac{7}{6}$

$\dfrac{7}{6} \div 2 = \dfrac{7}{6} \cdot \dfrac{1}{2} = \dfrac{7}{12}$

13. $\sqrt{10} \cdot \sqrt{5} = \sqrt{10 \cdot 5}$

$\qquad = \sqrt{50}$

$\qquad = \sqrt{25 \cdot 2}$

$\qquad = \sqrt{25} \cdot \sqrt{2}$

$\qquad = 5\sqrt{2}$

14. $\sqrt{50} + \sqrt{32} = \sqrt{25} \cdot \sqrt{2} + \sqrt{16} \cdot \sqrt{2}$

$\qquad = 5\sqrt{2} + 4\sqrt{2}$

$\qquad = (5 + 4)\sqrt{2}$

$\qquad = 9\sqrt{2}$

15. $\dfrac{6}{\sqrt{2}} = \dfrac{6}{\sqrt{2}} \cdot \dfrac{\sqrt{2}}{\sqrt{2}} = \dfrac{6\sqrt{2}}{\sqrt{4}} = \dfrac{6\sqrt{2}}{2} = 3\sqrt{2}$

16. The rational numbers are
$-7, -\dfrac{4}{5}, 0, 0.25, \sqrt{4}, \dfrac{22}{7}$.

17. Commutative property of addition

18. Distributive property of multiplication over addition

19. $3^3 \cdot 3^2 = 3^{3+2} = 3^5 = 243$

20. $\dfrac{4^6}{4^3} = 4^{6-3} = 4^3 = 64$

21. $8^{-2} = \dfrac{1}{8^2} = \dfrac{1}{64}$

22. $(3 \times 10^8)(2.5 \times 10^{-5}) = (3 \times 2.5) \times 10^{8-5}$

$\qquad = 7.5 \times 10^3$

$\qquad = 7500$

23. $\dfrac{49,000}{0.007} = \dfrac{4.9 \times 10^4}{7 \times 10^{-3}}$

$\qquad = \left(\dfrac{4.9}{7}\right) \times 10^{4-(-3)}$

$\qquad = 0.7 \times 10^7$

$\qquad = 7 \times 10^{-1} \times 10^7$

$\qquad = 7 \times 10^6$

24. $\dfrac{2.18 \times 10^{12}}{3 \times 10^8} = \left(\dfrac{2.18}{3}\right) \times 10^4$

$\qquad = 0.7267 \times 10^4$

$\qquad = \$7267$

25. $a_1 = 1, d = -5$

$1, 1 - 5 = -4, -4 - 5 = -9,$

$-9 - 5 = -14, -14 - 5 = -19,$

$-19 - 5 = -24$

$1, -4, -9, -14, -19, -24$

26. $a_1 = -2,\ d = 3$

$a_9 = -2 + (9-1)(3)$

$\quad = -2 + 8(3)$

$\quad = -2 + 24$

$\quad = 22$

27. $a_1 = 16,\ r = \dfrac{1}{2}$

$16,\ 16 \cdot \dfrac{1}{2} = 8,\ 8 \cdot \dfrac{1}{2} = 4,\ 4 \cdot \dfrac{1}{2} = 2,\ 2 \cdot \dfrac{1}{2} = 1,\ 1 \cdot \dfrac{1}{2} = \dfrac{1}{2}$

$16,\ 8,\ 4,\ 2,\ 1,\ \dfrac{1}{2}$

28. $a_1 = 5,\ r = 2$

$a_7 = 5(2)^{7-1}$

$\quad = 5(2)^6$

$\quad = 5(64)$

$\quad = 320$

Chapter 6
Algebra: Equations and Inequalities

Check Points 6.1

1. $8 + 6(x-3)^2 = 8 + 6(13-3)^2$
$$= 8 + 6(10)^2$$
$$= 8 + 6(100)$$
$$= 608$$

2. If $x = -5$, then $x^2 + 4x - 7 = (-5)^2 + 4(-5) - 7$
$$= 25 - 20 - 7$$
$$= -2$$

3. If $x = 5$ and $y = -1$, then
$$-3x^2 + 4xy - y^3 = -3(5)^2 + 4(5)(-1) - (-1)^3$$
$$= -3(25) - 20 - (-1)$$
$$= -75 - 20 + 1$$
$$= -94$$

4. In 2003 the graph shows a value of $400 million. This is best described by Model 3 as shown.
Model 1: $D = 236(1.5)^x$
$$= 236(1.5)^1$$
$$= 354$$
Model 2: $D = 127x + 239$
$$= 127(1) + 239$$
$$= 366$$
Model 3: $D = -54x^2 + 234x + 220$
$$= -54(1)^2 + 234(1) + 220$$
$$= -54 + 234 + 220$$
$$= 400$$

5. $7(2x-3) - 11x = 7 \cdot 2x - 7 \cdot 3 - 11x$
$$= 14x - 21 - 11x$$
$$= 3x - 21$$

6. $7(4x^2 + 3x) + 2(5x^2 + x) = 28x^2 + 21x + 10x^2 + 2x$
$$= 38x^2 + 23x$$

7. $6x + 4[7 - (x-2)] = 6x + 4[7 - x + 2]$
$$= 6x + 4[9 - x]$$
$$= 6x + 36 - 4x$$
$$= 2x + 36$$

Exercise Set 6.1

1. $5x + 7 = 5 \cdot 4 + 7 = 20 + 7 = 27$

3. $-7x - 5 = -7(-4) - 5 = 28 - 5 = 23$

5. $x^2 + 4 = 5^2 + 4 = 25 + 4 = 29$

7. $x^2 - 6 = (-2)^2 - 6 = 4 - 6 = -2$

9. $x^2 + 4x = (10)^2 + 4 \cdot 10 = 100 + 40 = 140$

11. $8x^2 + 17 = 8(5)^2 + 17$
$$= 8(25) + 17$$
$$= 200 + 17$$
$$= 217$$

13. $x^2 - 5x = (-11)^2 - 5(-11)$
$$= 121 + 55$$
$$= 176$$

15. $x^2 + 5x - 6 = 4^2 + 5 \cdot 4 - 6$
$$= 16 + 20 - 6$$
$$= 30$$

17. $4 + 5(x-7)^3 = 4 + 5(9-7)^3$
$$= 4 + 5(2)^3$$
$$= 4 + 5(8)$$
$$= 44$$

19. $x^2 - 3(x - y) = 2^2 - 3(2 - 8)$
$$= 4 - 3(-6)$$
$$= 4 + 18$$
$$= 22$$

21. $2x^2 - 5x - 6 = 2(-3)^2 - 5(-3) - 6$
$$= 2(9) - 5(-3) - 6$$
$$= 18 + 15 - 6$$
$$= 27$$

23. $-5x^2 - 4x - 11 = -5(-1)^2 - 4(-1) - 11$
$$= -5(1) - 4(-1) - 11$$
$$= -5 + 4 - 11$$
$$= -12$$

25. $3x^2 + 2xy + 5y^2 = 3(2)^2 + 2(2)(3) + 5(3)^2$
$$= 3(4) + 2(2)(3) + 5(9)$$
$$= 12 + 12 + 45$$
$$= 69$$

27. $-x^2 - 4xy + 3y^3 = -(-1)^2 - 4(-1)(-2) + 3(-2)^3$
$$= -(1) - 8 + 3(-8)$$
$$= -1 - 8 - 24$$
$$= -33$$

29. If $x = -2$ and $y = 4$ then
$$\frac{2x + 3y}{x + 1} = \frac{2(-2) + 3(4)}{-2 + 1} = \frac{-4 + 12}{-1} = \frac{8}{-1} = -8$$

31. $C = \dfrac{5}{9}(50 - 32) = \dfrac{5}{9}(18) = 10$
$10°C$ is equivalent to $50°F$.

33. $h = 4 + 60t - 16t^2 = 4 + 60(2) - 16(2)^2$
$$= 4 + 120 - 16(4) = 4 + 120 - 64$$
$$= 124 - 64 = 60$$
Two seconds after it is kicked, the ball's height is 60 feet.

35. $7x + 10x = 17x$

37. $5x^2 - 8x^2 = -3x^2$

39. $3(x + 5) = 3x + 15$

41. $4(2x - 3) = 8x - 12$

43. $5(3x + 4) - 4 = 5 \cdot 3x + 5 \cdot 4 - 4$
$$= 15x + 20 - 4$$
$$= 15x + 16$$

45. $5(3x - 2) + 12x = 5 \cdot 3x - 5 \cdot 2 + 12x$
$$= 15x - 10 + 12x$$
$$= 27x - 10$$

47. $7(3y - 5) + 2(4y + 3)$
$$= 7 \cdot 3y - 7 \cdot 5 + 2 \cdot 4y + 2 \cdot 3$$
$$= 21y - 35 + 8y + 6$$
$$= 29y - 29$$

49. $5(3y - 2) - (7y + 2) = 15y - 10 - 7y - 2$
$$= 8y - 12$$

51. $7 - 4\left[3 - (4y - 5)\right] = 7 - 4[3 - 4y + 5]$
$$= 7 - 4[8 - 4y]$$
$$= 7 - 32 + 16y$$
$$= 16y - 25$$

53. $18x^2 + 4 - \left[6(x^2 - 2) + 5\right]$
$$= 18x^2 + 4 - \left[6x^2 - 12 + 5\right]$$
$$= 18x^2 + 4 - \left[6x^2 - 7\right]$$
$$= 18x^2 + 4 - 6x^2 + 7$$
$$= 18x^2 - 6x^2 + 4 + 7$$
$$= (18 - 6)x^2 + 11 = 12x^2 + 11$$

55. $2(3x^2 - 5) - [4(2x^2 - 1) + 3]$
$$= 6x^2 - 10 - [8x^2 - 4 + 3]$$
$$= 6x^2 - 10 - [8x^2 - 1]$$
$$= 6x^2 - 10 - 8x^2 + 1$$
$$= -2x^2 - 9$$

57. $x - (x + 4) = x - x - 4 = -4$

59. $6(-5x) = -30x$

61. $5x - 2x = 3x$

63. $8x - (3x + 6) = 8x - 3x - 6 = 5x - 6$

65. $N = 17x^2 - 65.4x + 302.2$
$$= 17(4)^2 - 65.4(4) + 302.2$$
$$= 17(16) - 261.6 + 302.2$$
$$= 272 - 261.6 + 302.2$$
$$\approx 313$$
The formula models the graph's data very well.

67. $N = 17x^2 - 65.4x + 302.2$
$$= 17(10)^2 - 65.4(10) + 302.2$$
$$= 17(100) - 654 + 302.2$$
$$= 1700 - 654 + 302.2$$
$$\approx 1348$$
The formula predicts that there will be 1348 U.S. billionaires in 2010.

69. Model 1:

$N = 1.8x + 5.1$

$N = 1.8(2) + 5.1$

$N = 8.7$

Model 2:

$N = 5.6(1.2)^x$

$N = 5.6(1.2)^2$

$N = 8.064$

Model 3:

$N = 0.17x^2 + 0.95x + 5.68$

$N = 0.17(2)^2 + 0.95(2) + 5.68$

$N = 8.26$

Model 4:

$N = 0.09x^2 + 0.01x^3 + 1.1x + 5.64$

$N = 0.09(2)^2 + 0.01(2)^3 + 1.1(2) + 5.64$

$N = 8.28$

Model 1 best describes the data in 2000.

71. Model 3 is the model of degree 2.

$N = 0.17x^2 + 0.95x + 5.68$

$N = 0.17(5)^2 + 0.95(5) + 5.68$

$N = 14.68$

Model 3 best describes the data in 2003 very well.

73. a. $32,000

b. $C = 865x + 15,316$

$= 865(20) + 15,316$

$= \$32,616$

It describes the estimate from part (a) reasonably well.

c. $C = -2x^2 + 900x + 15,397$

$= -2(20)^2 + 900(20) + 15,397$

$= \$32,597$

It describes the estimate from part (a) reasonably well.

75. Model 1: $C = 865x + 15,316$

$= 865(25) + 15,316$

$= \$36,941$

Model 2: $C = -2x^2 + 900x + 15,397$

$= -2(25)^2 + 900(25) + 15,397$

$= \$36,647$

Model 1 best describes the 2005 value of $36,952.

83. d is true.

85. $\dfrac{0.5x + 5000}{x}$

a. $x = 100$

$\dfrac{0.5(100) + 5000}{100} = \$50.50$

$x = 1000$

$\dfrac{0.5(1000) + 5000}{1000} = \$5.50$

$x = 10,000$

$\dfrac{0.5(10,000) + 5000}{10,000} = \$1$

b. No; the business must produce at least 10,000 clocks each week to be competitive.

Check Points 6.2

1. $4x + 5 = 29$

$4x + 5 - 5 = 29 - 5$

$4x = 24$

$\dfrac{4x}{4} = \dfrac{24}{4}$

$x = 6$

Check:

$4x + 5 = 29$

$4(6) + 5 = 29$

$24 + 5 = 29$

$29 = 29$

The solution set is $\{6\}$.

2. $6(x - 3) - 10x = -10$

$6x - 18 - 10x = -10$

$-4x - 18 = -10$

$-4x - 18 + 18 = -10 + 18$

$-4x = 8$

$\dfrac{-4x}{-4} = \dfrac{8}{-4}$

$x = -2$

Check:

$6(-2 - 3) - 10(-2) = -10$

$6(-5) + 20 = -10$

$-30 + 20 = -10$

$-10 = -10$

The solution set is $\{-2\}$.

3.
$$2x+9=8x-3$$
$$2x+9-8x=8x-3-8x$$
$$-6x+9=-3$$
$$-6x+9-9=-3-9$$
$$-6x=-12$$
$$\frac{-6x}{-6}=\frac{-12}{-6}$$
$$x=2$$
The solution set is $\{2\}$.

4. $4(2x+1)-29=3(2x-5)$
$$8x+4-29=6x-15$$
$$8x-25=6x-15$$
$$8x-25-6x=6x-15-6x$$
$$2x-25=-15$$
$$2x-25+25=-15+25$$
$$2x=10$$
$$\frac{2x}{2}=\frac{10}{2}$$
$$x=5$$
The solution set is $\{5\}$.

5.
$$\frac{x}{4}=\frac{2x}{3}+\frac{5}{6}$$
$$12\cdot\frac{x}{4}=12\cdot\left(\frac{2x}{3}+\frac{5}{6}\right)$$
$$12\cdot\frac{x}{4}=12\cdot\frac{2x}{3}+12\cdot\frac{5}{6}$$
$$3\cdot x=4\cdot2x+2\cdot5$$
$$3x=8x+10$$
$$3x-8x=8x+10-8x$$
$$-5x=10$$
$$\frac{-5x}{-5}=\frac{10}{-5}$$
$$x=-2$$
The solution set is $\{-2\}$.

6.
$$D=\frac{10}{9}x+\frac{53}{9}$$
$$10=\frac{10}{9}x+\frac{53}{9}$$
$$9\cdot10=9\cdot\left(\frac{10}{9}x+\frac{53}{9}\right)$$
$$90=10x+53$$
$$37=10x$$
$$3.7=x$$
This is shown on the graph as the point $(3.7,\ 10)$.

7. $2l+2w=P$
$$2l+2w-2l=P-2l$$
$$2w=P-2l$$
$$\frac{2w}{2}=\frac{P-2l}{2}$$
$$w=\frac{P-2l}{2}$$

8.
$$T=D+pm$$
$$T-D=D-D+pm$$
$$T-D=pm$$
$$\frac{T-D}{p}=\frac{pm}{p}$$
$$\frac{T-D}{p}=m$$

9.
$$3x+7=3(x+1)$$
$$3x+7=3x+3$$
$$3x+7-3x=3x+3-3x$$
$$7=3$$
There is no solution, $\varnothing$.

10. $7x+9=9(x+1)-2x$
$$7x+9=9x+9-2x$$
$$7x+9=7x+9$$
$$9=9$$
The solution set is $\{x\,|\,x \text{ is a real number}\}$.

Exercise Set 6.2

1. $x - 7 = 3$
    $x - 7 + 7 = 3 + 7$
    $x = 10$
    The solution set is $\{10\}$.

3. $x + 5 = -12$
    $x + 5 - 5 = -12 - 5$
    $x = -17$
    The solution set is $\{-17\}$.

5. $\dfrac{x}{3} = 4$

    $3\left(\dfrac{x}{3}\right) = 3(4)$

    $x = 12$
    The solution set is $\{12\}$.

7. $5x = 45$
    $\dfrac{5x}{5} = \dfrac{45}{5}$
    $x = 9$

    The solution set is $\{9\}$.

9. $8x = -24$
    $\dfrac{8x}{8} = \dfrac{-24}{8}$
    $x = -3$
    The solution set is $\{-3\}$.

11. $-8x = 2$
    $\dfrac{-8x}{-8} = \dfrac{2}{-8}$
    $x = -\dfrac{1}{4}$

    The solution set is $\left\{-\dfrac{1}{4}\right\}$.

13. $5x + 3 = 18$
    $5x + 3 - 3 = 18 - 3$
    $5x = 15$
    $\dfrac{5x}{5} = \dfrac{15}{5}$
    $x = 3$
    The solution set is $\{3\}$.

15. $6x - 3 = 63$
    $6x - 3 + 3 = 63 + 3$
    $6x = 66$
    $\dfrac{6x}{6} = \dfrac{66}{6}$
    $x = 11$
    The solution set is $\{11\}$.

17. $4x - 14 = -82$
    $4x - 14 + 14 = -82 + 14$
    $4x = -68$
    $\dfrac{4x}{4} = \dfrac{-68}{4}$
    $x = -17$
    The solution set is $\{-17\}$.

19. $14 - 5x = -41$
    $14 - 5x - 14 = -41 - 14$
    $-5x = -55$
    $\dfrac{-5x}{-5} = \dfrac{-55}{-5}$
    $x = 11$
    The solution set is $\{11\}$.

21. $9(5x - 2) = 45$
    $45x - 18 = 45$
    $45x - 18 + 18 = 45 + 18$
    $45x = 63$
    $\dfrac{45x}{45} = \dfrac{63}{45}$
    $x = \dfrac{7}{5}$

    The solution set is $\left\{\dfrac{7}{5}\right\}$.

23. $5x - (2x - 10) = 35$
    $5x - 2x + 10 = 35$
    $3x + 10 = 35$
    $3x + 10 - 10 = 35 - 10$
    $3x = 25$
    $\dfrac{3x}{3} = \dfrac{25}{3}$
    $x = \dfrac{25}{3}$

    The solution set is $\left\{\dfrac{25}{3}\right\}$.

25. $3x + 5 = 2x + 13$
    $3x + 5 - 5 = 2x + 13 - 5$
    $3x = 2x + 8$
    $3x - 2x = 2x + 8 - 2x$
    $x = 8$
    The solution set is $\{8\}$.

27. $8x - 2 = 7x - 5$
    $8x - 2 + 2 = 7x - 5 + 2$
    $8x = 7x - 3$
    $8x - 7x = 7x - 3 - 7x$
    $x = -3$
    The solution set is $\{-3\}$.

29.
$$7x + 4 = x + 16$$
$$7x + 4 - 4 = x + 16 - 4$$
$$7x = x + 12$$
$$7x - x = x + 12 - x$$
$$6x = 12$$
$$\frac{6x}{6} = \frac{12}{6}$$
$$x = 2$$
The solution set is {2}.

31.
$$8y - 3 = 11y + 9$$
$$8y - 3 + 3 = 11y + 9 + 3$$
$$8y = 11y + 12$$
$$8y - 11y = 11y + 12 - 11y$$
$$-3y = 12$$
$$\frac{-3y}{-3} = \frac{12}{-3}$$
$$y = -4$$
The solution set is {−4}.

33.
$$2(4 - 3x) = 2(2x + 5)$$
$$8 - 6x = 4x + 10$$
$$8 - 6x - 8 = 4x + 10 - 8$$
$$-6x = 4x + 2$$
$$-6x - 4x = 4x + 2 - 4x$$
$$-10x = 2$$
$$\frac{-10x}{-10} = \frac{2}{-10}$$
$$x = -\frac{1}{5}$$
The solution set is $\left\{-\frac{1}{5}\right\}$.

35.
$$8(y + 2) = 2(3y + 4)$$
$$8y + 16 = 6y + 8$$
$$8y + 16 - 16 = 6y + 8 - 16$$
$$8y = 6y - 8$$
$$8y - 6y = 6y - 8 - 6y$$
$$2y = -8$$
$$\frac{2y}{2} = \frac{-8}{2}$$
$$y = -4$$
The solution set is {−4}.

37.
$$3(x + 1) = 7(x - 2) - 3$$
$$3x + 3 = 7x - 14 - 3$$
$$3x + 3 = 7x - 17$$
$$3x + 3 - 3 = 7x - 17 - 3$$
$$3x = 7x - 20$$
$$3x - 7x = 7x - 20 - 7x$$
$$-4x = -20$$
$$\frac{-4x}{-4} = \frac{-20}{-4}$$
$$x = 5$$
The solution set is {5}.

39.
$$5(2x - 8) - 2 = 5(x - 3) + 3$$
$$10x - 40 - 2 = 5x - 15 + 3$$
$$10x - 42 = 5x - 12$$
$$10x - 42 + 42 = 5x - 12 + 42$$
$$10x = 5x + 30$$
$$10x - 5x = 5x + 30 - 5x$$
$$5x = 30$$
$$\frac{5x}{5} = \frac{30}{5}$$
$$x = 6$$
The solution set is {6}.

41.
$$5(x - 2) - 2(2x + 1) = 2 + 5x$$
$$5x - 10 - 4x - 2 = 2 + 5x$$
$$x - 12 = 2 + 5x$$
$$x - 12 + 12 = 2 + 5x + 12$$
$$x = 14 + 5x$$
$$x - 5x = 14 + 5x - 5x$$
$$-4x = 14$$
$$\frac{-4x}{-4} = \frac{14}{-4}$$
$$x = -\frac{7}{2}$$
The solution set is $\left\{-\frac{7}{2}\right\}$.

43.
$$\frac{x}{3} + \frac{x}{2} = \frac{5}{6}$$
$$6\left(\frac{x}{3} + \frac{x}{2}\right) = 6\left(\frac{5}{6}\right)$$
$$6\left(\frac{x}{3}\right) + 6\left(\frac{x}{2}\right) = 6\left(\frac{5}{6}\right)$$
$$2x + 3x = 5$$
$$5x = 5$$
$$\frac{5x}{5} = \frac{5}{5}$$
$$x = 1$$
The solution set is {1}.

45.
$$\frac{x}{2} = 20 - \frac{x}{3}$$
$$6\left(\frac{x}{2}\right) = 6\left(20 - \frac{x}{3}\right)$$
$$6\left(\frac{x}{2}\right) = 6(20) - 6\left(\frac{x}{3}\right)$$
$$3x = 120 - 2x$$
$$3x + 2x = 120 - 2x + 2x$$
$$5x = 120$$
$$\frac{5x}{5} = \frac{120}{5}$$
$$x = 24$$
The solution set is $\{24\}$.

47.
$$\frac{3y}{4} - 3 = \frac{y}{2} + 2$$
$$4\left(\frac{3y}{4} - 3\right) = 4\left(\frac{y}{2} + 2\right)$$
$$4 \cdot \frac{3y}{4} - 4 \cdot 3 = 4 \cdot \frac{y}{2} + 4 \cdot 2$$
$$3y - 12 = 2y + 8$$
$$3y - 12 + 12 = 2y + 8 + 12$$
$$3y = 2y + 20$$
$$3y - 2y = 2y + 20 - 2y$$
$$y = 20$$
The solution set is $\{20\}$.

49.
$$\frac{3x}{5} - x = \frac{x}{10} - \frac{5}{2}$$
$$10\left(\frac{3x}{5} - x\right) = 10\left(\frac{x}{10} - \frac{5}{2}\right)$$
$$10\left(\frac{3x}{5}\right) - 10(x) = 10\left(\frac{x}{10}\right) - 10\left(\frac{5}{2}\right)$$
$$6x - 10x = x - 25$$
$$-4x = x - 25$$
$$-4x - x = x - 25 - x$$
$$-5x = -25$$
$$\frac{-5x}{-5} = \frac{-25}{-5}$$
$$x = 5$$
The solution set is $\{5\}$.

51. $A = LW$ for L
$$\frac{A}{W} = \frac{LW}{W}$$
$$\frac{A}{W} = L \text{ or } L = \frac{A}{W}$$

53. $A = \frac{1}{2}bh$ for b
$$2(A) = 2\left(\frac{1}{2}bh\right)$$
$$2A = bh$$
$$\frac{2A}{h} = \frac{bh}{h}$$
$$\frac{2A}{h} = b \text{ or } b = \frac{2A}{h}$$

55. $I = Prt$ for P
$$\frac{I}{rt} = \frac{Prt}{rt}$$
$$\frac{I}{rt} = P \text{ or } P = \frac{I}{rt}$$

57. $E = mc^2$ for m
$$\frac{E}{c^2} = \frac{mc^2}{c^2}$$
$$\frac{E}{c^2} = m \text{ or } m = \frac{E}{c^2}$$

59. $y = mx + b$ for m
$$y - b = mx + b - b$$
$$y - b = mx$$
$$\frac{y - b}{x} = \frac{mx}{x}$$
$$\frac{y - b}{x} = m \text{ or } m = \frac{y - b}{x}$$

61. $A = \frac{1}{2}(a + b)$ for a
$$2 \cdot A = 2 \cdot \frac{1}{2}(a + b)$$
$$2A = a + b$$
$$2A - b = a + b - b$$
$$2A - b = a \text{ or } a = 2A - b$$

63. $S = P + Prt$ for r

$S - P = P + Prt - P$

$S - P = Prt$

$\dfrac{S - P}{Pt} = \dfrac{Prt}{Pt}$

$\dfrac{S - P}{Pt} = r$ or $r = \dfrac{S - P}{Pt}$

65. $Ax + By = C$ for x

$Ax + By = C$

$Ax = C - By$

$x = \dfrac{C - By}{A}$

67. $a_n = a_1 + (n-1)$ for n

$a_n = a_1 + (n-1)d$

$a_n - a_1 = dn - d$

$a_n - a_1 + d = dn$

$\dfrac{a_n - a_1 + d}{d} = n$

69. $3x - 7 = 3(x+1)$

$3x - 7 = 3x + 3$

$3x - 7 - 3x = 3x + 3 - 3x$

$-7 = 3$

The statement is false. The solution set is { }.

71. $2(x+4) = 4x + 5 - 2x + 3$

$2x + 8 = 2x + 8$

$2x - 8 - 2x = 2x + 8 - 2x$

$8 = 8$

The statement is true. The solution is $\{x \mid x \text{ is a real number}\}$.

73. $7 + 2(3x - 5) = 8 - 3(2x + 1)$

$7 + 6x - 10 = 8 - 6x - 3$

$6x - 3 = 5 - 6x$

$6x + 6x - 3 = 5 - 6x + 6x$

$12x - 3 = 5$

$12x - 3 + 3 = 5 + 3$

$12x = 8$

$\dfrac{12x}{12} = \dfrac{8}{12}$

$x = \dfrac{2}{3}$

The solution set is $\left\{\dfrac{2}{3}\right\}$.

75. $4x + 1 - 5x = 5 - (x + 4)$

$-x + 1 = 5 - x - 4$

$-x + 1 = 1 - x$

$-x + 1 + x = 1 - x + x$

$1 = 1$

The statement is true. The solution set is $\{x \mid x \text{ is a real number}\}$.

77. $4(x + 2) + 1 = 7x - 3(x - 2)$

$4x + 8 + 1 = 7x - 3x + 6$

$4x + 9 = 4x + 6$

$4x - 4x + 9 = 4x - 4x + 6$

$9 = 6$

The statement is false. The solution set is { }.

79. $3 - x = 2x + 3$

$3 - x + x = 2x + x + 3$

$3 = 3x + 3$

$3 - 3 = 3x + 3 - 3$

$0 = 3x$

$\dfrac{0}{3} = \dfrac{3x}{3}$

$0 = x$

The solution set is $\{0\}$.

81.
$$\frac{x}{3} + 2 = \frac{x}{3}$$
$$6\left(\frac{x}{3} + 2\right) = 6\left(\frac{x}{3}\right)$$
$$2x + 12 = 2x$$
$$2x + 12 - 2x = 2x - 2x$$
$$12 = 0$$
The statement is false. The solution set is { }.

83.
$$\frac{x}{2} - \frac{x}{4} + 4 = x + 4$$
$$4\left(\frac{x}{2} - \frac{x}{4} + 4\right) = 4(x + 4)$$
$$4\left(\frac{x}{2}\right) - 4\left(\frac{x}{4}\right) + 16 = 4x + 16$$
$$2x - x + 16 = 4x + 16$$
$$x + 16 = 4x + 16$$
$$x - x + 16 = 4x - x + 16$$
$$16 = 3x + 16$$
$$16 - 16 = 3x + 16 - 16$$
$$0 = 3x$$
$$\frac{0}{3} = \frac{3x}{3}$$
$$0 = x$$
The solution set is $\{0\}$.

85. Solve: $4(x - 2) + 2 = 4x - 2(2 - x)$
$$4x - 8 + 2 = 4x - 4 + 2x$$
$$4x - 6 = 6x - 4$$
$$-2x - 6 = -4$$
$$-2x = 2$$
$$x = -1$$
Now, evaluate $x^2 - x$ for $x = -1$:
$$x^2 - x = (-1)^2 - (-1)$$
$$= 1 - (-1) = 1 + 1 = 2$$

87. Solve for x.
$$\frac{x}{5} - 2 = \frac{x}{3}$$
$$15 \cdot \left(\frac{x}{5} - 2\right) = 15 \cdot \frac{x}{3}$$
$$3x - 30 = 5x$$
$$-2x = 30$$
$$x = -15$$
Solve for y.
$$-2y - 10 = 5y + 18$$
$$-2y - 5y = 18 + 10$$
$$-7y = 28$$
$$y = -4$$
Evaluate
$$x^2 - (xy - y) = (-15)^2 - [(-15)(-4) - (-4)]$$
$$= 161$$

89.
$$\left[(3 + 6)^2 \div 3\right] \cdot 4 = -54x$$
$$\left(9^2 \div 3\right) \cdot 4 = -54x$$
$$(81 \div 3) \cdot 4 = -54x$$
$$27 \cdot 4 = -54x$$
$$108 = -54x$$
$$-2 = x$$
The solution set is $\{-2\}$.

91.
$$5 - 12x = 8 - 7x - \left[6 \div 3\left(2 + 5^3\right) + 5x\right]$$
$$5 - 12x = 8 - 7x - \left[6 \div 3(2 + 125) + 5x\right]$$
$$5 - 12x = 8 - 7x - \left[6 \div 3 \cdot 127 + 5x\right]$$
$$5 - 12x = 8 - 7x - \left[2 \cdot 127 + 5x\right]$$
$$5 - 12x = 8 - 7x - \left[254 + 5x\right]$$
$$5 - 12x = 8 - 7x - 254 - 5x$$
$$5 - 12x = -12x - 246$$
$$5 = -246$$
The final statement is a contradiction, so the equation has no solution. The solution set is $\varnothing$.

93.
$$0.7x + 0.4(20) = 0.5(x + 20)$$
$$0.7x + 8 = 0.5x + 10$$
$$0.2x + 8 = 10$$
$$0.2x = 2$$
$$x = 10$$
The solution set is $\{10\}$.

95. $4x+13-\left\{2x-\left[4\left(x-3\right)-5\right]\right\}=2(x-6)$

$4x+13-\left\{2x-\left[4x-12-5\right]\right\}=2x-12$

$4x+13-\left\{2x-\left[4x-17\right]\right\}=2x-12$

$4x+13-\left\{2x-4x+17\right\}=2x-12$

$4x+13-\left\{-2x+17\right\}=2x-12$

$4x+13+2x-17=2x-12$

$6x-4=2x-12$

$4x-4=-12$

$4x=-8$

$x=-2$

The solution set is $\{-2\}$.

97. $H=10x+59$

$179=10x+59$

$120=10x$

$12=x$

12 years after 2001 or 2013.

99. $\dfrac{n}{2}+80=2T$

$\dfrac{n}{2}+80=2(70)$

$\dfrac{n}{2}+80=140$

$n+160=280$

$n=120$ chirps per minute

101. $p=15+\dfrac{5d}{11}$

$201=15+\dfrac{5d}{11}$

$11(201)=11\left(15+\dfrac{5d}{11}\right)$

$2211=165+5d$

$-5d=-2046$

$d=409.2$ ft

103. $W=-x+140$

$127=-x+140$

$x=140-127$

$x=13$

13 years after 1998, or 2011.

105. $W=\dfrac{1}{4}x+159$

$163=\dfrac{1}{4}x+159$

$652=x+636$

$x=16$

16 years after 1998, or 2014.

107. $\left(\dfrac{1}{4}x+159\right)-(-x+140)=49$

$\dfrac{1}{4}x+159+x-140=49$

$x+636+4x-560=196$

$5x+76=196$

$5x=120$

$x=24$

24 years after 1998, or 2022.

117. c is true

Check Points 6.3

1. Let x = the number.
$$6x - 4 = 68$$
$$6x - 4 + 4 = 68 + 4$$
$$6x = 72$$
$$\frac{6x}{6} = \frac{72}{6}$$
$$x = 12$$

2. Let x = the number of years after 2004 that it will take until Americans will purchase 79.9 million gallons of organic milk.
$$40.7 + 5.6x = 79.9$$
$$5.6x = 79.9 - 40.7$$
$$5.6x = 39.2$$
$$x = \frac{39.2}{5.6}$$
$$x = 7$$
Americans will purchase 79.9 million gallons of organic milk 7 years after 2004, or 2011.

3. Let x = the number of football injuries
Let $x + 0.6$ = the number of basketball injuries
Let $x + 0.3$ = the number of bicycling injuries
$$x + (x + 0.6) + (x + 0.3) = 3.9$$
$$x + x + 0.6 + x + 0.3 = 3.9$$
$$3x + 0.9 = 3.9$$
$$3x = 3$$
$$x = 1$$
$$x = 1$$
$$x + 0.6 = 1 + 0.6 = 1.6$$
$$x + 0.3 = 1 + 0.3 = 1.3$$
In 2004 there were 1 million football injuries, 1.6 million basketball injuries, and 1.3 million bicycling injuries.

4. Let x = the number of minutes at which the costs of the two plans are the same.

$$\overbrace{15 + 0.08x}^{\text{Plan A}} = \overbrace{3 + 0.12x}^{\text{Plan B}}$$
$$15 + 0.08x - 15 = 3 + 0.12x - 15$$
$$0.08x = 0.12x - 12$$
$$0.08x - 0.12x = 0.12x - 12 - 0.12x$$
$$-0.04x = -12$$
$$\frac{-0.04x}{-0.04} = \frac{-12}{-0.04}$$
$$x = 300$$
The two plans are the same at 300 minutes.

Exercise Set 6.3

1. Let x = the number
$$5x - 4 = 26$$
$$5x = 30$$
$$x = 6$$
The number is 6.

3. Let x = the number
$x + 26$ = the other number
$$x + (x + 26) = 64$$
$$x + x + 26 = 64$$
$$2x + 26 = 64$$
$$2x = 38$$
$$x = 19$$
If $x = 19$, then $x + 26 = 45$.
The numbers are 19 and 45.

5. Let L = the life expectancy of an American man.
y = the number of years after 1900.
$$L = 55 + 0.2y$$
$$85 = 55 + 0.2y$$
$$30 = 0.2y$$
$$150 = y$$
The life expectancy will be 85 years in the year $1900 + 150 = 2050$.

7. $$1.7x + 39.8 = 44.9 + 8.5$$
$$1.7x + 39.8 = 53.4$$
$$1.7x = 13.6$$
$$\frac{1.7x}{1.7} = \frac{13.6}{1.7}$$
$$x = 8$$
The number of Americans without health insurance will exceed 44.9 million by 8.5 million 8 years after 2000, or 2008.

9. Let x = the number of years after 2000
    $$10,600,000 - 28,000x = 10,200,000 - 12,000x$$
    $$-16,000x = -400,000$$
    $$x = 25$$
    The countries will have the same population 25 years after the year 2000, or the year 2025.
    $$10,200,000 - 12,000x = 10,200,000 - 12,000(25)$$
    $$= 10,200,000 - 300,000$$
    $$= 9,900,000$$
    The population in the year 2025 will be 9,900,000.

11. Let x = the number of births (in thousands)
    Let $x - 229$ = the number of deaths (in thousands).
    $$x + (x - 229) = 521$$
    $$x + x - 229 = 521$$
    $$2x - 229 = 521$$
    $$2x = 750$$
    $$x = 375$$
    There are 375 thousand births and $375 - 229 = 146$ thousand deaths each day.

13. Let x = the number of hate crimes based on sexual orientation.
    $3x + 127$ = the number of hate crimes based on race.
    $$(3x + 127) + 1343 + x + 1026 + 33 = 7485$$
    $$3x + 127 + 1343 + x + 1026 + 33 = 7485$$
    $$4x + 2529 = 7485$$
    $$4x = 4956$$
    $$\frac{4x}{4} = \frac{4956}{4}$$
    $$x = 1239$$
    $$3x + 127 = 3844$$
    Thus, there were 3844 hate crimes based on race and 1239 based on sexual orientation.

15. Let x = the number of months.
    The cost for Club A: $25x + 40$
    The cost for Club B: $30x + 15$
    $$25x + 40 = 30x + 15$$
    $$-5x + 40 = 15$$
    $$-5x = -25$$
    $$x = 5$$
    The total cost for the clubs will be the same at 5 months. The cost will be
    $$25(5) + 40 = 30(5) + 15 = \$165$$

17. Let x = the number of uses.
    Cost without coupon book: $1.25x$
    Cost with coupon book: $15 + 0.75x$
    $$1.25x = 15 + 0.75x$$
    $$0.50x = 15$$
    $$x = 30$$
    The bus must be used 30 times in a month for the costs to be equal.

19. Let x = dollars of merchandise purchased.
    $$\overbrace{100 + 0.80x}^{\text{Plan A}} = \overbrace{40 + 0.90x}^{\text{Plan B}}$$
    $$0.80x - 0.90x = 40 - 100$$
    $$-0.10x = -60$$
    $$x = \$600$$
    $600 of merchandise must be purchased for the costs to be equal.
    The cost of each plan would be
    $$100 + 0.80(600) = \$580$$

21. Let x = number of hours
    $35x$ = labor cost
    $$35x + 63 = 448$$
    $$35x = 385$$
    $$x = 11$$
    It took 11 hours.

23. Let g = the gross amount of the paycheck
    Yearly Salary = $2(12)g + 750$
    $$33150 = 24g + 750$$
    $$32400 = 24g$$
    $$1350 = g$$
    The gross amount of each paycheck is $1350.

29. Let x = number of inches over 5 feet.
    $$100 + 5x = 135$$
    $$100 + 5x - 100 = 135 - 100$$
    $$5x = 35$$
    $$\frac{5x}{5} = \frac{35}{5}$$
    $$x = 7$$
    The height that corresponds to 135 pounds is $5'7''$.

31. Let x = current age of woman, then
$$3x = \text{current age of "uncle."}$$
$$2(x + 20) = 3x + 20$$
$$2x + 40 = 3x + 20$$
$$2x + 40 - 2x = 3x + 20 - 2x$$
$$40 = x + 20$$
$$40 - 20 = x + 20 - 20$$
$$20 = x$$
$$60 = 3x$$
The woman is 20 years and the "uncle" is 60 years.

33. Let x = mother's amount
$$2x = \text{boy's amount}$$
$$\frac{x}{2} = \text{girl's amount}$$
$$x + 2x + \frac{x}{2} = 14,000$$
$$\frac{7}{2}x = 14,000$$
$$x = \$4,000$$

The mother received \$4000, the boy received \$8000, and the girl received \$2000.

Check Points 6.4

1. a.
$$\frac{10}{x} = \frac{2}{3}$$
$$10 \cdot 3 = 2x$$
$$30 = 2x$$
$$\frac{30}{2} = \frac{2x}{2}$$
$$15 = x$$
The solution set is $\{15\}$.

b.
$$\frac{11}{910 - x} = \frac{2}{x}$$
$$11x = 2(910 - x)$$
$$11x = 1820 - 2x$$
$$11x + 2x = 1820 - 2x + 2x$$
$$13x = 1820$$
$$\frac{13x}{13} = \frac{1820}{13}$$
$$x = 140$$
The solution set is $\{140\}$.

2. Let x = tax on a \$112,500 house.
$$\frac{\$600}{\$45,000} = \frac{\$x}{\$112,500}$$
$$\frac{600}{45,000} = \frac{x}{112,500}$$
$$(600)(112,500) = 45,000x$$
$$67,500,000 = 45,000x$$
$$\frac{67,500,000}{45,000} = \frac{45,000x}{45,000}$$
$$1500 = x$$
The tax on the \$112,500 house is \$1500.

3. Let x = the number of deer in the refuge.
$$\frac{120 \text{ tagged deer}}{x} = \frac{25 \text{ tagged deer in sample}}{150 \text{ deer in sample}}$$
$$\frac{120}{x} = \frac{25}{150}$$
$$25x = (120)(150)$$
$$25x = 18,000$$
$$\frac{25x}{25} = \frac{18,000}{25}$$
$$x = 720$$
There are approximately 720 deer in the refuge.

4. Let x = the number of gallons.
$$\frac{11 \text{ minutes}}{x} = \frac{5 \text{ minutes}}{30 \text{ gallons}}$$
$$\frac{11}{x} = \frac{5}{30}$$
$$5x = (11)(30)$$
$$5x = 330$$
$$x = 66 \text{ gallons}$$
A shower lasting 11 minutes will use 66 gallons of water.

5. Let x = the distance (in feet) required to stop a car traveling at 100 mph.

$$\frac{200 \text{ feet}}{60^2 \text{ miles per hour}} = \frac{x}{100^2 \text{ miles per hour}}$$

$$\frac{200}{60^2} = \frac{x}{100^2}$$

$$\frac{200}{3600} = \frac{x}{10,000}$$

$$(200)(10,000) = 3600x$$

$$2,000,000 = 3600x$$

$$\frac{2,000,000}{3600} = \frac{3600x}{3600}$$

$$556 \approx x$$

Approximately 556 feet are needed to stop at 100 mph.

6. Let x = pounds per square inch when the volume is 22 cubic inches.

$$\frac{8 \text{ cubic inches}}{\underbrace{x}_{\substack{\text{corresponds to} \\ \text{22 cubic inches}}}} = \frac{\overbrace{22 \text{ cubic inches}}}{\underbrace{12 \text{ pounds per square inch}}_{\substack{\text{corresponds to} \\ \text{8 cubic inches}}}}$$

$$\frac{8}{x} = \frac{22}{12}$$

$$(8)(12) = 22x$$

$$96 = 22x$$

$$\frac{96}{22} = \frac{22x}{22}$$

$$4.36 \approx x$$

The pressure is about 4.36 pounds per square inch.

Exercise Set 6.4

1.
$$\frac{24}{x} = \frac{12}{7}$$
$$12x = 24 \cdot 7$$
$$12x = 168$$
$$\frac{12x}{12} = \frac{168}{12}$$
$$x = 14$$

3.
$$\frac{x}{6} = \frac{18}{4}$$
$$4x = 6 \cdot 18$$
$$4x = 108$$
$$\frac{4x}{4} = \frac{108}{4}$$
$$x = 27$$

5.
$$\frac{x}{3} = -\frac{3}{4}$$
$$4x = 3(-3)$$
$$4x = -9$$
$$\frac{4x}{4} = \frac{-9}{4}$$
$$x = -\frac{9}{4}$$

7.
$$\frac{-3}{8} = \frac{x}{40}$$
$$8x = -3(40)$$
$$8x = -120$$
$$\frac{8x}{8} = \frac{-120}{8}$$
$$x = -15$$

9.
$$\frac{x-2}{5} = \frac{3}{10}$$
$$10(x-2) = 3 \cdot 5$$
$$10x - 20 = 15$$
$$10x - 20 + 20 = 15 + 20$$
$$10x = 35$$
$$\frac{10x}{10} = \frac{35}{10}$$
$$x = \frac{7}{2}$$

11.
$$\frac{y+10}{10} = \frac{y-2}{4}$$
$$4(y+10) = 10(y-2)$$
$$4y + 40 = 10y - 20$$
$$4y + 40 - 40 = 10y - 20 - 40$$
$$4y = 10y - 60$$
$$4y - 10y = 10y - 60 - 10y$$
$$-6y = -60$$
$$\frac{-6y}{-6} = \frac{-60}{-6}$$
$$y = 10$$

13.
$$\frac{1\text{st } y}{1\text{st } x} = \frac{2\text{nd } y}{2\text{nd } x}$$
$$\frac{65}{5} = \frac{y}{12}$$
$$5y = 65 \cdot 12$$
$$5y = 780$$
$$y = 156$$

15.
$$\frac{1\text{st } y}{2\text{nd } x} = \frac{2\text{nd } y}{1\text{st } x}$$
$$\frac{12}{2} = \frac{y}{5}$$
$$2y = 12 \cdot 5$$
$$2y = 60$$
$$y = 30$$

17.
$$\frac{x}{a} = \frac{b}{c}$$
$$cx = ab$$
$$x = \frac{ab}{c}$$

19.
$$\frac{a+b}{c} = \frac{x}{d}$$
$$cx = (a+b)d$$
$$x = \frac{(a+b)d}{c}$$

21.
$$\frac{x+a}{a} = \frac{b+c}{c}$$
$$c(x+a) = (b+c)a$$
$$cx + ac = ab + ac$$
$$cx = ab$$
$$x = \frac{ab}{c}$$

23. Let x = tax.
$$\frac{725}{65,000} = \frac{x}{100,000}$$
$$65,000x = 725 \cdot 100,000$$
$$65,000x = 72,500,000$$
$$\frac{65,000x}{65,000} = \frac{72,500,000}{65,000}$$
$$x \approx 1115.38$$
The tax on a property with an assessed value of $100,000 is $1115.38.

25. Let x = total number of fur seal pups in this rookery.
$$\frac{218}{900} = \frac{4963}{x}$$
$$218x = 900 \cdot 4963$$
$$218x = 4,466,700$$
$$\frac{218x}{218} = \frac{4,466,700}{218}$$
$$x \approx 20,489$$
There were an estimated 20,489 fur seal pups in this rookery.

27. Let x = amount to pay in monthly child support.
$$\frac{1}{40} = \frac{x}{38,000}$$
$$1 \cdot 38,000 = 40x$$
$$\frac{38,000}{40} = \frac{40x}{40}$$
$$950 = x$$
The father should pay $950 each month.

29. Let x = moon weight of a person who weighs 186 pounds on earth.
$$\frac{60}{360} = \frac{x}{186}$$
$$360x = 60 \cdot 186$$
$$360x = 11,160$$
$$x = 31$$
A person who weighs 186 pounds on Earth will weigh 31 pounds on the moon.

31. Let x = height of the critter.
$$\frac{67}{10} = \frac{x}{23}$$
$$10x = 67 \cdot 23$$
$$10x = 1541$$
$$\frac{10x}{10} = \frac{1541}{10}$$
$$x = 154.1$$
The critter was 154.1 inches tall or about 12 ft 10 in.

33. Let x = Mr. Wadlow's weight.

$$\frac{x}{(107)^3} = \frac{170}{(70)^3}$$

$$\frac{x}{1,225,043} = \frac{170}{343,000}$$

$$343,000x = 1,225,043 \cdot 170$$

$$343,000x = 208,257,310$$

$$\frac{343,000x}{343,000} = \frac{208,257,310}{343,000}$$

$$x \approx 607$$

Robert Wadlow's weight was about 607 pounds.

35. Let x = the stopping distance.

$$\frac{45^2}{67.5} = \frac{60^2}{x}$$

$$\frac{2025}{67.5} = \frac{3600}{x}$$

$$2025x = 3600 \cdot 67.5$$

$$2025x = 243,000$$

$$x = 120$$

The stopping distance is 120 feet.

37. Let x = water temperature at 5000 meters.

$$\frac{4.4}{5000} = \frac{x}{1000}$$

$$5000x = 4.4 \cdot 1000$$

$$5000x = 4400$$

$$x = 0.88$$

The water temperature is $0.88°$ Celsius at a depth of 5000 meters.

39. Let x = pressure when the volume is 40 cubic centimeters.

$$\frac{32}{x} = \frac{40}{8}$$

$$40x = 32(8)$$

$$40x = 256$$

$$\frac{40x}{40} = \frac{256}{40}$$

$$x = 6.4$$

The pressure is 6.4 pounds.

41. Let x = loudness in decibels.

$$\frac{28}{4^2} = \frac{x}{8^2}$$

$$\frac{28}{16} = \frac{x}{64}$$

$$16x = 28 \cdot 64$$

$$16x = 1792$$

$$x = 112$$

The loudness at 4 feet is 112 decibels.

49.

$$\frac{\text{front}}{\text{rear}} = \frac{5}{1}$$

$$\frac{60}{x} = \frac{5}{1} \quad \text{or} \quad \frac{y}{20} = \frac{5}{1}$$

$$5x = 60 \qquad\qquad 1y = 5 \cdot 20$$

$$x = 12 \qquad\qquad y = 100$$

Replace the rear sprocket with one that has 12 teeth or replace the front sprocket with one that has 100 teeth.

51. Let p = the initial pressure and
Let x = the pressure after the wind velocity doubles
Let v = the initial wind velocity, then
    $2v$ = double the wind velocity

$$\frac{p}{v^2} = \frac{x}{(2v)^2}$$

$$\frac{p}{v^2} = \frac{x}{4v^2}$$

$$v^2 \cdot x = p \cdot 4v^2$$

$$\frac{v^2 \cdot x}{v^2} = \frac{p \cdot 4v^2}{v^2}$$

$$x = 4p$$

If the wind velocity doubles, the pressure quadruples.

Check Points 6.5

1. **a.** $x < 4$
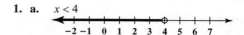

b. $x \geq -2$

c. $-4 \leq x < 1$

2. $5x - 3 \leq 17$
$$5x - 3 + 3 \leq 17 + 3$$
$$5x \leq 20$$
$$\frac{5x}{5} \leq \frac{20}{5}$$
$$x \leq 4$$
$$\{x \mid x \leq 4\}$$

3. **a.** $\frac{1}{4}x < 2$
$$4 \cdot \frac{1}{4}x < 4 \cdot 2$$
$$x < 8$$
$$\{x \mid x < 8\}$$

b. $-6x < 18$
$$\frac{-6x}{-6} > \frac{18}{-6}$$
$$x > -3$$
$$\{x \mid x > -3\}$$

4. $7x - 3 > 13x + 33$
$$7x - 3 + 3 > 13x + 33 + 3$$
$$7x > 13x + 36$$
$$7x - 13x > 13x + 36 - 13x$$
$$-6x > 36$$
$$\frac{-6x}{-6} < \frac{36}{-6}$$
$$x < -6$$
$$\{x \mid x < -6\}$$

5. $2(x - 3) - 1 \leq 3(x + 2) - 14$
$$2x - 6 - 1 \leq 3x + 6 - 14$$
$$2x - 7 \leq 3x - 8$$
$$2x - 7 + 7 \leq 3x - 8 + 7$$
$$2x \leq 3x - 1$$
$$2x - 3x \leq 3x - 1 - 3x$$
$$-x \leq -1$$
$$\frac{-x}{-1} \geq \frac{-1}{-1}$$
$$x \geq 1$$
$$\{x \mid x \geq 1\}$$

6. $1 \leq 2x + 3 < 11$
$$-2 \leq 2x < 8$$
$$-1 \leq x < 4$$
The solution set is $\{x \mid -1 \leq x < 4\}$ or $[-1, 4)$.

7. Let x = your grade on the final exam.
$$\frac{82 + 74 + 78 + x + x}{5} \geq 80$$
$$\frac{234 + 2x}{5} \geq 80$$
$$5\left(\frac{234 + 2x}{5}\right) \geq 5(80)$$
$$234 + 2x \geq 400$$
$$234 + 2x - 234 \geq 400 - 234$$
$$2x \geq 166$$
$$\frac{2x}{2} \geq \frac{166}{2}$$
$$x \geq 83$$
You need at least an 83% on the final to get a B in the course.

Exercise Set 6.5

1. $x > 6$
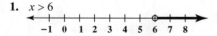

3. $x < -4$

5. $x \geq -3$

7. $x \le 4$

9. $-2 < x \le 5$

11. $-1 < x < 4$

13. $x - 3 > 2$

$x - 3 + 3 > 2 + 3$

$x > 5$

$\{x | x > 5\}$

15. $x + 4 \le 9$

$x + 4 - 4 \le 9 - 4$

$x \le 5$

$\{x | x \le 5\}$

17. $x - 3 < 0$

$x - 3 + 3 < 0 + 3$

$x < 3$

$\{x | x < 3\}$

19. $4x < 20$

$\dfrac{4x}{4} < \dfrac{20}{4}$

$x < 5$

$\{x | x < 5\}$

21. $3x \ge -15$

$\dfrac{3x}{3} \ge \dfrac{-15}{3}$

$x \ge -5$

$\{x | x \ge -5\}$

23. $2x - 3 > 7$

$2x - 3 + 3 > 7 + 3$

$2x > 10$

$\dfrac{2x}{2} > \dfrac{10}{2}$

$x > 5$

$\{x | x > 5\}$

25. $3x + 3 < 18$

$3x + 3 - 3 < 18 - 3$

$3x < 15$

$\dfrac{3x}{3} < \dfrac{15}{3}$

$x < 5$

$\{x | x < 5\}$

27. $\dfrac{1}{2} x < 4$

$2 \cdot \dfrac{1}{2} x < 2 \cdot 4$

$x < 8$

$\{x | x < 8\}$

29. $\dfrac{x}{3} > -2$

$3 \cdot \dfrac{x}{3} > 3 \cdot (-2)$

$x > -6$

$\{x | x > -6\}$

31. $-3x < 15$

$$\frac{-3x}{-3} > \frac{15}{-3}$$

$$x > -5$$

$$\{x \mid x > -5\}$$

33. $-3x \geq -15$

$$\frac{-3x}{-3} \leq \frac{-15}{-3}$$

$$x \leq 5$$

$$\{x \mid x \leq 5\}$$

35. $3x + 4 \leq 2x + 7$

$$3x + 4 - 4 \leq 2x + 7 - 4$$

$$3x \leq 2x + 3$$

$$3x - 2x \leq 2x + 3 - 2x$$

$$x \leq 3$$

$$\{x \mid x \leq 3\}$$

37. $5x - 9 < 4x + 7$

$$5x - 9 + 9 < 4x + 7 + 9$$

$$5x < 4x + 16$$

$$5x - 4x < 4x + 16 - 4x$$

$$x < 16$$

$$\{x \mid x < 16\}$$

39. $-2x - 3 < 3$

$$-2x - 3 + 3 < 3 + 3$$

$$-2x < 6$$

$$\frac{-2x}{-2} > \frac{6}{-2}$$

$$x > -3$$

$$\{x \mid x > -3\}$$

41. $3 - 7x \leq 17$

$$3 - 7x - 3 \leq 17 - 3$$

$$-7x \leq 14$$

$$\frac{-7x}{-7} \geq \frac{14}{-7}$$

$$x \geq -2$$

$$\{x \mid x \geq -2\}$$

43. $-x < 4$

$$\frac{-x}{-1} > \frac{4}{-1}$$

$$x > -4$$

$$\{x \mid x > -4\}$$

45. $5 - x \leq 1$

$$5 - x - 5 \leq 1 - 5$$

$$-x \leq -4$$

$$\frac{-x}{-1} \geq \frac{-4}{-1}$$

$$x \geq 4$$

$$\{x \mid x \geq 4\}$$

47. $2x - 5 > -x + 6$

$$2x - 5 + 5 > -x + 6 + 5$$

$$2x > -x + 11$$

$$2x + x > -x + 11 + x$$

$$3x > 11$$

$$\frac{3x}{3} > \frac{11}{3}$$

$$x > \frac{11}{3}$$

$$\left\{x \mid x > \frac{11}{3}\right\}$$

49.
$$2x-5<5x-11$$
$$2x-5+5<5x-11+5$$
$$2x<5x-6$$
$$2x-5x<5x-6-5x$$
$$-3x<-6$$
$$\frac{-3x}{-3}>\frac{-6}{-3}$$
$$x>2$$
$$\{x\,|\,x>2\}$$

51. $3(x+1)-5<2x+1$
$$3x+3-5<2x+1$$
$$3x-2<2x+1$$
$$3x-2+2<2x+1+2$$
$$3x<2x+3$$
$$3x-2x<2x+3-2x$$
$$x<3$$
$$\{x\,|\,x<3\}$$

53.
$$8x+3>3(2x+1)-x+5$$
$$8x+3>6x+3-x+5$$
$$8x+3>5x+8$$
$$8x+3-3>5x+8-3$$
$$8x>5x+5$$
$$8x-5x>5x+5-5x$$
$$3x>5$$
$$\frac{3x}{3}>\frac{5}{3}$$
$$x>\frac{5}{3}$$
$$\left\{x\,\middle|\,x>\frac{5}{3}\right\}$$

55.
$$\frac{x}{4}-\frac{3}{2}\le\frac{x}{2}+1$$
$$\frac{4x}{4}-\frac{4\cdot3}{2}\le\frac{4\cdot x}{2}+4\cdot1$$
$$x-6\le2x+4$$
$$-x\le10$$
$$x\ge-10$$
The solution set is $\{x\,|\,x\ge-10\}$.

57. $1-\frac{x}{2}>4$
$$-\frac{x}{2}>3$$
$$x<-6$$
The solution set is $\{x\,|\,x<-6\}$.

59. $6<x+3<8$
$$6-3<x+3-3<8-3$$
$$3<x<5$$
The solution set is $\{x\,|\,3<x<5\}$.

61. $-3\le x-2<1$
$$-1\le x<3$$
The solution set is $\{x\,|\,-1\le x<3\}$.

63. $-11<2x-1\le-5$
$$-10<2x\le-4$$
$$-5<x\le-2$$
The solution set is $\{x\,|\,-5<x\le-2\}$.

65. $-3\le\frac{2}{3}x-5<-1$
$$2\le\frac{2}{3}x<4$$
$$3\le x<6$$
The solution set is $\{x\,|\,3\le x<6\}$.

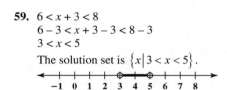
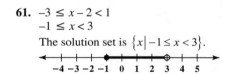
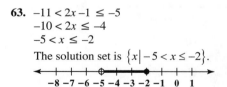

67. $Ax + By > C$

$$Ax > C - By$$

$$x > \frac{C - By}{A}$$

69. $Ax + By > C$

$$Ax > C - By$$

$$x < \frac{C - By}{A}$$

71. $\{x \mid x + 5 \geq 2x\}$

73. $\{x \mid 2(4 + x) \leq 36\}$

75. $\left\{x \mid \dfrac{3x}{5} + 4 \leq 34\right\}$

77. $\{x \mid 0 < x < 4\}$

79. intimacy $\geq$ passion or
    passion $\leq$ intimacy

81. commitment > passion or
    passion < commitment

83. The maximum level of intensity for passion is 9.
    This occurs after 3 years.

85. $0.4x + 8.8 > 20$

$$0.4x > 11.2$$

$$x > 28 \text{ years}$$

Years after 1984 + 28 or 2012.

87. a. $\dfrac{88 + 78 + 86 + 100}{4} = \dfrac{352}{4} = 88$

The average is 88. An A is not possible.

b. $\dfrac{88 + 78 + 86 + x}{4} \geq 80$

$$\frac{252 + x}{4} \geq 80$$

$$4 \cdot \frac{252 + x}{4} \geq 80 \cdot 4$$

$$252 + x \geq 320$$

$$252 + x - 252 \geq 320 - 252$$

$$x \geq 68$$

You must get at least 68 on the final to earn a B
in the course.

89. Let x = number of miles driven.

$$80 + 0.25x \leq 400$$

$$80 + 0.25x - 80 \leq 400 - 80$$

$$0.25x \leq 320$$

$$\frac{0.25x}{0.25} \leq \frac{320}{0.25}$$

$$x \leq 1280$$

You can drive at most 1280 miles.

91. Let x = number of cement bags.

$$245 + 95x \leq 3000$$

$$245 + 95x - 245 \leq 3000 - 245$$

$$95x \leq 2755$$

$$\frac{95x}{95} \leq \frac{2755}{95}$$

$$x \leq 29$$

At most 29 bags can be safely lifted.

93. $28 \leq 20 + 0.40(x - 60) \leq 40$

$$28 \leq 20 + 0.40x - 24 \leq 40$$

$$28 \leq 0.40x - 4 \leq 40$$

$$32 \leq 0.40x \leq 44$$

$$80 \leq x \leq 110$$

Between 80 and 110 ten minutes, inclusive.

99. Let x = the number of packages produced

$$5.50x > 3000 + 3.00x$$

$$5.50x - 3.00x > 3000$$

$$2.5x > 3000$$

$$2.5x > 3000$$

$$\frac{2.5x}{2.5} > \frac{3000}{2.5}$$

$$x > 1200$$

The number of packages produced must exceed
1200 for the company to generate a profit.

Check Points 6.6

1. $(x+5)(x+6) = x \cdot x + x \cdot 6 + 5 \cdot x + 5 \cdot 6$
$$= x^2 + 6x + 5x + 30$$
$$= x^2 + 11x + 30$$

2. $(7x+5)(4x-3) = 7x \cdot 4x + 7x(-3) + 5 \cdot 4x + 5(-3)$
$$= 28x^2 - 21x + 20x - 15$$
$$= 28x^2 - x - 15$$

3. $x^2 + 5x + 6 = (x+2)(x+3)$

4. $x^2 + 3x - 10 = (x+5)(x-2)$

5. $5x^2 - 14x + 8 = (5x-4)(x-2)$

6. $6y^2 + 19y - 7 = (3y-1)(2y+7)$

7. $(x+6)(x-3) = 0$
$x+6 = 0$ or $x-3 = 0$
$x = -6$ $x = 3$
The solution set is $\{-6,\ 3\}$.

8. $\qquad x^2 - 6x = 16$
$x^2 - 6x - 16 = 16 - 16$
$x^2 - 6x - 16 = 0$
$(x+2)(x-8) = 0$
$x+2 = 0$ or $x-8 = 0$
$x = -2$ $x = 8$
The solution set is $\{-2,\ 8\}$.

9. $\qquad 2x^2 + 7x - 4 = 0$
$(2x-1)(x+4) = 0$
$2x-1 = 0$ or $x+4 = 0$
$2x = 1$ $x = -4$
$x = \dfrac{1}{2}$
The solution set is $\left\{-4,\ \dfrac{1}{2}\right\}$.

10. $8x^2 + 2x - 1 = 0$
$$x = \frac{-b \pm \sqrt{b^2 - 4ac}}{2a}$$
$$x = \frac{-(2) \pm \sqrt{(2)^2 - 4(8)(-1)}}{2(8)}$$
$$x = \frac{-2 \pm \sqrt{4+32}}{16}$$
$$x = \frac{-2 \pm \sqrt{36}}{16}$$
$$x = \frac{-2 \pm 6}{16}$$
$x = \dfrac{-2+6}{16}$ or $x = \dfrac{-2-6}{16}$
$x = \dfrac{4}{16}$ $x = \dfrac{-8}{16}$
$x = \dfrac{1}{4}$ $x = -\dfrac{1}{2}$
The solution set is $\left\{-\dfrac{1}{2},\ \dfrac{1}{4}\right\}$.

11. $\qquad\qquad 2x^2 = 6x - 1$
$2x^2 - 6x + 1 = 0$
$$x = \frac{-b \pm \sqrt{b^2 - 4ac}}{2a}$$
$$x = \frac{-(-6) \pm \sqrt{(-6)^2 - 4(2)(1)}}{2(2)}$$
$$x = \frac{6 \pm \sqrt{36-8}}{4}$$
$$x = \frac{6 \pm \sqrt{28}}{4}$$
$$x = \frac{6 \pm 2\sqrt{7}}{4}$$
$$x = \frac{2\left(3 \pm \sqrt{7}\right)}{4}$$
$$x = \frac{3 \pm \sqrt{7}}{2}$$
$x = \dfrac{3+\sqrt{7}}{2}$ or $x = \dfrac{3-\sqrt{7}}{2}$
The solution set is $\left\{\dfrac{3+\sqrt{7}}{2},\ \dfrac{3-\sqrt{7}}{2}\right\}$.

12. $P = 0.01A^2 + 0.05A + 107$

$115 = 0.01A^2 + 0.05A + 107$

$0 = 0.01A^2 + 0.05A - 8$

$a = 0.01, \quad b = 0.05, \quad c = -8$

$A = \dfrac{-b \pm \sqrt{b^2 - 4ac}}{2a}$

$A = \dfrac{-(0.05) \pm \sqrt{(0.05)^2 - 4(0.01)(-8)}}{2(0.01)}$

$A = \dfrac{-0.05 \pm \sqrt{0.3225}}{0.02}$

$A \approx \dfrac{-0.05 + \sqrt{0.3225}}{0.02} \qquad A \approx \dfrac{-0.05 - \sqrt{0.3225}}{0.02}$

$A \approx 26 \qquad\qquad\qquad A \approx -31$

Age cannot be negative, reject the negative answer.
Thus, a woman whose normal systolic blood
pressure is 115 mm Hg is 26 years old.

Exercise Set 6.6

1. $(x+3)(x+5) = x^2 + 5x + 3x + 15$
$\qquad\qquad\quad = x^2 + 8x + 15$

3. $(x-5)(x+3) = x^2 + 3x - 5x - 15$
$\qquad\qquad\quad = x^2 - 2x - 15$

5. $(2x-1)(x+2) = 2x^2 + 4x - 1x - 2$
$\qquad\qquad\qquad = 2x^2 + 3x - 2$

7. $(3x-7)(4x-5) = 12x^2 - 15x - 28x + 35$
$\qquad\qquad\qquad\ = 12x^2 - 43x + 35$

9. $x^2 + 5x + 6 = (x+2)(x+3)$
Check: $(x+2)(x+3)$
$\qquad = x^2 + 3x + 2x + 6$
$\qquad = x^2 + 5x + 6$

11. $x^2 - 2x - 15 = (x-5)(x+3)$
Check: $(x-5)(x+3)$
$\qquad = x^2 + 3x - 5x - 15$
$\qquad = x^2 - 2x - 15$

13. $x^2 - 8x + 15 = (x-3)(x-5)$
Check: $(x-3)(x-5)$
$\qquad = x^2 - 5x - 3x + 15$
$\qquad = x^2 - 8x + 15$

15. $x^2 - 9x - 36 = (x-12)(x+3)$
Check: $(x-12)(x+3)$
$\qquad = x^2 + 3x - 12x - 36$
$\qquad = x^2 - 9x - 36$

17. $x^2 - 8x + 32$ is prime.

19. $x^2 + 17x + 16 = (x+16)(x+1)$
Check: $(x+16)(x+1)$
$\qquad = x^2 + x + 16x + 16$
$\qquad = x^2 + 17x + 16$

21. $2x^2 + 7x + 3 = (2x+1)(x+3)$
Check: $(2x+1)(x+3)$
$\qquad = 2x^2 + 6x + x + 3$
$\qquad = 2x^2 + 7x + 3$

23. $2x^2 - 17x + 30 = (2x-5)(x-6)$
Check: $(2x-5)(x-6)$
$\qquad = 2x^2 - 12x - 5x + 30$
$\qquad = 2x^2 - 17x + 30$

25. $3x^2 - x - 2 = (3x+2)(x-1)$
Check: $(3x+2)(x-1)$
$\qquad = 3x^2 - 3x + 2x - 2$
$\qquad = 3x^2 - x - 2$

27. $3x^2 - 25x - 28 = (3x-28)(x+1)$
Check: $(3x-28)(x+1)$
$\qquad = 3x^2 + 3x - 28x - 28$
$\qquad = 3x^2 - 25x - 28$

29. $6x^2 - 11x + 4 = (2x-1)(3x-4)$
Check: $(2x-1)(3x-4)$
$\qquad = 6x^2 - 8x - 3x + 4$
$\qquad = 6x^2 - 11x + 4$

31. $4x^2 + 16x + 15 = (2x+5)(2x+3)$
Check: $(2x+5)(2x+3)$
$\qquad = 4x^2 + 6x + 10x + 15$
$\qquad = 4x^2 + 16x + 15$

33. $(x-8)(x+3) = 0$
$x - 8 = 0 \quad$ or $\quad x + 3 = 0$
$\qquad x = 8 \qquad\qquad x = -3$
The solution set is $\{-3, 8\}$.

35. $(4x+5)(x-2)=0$

$4x+5=0$ or $x-2=0$

$4x=-5$ $x=2$

$x=-\dfrac{5}{4}$

The solution set is $\left\{-\dfrac{5}{4},2\right\}$.

37. $x^2+8x+15=0$

$(x+5)(x+3)=0$

$x+5=0$ or $x+3=0$

$x=-5$ $x=-3$

The solution set is $\{-5,-3\}$.

39. $x^2-2x-15=0$

$(x-5)(x+3)=0$

$x-5=0$ or $x+3=0$

$x=5$ $x=-3$

The solution set is $\{-3,5\}$.

41. $x^2-4x=21$

$x^2-4x-21=0$

$(x+3)(x-7)=0$

$x+3=0$ or $x-7=0$

$x=-3$ $x=7$

The solution set is $\{-3,7\}$.

43. $x^2+9x=-8$

$x^2+9x+8=0$

$(x+8)(x+1)=0$

$x+8=0$ or $x+1=0$

$x=-8$ $x=-1$

The solution set is $\{-8,-1\}$.

45. $x^2-12x=-36$

$x^2-12x+36=0$

$(x-6)(x-6)=0$

$x-6=0$ or $x-6=0$

$x=6$ $x=6$

The solution set is $\{6\}$.

47. $2x^2=7x+4$

$2x^2-7x-4=0$

$(2x+1)(x-4)=0$

$2x+1=0$ or $x-4=0$

$2x=-1$ $x=4$

$x=-\dfrac{1}{2}$

The solution set is $\left\{-\dfrac{1}{2},4\right\}$.

49. $5x^2+x=18$

$5x^2+x-18=0$

$(5x-9)(x+2)=0$

$5x-9=0$ or $x+2=0$

$5x=9$ $x=-2$

$x=\dfrac{9}{5}$

The solution set is $\left\{-2,\dfrac{9}{5}\right\}$.

51. $x(6x+23)+7=0$

$6x^2+23x+7=0$

$(2x+7)(3x+1)=0$

$2x+7=0$ or $3x+1=0$

$2x=-7$ $3x=-1$

$x=-\dfrac{7}{2}$ $x=-\dfrac{1}{3}$

The solution set is $\left\{-\dfrac{7}{2},-\dfrac{1}{3}\right\}$.

53. $x^2+8x+15=0$

$x=\dfrac{-b\pm\sqrt{b^2-4ac}}{2a}$

$x=\dfrac{-8\pm\sqrt{8^2-4(1)(15)}}{2(1)}$

$x=\dfrac{-8\pm\sqrt{4}}{2}$

$x=\dfrac{-8\pm2}{2}$

$x=\dfrac{-8-2}{2}$ or $x=\dfrac{-8+2}{2}$

$x=-5$ $x=-3$

The solution set is $\{-5,-3\}$.

55. $x^2 + 5x + 3 = 0$

$x = \dfrac{-b \pm \sqrt{b^2 - 4ac}}{2a}$

$x = \dfrac{-5 \pm \sqrt{5^2 - 4(1)(3)}}{2(1)}$

$x = \dfrac{-5 \pm \sqrt{13}}{2}$

The solution set is $\left\{ \dfrac{-5 - \sqrt{13}}{2}, \dfrac{-5 + \sqrt{13}}{2} \right\}$.

57. $x^2 + 4x = 6$

$x^2 + 4x - 6 = 0$

$x = \dfrac{-b \pm \sqrt{b^2 - 4ac}}{2a}$

$x = \dfrac{-4 \pm \sqrt{4^2 - 4(1)(-6)}}{2(1)}$

$x = \dfrac{-4 \pm \sqrt{40}}{2}$

$x = \dfrac{-4 \pm 2\sqrt{10}}{2}$

$x = -2 \pm \sqrt{10}$

The solution set is $\left\{ -2 - \sqrt{10}, \ -2 + \sqrt{10} \right\}$.

59. $x^2 + 4x - 7 = 0$

$x = \dfrac{-b \pm \sqrt{b^2 - 4ac}}{2a}$

$x = \dfrac{-4 \pm \sqrt{4^2 - 4(1)(-7)}}{2(1)}$

$x = \dfrac{-4 \pm \sqrt{44}}{2}$

$x = \dfrac{-4 \pm 2\sqrt{11}}{2}$

$x = -2 \pm \sqrt{11}$

The solution set is $\left\{ -2 - \sqrt{11}, \ -2 + \sqrt{11} \right\}$.

61. $x^2 - 3x = 18$

$x^2 - 3x - 18 = 0$

$x = \dfrac{-b \pm \sqrt{b^2 - 4ac}}{2a}$

$x = \dfrac{-(-3) \pm \sqrt{(-3)^2 - 4(1)(-18)}}{2(1)}$

$x = \dfrac{3 \pm \sqrt{81}}{2}$

$x = \dfrac{3 \pm 9}{2}$

$x = \dfrac{3 - 9}{2}$ or $x = \dfrac{3 + 9}{2}$

$x = -3$ $x = 6$

The solution set is $\{-3, 6\}$.

63. $6x^2 - 5x - 6 = 0$

$x = \dfrac{-b \pm \sqrt{b^2 - 4ac}}{2a}$

$x = \dfrac{-(-5) \pm \sqrt{(-5)^2 - 4(6)(-6)}}{2(6)}$

$x = \dfrac{5 \pm \sqrt{169}}{12}$

$x = \dfrac{5 \pm 13}{12}$

$x = \dfrac{5 + 13}{12}$ or $x = \dfrac{5 - 13}{12}$

$x = \dfrac{18}{12}$ $x = \dfrac{-8}{12}$

$x = \dfrac{3}{2}$ $x = -\dfrac{2}{3}$

The solution set is $\left\{ \dfrac{3}{2}, \ -\dfrac{2}{3} \right\}$.

65. $x^2 - 2x - 10 = 0$

$x = \dfrac{-b \pm \sqrt{b^2 - 4ac}}{2a}$

$x = \dfrac{-(-2) \pm \sqrt{(-2)^2 - 4(1)(-10)}}{2(1)}$

$x = \dfrac{2 \pm \sqrt{44}}{2}$

$x = \dfrac{2 \pm 2\sqrt{11}}{2}$

$x = 1 \pm \sqrt{11}$

The solution set is $\left\{ 1 - \sqrt{11}, \ 1 + \sqrt{11} \right\}$.

67.
$$x^2 - x = 14$$
$$x^2 - x - 14 = 0$$
$$x = \frac{-b \pm \sqrt{b^2 - 4ac}}{2a}$$
$$x = \frac{-(-1) \pm \sqrt{(-1)^2 - 4(1)(-14)}}{2(1)}$$
$$x = \frac{1 \pm \sqrt{57}}{2}$$

The solution set is $\left\{ \dfrac{1-\sqrt{57}}{2}, \ \dfrac{1+\sqrt{57}}{2} \right\}$.

69. $6x^2 + 6x + 1 = 0$
$$x = \frac{-b \pm \sqrt{b^2 - 4ac}}{2a}$$
$$x = \frac{-6 \pm \sqrt{6^2 - 4(6)(1)}}{2(6)}$$
$$x = \frac{-6 \pm \sqrt{12}}{12}$$
$$x = \frac{-6 \pm 2\sqrt{3}}{12}$$
$$x = \frac{-3 \pm \sqrt{3}}{6}$$

The solution set is $\left\{ \dfrac{-3-\sqrt{3}}{6}, \ \dfrac{-3+\sqrt{3}}{6} \right\}$.

71.
$$4x^2 = 12x - 9$$
$$4x^2 - 12x + 9 = 0$$
$$x = \frac{-b \pm \sqrt{b^2 - 4ac}}{2a}$$
$$x = \frac{-(-12) \pm \sqrt{(-12)^2 - 4(4)(9)}}{2(4)}$$
$$x = \frac{12 \pm \sqrt{0}}{8}$$
$$x = \frac{12}{8} = \frac{3}{2}$$

The solution set is $\left\{ \dfrac{3}{2} \right\}$.

73.
$$\frac{3x^2}{4} - \frac{5x}{2} - 2 = 0$$
$$3x^2 - 10x - 8 = 0$$
$$(x - 4)(3x + 2) = 0$$
$$x - 4 = 0 \quad \text{or} \quad 3x + 2 = 0$$
$$x = 4 \qquad\qquad x = -\frac{2}{3}$$

The solution set is $\left\{ -\dfrac{2}{3}, 4 \right\}$.

75. $(x-1)(3x+2) = -7(x-1)$
$$3x^2 - x - 2 = -7x + 7$$
$$3x^2 + 6x - 9 = 0$$
$$x^2 + 2x - 3 = 0$$
$$(x-1)(x+3) = 0$$
$$x - 1 = 0 \quad \text{or} \quad x + 3 = 0$$
$$x = 1 \qquad\qquad x = -3$$

The solution set is $\{-3, 1\}$.

77. $(2x-6)(x+2) = 5(x-1) - 12$
$$(2x-6)(x+2) = 5x - 5 - 12$$
$$2x^2 - 2x - 12 = 5x - 17$$
$$2x^2 - 7x + 5 = 0$$
$$(x-1)(2x-5) = 0$$
$$x - 1 = 0 \quad \text{or} \quad 2x - 5 = 0$$
$$x = 1 \qquad\qquad x = \frac{5}{2}$$

The solution set is $\left\{ 1, \dfrac{5}{2} \right\}$.

79. $2x^2 - 9x - 3 = 9 - 9x$
$$2x^2 = 12$$
$$x^2 = 6$$
$$x = \pm\sqrt{6}$$

The solution set is $\left\{ \pm\sqrt{6} \right\}$.

81. Let x = the number.

$$x^2 - (6 + 2x) = 0$$
$$x^2 - 2x - 6 = 0$$

Apply the quadratic formula.

$$a = 1 \quad b = -2 \quad c = -6$$

$$x = \frac{-(-2) \pm \sqrt{(-2)^2 - 4(1)(-6)}}{2(1)}$$

$$= \frac{2 \pm \sqrt{4 - (-24)}}{2}$$

$$= \frac{2 \pm \sqrt{28}}{2}$$

$$= \frac{2 \pm \sqrt{4 \cdot 7}}{2} = \frac{2 \pm 2\sqrt{7}}{2} = 1 \pm \sqrt{7}$$

We disregard $1 - \sqrt{7}$ because it is negative, and we are looking for a positive number. Thus, the number is $1 + \sqrt{7}$.

83. $N = \dfrac{t^2 - t}{2}$

$$36 = \frac{t^2 - t}{2}$$

$$72 = t^2 - t$$

$$0 = t^2 - t - 72$$

$$0 = (t + 8)(t - 9)$$

$$t + 8 = 0 \quad \text{or} \quad t - 9 = 0$$
$$t = -8 \qquad\qquad t = 9$$

Thus, the league has 9 teams.

85. $P = 0.005x^2 - 0.37x + 11.8$

$$P = 0.005(70)^2 - 0.37(70) + 11.8$$

$$P = 10.4$$

The formula models the data for 2000 exactly.

87. $P = 0.005x^2 - 0.37x + 11.8$

$$16 = 0.005x^2 - 0.37x + 11.8$$

$$0 = 0.005x^2 - 0.37x - 4.2$$

$$x = \frac{-b \pm \sqrt{b^2 - 4ac}}{2a}$$

$$x = \frac{-(-0.37) \pm \sqrt{(-0.37)^2 - 4(0.005)(-4.2)}}{2(0.005)}$$

$$x = \frac{0.37 \pm \sqrt{0.2209}}{0.01}$$

$$x \approx -10 \quad \text{or} \quad x \approx 84$$

Reject -10. The model projects that 16% of the U.S. population will be foreign-born 84 years after 1930, or 2014.

89. $N = 0.013x^2 - 1.19x + 28.24$, for $N = 3$

$$3 = 0.013x^2 - 1.19x + 28.24$$

$$0 = 0.013x^2 - 1.19x + 28.24 - 3$$

$$0 = 0.013x^2 - 1.19x + 25.24$$

$$x = \frac{-b \pm \sqrt{b^2 - 4ac}}{2a}$$

$$x = \frac{-(-1.19) \pm \sqrt{(-1.19)^2 - 4(0.013)(25.24)}}{2(0.013)}$$

$$x = \frac{1.19 \pm \sqrt{0.10362}}{0.026}$$

$$x \approx \frac{1.19 \pm 0.3219}{0.026}$$

$$x \approx \frac{1.19 + 0.3219}{0.026} \quad \text{or} \quad x \approx \frac{1.19 - 0.3219}{0.026}$$

$$x \approx \frac{1.5119}{0.026} \qquad\qquad x \approx \frac{0.8681}{0.026}$$

$$x \approx 58 \qquad\qquad\qquad x \approx 33$$

33-year-olds and 58-year-olds are expected to be involved in 3 fatal crashes per 100 million miles driven. The formula models the data quite well.

97. $x^2 + bx + 15$

$$(x + 3)(x + 5) = x^2 + 8x + 15$$

$$(x + 1)(x + 15) = x^2 + 16x + 15$$

Therefore, $b = 8, 16$.

99. $x^{2n} + 20x^n + 99 = (x^n + 11)(x^n + 9)$

Chapter 6 Review Exercises

1. $6x + 9 = 6 \cdot 4 + 9 = 24 + 9 = 33$

2. $7x^2 + 4x - 5 = 7(-2)^2 + 4(-2) - 5$
   $$= 7(4) + 4(-2) - 5$$
   $$= 28 - 8 - 5$$
   $$= 15$$

3. $6 + 2(x - 8)^3 = 6 + 2(5 - 8)^3$
   $$= 6 + 2(-3)^3$$
   $$= 6 + 2(-27)$$
   $$= 6 - 54$$
   $$= -48$$

4. Since $2002 - 1997 = 5$, let $x = 5$.
   $N = -26x^2 + 143x + 740$
   $$= -26(5)^2 + 143(5) + 740$$
   $$= -26(25) + 143(5) + 740$$
   $$= -650 + 715 + 740$$
   $$= 805$$
   805 million CDs were sold in 2002. According to the model, CD sales were 800 million. The model fits the data fairly well, but there is a slight overestimation.

5. $5(2x - 3) + 7x = 10x - 15 + 7x$
   $$= 17x - 15$$

6. $3(4y - 5) - (7y - 2) = 12y - 15 - 7y + 2$
   $$= 5y - 13$$

7. $2(x^2 + 5x) + 3(4x^2 - 3x) = 2x^2 + 10x + 12x^2 - 9x$
   $$= 14x^2 + x$$

8. $\quad 4x + 9 = 33$
   $4x + 9 - 9 = 33 - 9$
   $\quad\quad 4x = 24$
   $\quad\quad \dfrac{4x}{4} = \dfrac{24}{4}$
   $\quad\quad\quad x = 6$
   The solution set is $\{6\}$.

9. $\quad 5x - 3 = x + 5$
   $5x - 3 + 3 = x + 5 + 3$
   $\quad\quad 5x = x + 8$
   $5x - x = x + 8 - x$
   $\quad\quad 4x = 8$
   $\quad\quad \dfrac{4x}{4} = \dfrac{8}{4}$
   $\quad\quad\quad x = 2$
   The solution set is $\{2\}$.

10. $\quad\quad 3(x + 4) = 5x - 12$
    $\quad\quad 3x + 12 = 5x - 12$
    $3x + 12 - 12 = 5x - 12 - 12$
    $\quad\quad\quad 3x = 5x - 24$
    $\quad 3x - 5x = 5x - 24 - 5x$
    $\quad\quad\quad -2x = -24$
    $\quad\quad\quad \dfrac{-2x}{-2} = \dfrac{-24}{-2}$
    $\quad\quad\quad\quad x = 12$
    The solution set is $\{12\}$.

11. $\quad 2(x - 2) + 3(x + 5) = 2x - 2$
    $\quad 2x - 4 + 3x + 15 = 2x - 2$
    $\quad\quad\quad 5x + 11 = 2x - 2$
    $\quad 5x + 11 - 11 = 2x - 2 - 11$
    $\quad\quad\quad 5x = 2x - 13$
    $\quad\quad 5x - 2x = 2x - 13 - 2x$
    $\quad\quad\quad 3x = -13$
    $\quad\quad\quad \dfrac{3x}{3} = \dfrac{-13}{3}$
    $\quad\quad\quad x = -\dfrac{13}{3}$
    The solution set is $\left\{ -\dfrac{13}{3} \right\}$.

12.
$$\frac{2x}{3} = \frac{x}{6} + 1$$
$$6\left(\frac{2x}{3}\right) = 6\left(\frac{x}{6} + 1\right)$$
$$4x = x + 6$$
$$4x - x = x + 6 - x$$
$$3x = 6$$
$$\frac{3x}{3} = \frac{6}{3}$$
$$x = 2$$
The solution set is $\{2\}$.

13.
$$7x + 5 = 5(x + 3) + 2x$$
$$7x + 5 = 5x + 15 + 2x$$
$$7x + 5 = 7x + 15$$
$$5 = 15$$
This is a false statement. The solution set is $\{\ \}$.

14.
$$7x + 13 = 2(2x - 5) + 3x + 23$$
$$7x + 13 = 4x - 10 + 3x + 23$$
$$7x + 13 = 7x + 13$$
$$13 = 13$$
This is a true statement. The solution set is $\{x | x \text{ is a real number}\}$.

15.
$$Ax - By = C$$
$$Ax = By + C$$
$$\frac{Ax}{A} = \frac{By + C}{A}$$
$$x = \frac{By + C}{A}$$

16.
$$A = \frac{1}{2}bh$$
$$2A = bh$$
$$\frac{2A}{b} = \frac{bh}{b}$$
$$\frac{2A}{b} = h$$
$$h = \frac{2A}{b}$$

17.
$$A = \frac{B + C}{2}$$
$$2A = B + C$$
$$2A - C = B$$
$$B = 2A - C$$

18.
$$vt + gt^2 = s$$
$$gt^2 = s - vt$$
$$\frac{gt^2}{t^2} = \frac{s - vt}{t^2}$$
$$g = \frac{s - vt}{t^2}$$

19.
$$P = 4.6x + 45.8$$
$$101 = 4.6x + 45.8$$
$$55.2 = 4.6x$$
$$12 = x$$
The model projects that the average prescription-drug price will be \$101 12 years after 2000, or 2012.

20.
$$7x - 1 = 5x + 9$$
$$7x - 1 + 1 = 5x + 9 + 1$$
$$7x = 5x + 10$$
$$7x - 5x = 5x + 10 - 5x$$
$$2x = 10$$
$$\frac{2x}{2} = \frac{10}{2}$$
$$x = 5$$
The number is 5.

21. Let x = the number of calories in Burger King's Chicken Caesar.
$x + 125$ = the number of calories in Taco Bell's Express Taco Salad.
$x + 95$ = the number of calories in Wendy's Mandarin Chicken Salad.
$$x + (x + 125) + (x + 95) = 1705$$
$$3x + 220 = 1705$$
$$3x = 1485$$
$$x = 495$$
$$x + 125 = 495 + 125 = 620$$
$$x + 95 = 495 + 95 = 590$$
There are 495 calories in the Chicken Caesar, 620 calories in the Express Taco Salad, and 590 calories in the Mandarin Chicken Salad.

22. Let E = the number of endangered plant species, and let x represent the number of years since 1998.
$$E = 6.4x + 567$$
$$663 = 6.4x + 567$$
$$96 = 6.4x$$
$$15 = x$$
There will be 663 endangered species in the United States in the year $1998 + 15 = 2013$.

23. Let x = the number of minutes at which the costs of the two plans are the same.

$$\overbrace{15+0.05x}^{\text{first plan}} = \overbrace{5+0.07x}^{\text{other plan}}$$

$$15+0.05x-15 = 5+0.07x-15$$

$$0.05x = 0.07x-10$$

$$0.05x-0.07x = 0.07x-10-0.07x$$

$$-0.02x = -10$$

$$\frac{-0.02x}{-0.02} = \frac{-10}{-0.02}$$

$$x = 500$$

The two plans are the same at 500 minutes.

24. Let x = the daily oil consumption in China
$3x+1$ = t daily oil consumption in the U.S.

$$x+(3x+1) = 27$$

$$4x+1 = 27$$

$$4x = 26$$

$$x = 6.5$$

$$3x+1 = 20.5$$

The U.S. consumes 20.5 million barrels per day and China consumes 6.5 million barrels per day.

25.
$$\frac{3}{x} = \frac{15}{25}$$

$$3 \cdot 25 = x \cdot 15$$

$$75 = 15x$$

$$\frac{75}{15} = \frac{15x}{15}$$

$$5 = x$$

The solution set is $\{5\}$.

26.
$$\frac{-7}{5} = \frac{91}{x}$$

$$-7 \cdot x = 5 \cdot 91$$

$$-7x = 455$$

$$\frac{-7x}{-7} = \frac{455}{-7}$$

$$x = -65$$

The solution set is $\{-65\}$.

27.
$$\frac{x+2}{3} = \frac{4}{5}$$

$$5(x+2) = 3 \cdot 4$$

$$5x+10 = 12$$

$$5x+10-10 = 12-10$$

$$5x = 2$$

$$\frac{5x}{5} = \frac{2}{5}$$

$$x = \frac{2}{5}$$

The solution set is $\left\{\dfrac{2}{5}\right\}$.

28.
$$\frac{5}{x+7} = \frac{3}{x+3}$$

$$5(x+3) = 3(x+7)$$

$$5x+15 = 3x+21$$

$$5x+15-15 = 3x+21-15$$

$$5x = 3x+6$$

$$5x-3x = 3x+6-3x$$

$$2x = 6$$

$$\frac{2x}{2} = \frac{6}{2}$$

$$x = 3$$

The solution set is $\{3\}$.

29. Let x = number of teachers

$$\frac{3}{50} = \frac{x}{5400}$$

$$50 \cdot x = 3 \cdot 5400$$

$$50x = 16,200$$

$$\frac{50x}{50} = \frac{16,200}{50}$$

$$x = 324$$

There should be 324 teachers for 5400 students.

30. Let x = number of trout in lake

$$\frac{32}{82} = \frac{112}{x}$$

$$32x = 82 \cdot 112$$

$$32x = 9184$$

$$\frac{32x}{32} = \frac{9184}{32}$$

$$x = 287$$

There are 287 trout in the lake.

31. Let x = the dollar amount of the electric bill.

$$\frac{1400}{98} = \frac{2200}{x}$$
$$1400 \cdot x = 98 \cdot 2200$$
$$1400x = 215,600$$
$$\frac{1400x}{1400} = \frac{215,600}{1400}$$
$$x = 154$$

The electric bill is \$154.

32. Let x = feet the object will fall in 10 seconds.

$$\frac{144}{3^2} = \frac{x}{10^2}$$
$$\frac{144}{9} = \frac{x}{100}$$
$$144 \cdot 100 = 9x$$
$$14,400 = 9x$$
$$\frac{14,400}{9} = \frac{9x}{9}$$
$$1600 = x$$

The object will fall 1600 feet in 10 seconds.

33. Let x = hours needed when driving 40 mph.

$$\frac{50}{x} = \frac{40}{4}$$
$$50 \cdot 4 = 40x$$
$$200 = 40x$$
$$\frac{200}{40} = \frac{40x}{40}$$
$$5 = x$$

At a rate of 40 mph, it will take 5 hours.

34.
$$2x - 5 < 3$$
$$2x - 5 + 5 < 3 + 5$$
$$2x < 8$$
$$\frac{2x}{2} < \frac{8}{2}$$
$$x < 4$$
$$\{x \mid x < 4\}$$

35. $\dfrac{x}{2} > -4$

$$2 \cdot \frac{x}{2} > 2(-4)$$
$$x > -8$$
$$\{x \mid x > -8\}$$

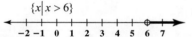

36.
$$3 - 5x \le 18$$
$$3 - 5x - 3 \le 18 - 3$$
$$-5x \le 15$$
$$\frac{-5x}{-5} \ge \frac{15}{-5}$$
$$x \ge -3$$
$$\{x \mid x \ge -3\}$$

```
 +--+--+--+--+--+--+--●--+--+--+
-10 -9 -8 -7 -6 -5 -4 -3 -2 -1
```

37.
$$4x + 6 < 5x$$
$$4x + 6 - 6 < 5x - 6$$
$$4x < 5x - 6$$
$$4x - 5x < 5x - 6 - 5x$$
$$-x < -6$$
$$\frac{-x}{-1} > \frac{-6}{-1}$$
$$x > 6$$
$$\{x \mid x > 6\}$$

```
 +--+--+--+--+--+--+--+--+⊕--+
-2 -1  0  1  2  3  4  5  6  7
```

38.
$$6x - 10 \ge 2(x + 3)$$
$$6x - 10 + 10 \ge 2x + 6 + 10$$
$$6x \ge 2x + 16$$
$$6x - 2x \ge 2x + 16 - 2x$$
$$4x \ge 16$$
$$\frac{4x}{4} \ge \frac{16}{4}$$
$$x \ge 4$$
$$\{x \mid x \ge 4\}$$

```
 +--+--+--+--+--+--●--+--+--+
-2 -1  0  1  2  3  4  5  6  7
```

39. $4x + 3(2x - 7) \le x - 3$

$4x + 6x - 21 \le x - 3$

$10x - 21 \le x - 3$

$10x - 21 + 21 \le x - 3 + 21$

$10x \le x + 18$

$10x - x \le x + 18 - x$

$9x \le 18$

$\dfrac{9x}{9} \le \dfrac{18}{9}$

$x \le 2$

$\{x \mid x \le 2\}$

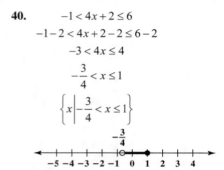

40. $-1 < 4x + 2 \le 6$

$-1 - 2 < 4x + 2 - 2 \le 6 - 2$

$-3 < 4x \le 4$

$-\dfrac{3}{4} < x \le 1$

$\left\{ x \mid -\dfrac{3}{4} < x \le 1 \right\}$

41. Let x = score on third test.

$\dfrac{42 + 74 + x}{3} \ge 60$

$\dfrac{116 + x}{3} \ge 60$

$3 \cdot \dfrac{116 + x}{3} \ge 3 \cdot 60$

$116 + x \ge 180$

$116 + x - 116 \ge 180 - 116$

$x \ge 64$

The score on the third test must be at least 64.

42. $(x + 9)(x - 5) = x^2 - 5x + 9x - 45$

$= x^2 + 4x - 45$

43. $(4x - 7)(3x + 2) = 12x^2 + 8x - 21x - 14$

$= 12x^2 - 13x - 14$

44. $x^2 - x - 12 = (x - 4)(x + 3)$

45. $x^2 - 8x + 15 = (x - 5)(x - 3)$

46. $x^2 + 2x + 3$ is prime.

47. $3x^2 - 17x + 10 = (3x - 2)(x - 5)$

48. $6x^2 - 11x - 10 = (3x + 2)(2x - 5)$

49. $3x^2 - 6x - 5$ is prime.

50. $x^2 + 5x - 14 = 0$

$(x + 7)(x - 2) = 0$

$x + 7 = 0 \quad \text{or} \quad x - 2 = 0$

$x = -7 \qquad\qquad x = 2$

The solution set is $\{-7, 2\}$.

51. $x^2 - 4x = 32$

$x^2 - 4x - 32 = 0$

$(x - 8)(x + 4) = 0$

$x - 8 = 0 \quad \text{or} \quad x + 4 = 0$

$x = 8 \qquad\qquad x = -4$

The solution set is $\{-4, 8\}$.

52. $2x^2 + 15x - 8 = 0$

$(2x - 1)(x + 8) = 0$

$2x - 1 = 0 \quad \text{or} \quad x + 8 = 0$

$2x = 1 \qquad\qquad x = -8$

$x = \dfrac{1}{2}$

The solution set is $\left\{ -8, \dfrac{1}{2} \right\}$.

53. $3x^2 = -21x - 30$

$3x^2 + 21x + 30 = 0$

$(3x + 6)(x + 5) = 0$

$3x + 6 = 0 \quad \text{or} \quad x + 5 = 0$

$3x = -6 \qquad\qquad x = -5$

$x = -2$

The solution set is $\{-5, -2\}$.

54. $x^2 - 4x + 3 = 0$

$x = \dfrac{-b \pm \sqrt{b^2 - 4ac}}{2a}$

$x = \dfrac{-(-4) \pm \sqrt{(-4)^2 - 4(1)(3)}}{2(1)}$

$x = \dfrac{4 \pm \sqrt{4}}{2}$

$x = \dfrac{4 \pm 2}{2}$

$x = \dfrac{4 - 2}{2}$ or $x = \dfrac{4 + 2}{2}$

$x = 1$ $x = 3$

The solution set is $\{1, 3\}$.

55. $x^2 - 5x = 4$

$x^2 - 5x - 4 = 0$

$x = \dfrac{-b \pm \sqrt{b^2 - 4ac}}{2a}$

$x = \dfrac{-(-5) \pm \sqrt{(-5)^2 - 4(1)(-4)}}{2(1)}$

$x = \dfrac{5 \pm \sqrt{41}}{2}$

The solution set is $\left\{ \dfrac{5 - \sqrt{41}}{2}, \dfrac{5 + \sqrt{41}}{2} \right\}$.

56. $2x^2 + 5x - 3 = 0$

$x = \dfrac{-b \pm \sqrt{b^2 - 4ac}}{2a}$

$x = \dfrac{-5 \pm \sqrt{5^2 - 4(2)(-3)}}{2(2)}$

$x = \dfrac{-5 \pm \sqrt{49}}{4}$

$x = \dfrac{-5 \pm 7}{4}$

$x = \dfrac{-5 + 7}{4}$ or $x = \dfrac{-5 - 7}{4}$

$x = \dfrac{1}{2}$ $x = -3$

The solution set is $\left\{ -3, \dfrac{1}{2} \right\}$.

57. $3x^2 - 6x = 5$

$3x^2 - 6x - 5 = 0$

$x = \dfrac{-b \pm \sqrt{b^2 - 4ac}}{2a}$

$x = \dfrac{-(-6) \pm \sqrt{(-6)^2 - 4(3)(-5)}}{2(3)}$

$x = \dfrac{6 \pm \sqrt{96}}{6}$

$x = \dfrac{6 \pm 4\sqrt{6}}{6}$

$x = \dfrac{3 \pm 2\sqrt{6}}{3}$

The solution set is $\left\{ \dfrac{3 - 2\sqrt{6}}{3}, \dfrac{3 + 2\sqrt{6}}{3} \right\}$.

58. $P = -0.035x^2 + 0.65x + 7.6$

$0 = -0.035x^2 + 0.65x + 7.6$

$x = \dfrac{-b \pm \sqrt{b^2 - 4ac}}{2a}$

$x = \dfrac{-(0.65) \pm \sqrt{(0.65)^2 - 4(-0.035)(7.6)}}{2(-0.035)}$

$x \approx 27$ $x \approx -8$ (rejected)

If this trend continues, corporations will pay no taxes 27 years after 1985, or 2012.

Chapter 6 Test

1. $x^3 - 4(x-1)^2 = (-2)^3 - 4(-2-1)^2$
$= -8 - 4(-3)^2$
$= -8 - 4(9)$
$= -8 - 4(9)$
$= -44$

2. $5(3x-2) - (x-6) = 15x - 10 - x + 6$
$= 14x - 4$

3. $F = 24t^2 - 260t + 816$
$= 24(10)^2 - 260(10) + 816$
$= 24(100) - 2600 + 816$
$= 2400 - 2600 + 816 = 616$
There were 616 convictions in 2000.

4. $12x + 4 = 7x - 21$
$5x = -25$
$x = -5$
The solution set is $\{-5\}$.

5. $3(2x-4) = 9 - 3(x+1)$
$6x - 12 = 9 - 3x - 3$
$6x - 12 = 6 - 3x$
$6x - 12 + 12 = 6 - 3x + 12$
$6x = -3x + 18$
$6x + 3x = -3x + 18 + 3x$
$9x = 18$
$\dfrac{9x}{9} = \dfrac{18}{9}$
$x = 2$
The solution set is $\{2\}$.

6. $3(x-4) + x = 2(6+2x)$
$3x - 12 + x = 12 + 4x$
$4x - 12 = 4x + 12$
$-12 = 12$
This is a false statement. The solution set is $\{\ \}$.

7. $\dfrac{x}{5} - 2 = \dfrac{x}{3}$
$15\left(\dfrac{x}{5} - 2\right) = 15\left(\dfrac{x}{3}\right)$
$3x - 30 = 5x$
$-2x = 30$
$x = -15$
The solution set is $\{-15\}$.

8. $By - Ax = A$
$By = Ax + A$
$\dfrac{By}{B} = \dfrac{Ax + A}{B}$
$y = \dfrac{Ax + A}{B}$

9. $D = 0.12x + 5.44$
$6.4 = 0.12x + 5.44$
$0.96 = 0.12x$
$\dfrac{0.96}{0.12} = \dfrac{0.12x}{0.12}$
$8 = x$
The model projects that there will be 6.4 million children in the U.S. with physical disabilities 8 years after 2000, or 2008.

10. Let x = the number.
$5x - 9 = 310$
$5x - 9 + 9 = 310 + 9$
$5x = 319$
$\dfrac{5x}{5} = \dfrac{319}{5}$
$x = 63.8$
The number is 63.8.

11. Let x = the amount earned by a preschool teacher
Let $x + 22{,}870$ = the amount earned by a fitness instructor
$x + (x + 22{,}870) = 79{,}030$
$x + x + 22{,}870 = 79{,}030$
$2x + 22{,}870 = 79{,}030$
$2x + 22{,}870 - 22{,}870 = 79{,}030 - 22{,}870$
$2x = 79{,}030 - 22{,}870$
$2x = 56{,}160$
$\dfrac{2x}{2} = \dfrac{56{,}160}{2}$
$x = 28{,}080$
A preschool teacher makes $28,080 and a fitness trainer makes $28,080 + $22,870 = $50,950

12. Let x = the number of years since the car was purchased

Value $= \$13,805 - \$1820x$

$4705 = 13,805 - 1820x$

$1820x = 9100$

$x = 5$

The car will have a value of $4705 in 5 years.

13. Let x = the number of prints.

Photo Shop A: $0.11x + 1.60$

Photo Shop B: $0.13x + 1.20$

$0.13x + 1.20 = 0.11x + 1.60$

$0.02x + 1.20 = 1.60$

$0.02x = 0.40$

$x = 20$

The cost will be the same for 20 prints. That common price is $0.11(20) + 1.60 = 0.13(20) + 1.20$

$= \$3.80$

14. $\dfrac{5}{8} = \dfrac{x}{12}$

$8 \cdot x = 5 \cdot 12$

$8x = 60$

$\dfrac{8x}{8} = \dfrac{60}{8}$

$x = 7.5$

The solution set is $\{7.5\}$.

15. $\dfrac{x+5}{8} = \dfrac{x+2}{5}$

$5(x+5) = 8(x+2)$

$5x + 25 = 8x + 16$

$5x + 25 - 25 = 8x + 16 - 25$

$5x = 8x - 9$

$5x - 8x = 8x - 9 - 8x$

$-3x = -9$

$\dfrac{-3x}{-3} = \dfrac{-9}{-3}$

$x = 3$

The solution set is $\{3\}$.

16. Let x = number of elk in the park.

$\dfrac{5}{150} = \dfrac{200}{x}$

$5x = 150 \cdot 200$

$5x = 30,000$

$\dfrac{5x}{5} = \dfrac{30,000}{5}$

$x = 6000$

There are 6000 elk in the park.

17. Let x = pressure.

$\dfrac{25}{60} = \dfrac{x}{330}$

$60x = 25 \cdot 330$

$60x = 8250$

$\dfrac{60x}{60} = \dfrac{8250}{60}$

$x = 137.5$

The pressure will be 137.5 pounds per square inch.

18. Let x = time.

$\dfrac{6}{30} = \dfrac{x}{20}$

$30x = 6 \cdot 20$

$30x = 120$

$x = 4$ hours

It would take 30 people 4 hours to do the job.

19. $6 - 9x \geq 33$

$6 - 9x - 6 \geq 33 - 6$

$-9x \geq 27$

$\dfrac{-9x}{-9} \leq \dfrac{27}{-9}$

$x \leq -3$

$\{x \mid x \leq -3\}$

20.
$$4x - 2 > 2(x + 6)$$
$$4x - 2 > 2x + 12$$
$$4x - 2 + 2 > 2x + 12 + 2$$
$$4x > 2x + 14$$
$$4x - 2x > 2x + 14 - 2x$$
$$2x > 14$$
$$\frac{2x}{2} > \frac{14}{2}$$
$$x > 7$$
$$\{x \mid x > 7\}$$

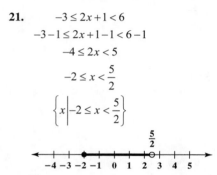

21.
$$-3 \le 2x + 1 < 6$$
$$-3 - 1 \le 2x + 1 - 1 < 6 - 1$$
$$-4 \le 2x < 5$$
$$-2 \le x < \frac{5}{2}$$
$$\left\{ x \,\middle|\, -2 \le x < \frac{5}{2} \right\}$$

22. Let x = grade on 4th examination.
$$\frac{76 + 80 + 72 + x}{4} \ge 80$$
$$\frac{228 + x}{4} \ge 80$$
$$4 \cdot \frac{228 + x}{4} \ge 80 \cdot 4$$
$$228 + x \ge 320$$
$$228 + x - 228 \ge 320 - 228$$
$$x \ge 92$$
The student must earn at least a 92 to receive a B.

23. $(2x - 5)(3x + 4) = 6x^2 + 8x - 15x - 20$
$$= 6x^2 - 7x - 20$$

24. $2x^2 - 9x + 10 = (2x - 5)(x - 2)$

25.
$$x^2 + 5x = 36$$
$$x^2 + 5x - 36 = 0$$
$$(x + 9)(x - 4) = 0$$
$$x + 9 = 0 \quad \text{or} \quad x - 4 = 0$$
$$x = -9 \qquad \qquad x = 4$$
The solution set is $\{-9, 4\}$.

26.
$$2x^2 + 4x = -1$$
$$2x^2 + 4x + 1 = 0$$
$$x = \frac{-b \pm \sqrt{b^2 - 4ac}}{2a}$$
$$x = \frac{-4 \pm \sqrt{4^2 - 4(2)(1)}}{2(2)}$$
$$x = \frac{-4 \pm \sqrt{8}}{4}$$
$$x = \frac{-4 \pm 2\sqrt{2}}{4}$$
$$x = \frac{-2 \pm \sqrt{2}}{2}$$
The solution set is $\left\{ \dfrac{-2 - \sqrt{2}}{2}, \dfrac{-2 + \sqrt{2}}{2} \right\}$.

27.
$$43x + 575 = 1177$$
$$43x = 602$$
$$x = 14$$
The system's income will be \$1177 billion 14 years after 2004, or 2018.

28.
$$B = 0.07x^2 + 47.4x + 500$$
$$1177 = 0.07x^2 + 47.4x + 500$$
$$0 = 0.07x^2 + 47.4x - 677$$
$$0 = 0.07x^2 + 47.4x - 677$$
$$x = \frac{-b \pm \sqrt{b^2 - 4ac}}{2a}$$
$$x = \frac{-(47.4) \pm \sqrt{(47.4)^2 - 4(0.07)(-677)}}{2(0.07)}$$
$$x \approx 14, \quad x \approx -691 \ (\text{rejected})$$
The system's income will be \$1177 billion 14 years after 2004, or 2018.

29. The formulas model the data quite well.

Chapter 7
Algebra: Graphs, Functions, and Linear Systems

Check Points 7.1

1.

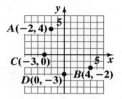

2.

x	$y = x^2 - 1$	(x, y)
-3	$y = (-3)^2 - 1 = 9 - 1 = 8$	$(-3, 8)$
-2	$y = (-2)^2 - 1 = 4 - 1 = 3$	$(-2, 3)$
-1	$y = (-1)^2 - 1 = 1 - 1 = 0$	$(-1, 0)$
0	$y = (0)^2 - 1 = 0 - 1 = -1$	$(0, -1)$
1	$y = (1)^2 - 1 = 1 - 1 = 0$	$(1, 0)$
2	$y = (2)^2 - 1 = 4 - 1 = 3$	$(2, 3)$
3	$y = (3)^2 - 1 = 9 - 1 = 8$	$(3, 8)$

3. a.

Without the coupon book

x	$y = 2x$	(x, y)
0	$y = 2(0) = 0$	$(0, 0)$
2	$y = 2(2) = 4$	$(2, 4)$
4	$y = 2(4) = 8$	$(4, 8)$
6	$y = 2(6) = 12$	$(6, 12)$
8	$y = 2(8) = 16$	$(8, 16)$
10	$y = 2(10) = 20$	$(10, 20)$
12	$y = 2(12) = 24$	$(12, 24)$

With the coupon book

x	$y = 10 + x$	(x, y)
0	$y = 10 + 0 = 10$	$(0, 10)$
2	$y = 10 + 2 = 12$	$(2, 12)$
4	$y = 10 + 4 = 14$	$(4, 14)$
6	$y = 10 + 6 = 16$	$(6, 16)$
8	$y = 10 + 8 = 18$	$(8, 18)$
10	$y = 10 + 10 = 20$	$(10, 20)$
12	$y = 10 + 12 = 22$	$(12, 22)$

b.

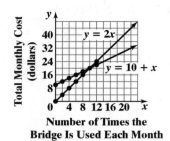

**Number of Times the
Bridge Is Used Each Month**

c.   The graphs intersect at $(10, 20)$. This means that if the bridge is used ten times in a month, the total monthly cost is $20 with or without the coupon book.

4. a.   Women's earnings as a percentage of men's was about 73% in 2000.

b. $f(x) = 0.012x^2 - 0.16x + 60$

   $f(40) = 0.012(40)^2 - 0.16(40) + 60 = 72.8\%$

5.

x	$f(x) = 2x$	(x, y) or $(x, f(x))$
-2	$f(-2) = 2(-2) = -4$	$(-2, -4)$
-1	$f(-1) = 2(-1) = -2$	$(-1, -2)$
0	$f(0) = 2(0) = 0$	$(0, 0)$
1	$f(1) = 2(1) = 2$	$(1, 2)$
2	$f(2) = 2(2) = 4$	$(2, 4)$

x	$g(x) = 2x - 3$	(x, y) or $(x, f(x))$
-2	$g(-2) = 2(-2) - 3 = -7$	$(-2, -7)$
-1	$g(-1) = 2(-1) - 3 = -5$	$(-1, -5)$
0	$g(0) = 2(0) - 3 = -3$	$(0, -3)$
1	$g(1) = 2(1) - 3 = -1$	$(1, -1)$
2	$g(2) = 2(2) - 3 = 1$	$(2, 1)$

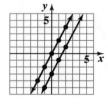

The graph of g is the graph of f shifted vertically down 3 units.

6. a. y is a function of x.

b. y is a function of x.

c. y is not a function of x. Two values of y correspond to an x-value.

7. a.   The concentration is increasing from 0 to 3 hours.

b.   The concentration is decreasing from 3 to 13 hours.

c.   The maximum concentration of 0.05 mg per 100 ml occurs after 3 hours.

d.   None of the drug is left in the body.

e.   The graph defines y as a function of x because no vertical line intersects the graph in more than one point.

Exercise Set 7.1

1.

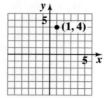

3.

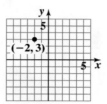

5.

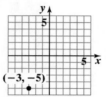

7.

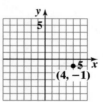

9.

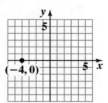

11.

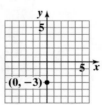

13.

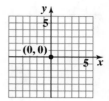

15.

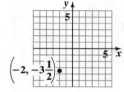

17.

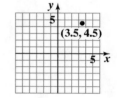

19.

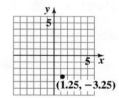

21.

x	-3	-2	-1	0	1	2	3
$y = x^2 - 2$	7	2	-1	-2	-1	2	7

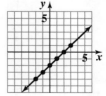

23.

x	-3	-2	-1	0	1	2	3
$y = x - 2$	-5	-4	-3	-2	-1	0	1

25.

x	-3	-2	-1	0	1	2	3
$y = 2x+1$	-5	-3	-1	1	3	5	7

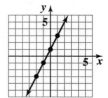

27.

x	-3	-2	-1	0	1	2	3
$y = -\frac{1}{2}x$	$\frac{3}{2}$	1	$\frac{1}{2}$	0	$-\frac{1}{2}$	-1	$-\frac{3}{2}$

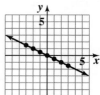

29.

x	-3	-2	-1	0	1	2	3
$y = x^3$	-27	-8	-1	0	1	8	27

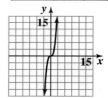

31.

x	-3	-2	-1	0	1	2	3
$y = \lvert x\rvert +1$	4	3	2	1	2	3	4

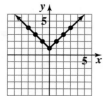

33. $f(x) = x - 4$

    a. $f(8) = 8 - 4 = 4$

    b. $f(1) = 1 - 4 = -3$

35. $f(x) = 3x - 2$

    a. $f(7) = 3(7) - 2 = 21 - 2 = 19$

    b. $f(0) = 3(0) - 2 = 0 - 2 = -2$

37. $f(x) = x^2 + 1$

    a. $f(2) = (2)^2 + 1 = 4 + 1 = 5$

    b. $f(-2) = (-2)^2 + 1 = 4 + 1 = 5$

39. $f(x) = 3x^2 + 5$

    a. $f(4) = 3(4)^2 + 5$
$$= 3(16) + 5$$
$$= 48 + 5$$
$$= 53$$

    b. $f(-1) = 3(-1)^2 + 5 = 3 + 5 = 8$

41. $f(x) = 2x^2 + 3x - 1$

    a. $f(3) = 2(3)^2 + 3(3) - 1$
$$= 2(9) + 9 - 1$$
$$= 18 + 9 - 1$$
$$= 26$$

    b. $f(-4) = 2(-4)^2 + 3(-4) - 1$
$$= 2(16) - 12 - 1$$
$$= 32 - 12 - 1$$
$$= 19$$

43.

x	$f(x) = x^2 - 1$
−2	3
−1	0
0	−1
1	0
2	3

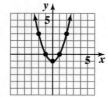

45.

x	$f(x) = x - 1$
−2	−3
−1	−2
0	−1
1	0
2	1

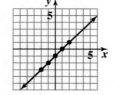

47.

x	$f(x) = (x-2)^2$
0	4
1	1
2	0
3	1
4	4

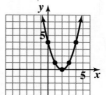

49.

x	$f(x) = x^3 + 1$
−3	−26
−2	−7
−1	0
0	1
1	2

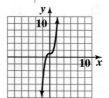

51. y is a function of x.

53. y is a function of x.

55. y is not a function of x. Two values of y correspond to an x-value.

57. y is a function of x.

59. $g(1) = 3(1) - 5 = 3 - 5 = -2$

$f(g(1)) = f(-2) = (-2)^2 - (-2) + 4$

$\qquad = 4 + 2 + 4 = 10$

61. $\sqrt{3-(-1)}-(-6)^2+6\div(-6)\cdot4$

$=\sqrt{3+1}-36+6\div(-6)\cdot4$

$=\sqrt{4}-36+-1\cdot4$

$=2-36+-4$

$=-34+-4$

$=-38$

63.

$y = 2x + 4$

65.

$y = 3 - x^2$

67. Point A is (91, 125). This means that in 1991, 125,000 acres were used for opium-poppy cultivation.

69. 200 thousand acres were used for opium-poppy cultivation in 2000 and 2003.

71. Opium cultivation was at a minimum in 2001 when approximately 25,000 acres were used.

73. Opium cultivation did not change between 1991 and 1992.

75. a. $W(20) = 13.2(20) + 443 = 707$. Approximately 707,000 bachelor's degrees were awarded to women in 2000. This is represented as (20, 707) on the graph.

b. $W(20) = 707$ is a slight overestimation of the actual data shown by the bar graph.

77. a. $W(10) - M(10) = [13.2(10) + 443] - [3.5(10) + 472] = 575 - 507 = 68$

Approximately 68,000 more bachelor's degrees were awarded to women than to men in 1990. The points on the graph with first coordinate 10 are 68 units apart.

b. $W(10) - M(10) = 68$ does not model the actual data shown by the bar graph very well. The actual difference shown in the graph is $555 - 500 = 55$.

79. a. The number of hate crimes reached a maximum in 2001. There were approximately 9200 hate crimes in 2001.

b. $f(1) = 339(1)^3 - 2085(1)^2 + 2814(1) + 8123 = 9191$ hate crimes

81. $f(2) = 339(2)^3 - 2085(2)^2 + 2814(2) + 8123 = 8123$ hate crimes
This is represented on the graph by the point (2, 8123).

89. $f(-1) + g(-1) = 1 + (-3) = -2$

91. $f(g(-1)) = f(-3) = 1$

Check Points 7.2

1. Find the *x*-intercept by setting $y = 0$
$$2x + 3(0) = 6$$
$$2x = 6$$
$$x = 3; \text{ resulting point } (3, 0)$$
Find the *y*-intercept by setting $x = 0$
$$2(0) + 3y = 6$$
$$3y = 6$$
$$y = 2; \text{ resulting point } (0, 2)$$
Find a checkpoint by substituting any value.
$$2(1) + 3y = 6$$
$$2 + 3y = 6$$
$$3y = 4$$
$$y = \frac{4}{3}; \text{ resulting point } \left(1, \frac{4}{3}\right)$$

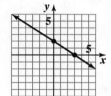

2. a. $m = \dfrac{-2 - 4}{-4 - (-3)} = \dfrac{-6}{-1} = 6$

b. $m = \dfrac{5 - (-2)}{-1 - 4} = \dfrac{7}{-5} = -\dfrac{7}{5}$

3. Step 1. Plot the *y*-intercept of (0, 1)

Step 2. Obtain a second point using the slope *m*.
$$m = \frac{3}{5} = \frac{\text{Rise}}{\text{Run}}$$
Starting from the *y*-intercept move up 3 units and move 5 units to the right. This puts the second point at (3, 6).

Step 3. Draw the line through the two points.

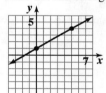

4. Solve for *y*.
$$3x + 4y = 0$$
$$4y = -3x + 0$$
$$\frac{4y}{4} = \frac{-3x}{4} + \frac{0}{4}$$
$$y = \frac{-3}{4}x + 0$$
$m = \dfrac{-3}{4}$ and the *y*-intercept is (0, 0)

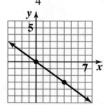

5. Draw horizontal line that intersects the y-axis at 3.

6. Draw vertical line that intersects the x-axis at -2.

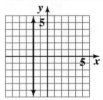

7. The two points shown on the line segment for men are (1990, 9.0) and (2003, 12.5).

$$m = \frac{12.5 - 9.0}{2003 - 1990} = \frac{3.5}{13} \approx 0.27$$

The slope indicates that the number of men living alone is projected to increase by 0.27 million each year.

8. Find slope by using the endpoints of the line segment.

$$m = \frac{3.9 - 1}{13 - 0} = \frac{2.9}{13} \approx 0.22$$

The value of b is the y-intercept, or 1.
Thus, $S(x) = mx + b$ becomes $S(x) = 0.22x + 1$.

Exercise Set 7.2

1. Find the x-intercept by setting $y = 0$
$x - y = 3$
$x - 0 = 3$
    $x = 3$; resulting point (3, 0)
Find the y-intercept by setting $x = 0$
$0 - y = 3$
  $-y = 3$
    $y = -3$; resulting point (0, -3)

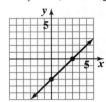

3. Find the x-intercept by setting $y = 0$
$3x - 4(0) = 12$
  $3x = 12$
    $x = 4$; resulting point (4, 0)
Find the y-intercept by setting $x = 0$
$3(0) - 4y = 12$
  $-4y = 12$
    $y = -3$; resulting point (0, -3)

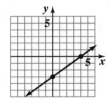

5. Find the x-intercept by setting $y = 0$
$2x + 0 = 6$
  $2x = 6$
    $x = 3$; resulting point (3, 0)
Find the y-intercept by setting $x = 0$
$2(0) + y = 6$
    $y = 6$; resulting point (0, 6)

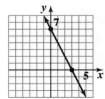

7. Find the x-intercept by setting $y = 0$

$$5x = 3(0) - 15$$
$$5x = -15$$
$$x = -3; \text{ resulting point } (-3, 0)$$

Find the y-intercept by setting $x = 0$

$$5(0) = 3y - 15$$
$$0 = 3y - 15$$
$$-3y = -15$$
$$y = 5; \text{ resulting point } (0, 5)$$

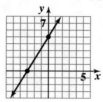

9. $m = \dfrac{5 - 6}{3 - 2} = \dfrac{-1}{1} = -1$; line falls.

11. $m = \dfrac{2 - 1}{2 - (-2)} = \dfrac{1}{4}$; line rises.

13. $m = \dfrac{-1 - 4}{-1 - (-2)} = \dfrac{-5}{1} = -5$; line falls.

15. $m = \dfrac{-2 - 3}{5 - 5} = \dfrac{-5}{0}$;

Slope undefined. Line is vertical.

17. $m = \dfrac{8 - 0}{0 - 2} = \dfrac{8}{-2} = -4$; line falls.

19. $m = \dfrac{1 - 1}{-2 - 5} = \dfrac{0}{-7} = 0$; line is horizontal.

21. $y = 2x + 3$

Slope: 2, y-intercept: 3

Plot point $(0, 3)$ and second point using

$$m = \frac{2}{1} = \frac{\text{rise}}{\text{run}}$$

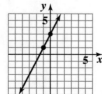

23. $y = -2x + 4$

Slope: -2, y-intercept: 4

Plot point $(0, 4)$ and second point using

$$m = \frac{-2}{1} = \frac{\text{rise}}{\text{run}}$$

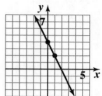

25. $y = \dfrac{1}{2}x + 3$

Slope: $\dfrac{1}{2}$, y-intercept: 3

Plot point $(0, 3)$ and second point using

$$m = \frac{1}{2} = \frac{\text{rise}}{\text{run}}.$$

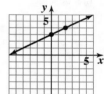

27. $f(x) = \dfrac{2}{3}x - 4$

Slope: $\dfrac{2}{3}$, y-intercept: -4

Plot point $(0, -4)$ and second point using

$$m = \frac{2}{3} = \frac{\text{rise}}{\text{run}}.$$

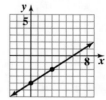

29. $y = -\dfrac{3}{4}x + 4$

Slope: $-\dfrac{3}{4}$, y-intercept: 4

Plot point (0, 4) and second point using

$m = \dfrac{-3}{4} = \dfrac{\text{rise}}{\text{run}}$.

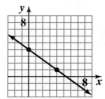

31. $f(x) = -\dfrac{5}{3}x$ or $f(x) = -\dfrac{5}{3}x + 0$

Slope: $-\dfrac{5}{3}$, y-intercept: 0

Plot point (0, 0) and second point using

$m = \dfrac{-5}{3} = \dfrac{\text{rise}}{\text{run}}$.

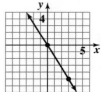

33. a. $3x + y = 0$

$y = -3x$ or $y = -3x + 0$

b. Slope $= -3$
y-intercept $= 0$

c.

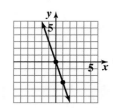

35. a. $3y = 4x$

$y = \dfrac{4}{3}x$ or $y = \dfrac{4}{3}x + 0$

b. Slope $= \dfrac{4}{3}$
y-intercept $= 0$

c.

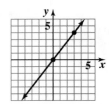

37. a. $2x + y = 3$

$y = -2x + 3$

b. Slope $= -2$
y-intercept $= 3$

c.

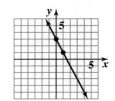

39. a. $7x + 2y = 14$

$2y = -7x + 14$

$y = -\dfrac{7}{2}x + 7$

b. Slope $= -\dfrac{7}{2}$
y-intercept $= 7$

c.

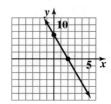

41. $y = 4$

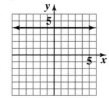

43. $y = -2$

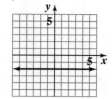

45. $x = 2$

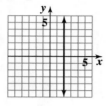

47. $x + 1 = 0$ or $x = -1$

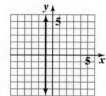

49. $m = \dfrac{0-a}{b-0} = \dfrac{-a}{b} = -\dfrac{a}{b}$

Since a and b are both positive, $-\dfrac{a}{b}$ is

negative. Therefore, the line falls.

51. $m = \dfrac{(b+c)-b}{a-a} = \dfrac{c}{0}$

The slope is undefined.
The line is vertical.

53. $Ax + By = C$

$By = -Ax + C$

$y = -\dfrac{A}{B}x + \dfrac{C}{B}$

The slope is $-\dfrac{A}{B}$ and the $y-$ intercept is $\dfrac{C}{B}$.

55.
$-3 = \dfrac{4-y}{1-3}$

$-3 = \dfrac{4-y}{-2}$

$6 = 4 - y$

$2 = -y$

$-2 = y$

57. $m_1,\ m_3,\ m_2,\ m_4$

59. a. $m = \dfrac{1.36-1.51}{2002-2000} \approx -0.075$

The slope indicates that between 2000 and
2002, the average priced decreased by 7.5 cents
per year.

b. $m = \dfrac{2.08-1.36}{2005-2002} \approx 0.24$

The slope indicates that between 2002 and
2005, the average priced increased by 24 cents
per year.

61. a. The y-intercept is 33. In 1999 (0 years after
1999), there were 33 million HIV/AIDS cases.

b. $m = \dfrac{41-33}{4-0} = 2$

The number of HIV/AIDS cases increased by 2
million each year.

c. $y = mx + b$

$y = 2x + 33$

d. $y = 2(11) + 33 = 55$ million HIV/AIDS cases

73. First, find the slope using the points
$(0, 32)$ and $(100, 212)$.

$m = \dfrac{212-32}{100-0} = \dfrac{180}{100} = \dfrac{9}{5}$

The slope is $\dfrac{9}{5}$. We also know that the y-

intercept is 32. Using slope-intercept form, we

have $F = \dfrac{9}{5}C + 32$.

Check Points 7.3

1. Replace x with -4 and y with 3.

$$x + 2y = 2 \qquad\qquad x - 2y = 6$$
$$-4 + 2(3) = 2 \qquad -4 - 2(3) = 6$$
$$-4 + 6 = 2 \qquad\qquad -4 - 6 = 6$$
$$2 = 2 \ \text{true} \qquad\quad -10 = 6 \ \text{false}$$

The pair $(-4, 3)$ does not satisfy both equations. Therefore it is not a solution of the system.

2.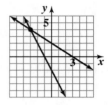

Check coordinates of intersection:

$$2x + 3y = 6 \qquad\qquad 2x + y = -2$$
$$2(-3) + 3(4) = 6 \qquad 2(-3) + (4) = -2$$
$$-6 + 12 = 6 \qquad\qquad -6 + 2 = -2$$
$$6 = 6, \ \text{true} \qquad\quad -2 = -2, \text{true}$$

The solution set is $\{(-3, 4)\}$.

3. Step 1. Solve one of the equations for one variable: $y = 3x - 7$

   Step 2. Substitute into the other equation:
   $$5x - 2y = 8$$
   $$5x - 2\overbrace{(3x - 7)}^{y} = 8$$

   Step 3. Solve: $5x - 2(3x - 7) = 8$
   $$5x - 6x + 14 = 8$$
   $$-x + 14 = 8$$
   $$-x = -6$$
   $$x = 6$$

   Step 4. Back-substitute the obtained value into the equation from step 1:
   $$y = 3x - 7$$
   $$y = 3(6) - 7$$
   $$y = 11$$

   Step 5. Check (6, 11) in both equations:
   $$y = 3x - 7 \qquad\qquad 5x - 2y = 8$$
   $$11 = 3(6) - 7 \qquad 5(6) - 2(11) = 8$$
   $$11 = 11, \text{true} \qquad\qquad 8 = 8, \text{true}$$

   The solution set is $\{(6, 11)\}$.

4. Step 1.   Solve one of the equations for one variable:

$x - y = 3$

$x = y + 3$

Step 2.   Substitute into the other equation:

$3x + 2y = -1$

$$3\overbrace{(y+3)}^{x} + 2y = -1$$

Step 3.   Solve: $3(y+3) + 2y = -1$

$3y + 9 + 2y = -1$

$5y + 9 = -1$

$5y = -10$

$y = -2$

Step 4.   Back-substitute the obtained value into the equation from step 1:

$x = y + 3$

$x = -2 + 3$

$x = 1$

Step 5.   Check $(1, -2)$ in both equations:

$$x - y = 3 \qquad\qquad 3x + 2y = -1$$
$$1 - (-2) = 3 \qquad 3(1) + 2(-2) = -1$$
$$3 = 3, \text{ true} \qquad\qquad -1 = -1, \text{ true}$$

The solution set is $\{(1, -2)\}$.

5. Rewrite one or both equations:

$$4x + 5y = 3 \xrightarrow{\text{No change}} \quad 4x + 5y = \;\; 3$$
$$2x - 3y = 7 \xrightarrow{\text{Mult. by } -2} \underline{-4x + 6y = -14}$$
$$11y = -11$$
$$y = -1$$

Back-substitute into either equation:

$4x + 5y = 3$

$4x + 5(-1) = 3$

$4x - 5 = 3$

$4x = 8$

$x = 2$

Checking confirms the solution set is $\{(2, -1)\}$.

6. Rewrite both equations in the form $Ax + By = C$:

$3x = 2 - 4y \quad \rightarrow \quad 3x + 4y = 2$

$5y = -1 - 2x \quad \rightarrow \quad 2x + 5y = -1$

Rewrite with opposite coefficients, then add and solve:

$$3x + 4y = 2 \quad \xrightarrow{\text{Mult. by 2}} \quad 6x + 8y = 4$$
$$2x + 5y = -1 \quad \xrightarrow{\text{Mult. by } -3} \quad \underline{-6x - 15y = 3}$$
$$-7y = 7$$
$$y = -1$$

Back-substitute into either equation:

$$3x = 2 - 4y$$
$$3x = 2 - 4(-1)$$
$$3x = 6$$
$$x = 2$$

Checking confirms the solution set is $\{(2, -1)\}$.

7. Rewrite with a pair of opposite coefficients, then add:

$$x + 2y = 4 \quad \xrightarrow{\text{Mult. by } -3} \quad -3x - 6y = -12$$
$$3x + 6y = 13 \quad \xrightarrow{\text{No change}} \quad \underline{3x + 6y = 13}$$
$$0 = 1$$

The statement $0 = 1$ is false which indicates that the system has no solution. The solution set is the empty set, $\varnothing$.

8. Substitute $4x - 4$ for y in the other equation:

$$8x - 2\overbrace{(4x - 4)}^{y} = 8$$
$$8x - 8x + 8 = 8$$
$$8 = 8$$

The statement $8 = 8$ is true which indicates that the system has infinitely many solutions. The solution set is $\{(x, y) | y = 4x - 4\}$ **or** $\{(x, y) | 8x - 2y = 8\}$.

9. **a.** $C(x) = 300,000 + 30x$

   b. $R(x) = 80x$

   c. $R(x) = C(x)$
   $$80x = 300,000 + 30x$$
   $$50x = 300,000$$
   $$x = 6000$$
   $$C(6000) = 300,000 + 30(6000) = 480,000$$
   Break even point (6000, 480000)
   The company will need to make 6000 pairs of shoes and earn \$480,000 to break even.

Exercise Set 7.3

1. Replace x with 2 and y with 3.

$x + 3y = 11$	$x - 5y = -13$
$2 + 3(3) = 11$	$2 - 5(3) = 13$
$2 + 9 = 11$	$2 - 15 = 13$
$11 = 11$, true	$13 = 13$, true

   The pair (2, 3) is a solution of the system.

3. Replace x with 2 and y with 5.
   $$2x + 3y = 17$$
   $$2(2) + 3(5) = 17$$
   $$4 + 15 = 17$$
   $$19 = 17, \text{ false.}$$
   The pair (2, 5) is not a solution of the system.

5.

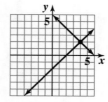

Check coordinates of intersection:

$x + y = 6$ $x - y = 2$

$4 + 2 = 6$ $4 - 2 = 2$

     $6 = 6$, true $2 = 2$, true

The solution set is $\{(4, 2)\}$.

7.

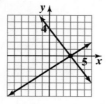

Check coordinates of intersection:

$2x - 3y = 6$ $4x + 3y = 12$

$2(3) - 3(0) = 6$ $4(3) + (0) = 12$

     $6 = 6$, true $12 = 12$, true

The solution set is $\{(3, 0)\}$.

9.

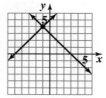

Check coordinates of intersection:

$y = x + 5$ $y = -x + 3$

$4 = -1 + 5$ $4 = -(-1) + 3$

$4 = 4$, true $4 = 4$, true

The solution set is $\{(-1, 4)\}$.

11.

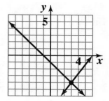

Check coordinates of intersection:

$y = -x - 1$ $4x - 3y = 24$

$-4 = -(3) - 1$ $4(3) - 3(-4) = 24$

$-4 = -4$, true $24 = 24$, true

The solution set is $\{(3, -4)\}$.

13. $y = 3x$ $x + y = 4$

$x + 3x = 4$

     $4x = 4$

       $x = 1$

$y = 3(1) = 3$

The proposed solution is $(1, 3)$

Check: $3 = 3(1)$ $1 + 3 = 4$

        $3 = 3$, true $4 = 4$, true

The pair $(1, 3)$ satisfies both equations.

The system's solution set is $\{(1, 3)\}$.

15. $y = 2x - 9$ $x + 3y = 8$

$x + 3(2x - 9) = 8$

   $x + 6x - 27 = 8$

        $7x = 35$

         $x = 5$

$y = 2(5) - 9 = 1$

The proposed solution is $(5, 1)$.

Check:

$1 = 2(5) - 9$ $5 + 3(1) = 8$

$1 = 10 - 9$ $5 + 3 = 8$

$1 = 1$, true $8 = 8$, true

The pair $(5, 1)$ satisfies both equations.

The system's solution set is $\{(5, 1)\}$.

17. $x + 3y = 5$

$x = 5 - 3y \quad 4x + 5y = 13$

$4(5 - 3y) + 5y = 13$

$20 - 12y + 5y = 13$

$20 - 7y = 13$

$-7y = -7$

$y = 1$

$x = 5 - 3(1) = 2$

The proposed solution is (2, 1).

Check:

$2 + 3(1) = 5 \qquad 4(2) + 5(1) = 13$

$5 = 5$, true $\qquad 8 + 5 = 13$

$13 = 13$, true

The pair (2, 1) satisfies both equations.

The system's solution set is $\{(2, 1)\}$.

19. $2x - y = -5$

$y = 2x + 5 \quad x + 5y = 14$

$x + 5(2x + 5) = 14$

$x + 10x + 25 = 14$

$11x = -11$

$x = -1$

$y = 2(-1) + 5 = -2 + 5 = 3$

The proposed solution is (−1, 3).

Check:

$2(-1) - 3 = -5 \qquad -1 + 5(3) = 14$

$-2 - 3 = -5 \qquad -1 + 15 = 14$

$-5 = -5$, true $\qquad 14 = 14$, true

The pair (− 1, 3) satisfies both equations.

The system's solution set is $\{(-1, 3)\}$.

21. $2x - y = 3$

$y = 2x - 3 \quad 5x - 2y = 10$

$5x - 2(2x - 3) = 10$

$5x - 4x + 6 = 10$

$x = 4$

$y = 2(4) - 3 = 8 - 3 = 5$

The proposed solution is (4, 5).

Check:

$2(4) - 5 = 3 \qquad 5(4) - 2(5) = 10$

$8 - 5 = 3 \qquad 20 - 10 = 10$

$3 = 3$, true $\qquad 10 = 10$, true

The pair (4, 5) satisfies both equations.

The system's solution set is $\{(4, 5)\}$.

23. $x + 8y = 6$

$x = 6 - 8y \quad 2x + 4y = -3$

$2(6 - 8y) + 4y = -3$

$12 - 16y + 4y = -3$

$-12y = -15$

$\dfrac{-12y}{-12} = \dfrac{-15}{-12}$

$y = \dfrac{15}{12} = \dfrac{5}{4}$

$x = 6 - 8\left(\dfrac{5}{4}\right) = 6 - 10 = -4$

The proposed solution is $\left(-4, \dfrac{5}{4}\right)$.

Check:

$-4 + 8\left(\dfrac{5}{4}\right) = 6 \quad 2(-4) + 4\left(\dfrac{5}{4}\right) = -3$

$-4 + 10 = 6 \qquad -8 + 5 = -3$

$6 = 6$, true $\qquad -3 = -3$, true

The pair $\left(-4, \dfrac{5}{4}\right)$ satisfies both equations.

The system's solution set is $\left\{\left(-4, \dfrac{5}{4}\right)\right\}$.

25. $x + y = 1$

$\dfrac{x - y = 3}{2x = 4}$

$x = 2$

$x + y = 1$

$2 + y = 1$

$y = -1$

Check: $2 + (-1) = 1 \quad 2 - (-1) = 3$

$1 = 1$, true $\qquad 3 = 3$, true

The solution set is $\{(2, -1)\}$.

27. $2x+3y=6$
$\underline{2x-3y=6}$
$4x=12$
$x=3$
$2x+3y=6$
$2\cdot3+3y=6$
$6+3y=6$
$3y=0$
$y=0$
Check:
$2(3)+3(0)=6 \qquad 2(3)-3(0)=6$
$6+0=6 \qquad\qquad 6-0=6$
$6=6$, true $\qquad\qquad 6=6$, true
The solution set is $\{(3, 0)\}$.

29. $x+2y=2$ Mult. by 3. $3x+6y=6$
$-4x+3y=25$ Mult. by -2. $\underline{8x-6y=-50}$
$11x=-44$
$x=-4$
$x+2y=2$
$-4+2y=2$
$2y=6$
$y=3$
Check:
$-4+2(3)=2 \qquad -4(-4)+3(3)=25$
$-4+6=2 \qquad\qquad 16+9=25$
$2=2$, true $\qquad\qquad 25=25$, true
The solution set is $\{(-4, 3)\}$.

31. $4x+3y=15$ Mult. by 5. $20x+15y=75$
$2x-5y=1$ Mult. by 3. $\underline{6x-15y=3}$
$26x=78$
$x=3$
$4x+3y=15$
$4\cdot3+3y=15$
$12+3y=15$
$3y=3$
$y=1$
Check:
$4(3)+3(1)=15 \qquad 2(3)-5(1)=1$
$12+3=15 \qquad\qquad 6-5=1$
$15=15$, true $\qquad 1=1$, true
The solution set is $\{(3, 1)\}$.

33. $3x-4y=11$ Mult. by 3. $9x-12y=33$
$2x+3y=-4$ Mult. by 4. $\underline{8x+12y=-16}$
$17x=17$
$x=1$
$2x+3y=-4$
$2\cdot1+3y=-4$
$2+3y=-4$
$3y=-6$
$y=-2$
Check:
$3(1)-4(-2)=11$
$3+8=11$
$11=11$, true
$2(1)+3(-2)=-4$
$2-6=-4$
$-4=-4$, true
The solution set is $\{(1, -2)\}$.

35. $2x=3y-4$ Rearrange and Mult. by 3. $6x-9y=-12$
$-6x+12y=6$ No change. $\underline{-6x+12y=6}$
$3y=-6$
$y=-2$
$2x=3y-4$
$2x=3(-2)-4$
$2x=-6-4$
$2x=-10$
$x=-5$
Check:
$2(-5)=3(-2)-4 \qquad -6(-5)+12(-2)=6$
$-10=-6-4 \qquad\qquad 30-24=6$
$-10=-10$, true $\qquad 6=6$, true
The solution set is $\{(-5, -2)\}$.

37. $x=9-2y$ $x+2y=13$
$(9-2y)+2y=13$
$9=13$ false
The system has no solution.
The solution set is the empty set, $\varnothing$.

39. $y = 3x - 5$ $21x - 35 = 7y$

$\qquad 21x - 35 = 7(3x - 5)$

$\qquad 21x - 35 = 21x - 35$

$\qquad 21x - 21x = 35 - 35$

$\qquad\qquad\quad 0 = 0,$ true

The system has infinitely many solutions.

The solution set is $\{(x, y) | y = 3x - 5\}$.

41. $3x - 2y = -5$ No change. $3x - 2y = -5$

$\quad\ 4x + y = 8$ Mult. by 2. $\underline{8x + 2y = 16}$

$\qquad\qquad\qquad\qquad\qquad\qquad\quad 11x = 11$

$\qquad\qquad\qquad\qquad\qquad\qquad\qquad x = 1$

$\quad 4x + y = 8$

$\quad 4(1) + y = 8$

$\qquad\qquad y = 4$

Check:

$3(1) - 2(4) = -5 \qquad 4(1) + (4) = 8$

$\quad\ 3 - 8 = -5 \qquad\qquad 4 + 4 = 8$

$\qquad\quad -5 = -5,$ true $\qquad\quad 8 = 8,$ true

The solution set is $\{(1, 4)\}$.

43. $x + 3y = 2$

$\quad x = 2 - 3y \qquad 3x + 9y = 6$

$\qquad\qquad\qquad 3(2 - 3y) + 9y = 6$

$\qquad\qquad\qquad\quad 6 - 9y + 9y = 6$

$\qquad\qquad\qquad\qquad\qquad 6 = 6$ true

The system has infinitely many solutions.

The solution set is $\{(x, y) | x + 3y = 2\}$.

45. The solution to a system of linear equations is the point of intersection of the graphs of the equations in the system. If $(6, 2)$ is a solution, then we need to find the lines that intersect at that point.

Looking at the graph, we see that the graphs of $x + 3y = 12$ and $x - y = 4$ intersect at the point $(6, 2)$. Therefore, the desired system of equations is

$x + 3y = 12$ or $y = -\dfrac{1}{3}x + 4$

$x - y = 4 \qquad\qquad y = x - 4$

47. $5ax + 4y = 17$

$\quad\ ax + 7y = 22$

Multiply the second equation by -5 and add the equations.

$\qquad 5ax + 4y = 17$

$\underline{\ -5ax - 35y = -110}$

$\qquad\qquad -31y = -93$

$\qquad\qquad\qquad y = 3$

Back-substitute into one of the original equations to solve for x.

$\qquad ax + 7y = 22$

$\qquad ax + 7(3) = 22$

$\qquad ax + 21 = 22$

$\qquad\qquad ax = 1$

$\qquad\qquad\ x = \dfrac{1}{a}$

The solution is $\left(\dfrac{1}{a}, 3\right)$.

49. $f(-2) = 11 \quad \rightarrow \quad -2m + b = 11$

$\quad\ f(3) = -9 \quad \rightarrow \quad 3m + b = -9$

We need to solve the resulting system of equations:

$-2m + b = 11$

$\quad 3m + b = -9$

Subtract the two equations:

$\quad -2m + b = 11$

$\underline{\quad\ 3m + b = -9}$

$\qquad\ -5m = 20$

$\qquad\qquad m = -4$

Back-substitute into one of the original equations to solve for b.

$\qquad -2m + b = 11$

$\quad -2(-4) + b = 11$

$\qquad\quad 8 + b = 11$

$\qquad\qquad\quad b = 3$

Therefore, $m = -4$ and $b = 3$.

51. At the break-even point, $R(x) = C(x)$.

$10000 + 30x = 50x$

$\qquad 10000 = 20x$

$\qquad 10000 = 20x$

$\qquad\quad 500 = x$

Five hundred radios must be produced and sold to break-even.

53. $R(x) = 50x$

$R(200) = 50(200) = 10000$

$C(x) = 10000 + 30x$

$C(200) = 10000 + 30(200)$

$\quad = 10000 + 6000 = 16000$

$R(200) - C(200) = 10000 - 16000$

$\quad\quad = -6000$

This means that if 200 radios are produced and sold the company will lose $6,000.

55. a. $P(x) = R(x) - C(x)$

$\quad = 50x - (10000 + 30x)$

$\quad = 50x - 10000 - 30x$

$\quad = 20x - 10000$

$P(x) = 20x - 10000$

b. $P(10000) = 20(10000) - 10000$

$\quad = 200000 - 10000 = 190000$

If 10,000 radios are produced and sold the profit will be $190,000.

57. a. The cost function is:

$C(x) = 18,000 + 20x$

b. The revenue function is:

$R(x) = 80x$

c. At the break-even point, $R(x) = C(x)$.

$80x = 18000 + 20x$

$60x = 18000$

$x = 300$

$R(x) = 80x$

$R(300) = 80(300)$

$\quad = 24,000$

When approximately 300 canoes are produced the company will break-even with cost and revenue at $24,000.

59. a. The cost function is:

$C(x) = 30000 + 2500x$

b. The revenue function is:

$R(x) = 3125x$

c. At the break-even point, $R(x) = C(x)$.

$3125x = 30000 + 2500x$

$625x = 30000$

$x = 48$

After 48 sold out performances, the investor will break-even. ($150,000)

61. a. $m = \dfrac{25.3 - 38}{17 - 0} = \dfrac{-12.7}{17} \approx -0.75$

From the point $(0, 38)$ we have that the y-intercept is $b = 38$. Therefore, the equation of the line is $y = -0.75x + 38$.

b. $m = \dfrac{23 - 40}{17 - 0} = \dfrac{-17}{17} = -1$

From the point $(0, 40)$ we have that the y-intercept is $b = 40$. Therefore, the equation of the line is $y = -x + 40$.

c. To find the year when cigarette use is the same, we set the two equations equal to each other and solve for x.

$-0.75x + 38 = -x + 40$

$0.25x + 38 = 40$

$0.25x = 2$

$x = \dfrac{2}{0.25} = 8$

$y = -(8) + 40 = 32$

Cigarette use was the same for African Americans and Hispanics in 1993 (8 years after 1985). At that time, 32% of each group used cigarettes.

63. $x + 2y = 1980$

$2x + y = 2670$

Multiply the first equation by -2 and add to the second equation. Solve for y.

$-2x - 4y = -3960$

$2x + y = 2670$

$-3y = -1290$

$y = 430$

Substitute 430 for y in the second equation and solve for x.

$2x + 430 = 2670$

$2x = 2240$

$x = 1120$

There are 1120 calories in a pan pizza and 430 calories in a beef burrito

65. $x + y = 300 + 241$ or $x + y = 541$
$2x + 3y = 1257$
Multiply the first equation by -2 and add to the second equation. Solve for y.
$-2x - 2y = -1082$
$2x + 3y = 1257$
$y = 175$
Substitute 175 for y in the first equation and solve for x.
$x + 175 = 541$
$x = 366$
There are 366 mg in scrambled eggs and 175 mg in a Double Beef Whooper.

67. $x + y = 200$
$100x + 80y = 17000$
Multiply the first equation by -100 and add to the second equation. Solve for y.
$-100x - 100y = -20000$
$100x + 80y = 17000$
$-20y = -3000$
$y = 150$
Substitute 150 for y in the first equation and solve for x.
$x + 150 = 200$
$x = 50$
There are 50 rooms with kitchenettes and 150 rooms without.

79. x = number of people upstairs
y = number of people downstairs
The following system results:
$x - 1 = y + 1$
$x + 1 = 2y$
Eliminate x by multiplying the first equation by -1 and adding the resulting equations.

$-x + 1 = -y - 1$
$x + 1 = 2y$
$2 = y - 1$

Which gives $y = 3$. Thus $x = 5$.

Check Points 7.4

1. $2x - 4y \geq 8$
Graph the equation $2x - 4y = 8$ as a solid line.
Choose a test point that is not on the line.
Test $(0,0)$
$2x - 4y \geq 8$
$2(0) - 4(0) \geq 8$
$0 \geq 8$, false
Since the statement is false, shade the other half-plane.

2. $y > -\dfrac{3}{4}x$

Graph the equation $y = -\dfrac{3}{4}x$ as a dashed line.
Choose a test point that is not on the line.
Test $(1,1)$
$y > -\dfrac{3}{4}x$
$1 > -\dfrac{3}{4}(1)$
$1 > -\dfrac{3}{4}$, true
Since the statement is true, shade the half-plane containing the point.

$y > -\dfrac{3}{4}x$

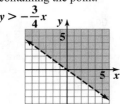

3. a. $y > 1$

Graph the equation $y = 1$ as a dashed line.
Choose a test point that is not on the line.

Test $(0, 0)$

$y > 1$

$0 > 1$, false

Since the statement is false, shade the other half-plane.

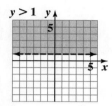

$y > 1$

b. Graph the equation $x = -2$ as a solid line.
Choose a test point that is not on the line.

Test $(0, 0)$

$x \leq -2$

$0 \leq -2$, false

Since the statement is false, shade the other half-plane.

$x \leq -2$

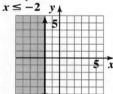

4. Test $T = 60$ and $P = 20$ in each inequality.

$T \geq 35$

$60 \geq 35$ True

$5T - 7P \geq 70$

$5(60) - 7(20) \geq 70$

$160 \geq 70$ True

$3T - 35P \leq -140$

$3(60) - 35(20) \leq -140$

$-520 \leq -140$ True

The point (60, 20) checks in all three equations.

5. $x + 2y > 4$

$2x - 3y \leq -6$

Graph the equation $x + 2y = 4$ as a dashed line.
Choose a test point that is not on the line.

Test $(0, 0)$

$x + 2y > 4$

$0 + 2(0) > 4$

$0 > 4$, false

Since the statement is false, shade the other half-plane.
Next, graph the equation $2x - 3y = -6$ as a solid line.
Choose a test point that is not on the line.

Test $(0, 0)$

$2x - 3y \leq -6$

$2(0) - 3(0) \leq -6$

$0 \leq -6$, false

Since the statement is false, shade the other half-plane.
The graph is the intersection (overlapping) of the two half-planes.

6. $x < 3$

$y \geq -1$

Graph the equation $x = 3$ as a dashed line.
Choose a test point that is not on the line.

Test $(0, 0)$

$x < 3$

$0 < 3$, true

Since the statement is true, shade the half-plane that contains the test point.
Next, graph the equation $y = -1$ as a solid line.

Choose a test point that is not on the line.

Test $(0, 0)$

$y \geq -1$

$0 \geq -1$, true

Since the statement is true, shade the half-plane that contains the test point.
The graph is the intersection (overlapping) of the two half-planes.

Exercise Set 7.4

1. To graph $x + y \geq 2$, begin by graphing $x + y = 2$ with a solid line because $\geq$ includes equality.
test point $(0, 0)$:

$x + y \geq 2$

$0 + 0 \geq 2$

$0 \geq 2$, false

Since the test point makes the inequality <u>false</u>, shade the half-plane <u>not containing</u> test point $(0, 0)$.

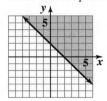

3. To graph $3x - y \geq 6$, begin by graphing $3x - y = 6$ with a solid line because $\geq$ includes equality.
test point $(0, 0)$:

$3x - y \geq 6$

$3(0) - 0 \geq 6$

$0 \geq 6$, false

Since the test point makes the inequality <u>false</u>, shade the half-plane <u>not containing</u> test point $(0, 0)$.

5. To graph $2x + 3y > 12$, begin by graphing $2x + 3y = 12$ with a dashed line because $>$ does not include equality.
test point $(0, 0)$:

$2x + 3y > 12$

$2(0) + 3(0) > 12$

$0 > 12$, false

Since the test point makes the inequality <u>false</u>, shade the half-plane <u>not containing</u> test point $(0, 0)$.

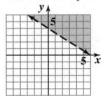

7. To graph $5x + 3y \leq -15$, begin by graphing $5x + 3y = -15$ with a solid line because $\leq$ includes equality.
test point $(0, 0)$:

$5x + 3y \leq -15$

$5(0) + 3(0) \leq -15$

$0 \leq -15$, false

Since the test point makes the inequality <u>false</u>, shade the half-plane <u>not containing</u> test point $(0, 0)$.

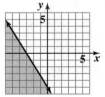

9. To graph $2y - 3x > 6$, begin by graphing
$2y - 3x = 6$ with a dashed line because $>$ does not
include equality.

test point $(0, 0)$:

$2y - 3x > 6$

$2(0) - 3(0) > 6$

$0 > 6$, false

Since the test point makes the inequality <u>false</u>,
shade the half-plane <u>not containing</u> test point $(0, 0)$.

11.
$$y > \frac{1}{3}x$$

Graph the equation $y = \frac{1}{3}$ with a dashed line.

Next, select a test point. We cannot use the
origin because it lies on the line. Use $(1,1)$.

$1 > \frac{1}{3}(1)$

$1 > \frac{1}{3}$

This is a true statement, so we know the point
$(1,1)$ lies in the shaded half-plane.

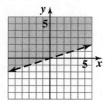

13. $y \le 3x + 2$

Graph the equation $y = 3x + 2$ with a solid line.
Next, use the origin as a test point.

$0 \le 3(0) + 2$

$0 \le 2$

This is a true statement. This means that the point
$(0,0)$ will fall in the shaded half-plane.

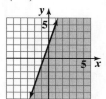

15.
$$y < -\frac{1}{4}x$$

Graph the equation $y = -\frac{1}{4}x$ with a dashed line.

Next, select a test point. We cannot use the origin
because it lies on the line. Use $(1,1)$.

$1 < -\frac{1}{4}(1)$

$1 < -\frac{1}{4}$

This is a false statement, so we know the point $(1,1)$
does not lie in the shaded half-plane.

17. $x \le 2$

Graph the equation $x = 2$ with a solid line.
Next, use the origin as a test point.

$x \le 2$

$0 \le 2$

This is a true statement, so we know the point $(0,0)$
lies in the shaded half-plane.

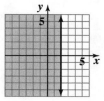

19. $y > -4$

Graph the equation $y = -4$ with a dashed line.

Next, use the origin as a test point.

$y > -4$

$0 > -4$

This is a true statement, so we know the point $(0,0)$ lies in the shaded half-plane.

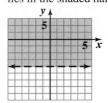

21. $y \geq 0$

Graph the equation $y = 0$ with a solid line.

Next, select a test point. We cannot use the origin because it lies on the line. Use $(1,1)$.

$y \geq 0$

$1 \geq 0$

This is a true statement, so we know the point $(1,1)$ lies in the shaded half-plane.

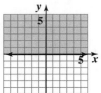

23. $3x + 6y \leq 6$

$2x + y \leq 8$

Graph the equations using the intercepts.

$3x + 6y = 6$ $2x + y = 8$

x – intercept = 2 x – intercept = 4

y – intercept = 1 y – intercept = 8

Use the origin as a test point to determine shading.

The solution set is the intersection of the shaded half-planes.

25. $2x + y < 3$

$x - y > 2$

Graph $2x + y = 3$ as a dashed line.

If $x = 0$, then $y = 3$ and if $y = 0$, then $x = \dfrac{3}{2}$.

Because (0, 0) makes the inequality true, shade the half-plane containing (0, 0).

Graph $x - y = 2$ as a dashed line.

If $x = 0$, then $y = -2$ and if $y = 0$, then $x = 2$.

Because (0, 0) makes the inequality false, shade the half-plane not containing (0, 0).

27. $2x + y < 4$

$x - y > 4$

Graph $2x + y = 4$ as a dashed line.

If $x = 0$, then $y = 4$ and if $y = 0$, then $x = 2$.

Because (0, 0) makes the inequality true, shade the half-plane containing (0, 0)

Graph $x - y = 4$ as a dashed line.

If $x = 0$, then $y = -4$. and if $y = 0$, then $x = 4$.

Because (0, 0) makes the inequality false, shade the half-plane not containing (0, 0).

29. $x \geq 2$

$y \leq 3$

Graph $x = 2$ as a solid line.

The points in the half-plane to the right of the line satisfy $x > 2$.

Graph $y = 3$ as a solid line.

The points in the half-plane below the line satisfy $y < 3$.

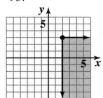

31. $x \le 5$
$y > -3$

Graph $x = 5$ as a solid line.
The points in the half-plane to the left of the line satisfy $x < 5$.
Graph $y = -3$ as a dashed line.
The points in the half-plane above the line satisfy $y > -3$.

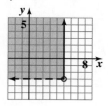

33. $x - y \le 1$
$x \ge 2$

Graph $x - y = 1$ as a solid line.
If $x = 0$, then $y = -1$ and if $y = 0$, then $x = 1$.
Because $(0, 0)$ satisfies the inequality, shade the half-plane containing $(0, 0)$.
Graph $x = 2$ as a solid line.
The points in the half-plane to the right of $x = 2$ satisfy the inequality $x \ge 2$.

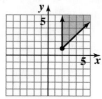

35. $y > 2x - 3$
$y < -x + 6$

Graph the equations using the intercepts.
$y = 2x - 3$ $\qquad$ $y = -x + 6$
$x-\text{intercept} = \dfrac{3}{2}$ $\quad$ $x-\text{intercept} = 6$
$y-\text{intercept} = -3$ $\qquad$ $y-\text{intercept} = 6$
Use the origin as a test point to determine shading.

The solution set is the intersection of the shaded half-planes.

37. $x + 2y \le 4$
$y \ge x - 3$

Graph the equations using the intercepts.
$x + 2y = 4$ $\qquad$ $y = x - 3$
$x-\text{intercept} = 4$ $\quad$ $x-\text{intercept} = 3$
$y-\text{intercept} = 2$ $\quad$ $y-\text{intercept} = -3$
Use the origin as a test point to determine shading.

The solution set is the intersection of the shaded half-planes.

39. $y \ge -2x + 4$

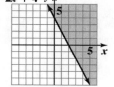

41. $x + y \le 4$
$3x + y \le 6$

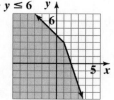

43. Find the union of solutions of
$y > \dfrac{3}{2}x - 2$ and $y < 4$.

45. a. The coordinates of point A are $(20,150)$.

This means that a 20 year-old person with a pulse rate of 150 beats per minute falls within the target zone.

b. $10 \le a \le 70$

$10 \le 20 \le 70$

    True

$150 \ge 0.7(220 - 20)$

      True

$150 \le 0.8(220 - 20)$

      True

Since point A makes all three inequalities true, it is a solution of the system.

47. $10 \le a \le 70$

$H \ge 0.6(220 - a)$

$H \le 0.7(220 - a)$

49. a. $50x + 150y > 2000$

b. Graph $50x + 150y$ as a dashed line using its x-intercept, $(40, 0)$, and its y-intercept, $\left(0, \dfrac{40}{3}\right)$.

Test $(0, 0)$:

$50(0) + 150(0) > 2000$?

      $0 > 2000$ false

Shade the half-plane not containing (0, 0).

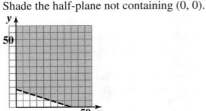

$50x + 150y > 2000$

c. Ordered pairs may vary.

51. a. $\text{BMI} = \dfrac{703W}{H^2} = \dfrac{703(200)}{72^2} \approx 27.1$

b. A 20 year old man with a BMI of 27.1 is classified as overweight.

59. $y > x - 3$

$y \le x$

61. The system $\begin{array}{l} 6x - y \le 24 \\ 6x - y > 24 \end{array}$ has no solution. The number $6x - y$ cannot both be less than or equal to 24 and greater than 24 at the same time.

63. The system $\begin{array}{l} 6x - y \le 24 \\ 6x - y \ge 24 \end{array}$ has infinitely many solutions. The solutions are all points on the line $6x - y = 24$.

Section 7.5

Check Point Exercises

1. The total profit is 25 times the number of bookshelves, x, plus 55 times the number of desks, y. The objective function is $z = 25x + 55y$

2. Not more than a total of 80 bookshelves and desks can be manufactured per day. This is represented by the inequality $x + y \le 80$.

3. Objective function: $z = 25x + 55y$

Constraints: $x + y \le 80$

$30 \le x \le 80$

$10 \le y \le 30$

4. Graph the constraints and find the corners, or vertices, of the region of intersection.

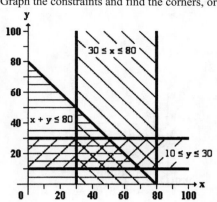
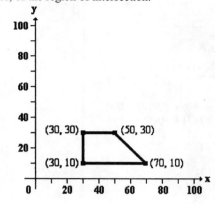

Find the value of the objective function at each corner of the graphed region.

Corner (x, y)	Objective Function $z = 25x + 55y$ z
(30, 10)	$z = 25(30) + 55(10)$ $= 750 + 550 = 1300$
(30, 30)	$z = 25(30) + 55(30)$ $= 750 + 1650 = 2400$
(50, 30)	$z = 25(50) + 55(30)$ $= 1250 + 1650 = 2900 \leftarrow$ Maximum
(70, 10)	$z = 25(70) + 55(10)$ $= 1750 + 550 = 2300$

The maximum value of z is 2900 and it occurs at the point (50, 30).
In order to maximize profit, 50 bookshelves and 30 desks must be produced each day for a profit of $2900.

Exercise Set 7.5

1. $z = 5x + 6y$

(1, 2): $5(1) + 6(2) = 5 + 12 = 17$
(2, 10): $5(2) + 6(10) = 10 + 60 = 70$
(7, 5): $5(7) + 6(5) = 35 + 30 = 65$
(8, 3): $5(8) + 6(3) = 40 + 18 = 58$
The maximum value is $z = 70$; the minimum value is $z = 17$.

3. $z = 40x + 50y$

(0, 0): $40(0) + 50(0) = 0 + 0 = 0$
(0, 8): $40(0) + 50(8) = 0 + 400 = 400$
(4, 9): $40(4) + 50(9) = 160 + 450 = 610$
(8, 0): $40(8) + 50(0) = 320 + 0 = 320$
The maximum value is $z = 610$; the minimum value is $z = 0$.

5. a.

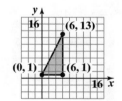

b. at (0, 1) $z = 0 + 1 = 1$
at (6, 13) $z = 6 + 13 = 19$
at (6, 1) $z = 6 + 1 = 7$

c. Maximum = 19
occurs at $x = 6$ and $y = 13$

7. a.

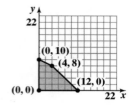

b. at $(0, 10)$ $z = 6(0) + 10(10) = 100$
at $(4, 8)$ $z = 6(4) + 10(8) = 104$
at $(12, 0)$ $z = 6(12) + 10(0) = 72$
at $(0, 0)$ $z = 6(0) + 10(0) = 0$

c. Maximum = 104
occurs at $x = 4$ and $y = 8$

9. $z = 5x - 2y$
$0 \leq x \leq 5$
$0 \leq y \leq 3$
$x + y \geq 2$

a.

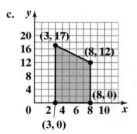

b. $(0, 3): z = 5(0) - 2(3) = -6$
$(0, 2): z = 5(0) - 2(2) = -4$
$(2, 0): z = 5(2) - 2(0) = 10$
$(5, 0): z = 5(5) - 2(0) = 25$
$(5, 3): z = 5(5) - 2(3) = 19$

c. The maximum value is 25 at $x = 5$ and $y = 0$.

11. $z = 10x + 12y$
$x \geq 0, y \geq 0$
$x + y \leq 7$
$2x + y \leq 10$
$2x + 3y \leq 18$

a.

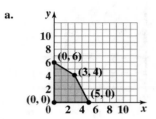

b. $(0, 6): z = 10(0) + 12(6) = 72$
$(0, 0): z = 10(0) + 12(0) = 0$
$(5, 0): z = 10(5) + 12(0) = 50$
$(3, 4): z = 10(3) + 12(4)$
$= 30 + 48 = 78$

c. The maximum value is 78 at $x = 3$ and $y = 4$.

13. a. Let $x =$ number of hours spent tutoring and $y =$ number of hours spent as a teacher's aid. The objective is to maximize $z = 10x + 7y$.

b. The constraints are:
$x + y \leq 20$
$x \geq 3$
$x \leq 8$

c.

d. $(3, 0): 10(3) + 7(0) = 30 + 0 = 30$
$(3, 17): 10(3) + 7(17) = 30 + 119 = 149$
$(8, 12): 10(8) + 7(12) = 80 + 84 = 164$
$(8, 0): 10(8) + 7(0) = 80 + 0 = 80$

e. The student can earn the maximum amount per week by tutoring for 8 hours a week and working as a teacher's aid for 12 hours a week. The maximum that the student can earn each week is $164.

15. Let x = the number of cartons of food and
y = the number of cartons of clothing.
The constraints are:
$20x + 10y \le 8,000$ or $2x + y \le 8000$
$50x + 20y \le 19,000$ or $5x + 2y \le 1900$
Graph these inequalities in the first quadrant, since x
and y cannot be negative.

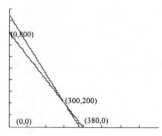

The quantity to be maximized is the number of
people helped, which is $12x + 5y$.
$(0, 0)$: $12(0) + 5(0) = 0 + 0 = 0$
$(0, 800)$: $12(0) + 5(800) = 0 + 4000 = 4000$
$(300, 200)$: $12(300) + 5(200) = 4600$
$(380, 0)$: $12(380) + 5(0) = 4500$
300 cartons of food and 200 cartons of clothing
should be shipped. This will help 4600 people.

17. Let x = number of students attending and
y = number of parents attending.
The constraints are
$x + y \le 150$
$2x \ge y$
or
$x + y \le 150$
$2x - y \ge 0$
Graph these inequalities in the first quadrant, since x
and y cannot be negative.

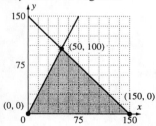

The quantity to be maximized is the amount of
money raised, which is $x + 2y$.
$(0, 0)$: $0 + 2(0) = 0 + 0 = 0$
$(50, 100)$: $50 + 2(100) = 50 + 200 = 250$
$(150, 0)$: $150 + 2(0) = 150 + 0 = 150$
50 students and 100 parents should attend.

Section 7.6

Check Point Exercises

1.

x	$f(x) = 3^x$
-2	$\dfrac{1}{9}$
-1	$\dfrac{1}{3}$
0	1
1	3
2	9

2. a. $f(x) = 0.073x + 2.316$
$f(21) = 0.073(21) + 2.316$
$f(21) \approx 3.8$

$g(x) = 2.569(1.017)^x$
$g(21) = 2.569(1.017)^{21}$
$g(21) \approx 3.7$
The exponential function, $g(x)$, serves as the
better model for this year.

b. $f(x) = 0.073x + 2.316$
$f(63) = 0.073(63) + 2.316$
$f(63) \approx 6.9$

$g(x) = 2.569(1.017)^x$
$g(63) = 2.569(1.017)^{63}$
$g(63) \approx 7.4$
$f(x)$ models this projection fairly well.
$g(x)$ overestimates this projection.

3. $R = 6e^{12.77x}$
$= 6e^{12.77(0.01)}$
$= 6.8\%$
The risk of a car accident with a blood alcohol
concentration of 0.01 is 6.8%.

4. $y = \log_3 x$ is equivalent to $x = 3^y$.

$x = 3^y$	y	(x, y)
$\dfrac{1}{9}$	-2	$\left(\dfrac{1}{9}, -2\right)$
$\dfrac{1}{3}$	-1	$\left(\dfrac{1}{3}, -1\right)$
1	0	$(1, 0)$
3	1	$(3, 1)$
9	2	$(9, 2)$

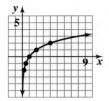

5. $f(x) = -11.6 + 13.4 \ln x$

$f(30) = -11.6 + 13.4 \ln 30$

$f(30) \approx 34°$

The function models the actual data extremely well.

6. Step 1. Since $a > 0$, the parabola opens upward $(a = 1)$.

Step 2. Find the vertex given $a = 1$ and $b = 6$.

x-coordinate of vertex $= \dfrac{-b}{2a} = \dfrac{-6}{2(1)} = \dfrac{-6}{2} = -3$

y-coordinate of vertex $= (-3)^2 + 6(-3) + 5 = 9 - 18 + 5 = -4$

Thus, the vertex is the point $(-3, -4)$.

Step 3. Replace y with 0 and solve the equation for x by factoring.

$x^2 + 6x + 5 = 0$

$(x + 5)(x + 1) = 0$

$x + 5 = 0$ or $x + 1 = 0$

$x = -5$ $x = -1$

Thus the x-intercepts are -5 and -1, , which are located at the points $(-5, 0)$ and $(-1, 0)$.

Step 4. Replace x with 0 and solve the equation for y.

$y = x^2 + 6x + 5$

$y = (0)^2 + 6(0) + 5$

$y = 5$

Thus the y-intercept is 5, which is located at the point $(0, 5)$.

Steps 5 and 6.   Plot the intercepts and the vertex. Connect these points with a smooth curve.

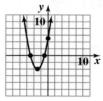

7. Use $x = \dfrac{-b}{2a}$ to find the age at which the minimum occurs.

$x = \dfrac{-(-36)}{2(0.4)} = 45$ years

$f(x) = 0.4x^2 - 36x + 1000$

$f(45) = 0.4(45)^2 - 36(45) + 1000$

$f(45) = 190$

The 45-year old age group has 190 accidents per 50 million miles driven.

Exercise Set 7.6

1.

x	$y = 4^x$
-2	$\dfrac{1}{16}$
-1	$\dfrac{1}{4}$
0	1
1	4
2	16

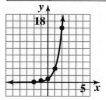

3.

x	$y = 2^{x+1}$
-2	$\dfrac{1}{2}$
-1	1
0	2
1	4
2	8

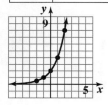

5.

x	$y = 3^{x-1}$
-2	$\dfrac{1}{27}$
-1	$\dfrac{1}{9}$
0	$\dfrac{1}{3}$
1	1
2	3

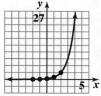

7. a. $y = \log_4 x$ is equivalent to $x = 4^y$.

b.

$x = 4^y$	y
$\dfrac{1}{16}$	-2
$\dfrac{1}{4}$	-1
1	0
4	1
16	2

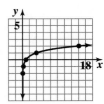

9. a. $a > 0$, thus the parabola opens upward.

b. x-coordinate: $x = \dfrac{-b}{2a} = \dfrac{-8}{2(1)} = -4$

y-coordinate: $y = x^2 + 8x + 7$
$= (-4)^2 + 8(-4) + 7$
$= -9$
vertex: $(-4, -9)$

c. x-intercepts: $y = x^2 + 8x + 7$
$0 = x^2 + 8x + 7$
$0 = (x + 7)(x + 1)$

$x + 7 = 0$ or $x + 1 = 0$
$x = -7$ $x = -1$

d. y-intercept: $y = x^2 + 8x + 7$
$y = 0^2 + 8(0) + 7$
$y = 7$

e.

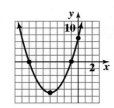

11. a. $a > 0$, thus the parabola opens upward.

b. x-coordinate: $x = \dfrac{-b}{2a} = \dfrac{-(-2)}{2(1)} = 1$

y-coordinate: $f(x) = x^2 - 2x - 8$
$f(1) = (1)^2 - 2(1) - 8$
$= -9$
vertex: $(1, -9)$

c. x-intercepts: $f(x) = x^2 - 2x - 8$
$0 = x^2 - 2x - 8$
$0 = (x + 2)(x - 4)$

$x + 2 = 0$ or $x - 4 = 0$
$x = -2$ $x = 4$

d. y-intercept: $f(x) = x^2 - 2x - 8$
$f(0) = 0^2 - 2(0) - 8$
$y = -8$

e.

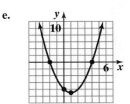

13. a. $a < 0$, thus the parabola opens downward.

b. x-coordinate: $x = \dfrac{-b}{2a} = \dfrac{-4}{2(-1)} = 2$

y-coordinate: $y = -x^2 + 4x - 3$
$= -(2)^2 + 4(2) - 3$
$= 1$
vertex: $(2, 1)$

c. x-intercepts: $y = -x^2 + 4x - 3$
$0 = -x^2 + 4x - 3$
$0 = x^2 - 4x + 3$
$0 = (x - 3)(x - 1)$

$x - 3 = 0$ or $x - 1 = 0$
$x = 3$ $x = 1$

d. y-intercept: $y = -x^2 + 4x - 3$
$y = -0^2 + 4(0) - 3$
$y = -3$

e.

15. a.

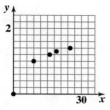

    b. logarithmic

17. a.

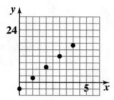

    b. linear

19. a.

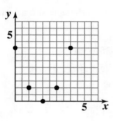

    b. quadratic

21. a.

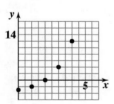

    b. exponential

23.

x	$f(x) = \left(\frac{1}{2}\right)^x$
-2	4
-1	2
0	1
1	$\dfrac{1}{2}$
2	$\dfrac{1}{4}$

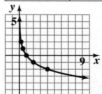

The graph is decreasing, although the rate of decrease is slowing down.

25.

$x = \left(\frac{1}{2}\right)^y$	y
4	-2
2	-1
1	0
$\dfrac{1}{2}$	1
$\dfrac{1}{4}$	2

The graph is decreasing, although the rate of decrease is slowing down.

27. a. $a < 0$, thus the parabola opens downward.

b. x-coordinate: $x = \dfrac{-b}{2a} = \dfrac{-4}{2(-2)} = 1$

y-coordinate: $f(x) = -2x^2 + 4x + 5$
$$f(1) = -2(1)^2 + 4(1) + 5$$
$$= 7$$
vertex: $(1, 7)$

c. x-intercepts: $f(x) = -2x^2 + 4x + 5$
$$0 = -2x^2 + 4x + 5$$
$$0 = 2x^2 - 4x - 5$$
$$x = \frac{-b \pm \sqrt{b^2 - 4ac}}{2a}$$
$$x = \frac{-(-4) \pm \sqrt{(-4)^2 - 4(2)(-5)}}{2(2)}$$
$$x \approx -0.9 \quad \text{or} \quad x \approx 2.9$$

d. y-intercept: $f(x) = -2x^2 + 4x + 5$
$$f(0) = -2(0)^2 + 4(0) + 5$$
$$f(0) = 5$$

e.

29. $y = (x - 3)^2 + 2$
$$y = x^2 - 6x + 9 + 2$$
$$y = x^2 - 6x + 11$$

x-coordinate: $x = \dfrac{-b}{2a} = \dfrac{-(-6)}{2(1)} = 3$

y-coordinate: $y = x^2 - 6x + 11$
$$y = 3^2 - 6(3) + 11$$
$$y = 2$$
vertex: $(3, 2)$

31. a. An exponential function was used because the graph is increasing more and more rapidly.

b. $f(x) = 1.402(1.078)^x$

c. $f(37) = 1.402(1.078)^{37}$
$$f(37) = 22.6\%$$
The function models the actual data value of 20% fairly well.

33. $f(45) = 0.16(45) + 1.43 = 8.63$

$g(45) = 1.8e^{0.04(45)} \approx 10.9$

The linear function is a better model for the graph's value of 8.9.

35. $f(x) = 13.4 + 46.3 \ln x$
$$f(3) = 13.4 + 46.3 \ln 3$$
$$f(3) \approx 64\%$$
The function describes the actual data value of 63% very well.

37. a. $f(x) = 62 + 35 \log(x - 4)$
$$f(13) = 62 + 35 \log(13 - 4)$$
$$f(13) \approx 95.3\%$$

b. A logarithmic function was used because height increases rapidly at first and then more slowly.

39. a. A quadratic function was used because data values decrease then increase.

b. $f(x) = 0.22x^2 - 0.49x + 7.68$

c. The minimum occurs at the vertex.

x-coordinate: $x = \dfrac{-b}{2a} = \dfrac{-(-.049)}{2(0.22)} \approx 1$

Thus the minimum occurs 1 year after 1999, or 2000.
Substitute the x-coordinate of the vertex into the function to find the y-coordinate.
$$f(x) = 0.22x^2 - 0.49x + 7.68$$
$$f(1) = 0.22(1)^2 - 0.49(1) + 7.68$$
$$f(1) \approx 7.4 \text{ million}$$
The function models the data in the graph extremely well.

47. Statement **c.** is true. In 2012 Uganda's population will exceed Canada's. Uganda's population will be $A = 25.6e^{0.03(9)} \approx 33.5$ million, while Canada's population will be $A = 32.2e^{0.003(9)} \approx 33.1$ million. Statement **a.** is false. In 2003, Uganda's population was not ten times that of Canada's. In fact, Uganda's population was smaller than Canada's. However, Uganda's rate of growth was ten times that of Canada's.
Statement **b.** is false. In 2003, Canada's population exceeded Uganda's by $32.2 - 25.6 = 6.6$ million, not 660,000.
Statement **d.** is false since statement **c.** is true.

49. There is only one x-intercept because the vertex on the x-axis. The vertex is the x-intercept.

Chapter 7 Review Exercises

1.

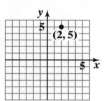

2.

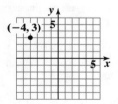

3.

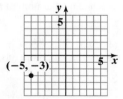

4.

5.

x	$y = 2x - 2$
-3	-8
-2	-6
-1	-4
0	-2
1	0
2	2
3	4

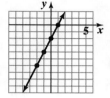

6.

x	$y = |x| + 2$
-3	5
-2	4
-1	3
0	2
1	3
2	4
3	5

7.

x	$y = x$
–3	–3
–2	–2
–1	–1
0	0
1	1
2	2
3	3

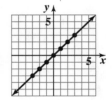

8. $f(x) = 4x + 11$

$f(-2) = 4(-2) + 11 = -8 + 11 = 3$

9. $f(x) = -7x + 5$

$f(-3) = -7(-3) + 5 = 21 + 5 = 26$

10. $f(x) = 3x^2 - 5x + 2$

$f(4) = 3(4)^2 - 5(4) + 2 = 48 - 20 + 2 = 30$

11. $f(x) = -3x^2 + 6x + 8$

$f(-4) = -3(-4)^2 + 6(-4) + 8$

$\quad\quad = -48 - 24 + 8 = -64$

12.

x	$f(x) = \frac{1}{2}|x|$
–6	3
–4	2
–2	1
0	0
2	1
4	2
6	3

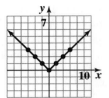

13.

x	$f(x) = x^2 - 2$
–2	2
–1	–1
0	–2
1	–1
2	2

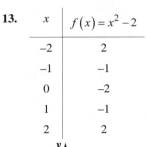

14. The graph passes the vertical line test. Thus y is a function of x.

15. The graph does not pass the vertical line test. Thus y is not a function of x.

16. a. $f(60) \approx 3.1$; This means about 3.1% of the U.S. population was made up of Jewish Americans in 1960.

b. The percentage of Jewish Americans in the U.S. population reached a maximum of about 3.7% in 1940.

c. The percentage of Jewish Americans in the U.S. population reached a minimum of about 1.3% in 1900.

d. The graph represents a function because it passes the vertical line test.

e. Generally the percentage rises until 1940 then it decreases.

17. $2x + y = 4$
x-intercept is 2; y-intercept is 4.

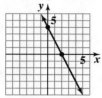

18. $2x - 3y = 6$
x-intercept is 3; y-intercept is -2.

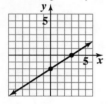

19. $5x - 3y = 15$
x-intercept is 3; y-intercept is -5.

20. Slope $= \dfrac{1-2}{5-3} = -\dfrac{1}{2}$; line falls

21. Slope $= \dfrac{-4-2}{-3-(-1)} = \dfrac{-6}{-2} = 3$; line rises

22. Slope $= \dfrac{4-4}{6-(-3)} = 0$; line horizontal

23. Slope $= \dfrac{-3-3}{5-5} = \dfrac{-6}{0}$ is undefined,
vertical line

24. $y = 2x - 4$; Slope: 2, y-intercept: -4
Plot point $(0, -4)$ and second point using
$m = \dfrac{2}{1} = \dfrac{\text{rise}}{\text{run}}$.

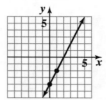

25. $y = -\dfrac{2}{3}x + 5$; Slope: $-\dfrac{2}{3}$, y-intercept: 5
Plot point $(0, 5)$ and second point using
$m = \dfrac{-2}{3} = \dfrac{\text{rise}}{\text{run}}$.

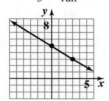

26. $y = \dfrac{3}{4}x - 2$; Slope: $\dfrac{3}{4}$, y-intercept: -2
Plot point $(0, -2)$ and second point using
$m = \dfrac{3}{4} = \dfrac{\text{rise}}{\text{run}}$.

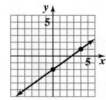

27. $y = \dfrac{1}{2}x + 0$; Slope: $\dfrac{1}{2}$, y-intercept: 0
Plot point $(0, 0)$ and second point using
$m = \dfrac{1}{2} = \dfrac{\text{rise}}{\text{run}}$.

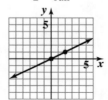

28. a. $2x + y = 0$
$y = -2x$

    b. Slope $= -2$
y-intercept $= 0$

    c.

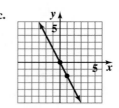

29. a. $3y = 5x$
$y = \dfrac{5}{3}x$

    b. Slope $= \dfrac{5}{3}$
y-intercept $= 0$

    c.

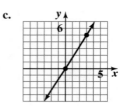

30. a. $3x + 2y = 4$
$2y = -3x + 4$
$y = -\dfrac{3}{2}x + 2$

    b. Slope $= -\dfrac{3}{2}$
y-intercept $= 2$

    c.

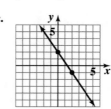

31. $x = 3$

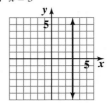

32. $y = -4$

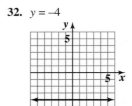

33. $x + 2 = 0$ or $x = -2$

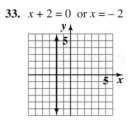

34. a. The y-intercept is 2970. This represents that in 1990 (0 years after 1990), the average credit card debt was $2970.

    b. $m = \dfrac{9312 - 2970}{14 - 0} = 453$

    This represents that the average credit card debt increased $453 per year.

    c. $y = mx + b$
$y = 453x + 2970$

    d. 2010 is 20 years after 1990.
$y = 453x + 2970$
$y = 453(20) + 2970$
$y = \$12{,}030$

35. The intersection is (2, 3).
    Check: $2 + 3 = 5$ $3(2) - 3 = 3$
             $5 = 5$ true $6 - 3 = 3$
                               $3 = 3$ true

The solution set is $\{(2, 3)\}$.

36. The intersection is (–2, –3).
Check: $2(-2)-(-3)=-1$ $-2-3=-5$
        $-4+3=-1$ $-5=-5$ true
              $-1=-1$ true
The solution set is $\{(-2, -3)\}$.

37. The intersection is (3, 2).
Check: $2=-3+5$ $2(3)-2=4$
        $2=2$ true $6-2=4$
                  $4=4$ true
The solution set is $\{(3, 2)\}$.

38. $x=3y+10$ $2x+3y=2$
$2(3y+10)+3y=2$
$6y+20+3y=2$
$9y=-18$
$y=-2$
$x=3(-2)+10=-6+10=4$
The solution set is $\{(4, -2)\}$.

39. $y=4x+1$ $3x+2y=13$
$3x+2(4x+1)=13$
$3x+8x+2=13$
$11x=11$
$x=1$
$y=4(1)+1=5$
The solution set is $\{(1, 5)\}$.

40. $x+4y=14$
$x=14-4y$ $2x-y=1$
$2(14-4y)-y=1$
$28-8y-y=1$
$-9y=-27$
$y=3$
$x=14-4(3)=2$
The solution set is $\{(2, 3)\}$.

41. $x+2y=-3$ No change. $x+2y=-3$
$x-y=-12$ Multiply by -1. $\underline{-x+\ \ y=12}$
                  $3y=9$
                  $y=3$
$x-y=-12$
$x-3=-12$
$x=-9$
The solution set is $\{(-9, 3)\}$.

42. $2x-y=2$ Mult. by 2. $4x-2y=4$
$x+2y=11$ No change $\underline{x+2y=11}$
                $5x=15$
                $x=3$
$x+2y=11$
$3+2y=11$
$2y=8$
$y=4$
The solution set is $\{(3, 4)\}$.

43. $5x+3y=1$ Mult. by 3. $15x+9y=3$
$3x+4y=-6$ Mult. by -5. $\underline{-15x-20y=30}$
                  $-11y=33$
                  $y=-3$
$5x+3y=1$
$5x+3(-3)=1$
$5x=10$
$x=2$
The solution set is $\{(2, -3)\}$.

44. $y=-x+4$ $3x+3y=-6$
$3x+3(-x+4)=-6$
$3x-3x+12=-6$
$12=-6$, false
There is no solution or $\{\ \}$.

45. $3x+y=8$
$y=8-3x$ $2x-5y=11$
$2x-5(8-3x)=11$
$2x-40+15x=11$
$17x=51$
$x=3$
$y=8-3(3)=-1$
The solution set is $\{(3, -1)\}$.

46. $3x-2y=6$ Mult. by -2. $-6x+4y=-12$
$6x-4y=12$ No change. $\underline{6x-4y=12}$
                $0=0$

The system has infinitely many solutions.
The solution set is $\{(x, y)|3x-2y=6\}$.

47. a. $C(x) = 60,000 + 200x$

b. $R(x) = 450x$

c. $450x = 60000 + 200x$
$250x = 60000$
$x = 240$
$450(240) = 108,000$
The company must make 240 desks at a cost of $108,000 to break even.

48. Let x = the number of years of healthy life expectancy of people in Japan.
Let y = the number of years of healthy life expectancy of people in Switzerland.
The system to solve is: $\begin{aligned} x + y &= 146.4 \\ x - y &= 0.8 \end{aligned}$

Add the two equations:
$x + y = 146.4$
$\underline{x - y = 0.8}$
$\quad 2x = 147.2$
$\quad\; x = 73.6$
Back substitute to find y:
$x + y = 146.4$
$73.6 + y = 146.4$
$\quad\;\; y = 72.8$
The number of years of healthy life expectancy in Japan is 73.6 years and in Switzerland is 72.8 years.

49. To graph $x - 3y \leq 6$, begin by graphing $x - 3y = 6$ with a solid line because $\leq$ includes equality.
test point (0, 0):

$x - 3y \leq 6$
$0 - 3(0) \leq 6$
$0 \leq 6$, true

Since the test point makes the inequality <u>true</u>, shade the half-plane <u>containing</u> test point (0, 0).

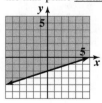

50. To graph $2x + 3y \geq 12$, begin by graphing $2x + 3y = 12$ with a solid line because $\geq$ includes equality.
test point (0, 0):

$2x + 3y \geq 12$
$2(0) + 3(0) \geq 12$
$0 \geq 12$, false

Since the test point makes the inequality <u>false</u>, shade the half-plane <u>not containing</u> test point (0, 0).

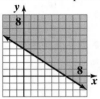

51. To graph $2x - 7y > 14$, begin by graphing $2x - 7y = 14$ with a dashed line because $>$ does not include equality.
test point (0, 0):

$2x - 7y > 14$
$2(0) - 7(0) > 14$
$0 > 14$, false

Since the test point makes the inequality <u>false</u>, shade the half-plane <u>not containing</u> test point (0, 0).

52. To graph $y > \frac{3}{5}x$, begin by graphing $y = \frac{3}{5}x$ as a dashed line passing through the origin with a slope of $\frac{3}{5}$, then shade above the line.

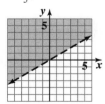

53. To graph $y \leq -\frac{1}{2}x + 2$, begin by graphing

$y = -\frac{1}{2}x + 2$ as a solid line passing through $(0, 2)$

with a slope of $\frac{-1}{2}$, then shade below the line.

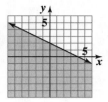

54. To graph $x \leq 2$, begin by graphing $x = 2$ as a solid

vertical line passing through $x = 2$, then shade to

the left of the line.

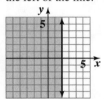

55. To graph $y > -3$, begin by graphing $y = -3$ as a

dashed horizontal line passing through $y = -3$, then

shade above the line.

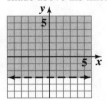

56. $3x - y \leq 6$

$x + y \geq 2$

Graph $3x - y = 6$ as a solid line.

Because $(0, 0)$ makes the inequality true, shade the

half-plane containing $(0,0)$.

Graph $x + y = 2$ as a solid line.

Because $(0, 0)$ makes the inequality false, shade the

half-plane not containing $(0, 0)$.

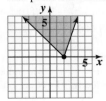

57. $x + y < 4$

$x - y < 4$

Graph $x + y = 4$ as a dashed line.

Because $(0, 0)$ makes the inequality true, shade the

half-plane containing $(0, 0)$.

Graph $x - y = 4$ as a dashed line.

Because $(0, 0)$ makes the inequality true, shade the

half-plane containing $(0, 0)$.

58. $x \leq 3$

$y > -2$

Graph $x = 3$ as a solid line.

The points in the half-plane to the left of the line

satisfy $x < 3$.

Graph $y = -2$ as a dashed line.

The points in the half-plane above the line satisfy $y > -2$.

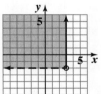

59. $4x + 6y = 24$

$y > 2$

Graph $4x + 6y = 24$ as a solid line.

Because $(0, 0)$ makes the inequality true, shade the

half-plane containing $(0, 0)$.

Graph $y = 2$ as a dashed line.

The points in the half-plane above the line satisfy $y > 2$.

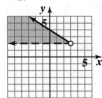

60. $x + y \leq 6$

$y \geq 2x - 3$

Graph $x + y = 6$ as a solid line.

Because $(0, 0)$ makes the inequality true, shade the half-plane containing $(0, 0)$.

Graph $y = 2x - 3$ as a solid line.

Because $(0, 0)$ makes the inequality true, shade the half-plane containing $(0, 0)$.

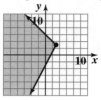

61. $y < -x + 4$

$y > x - 4$

Graph $y < -x + 4$ as a dashed line.

Because $(0, 0)$ makes the inequality true, shade the half-plane containing $(0, 0)$.

Graph $y = x - 4$ as a dashed line.

Because $(0, 0)$ makes the inequality true, shade the half-plane containing $(0, 0)$.

62. $z = 2x + 3y$

at $(1, 0)$ $z = 2(1) + 3(0) = 2$

at $\left(\dfrac{1}{2}, \dfrac{1}{2}\right)$ $z = 2\left(\dfrac{1}{2}\right) + 3\left(\dfrac{1}{2}\right) = \dfrac{5}{2}$

at $(2, 2)$ $z = 2(2) + 3(2) = 10$

at $(4, 0)$ $z = 2(4) + 3(0) = 8$

Maximum value of the objective function is 10.
Minimum value of the objective function is 2.

63. $z = 2x + 3y$

Constraints: $x \leq 6$

$y \leq 5$

$x + y \geq 2$

$x \geq 0$

$y \geq 0$

a.

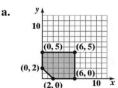

b.   at $(0, 2)$ $z = 2(0) + 3(2) = 6$

at $(0, 5)$ $z = 2(0) + 3(5) = 15$

at $(6, 5)$ $z = 2(6) + 3(5) = 27$

at $(6, 0)$ $z = 2(6) + 3(0) = 12$

at $(2, 0)$ $z = 2(2) + 3(0) = 4$

c.   The maximum value of the objective function is 27. It occurs at $x = 6$ and $y = 5$.
The minimum value of the objective function is 4. It occurs at $x = 2$ and $y = 0$.

64. a. $z = 500x + 350y$

b. $x + y \leq 200$
$x \geq 10$
$y \geq 80$

c.

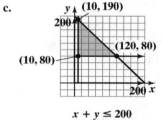

$x + y \leq 200$
$x \geq 10, y \geq 80$

d.

Vertex	Objective Function
	$z = 500x + 350y$
$(10, 80)$	$z = 500(10) + 350(80)$ $= 33,000$
$(10, 190)$	$z = 500(10) + 350(190)$ $= 71,500$
$(120, 80)$	$z = 500(120) + 350(80)$ $= 88,000$

e.   The company will make the greatest profit by producing 120 units of writing paper and 80 units of newsprint each day. The maximum daily profit is $88,000.

65.

x	$y = 2^x$
-2	$\frac{1}{4}$
-1	$\frac{1}{2}$
0	1
1	2
2	4

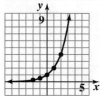

66.

x	$y = 2^{x+1}$
-2	$\frac{1}{2}$
-1	1
0	2
1	4
2	8

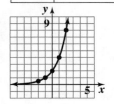

67. $y = \log_2 x$ is equivalent to $x = 2^y$.

$x = 2^y$	y
$\frac{1}{4}$	-2
$\frac{1}{2}$	-1
1	0
2	1
4	2

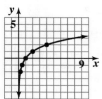

68. a. $a > 0$, thus the parabola opens upward.

b. x-coordinate: $x = \dfrac{-b}{2a} = \dfrac{-(-6)}{2(1)} = 3$

y-coordinate: $f(x) = x^2 - 6x - 7$
$f(3) = (3)^2 - 6(3) - 7$
$= -16$

vertex: $(3, -16)$

c. x-intercepts: $f(x) = x^2 - 6x - 7$
$0 = x^2 - 6x - 7$
$0 = (x+1)(x-7)$

$x + 1 = 0$ or $x - 7 = 0$
$x = -1$ $\quad x = 7$

d. y-intercept: $f(x) = x^2 - 6x - 7$
$f(0) = 0^2 - 6(0) - 7$
$y = -7$

e.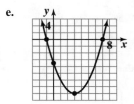

69. a. $a < 0$, thus the parabola opens downward.

b. x-coordinate: $x = \dfrac{-b}{2a} = \dfrac{-(-2)}{2(-1)} = -1$

y-coordinate: $f(x) = -x^2 - 2x + 3$
$f(-1) = -(-1)^2 - 2(-1) + 3$
$= 4$

vertex: $(-1, 4)$

c. *x*-intercepts: $f(x) = -x^2 - 2x + 3$

$$0 = -x^2 - 2x + 3$$
$$0 = x^2 + 2x - 3$$
$$0 = (x+3)(x-1)$$

$x + 3 = 0$ or $x - 1 = 0$
$x = -3$ $x = 1$

d. *y*-intercept: $f(x) = -x^2 - 2x + 3$

$$f(0) = -0^2 - 2(0) + 3$$
$$y = 3$$

e.

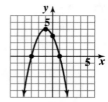

70. a. **Percentage of U.S. High School Seniors Taking Steroids**

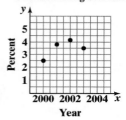

b. quadratic

71. a. **Hybrid Car Sales in U.S.**

b. exponential

72. a. **Percentage of U.S. Households with TVs with Cable Television**

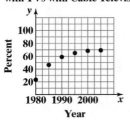

b. logarithmic

73. a. **Average Monthly Cellphone Bills**

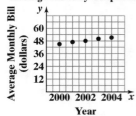

b. linear

74. $f(x) = 364(1.005)^x$

$$f(86) = 364(1.005)^{86}$$
$$\approx 559 \text{ ppm}$$

This is nearly double the preindustrial level.

75. $f(x) = 104.9e^{0.017x}$

$$f(41) = 104.9e^{0.017(41)}$$
$$\approx 210.7 \text{ million}$$

This is slightly more than twice the 2003 population.

76. a. The slope is –1.7. The means that the percentage of companies performing drug tests is decreasing by 1.7 each year.

b. The logarithmic model is better for 2003.
linear model: $f(x) = -1.7x + 74.5$
$$f(6) = -1.7(6) + 74.5$$
$$= 64.3$$
logarithmic model: $g(x) = -4.9 \ln x + 73.8$
$$g(6) = -4.9 \ln 6 + 73.8$$
$$= 65.0$$

Chapter 7 Test

1.

x	$y = |x| - 2$
-3	1
-2	0
-1	-1
0	-2
1	-1
2	0
3	1

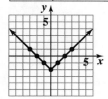

2. $f(-2) = 3(-2)^2 - 7(-2) - 5 = 12 + 14 - 5 = 21$

3. The graph does not pass the vertical line test. Thus y is not a function of x.

4. The graph passes the vertical line test. Thus y is a function of x.

5. a. Yes, it is a function. The graph passes the vertical line test.

   b. $f(15) = 0$ means that the eagle was on the ground after 15 seconds.

   c. The maximum height was 45 meters.

   d. The eagle was descending between second 3 and second 12.

6. Set $y = 0$ Set $x = 0$
$4x - 2 \cdot 0 = -8$ $4 \cdot 0 - 2y = -8$
      $4x = -8$ $-2y = -8$
        $x = -2$ $y = 4$
x-intercept is -2. y-intercept is 4.

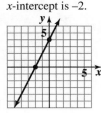

7. Slope $= \dfrac{-2-4}{-5-(-3)} = \dfrac{-6}{-2} = 3$

8. $y = \dfrac{2}{3}x - 1$

Slope: $\dfrac{2}{3}$, y-intercept: -1

Plot the point $(0, -1)$ and a second point using

$m = \dfrac{2}{3} = \dfrac{\text{rise}}{\text{run}}$.

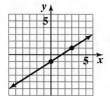

9. $f(x) = -2x + 3$

Slope: -2, y-intercept: 3
Plot the point $(0, 3)$ and a second point using

$m = \dfrac{-2}{1} = \dfrac{\text{rise}}{\text{run}}$.

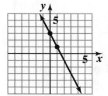

10. a. The y-intercept is 180. In 2000 (0 years after 2000), an American adult read a newspaper an average of 180 hours per year

    b. $m = \dfrac{176 - 180}{2 - 0} = -2$

The average number of hours decreased 2 hours per year.

    c. $y = mx + b$
$y = -2x + 180$

    d. $y = -2(12) + 180 = 156$ hours

11. The intersection is $(2, 4)$
Check: $2 + 4 = 6$ $4(2) - 4 = 4$
           $6 = 6$ true $8 - 4 = 4$
                            $4 = 4$ true
The solution set is $\{(2, 4)\}$.

12. $x = y + 4 \qquad 3x + 7y = -18$

$3(y + 4) + 7y = -18$

$3y + 12 + 7y = -18$

$10y = -30$

$y = -3$

$x = -3 + 4 = 1$

The solution set is $\{(1, -3)\}$.

13. $5x + 4y = 10 \xrightarrow{\text{Mult. by 3}} 15x + 12y = 30$

$3x + 5y = -7 \xrightarrow{\text{Mult. by } -5} \underline{-15x - 25y = 35}$

$-13y = 65$

$y = -5$

$5x + 4y = 10$

$5x + 4(-5) = 10$

$5x = 30$

$x = 6$

The solution set is $\{(6, -5)\}$.

14. a. $C(x) = 360,000 + 850x$

b. $R(x) = 1150x$

c. $1150x = 360000 + 850x$

$300x = 360000$

$x = 1200$

$1150(1200) = 1,380,000$

1200 computers need to be sold to make $1,380,00 for the company to break even.

15. $3x - 2y < 6$

Graph $3x - 2y = 6$ as a dashed line.

x-intercept:

$3x - 2 \cdot 0 = 6$

$3x = 6$

$x = 2$

y-intercept:

$3 \cdot 0 - 2y = 6$

$-2y = 6$

$y = -3$

Test point: (0, 0).

Is $3 \cdot 0 - 2 \cdot 0 < 6$?

$0 < 6$, true

Shade the half-plane containing (0, 0).

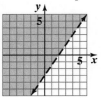

16. Graph $y = \frac{1}{2}x - 1$ as a solid line.

Use y-intercept of -1 and slope of $\frac{1}{2}$

Shade below this line.

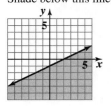

17. $y > -1$

Graph $y = -1$ as a dashed line.

Test point: (0, 0).

Is $0 > -1$?

$0 > -1$, true

Shade the half-plane containing (0, 0).

18. $2x - y \le 4$

$2x - y > -1$

Graph $2x - y = 4$ as a solid line.

Because (0, 0) makes the inequality true, shade the half-plane containing (0, 0).

Graph $2x - y = -1$ as a dashed line.

Because (0, 0) makes the inequality true, shade the half-plane containing (0, 0).

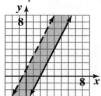

19. $z = 3x + 2y$

at $(2, 0)$ $z = 3(2) + 2(0) = 6$

at $(2, 6)$ $z = 3(2) + 2(6) = 18$

at $(6, 3)$ $z = 3(6) + 2(3) = 24$

at $(8, 0)$ $z = 3(8) + 2(0) = 24$

The maximum value of the objective function is 24.
The minimum value of the objective function is 6.

20.

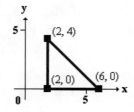

Objective function: $z = 3x + 5y$

at $(2, 0)$ $z = 3(2) + 5(0) = 6$

at $(6, 0)$ $z = 3(6) + 5(0) = 18$

at $(2, 4)$ $z = 3(2) + 5(4) = 26$

The maximum value of the objective function is 26.

21.

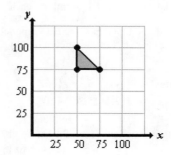

Objective function: $z = 200x + 250y$

Constraints: $x \geq 50$; $y \geq 75$; $x + y \leq 150$

Substitute vertices into objective function:

at $(50, 100)$ $z = 200(50) + 250(100) = 35,000$

at $(75, 75)$ $z = 200(75) + 250(75) = 33,750$

at $(50, 75)$ $z = 200(50) + 250(75) = 28,750$

The company will make the greatest profit by producing 50 regular jet skis and 100 deluxe jet skis each week. The maximum weekly profit is $35,000.

22.

x	$f(x) = 3^x$
-2	$\frac{1}{9}$
-1	$\frac{1}{3}$
0	1
1	3
2	9

23. $y = \log_2 x$ is equivalent to $x = 3^y$.

$x = 3^y$	y
$\frac{1}{9}$	-2
$\frac{1}{3}$	-1
1	0
3	1
9	2

24. x-coordinate: $x = \dfrac{-b}{2a} = \dfrac{-(-2)}{2(1)} = 1$

y-coordinate: $f(x) = x^2 - 2x - 8$

$f(1) = 1^2 - 2(1) - 8$

$y = -9$

vertex: $(1, -9)$

x-intercepts: $f(x) = x^2 - 2x - 8$

$$0 = x^2 - 2x - 8$$

$$0 = (x+2)(x-4)$$

$x + 2 = 0$ or $x - 4 = 0$

$x = -2$ $x = 4$

y-intercept: $f(x) = x^2 - 2x - 8$

$$f(0) = 0^2 - 2(0) - 8$$

$$y = -8$$

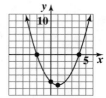

25. Plot the ordered pairs.

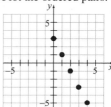

The values appear to belong to a linear function.

26. Plot the ordered pairs.

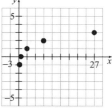

The values appear to belong to a logarithmic function.

27. Plot the ordered pairs.

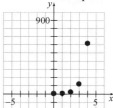

The values appear to belong to an exponential function.

28. Plot the ordered pairs.

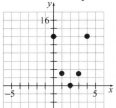

The values appear to belong to a quadratic function.

29. a. $f(x) = 110.5(1.13)^x$

$$f(3) = 110.5(1.13)^3$$

$f(3) \approx 159$ million

The function models the actual data in the graph extremely well.

b. $f(x) = 110.5(1.13)^x$

$$f(10) = 110.5(1.13)^{10}$$

$f(10) \approx 375$ million

30. a. $m = 15.3$, which means the number of Internet users is increasing by 15.3 million each year.

b.   A logarithmic function is the better model because the data values increase rapidly at first and then more slowly.

c.   linear model: $f(x) = 15.3x + 106.7$

$$f(2) = 15.3(2) + 106.7$$

$$= 137.3 \text{ million}$$

logarithmic model: $g(x) = 34.1 \ln x + 117.7$

$$g(2) = 34.1 \ln(2) + 117.7$$

$$\approx 141.3 \text{ million}$$

The logarithmic model serves better for 2001. This is consistent with the answer for part b.

Chapter 8
Consumer Mathematics and Financial Management

Check Points 8.1

1. Step 1: $\frac{1}{8} = 1 \div 8 = 0.125$
   Step 2: $0.125 \cdot 100 = 12.5$
   Step 3: 12.5%

2. $0.023 = 2.3\%$

3. **a.** $67\% = 0.67$

   b. $250\% = 2.50 = 2.5$

4. **a.** 6% of $\$1260 = 0.06 \times \$1260 = \$75.60$
   The tax paid is $\$75.60$

   b. $\$1260.00 + \$75.60 = \$1335.60$
   The total cost is $\$1335.60$

5. **a.** 35% of $\$380 = 0.35 \times \$380 = \$133$
   The discount is $\$133$

   b. $\$380 - \$133 = \$247$
   The sale price is $\$247$

6. Step 1. Determine the adjusted gross income.
   Adj. gross income = Gross income – Adjustments
   Adj. gross income = $\$40,000 - \$1000$
   $\qquad = \$39,000$

   Step 2. Determine the taxable income.
   Since the total deduction of $\$4800$ is less than the
   standard deduction of $\$5000$, use $\$5000$.
   Taxable inc. = Adj. gross inc– (Exempt.+Deduct.)
   Taxable inc. = $\$39,000 - (\$3200 + \$5000)$
   $\qquad = \$30,800$

   Step 3. Determine the income tax.
   Tax Computation
   $= 0.10(7300) + 0.15(29,700 - 7300)$
   $\qquad + 0.25(30,800 - 29,700)$
   $= \$4365$
   Income tax = Tax Computation – Tax credits
   Income tax = $\$4365 - \$0$
   $\qquad = \$4365$

7. **a.** Percent of increase $= \dfrac{\text{amount of increase}}{\text{original amount}}$

   $= \dfrac{4}{6} = 0.66\frac{2}{3} = 66\frac{2}{3}\%$

   b. Percent of decrease $= \dfrac{\text{amount of decrease}}{\text{original amount}}$

   $= \dfrac{4}{10} = 0.4 = 40\%$

8. Amount of decrease: $\$940 - \$611 = \$329$
   $\dfrac{\text{amount of decrease}}{\text{original amount}} = \dfrac{\$329}{\$940} = 0.35 = 35\%$
   There was a 35% decrease from 1998 to 1999.

9. Amount of increase: $12\% - 10\% = 2\%$
   $\dfrac{\text{amount of increase}}{\text{original amount}} = \dfrac{2\%}{10\%} = 0.2 = 20\%$
   There was a 20% increase for this episode.

10. **a.** 20% of $\$1200 = 0.20 \times \$1200 = \$240$
    Taxes for year 1 are $\$1200 - \$240 = \$960$
    20% of $\$960 = 0.20 \times \$960 = \$192$
    Taxes for year 2 are $\$960 + \$192 = \$1152$

    b. $\dfrac{\$1200 - \$1152}{\$1200} = \dfrac{\$48}{\$1200} = 0.04 = 4\%$
    Taxes for year 2 are 4% less than the original
    amount.

Exercise Set 8.1

1. $\dfrac{2}{5} = 2 \div 5 = 0.4 = 40\%$

3. $\dfrac{1}{4} = 1 \div 4 = 0.25 = 25\%$

5. $\dfrac{3}{8} = 3 \div 8 = 0.375 = 37.5\%$

7. $\dfrac{1}{40} = 1 \div 40 = 0.025 = 2.5\%$

9. $\dfrac{9}{80} = 9 \div 80 = 0.1125 = 11.25\%$

11. $0.59 = 59\%$

13. $0.3844 = 38.44\%$

15. $2.87 = 287\%$

17. $14.87 = 1487\%$

19. $100 = 10,000\%$

21. $72\% = 0.72$

23. $43.6\% = 0.436$

25. $130\% = 1.3$

27. $2\% = 0.02$

29. $\dfrac{1}{2}\% = 0.5\% = 0.005$

31. $\dfrac{5}{8}\% = 0.625\% = 0.00625$

33. $62\dfrac{1}{2}\% = 62.5\% = 0.625$

35. $A = PB$
$A = 0.03 \cdot 200$
$A = 6$

37. $A = PB$
$A = 0.18 \cdot 40$
$A = 7.2$

39. $A = PB$
$3 = 0.60 \cdot B$
$\dfrac{3}{0.60} = \dfrac{0.60B}{0.60}$
$5 = B$

41. $A = PB$
$40.8 = 0.24 \cdot B$
$\dfrac{40.8}{0.24} = \dfrac{0.24B}{0.24}$
$170 = B$

43. $A = PB$
$3 = P \cdot 15$
$\dfrac{3}{15} = \dfrac{P \cdot 15}{15}$
$0.2 = P$
$P = 20\%$

45. $A = PB$
$0.3 = P \cdot 2.5$
$\dfrac{0.3}{2.5} = \dfrac{P \cdot 2.5}{2.5}$
$0.12 = P$
$P = 12\%$

47. a. $(0.06)(16,800) = \$1,008$

b. $16,800 + 1008 = \$17,808$

49. a. $(0.12)(860) = \$103.20$

b. $860 - 103.20 = \$756.80$

51. Step 1. Determine the adjusted gross income.
Adj. gross income = Gross income – Adjustments
Adj. gross income = $75,000 – $4000
$\qquad\qquad\qquad = \$71,000$

Step 2. Determine the taxable income.
Since the total deduction of $35,200 is greater than the standard deduction of $5000, use $35,200.
Taxable inc. = Adj. gross inc– (Exempt.+Deduct.)
Taxable inc. = $71,000 – ($3200 + $35,200)
$\qquad\qquad\qquad = \$32,600$

Step 3. Determine the income tax.
Tax Computation
$= 0.10(7300) + 0.15(29,700 - 7300)$
$\quad + 0.25(32,600 - 29,700)$
$= \$4815$
Income tax = Tax Computation – Tax credits
Income tax = $4815 – $0
$\qquad\qquad\qquad = \$4815$

53. Step 1. Determine the adjusted gross income.
Adj. gross income = Gross income – Adjustments
Adj. gross income = $50,000 – $0
$\qquad\qquad\qquad = \$50,000$

Step 2. Determine the taxable income.
Since the total deduction of $6500 is less than the standard deduction of $7300, use $7300.
Taxable inc. = Adj. gross inc– (Exempt.+Deduct.)
Taxable inc. = $50,000 – ($3200 \cdot 3 + $7300)
$\qquad\qquad\qquad = \$33,100$

Step 3. Determine the income tax.
Tax Computation
$= 0.10(10,450) + 0.15(33,100 - 10,450)$
$= \$4442.50$
Income tax = Tax Computation – Tax credits
Income tax = $4442.50 – $2000
$\qquad\qquad\qquad = \$2442.50$

55. FICA tax
$= 0.0765(90,000) + 0.0145(100,000 - 90,000)$
$= \$7030$

57. FICA tax
$= 0.0765(90,000) + 0.0145(140,000 - 90,000)$
$= \$7610$
Since, this person is self-employed the FICA rate is doubled: $\$7610 \times 2 = \$15,220$

59. a. FICA tax $= 0.0765(20,000)$
$= \$1530$

b. Step 1. Determine the adjusted gross income.
Adj. gross income = Gross income − Adjustments
Adj. gross income $= \$20,000 - \$0 = \$20,000$

Step 2. Determine the taxable income.
The standard deduction is $5000.
Taxable inc. = Adj. gross inc−
(Exempt.+Deduct.)
Taxable inc. $= \$20,000 - (\$3200 + \$5000)$
$= \$11,800$

Step 3. Determine the income tax.
Tax Computation
$= 0.10(7300) + 0.15(11,800 - 7300)$
$= \$1405$
Income tax = Tax Computation − Tax credits
Income tax $= \$1405 - \$0$
$= \$1405$

c. $\dfrac{1530 + 1405}{20,000} \approx 0.147 = 14.7\%$

61. $\dfrac{974 - 624}{624} \approx 0.561 = 56.1\%$

63. $\dfrac{93 - 62}{62} = 0.5 = 50.0\%$

65. $\dfrac{840 - 714}{840} = 0.15 = 15\%$

67. Amount after first year
$= 10,000 - (0.3)(10,000)$
$= \$7000$
Amount after second year
$= 7000 + (0.4)(7000)$
$= \$9800$
Your adviser is not using percentages properly.

Actual change:
$\dfrac{10,000 - 9800}{10,000} = 0.02 = 2\%$ decrease.

77. Tax owed $= \dfrac{\$3.40}{\$100} \cdot \dfrac{\$78,500}{1} = \$2669$
Discount $= (0.03)(2669) = \$80.07$
Tax paid $= 2669 - 80.07 = \$2588.93$

Check Points 8.2

1. $I = Prt = (\$3000)(0.05)(1) = \$150$

2. $I = Prt = (\$2400)(0.07)(2) = \$336$

3. $A = P(1+rt) = 2040\left[1 + (0.075)\left(\dfrac{4}{12}\right)\right] = \$2091$

4. $A = P(1+rt)$
$6800 = 5000\left[1 + r(2)\right]$
$6800 = 5000 + 10,000r$
$1800 = 10,000r$
$\dfrac{1800}{10,000} = \dfrac{10,000r}{10,000}$
$0.18 = r$
$r = 18\%$

5. $A = P(1+rt)$
$4000 = P\left[1 + (0.08)\left(\frac{6}{12}\right)\right]$
$4000 = P(1.04)$
$\dfrac{4000}{1.04} = \dfrac{P(1.04)}{1.04}$
$3846.153 \approx P$
$P \approx \$3846.16$

6. a. $I = Prt = (5000)(0.12)(2) = 1200$
The loan's discount is $1200.

b. Amount received: $5000 - \$1200 = \$3800$

c. $I = Prt$
$1200 = (3800)(r)(2)$
$1200 = 7600r$
$\dfrac{1200}{7600} = \dfrac{7600r}{7600}$
$0.158 = r$
$r = 15.8\%$

Exercise Set 8.2

1. $I = (\$4000)(0.06)(1) = \$240$

3. $I = (\$180)(0.03)(2) = \$10.80$

5. $I = (\$5000)(0.085)\left(\dfrac{9}{12}\right) = \$318.75$

7. $I = (\$15,500)(0.11)\left(\dfrac{90}{360}\right) = \$426.25$

9. $A = P(1+rt) = 3000\left[1+(0.07)(2)\right] = \$3420$

11. $A = P(1+rt) = 26,000\left[1+(0.095)(5)\right] = \$38,350$

13. $A = P(1+rt) = 9000\left[1+(0.065)\left(\tfrac{8}{12}\right)\right] = \$9390$

15. $A = P(1+rt)$
$2150 = 2000\left[1+r(1)\right]$
$2150 = 2000 + 2000r$
$150 = 2000r$
$\dfrac{150}{2000} = \dfrac{2000r}{2000}$
$0.075 = r$
$r = 7.5\%$

17. $A = P(1+rt)$
$5900 = 5000\left[1+r(2)\right]$
$900 = 5000 + 10,000r$
$900 = 10,000r$
$\dfrac{900}{10,000} = \dfrac{10,000r}{10,000}$
$0.09 = r$
$r = 9\%$

19. $A = P(1+rt)$
$2840 = 2300\left[1+r\left(\tfrac{9}{12}\right)\right]$
$2840 = 2300 + 1725r$
$540 = 1725r$
$\dfrac{540}{1725} = \dfrac{1725r}{1725}$
$0.313 = r$
$r = 31.3\%$

21. $A = P(1+rt)$
$6000 = P\left[1+(0.08)(2)\right]$
$6000 = P(1.16)$
$\dfrac{6000}{1.16} = \dfrac{P(1.16)}{1.16}$
$5172.414 \approx P$
$P \approx \$5172.42$

23. $A = P(1+rt)$
$14,000 = P\left[1+(0.095)(6)\right]$
$14,000 = P(1.57)$
$\dfrac{14,000}{1.57} = \dfrac{P(1.57)}{1.57}$
$8917.197 \approx P$
$P \approx \$8917.20$

25. $A = P(1+rt)$
$5000 = P\left[1+(0.145)\left(\tfrac{9}{12}\right)\right]$
$5000 = P(1.10875)$
$\dfrac{5000}{1.10875} = \dfrac{P(1.10875)}{1.10875}$
$4509.583 \approx P$
$P \approx \$4509.59$

27. a. $I = Prt = (2000)(0.07)\left(\frac{8}{12}\right) = \$93.33$

  b. Amount received: $\$2000 - \$93.33 = \$1906.67$

  c.
  $$I = Prt$$
  $$93.33 = (1906.67)(r)\left(\tfrac{8}{12}\right)$$
  $$93.33 = 1271.113r$$
  $$\frac{93.33}{1271.113} = \frac{1271.113r}{1271.113}$$
  $$0.073 = r$$
  $$r = 7.3\%$$

29. a. $I = Prt = (12,000)(0.065)(2) = \$1560$

  b. Amount received: $\$12,000 - \$1560 = \$10,440$

  c.
  $$I = Prt$$
  $$1560 = (10,440)(r)(2)$$
  $$1560 = 20,880r$$
  $$\frac{1560}{20,880} = \frac{20,880r}{20,880}$$
  $$0.075 = r$$
  $$r = 7.5\%$$

31.
  $$A = P(1+rt)$$
  $$A = P + Prt$$
  $$A - P = Prt$$
  $$\frac{A-P}{Pt} = \frac{Prt}{Pt}$$
  $$\frac{A-P}{Pt} = r$$
  $$r = \frac{A-P}{Pt}$$

33.
  $$A = P(1+rt)$$
  $$\frac{A}{1+rt} = \frac{P(1+rt)}{1+rt}$$
  $$\frac{A}{1+rt} = P$$
  $$P = \frac{A}{1+rt}$$

35. a.
  $$I = Prt$$
  $$= (\$4000)(0.0825)\left(\frac{9}{12}\right)$$
  $$= \$247.50$$

  b. $\$4000 + \$247.50 = \$4247.50$

37.
  $$A = P(1+rt)$$
  $$2000 = 1400\left[1 + r(2)\right]$$
  $$2000 = 1400 + 2800r$$
  $$600 = 2800r$$
  $$\frac{600}{2800} = \frac{2800r}{2800}$$
  $$0.214 = r$$
  $$r = 21.4\%$$

39.
  $$A = P(1+rt)$$
  $$3000 = P\left[1 + (0.065)(2)\right]$$
  $$3000 = P(1.13)$$
  $$\frac{3000}{1.13} = \frac{P(1.13)}{1.13}$$
  $$2654.867 \approx P$$
  $$P \approx \$2654.87$$

45. a.
  $$A = P(1+rt)$$
  $$A = 5000\left[1 + (0.055)t\right]$$
  $$A = 5000 + 275t$$

  b. The slope is 275. This means the *rate of change* for the account is \$275 per year.

Check Points 8.3

1. a. $A = \$1000(1+0.04)^5 \approx \$1216.65$

  b. $\$1216.65 - \$1000 = \$216.65$

2. a. $A = \$4200\left(1 + \dfrac{0.04}{4}\right)^{4\cdot 10} \approx \$6253.23$

  b. $\$6253.23 - \$4200 = \$2053.23$

3. a. $A = P\left(1 + \dfrac{r}{n}\right)^{nt}$

$A = 10,000\left(1 + \dfrac{0.08}{4}\right)^{4(5)}$

$= \$14,859.47$

b. $A = Pe^{rt}$

$A = 10,000e^{0.08(5)}$

$= \$14,918.25$

4. $P = \dfrac{A}{\left(1 + \dfrac{r}{n}\right)^{nt}}$

$A = \$10,000, \ r = 0.06, \ n = 52, \ t = 8$

$P = \dfrac{10,000}{\left(1 + \dfrac{0.06}{52}\right)^{52 \cdot 8}} \approx \dfrac{10,000}{1.6156273} \approx \$6189.55$

5. a. $A = \$6000\left(1 + \dfrac{0.10}{12}\right)^{12 \cdot 1} \approx \$6628.28$

b. $A = P(1 + rt)$

$6628.28 = 6000\big[1 + (r)(1)\big]$

$6628.28 = 6000 + 6000r$

$628.28 = 6000r$

$\dfrac{628.28}{6000} = \dfrac{6000r}{6000}$

$0.105 \approx r$

$r \approx 10.5\%$

6. $Y = \left(1 + \dfrac{r}{n}\right)^{n} - 1$

$Y = \left(1 + \dfrac{0.08}{4}\right)^{4} - 1 \approx 0.0824 = 8.24\%$

Exercise Set 8.3

1. a. $A = \$10,000(1 + 0.04)^2 = \$10,816$

b. $\$10,816 - \$10,000 = \$816$

3. a. $A = \$3000\left(1 + \dfrac{0.05}{2}\right)^{2 \cdot 4}$

$= \$3000(1.025)^8$

$= \$3655.21$

b. $\$3655.21 - \$3000 = \$655.21$

5. a. $A = \$9500\left(1 + \dfrac{0.06}{4}\right)^{4 \cdot 5}$

$= \$9500(1.015)^{20}$

$= \$12,795.12$

b. $\$12,795.12 - \$9500 = \$3295.12$

7. a. $A = \$4500\left(1 + \dfrac{0.045}{12}\right)^{12 \cdot 3}$

$= \$4500(1.0038)^{36}$

$= \$5149.12$

b. $\$5149.12 - \$4500 = \$649.12$

9. a. $A = \$1500\left(1 + \dfrac{0.085}{360}\right)^{360 \cdot 2.5}$

$= \$1500(1.000236)^{900}$

$= \$1855.10$

b. $\$1855.10 - \$1500 = \$355.10$

11. a. $A = \$20,000\left(1 + \dfrac{0.045}{360}\right)^{360 \cdot 20}$

$= \$20,000(1.000125)^{7200}$

$= \$49,189.30$

b. $\$49,189.30 - \$20,000 = \$29,189.30$

13. a. $A = 10,000\left(1 + \dfrac{0.055}{2}\right)^{2(5)}$

$\approx \$13,116.51$

b. $A = 10,000\left(1 + \dfrac{0.055}{4}\right)^{4(5)}$

$\approx \$13,140.67$

c. $A = 10,000\left(1 + \dfrac{0.055}{12}\right)^{12(5)}$

$\approx \$13,157.04$

d. $A = 10,000e^{0.055(5)}$

$\approx \$13,165.31$

15. $A = 12,000\left(1 + \dfrac{0.07}{12}\right)^{12(3)}$

$\approx 14,795.11$ (7% yield)

$A = 12,000e^{0.0685(3)}$

$\approx 14,737.67$ (6.85% yield)

Investing $12,000 for 3 years at 7% compounded monthly yields the greater return.

17. $A = \$10,000$, $r = 0.06$, $n = 2$, $t = 3$

$P = \dfrac{10,000}{\left(1 + \frac{0.06}{2}\right)^{2 \cdot 3}} = \dfrac{10,000}{(1.03)^6} = \$8374.85$

19. $A = \$10,000$, $r = 0.095$, $n = 12$, $t = 3$

$P = \dfrac{10,000}{\left(1 + \frac{0.095}{12}\right)^{12 \cdot 3}} = \dfrac{10,000}{(1.00791667)^{36}} = \$7528.59$

21. a. $A = \$10,000\left(1 + \dfrac{0.045}{4}\right)^{4 \cdot 1}$

$= \$10,000(1.01125)^4$

$= \$10,457.65$

b. $A = P(1 + rt)$

$10,457.65 = 10,000\left[1 + r(1)\right]$

$10,457.65 = 10,000 + 10,000r$

$457.65 = 10,000r$

$\dfrac{457.65}{10,000} = \dfrac{10,000r}{10,000}$

$0.046 \approx r$

$r \approx 4.6\%$

23. $Y = \left(1 + \dfrac{0.06}{2}\right)^2 - 1 = 0.061 = 6.1\%$

25. $Y = \left(1 + \dfrac{0.06}{12}\right)^{12} - 1 \approx 0.062 = 6.2\%$

27. $Y = \left(1 + \dfrac{0.06}{1000}\right)^{1000} - 1 \approx 0.062 = 6.2\%$

29. $Y = \left(1 + \dfrac{0.08}{12}\right)^{12} - 1 \approx 0.0830 = 8.3\%$

$Y = \left(1 + \dfrac{0.0825}{1}\right)^1 - 1 \approx 0.0825 = 8.25\%$

8% compounded monthly is better..

31. $Y = \left(1 + \dfrac{0.055}{2}\right)^2 - 1 \approx 0.0558 = 5.6\%$

$Y = \left(1 + \dfrac{0.054}{360}\right)^{360} - 1 \approx 0.05548 = 5.5\%$

5.5% compounded semiannually is better.

33. $A = P(1 + r)^t$

$3P = P(1 + 0.05)^t$

$3 = (1.05)^t$

$t \approx 22.5$ years

35. $A = P(1 + r)^t$

$1.5P = P(1 + 0.10)^t$

$1.5 = (1.10)^t$

$t \approx 4.3$ years

37. $A = P(1 + r)^t$

$1.9P = P(1 + 0.08)^t$

$1.9 = (1.08)^t$

$t \approx 8.3$ years

39. a. $A = 24\left(1 + \dfrac{0.05}{12}\right)^{12 \cdot 380} \approx \$4,117,800,000$

b. $A = 24\left(1 + \dfrac{0.05}{360}\right)^{360 \cdot 380} \approx \$4,277,900,000$

41. $A = P\left(1 + \dfrac{r}{n}\right)^{nt}$

$A = \$10,000\left(1 + \dfrac{0.09}{12}\right)^{12 \cdot 21}$

$= \$10,000(1.0075)^{252}$

$= \$65,728.51$

43. $P = \dfrac{A}{\left(1 + \dfrac{r}{n}\right)^{nt}}$

$A = \$500,000$, $r = 0.09$, $n = 12$, $t = 65 - 30 = 35$

$P = \dfrac{500,000}{\left(1 + \frac{0.09}{12}\right)^{12 \cdot 35}} = \dfrac{500,000}{(1.0075)^{420}} = \$21,679.39$

51. $A = P\left(1 + \dfrac{r}{n}\right)^{nt}$

Start with $5000 in account for 2 years: $A = \$5000\left(1 + \dfrac{0.08}{12}\right)^{12 \cdot 2} = \$5864.44$

Then withdraw $1500, which leaves $5864.44 − $1500 = $4364.44

Leave $4364.44 in the account for 1 year: $A = \$4364.44\left(1 + \dfrac{0.08}{12}\right)^{12 \cdot 1} = \$4726.69$

Then add $2000, so now have $4726.69 + $2000 = $6726.69

Have $6726.69 in account for 3 years: $A = 6726.69\left(1 + \dfrac{0.08}{12}\right)^{12 \cdot 3} = \$8544.49$

Check Points 8.4

1. a. Value at end of year 1
$2000
Value at end of year 2
$2000(1 + 0.10) + \$2000 = \$4200$
Value at end of year 3
$4200(1 + 0.10) + \$2000 = \$6620$

b. $6620 − \$2000 \cdot 3 = \$620$

2. a. $A = \dfrac{P\left[(1+r)^t - 1\right]}{r}$

$A = \dfrac{3000\left[(1+0.08)^{40} - 1\right]}{0.08}$

$\approx \$777,170$

b. $777,170 − 40 \times \$3000 = \$657,170$

3. a. $A = \dfrac{P\left[\left(1 + \frac{r}{n}\right)^{nt} - 1\right]}{\frac{r}{n}}$

$A = \dfrac{100\left[\left(1 + \frac{0.095}{12}\right)^{12 \times 35} - 1\right]}{\frac{0.095}{12}}$

$\approx \$333,946$

b. $333,946 − \$100 \cdot 12 \cdot 35 = \$291,946$

4. a. $P = \dfrac{A\left(\frac{r}{n}\right)}{\left[\left(1 + \frac{r}{n}\right)^{nt} - 1\right]}$

$P = \dfrac{100,000\left(\frac{0.09}{12}\right)}{\left[\left(1 + \frac{0.09}{12}\right)^{12 \times 18} - 1\right]}$

$\approx \$187$

b. Deposits: $187 \times 18 \times 12 = \$40,392$
Interest: $100,000 − \$40,392 = \$59,608$

5. a. High price = $63.38,
Low price = $42.37

b. Dividend $= \$0.72 \cdot 3000 = \$2160$

c. Annual return for dividends alone = 1.5%
1.5% is much lower than the 3.5% bank rate.

d. Shares traded =
$72,032 \cdot 100 = 7,203,200$ shares

e. High price = $49.94,
Low price = $48.33

f. Price at close = $49.50

g. The price went up $0.03 per share.

h. Annual earnings per share $= \dfrac{\$49.50}{37} \approx \$1.34$

Exercise Set 8.4

1. a. $A = \dfrac{P\left[(1+r)^t - 1\right]}{r}$

$A = \dfrac{2000\left[(1+0.05)^{20} - 1\right]}{0.05}$

$\approx \$66,132$

b. $66,132 − 20 \times \$2000 = \$26,132$

3. a. $A = \dfrac{P\left[(1+r)^t - 1\right]}{r}$

$A = \dfrac{4000\left[(1+0.065)^{40} - 1\right]}{0.065}$

$\approx \$702,528$

b. $\$702,528 - 40 \times \$4000 = \$542,528$

5. a. $A = \dfrac{P\left[\left(1+\frac{r}{n}\right)^{nt} - 1\right]}{\frac{r}{n}}$

$A = \dfrac{50\left[\left(1+\frac{0.06}{12}\right)^{12\times30} - 1\right]}{\frac{0.06}{12}}$

$\approx \$50,226$

b. $\$50,226 - \$50 \cdot 12 \cdot 30 = \$32,226$

7. a. $A = \dfrac{P\left[\left(1+\frac{r}{n}\right)^{nt} - 1\right]}{\frac{r}{n}}$

$A = \dfrac{100\left[\left(1+\frac{0.045}{2}\right)^{2\times25} - 1\right]}{\frac{0.045}{2}}$

$\approx \$9076$

b. $\$9076 - \$100 \cdot 2 \cdot 25 = \$4076$

9. a. $A = \dfrac{P\left[\left(1+\frac{r}{n}\right)^{nt} - 1\right]}{\frac{r}{n}}$

$A = \dfrac{1000\left[\left(1+\frac{0.0625}{4}\right)^{4\times6} - 1\right]}{\frac{0.0625}{4}}$

$\approx \$28,850$

b. $\$28,850 - \$1000 \cdot 4 \cdot 6 = \$4850$

11. a. $P = \dfrac{A\left(\frac{r}{n}\right)}{\left[\left(1+\frac{r}{n}\right)^{nt} - 1\right]}$

$P = \dfrac{140,000\left(\frac{0.06}{1}\right)}{\left[\left(1+\frac{0.06}{1}\right)^{1\times18} - 1\right]}$

$\approx \$4530$

b. Deposits: $\$4530 \times 1 \times 18 = \$81,540$
Interest: $\$140,000 - \$81,540 = \$58,460$

13. a. $P = \dfrac{A\left(\frac{r}{n}\right)}{\left[\left(1+\frac{r}{n}\right)^{nt} - 1\right]}$

$P = \dfrac{200,000\left(\frac{0.045}{12}\right)}{\left[\left(1+\frac{0.045}{12}\right)^{12\times10} - 1\right]}$

$\approx \$1323$

b. Deposits: $\$1323 \times 12 \times 10 = \$158,760$
Interest: $\$200,000 - \$158,760 = \$41,240$

15. a. $P = \dfrac{A\left(\frac{r}{n}\right)}{\left[\left(1+\frac{r}{n}\right)^{nt} - 1\right]}$

$P = \dfrac{1,000,000\left(\frac{0.0725}{12}\right)}{\left[\left(1+\frac{0.0725}{12}\right)^{12\times40} - 1\right]}$

$\approx \$356$

b. Deposits: $\$356 \times 12 \times 40 = \$170,880$
Interest: $\$1,000,000 - \$170,880 = \$829,120$

17. a. $P = \dfrac{A\left(\frac{r}{n}\right)}{\left[\left(1+\frac{r}{n}\right)^{nt} - 1\right]}$

$P = \dfrac{20,000\left(\frac{0.035}{4}\right)}{\left[\left(1+\frac{0.035}{4}\right)^{4\times5} - 1\right]}$

$\approx \$920$

b. Deposits: $\$920 \times 4 \times 5 = \$18,400$
Interest: $\$20,000 - \$18,400 = \$1600$

19. a. High price = \$73.25,
Low price = \$45.44

b. Dividend = $\$1.20 \cdot 700 = \$840$

c. Annual return for dividends alone = 2.2%
2.2% is lower than a 3% bank rate.

d. Shares traded = $5915 \cdot 100 = 591,500$ shares

e. High price = \$56.38,
Low price = \$54.38

f. Price at close = \$55.50

g. The price went up \$1.25 per share.

h. Annual earnings per share $= \dfrac{\$55.50}{17}$
$\approx \$3.26$

21. a. Lump-Sum Deposit:

$$A = P(1+r)^t$$

$$A = 30,000(1+0.05)^{20}$$

$$\approx \$79,599$$

Periodic Deposit:

$$A = \frac{P\left[(1+r)^t - 1\right]}{r}$$

$$A = \frac{1500\left[(1+0.05)^{20} - 1\right]}{0.05}$$

$$\approx \$49,599$$

The lump-sum investment will have $79,599 − $49,599 = $30,000 more.

b. Lump-Sum Interest:
$79,599 − $30,000 = $49,599

Periodic Deposit Interest:
$49,599 − $30,000 = $19,599

The lump-sum investment will have $49,599 − $19,599 = $30,000 more.

23. a.

$$P = \frac{A\left(\frac{r}{n}\right)}{\left[\left(1+\frac{r}{n}\right)^{nt} - 1\right]}$$

$$P = \frac{1,000,000\left(\frac{0.08}{12}\right)}{\left[\left(1+\frac{0.08}{12}\right)^{12\times40} - 1\right]}$$

$$\approx \$287$$

b. Adjusted gross income *with* IRA:
Adj. gross income = $50,000 − $287·12

$$= \$46,556$$

Taxable income *with* IRA:
Taxable inc = $46,556 − ($3200 + $5000)

$$= \$38,356$$

Income tax *with* IRA:

$$= 0.10(7300) + 0.15(29,700 - 7300)$$

$$+ 0.25(38,356 - 29,700)$$

$$= \$6254$$

Adjusted gross income *without* IRA:
Adj. gross income = $50,000 − $0

$$= \$50,000$$

Taxable income *without* IRA:
Taxable inc = $50,000 − ($3200 + $5000)

$$= \$41,800$$

Income tax *without* IRA:

$$= 0.10(7300) + 0.15(29,700 - 7300)$$

$$+ 0.25(41,800 - 29,700)$$

$$= \$7115$$

c. Percent of gross income *with* IRA:

$$\frac{6254}{50,000} \approx 12.5\%$$

Percent of gross income *without* IRA:

$$\frac{7115}{50,000} \approx 14.2\%$$

25.

$$A = \frac{P\left[(1+r)^t - 1\right]}{r}$$

$$Ar = P\left[(1+r)^t - 1\right]$$

$$\frac{Ar}{\left[(1+r)^t - 1\right]} = \frac{P\left[(1+r)^t - 1\right]}{\left[(1+r)^t - 1\right]}$$

$$\frac{Ar}{(1+r)^t - 1} = P$$

$$P = \frac{Ar}{(1+r)^t - 1}$$

This formula describes the deposit necessary at the end of each year that yields *A* dollars after *t* years with interest rate *r* compounded annually.

27. a.

$$A = \frac{P\left[(1+r)^t - 1\right]}{r}$$

$$A = \frac{2000\left[(1+0.075)^5 - 1\right]}{0.075}$$

$$\approx \$11,617$$

b. $11,617 − 5 × $2000 = $1617

29. a.

$$A = \frac{P\left[\left(1+\frac{r}{n}\right)^{nt} - 1\right]}{\frac{r}{n}}$$

$$A = \frac{50\left[\left(1+\frac{0.055}{12}\right)^{12\times40} - 1\right]}{\frac{0.055}{12}}$$

$$\approx \$87,052$$

b. $87,052 − $50·12·40 = $63,052

31. a. $A = \dfrac{P\left[\left(1+\frac{r}{n}\right)^{nt} - 1\right]}{\frac{r}{n}}$

$A = \dfrac{10{,}000\left[\left(1+\frac{0.105}{4}\right)^{4\times10} - 1\right]}{\frac{0.105}{4}}$

$\approx \$693{,}031$

b. $\$693{,}031 - \$10{,}000 \cdot 4 \cdot 10 = \$293{,}031$

33. a. $P = \dfrac{A\left(\frac{r}{n}\right)}{\left[\left(1+\frac{r}{n}\right)^{nt} - 1\right]}$

$P = \dfrac{3500\left(\frac{0.05}{2}\right)}{\left[\left(1+\frac{0.05}{2}\right)^{2\times4} - 1\right]}$

$\approx \$401$

b. Deposits: $\$401 \times 2 \times 4 = \$3208$
Interest: $\$3500 - \$3208 = \$292$

35. $P = \dfrac{A\left(\frac{r}{n}\right)}{\left[\left(1+\frac{r}{n}\right)^{nt} - 1\right]}$

$P = \dfrac{4{,}000{,}000\left(\frac{0.085}{12}\right)}{\left[\left(1+\frac{0.085}{12}\right)^{12\times45} - 1\right]}$

$\approx \$641$

You must invest \$641 per month.

Amount from interest:
$\$4{,}000{,}000 - \$641 \cdot 12 \cdot 45 = \$3{,}653{,}860$

Check Points 8.5

1. a. Amount financed $= \$14{,}000 - \$280 = \$13{,}720$

b. Total installment price $= 60 \cdot \$315 + \$280 = \$19{,}180$

c. Finance charge $= \$19{,}180 - \$14{,}000 = \$5180$

2. Find the finance charge per \$100 financed: $\dfrac{\$5180}{\$13{,}720} \cdot \$100 = \$37.76$

For 60 monthly payments, \$37.76 is closest to \$38.06 in table 8.5. Therefore the APR is 13.5%.

3. a. $u = \dfrac{kRV}{100+V}$, where $k = 60 - 24 = 36$, $R = \$315$, $V = \$22.17$

V is found by looking up the APR from Check Point 2 in the "36-payments" row of table 8.5.

Interest saved $= u = \dfrac{36(\$315)(\$22.17)}{100+22.17} \approx \$2057.85$

b. Payoff amount = (Payment number 24) + (Total of remaining payments after payment 24) – (Interest saved)
$= \$315 + 36 \cdot \$315 - \$2057.85 = \$9597.15$

4. a. $u = \dfrac{k(k+1)}{n(n+1)} \cdot F$, where $k = 36$, $n = 60$, $F = \$5180$ (Computed in Check Point 1)

Interest saved $= u = \dfrac{36(36+1)}{60(60+1)} \cdot 5180 \approx \$1885.18$

b. Payoff amount = (Payment number 24) + (Total of remaining payments after payment 24) – (Interest saved)
$= \$315 + 36 \cdot \$315 - \$1885.18 = \$9769.82$

5. a. The interest is $I = Prt = (\$4720 - \$1000) \times 0.016 \times 1 = \$3720 \times 0.016 \times 1 = \$59.52$

 b. New balance $= \$3720 + \$59.52 + \$1025 + \$45 = \$4849.52$

 c. Minimum monthly payment $= \dfrac{\text{balance owed}}{36} = \dfrac{\$4849.52}{36} \approx \$135$.

6. a. Unpaid Balance Method: $I = Prt = (\$6800 - \$500) \times 0.018 \times 1 = \$6300 \times 0.018 \times 1 = \$113.40$

 b. Previous Balance Method: $I = Prt = \$6800 \times 0.018 \times 1 = \$122.40$

 c. Average daily balance $= \dfrac{(\$6800)(7) + (\$6300)(24)}{31} \approx \$6412.90$
 Average Daily Balance Method: $I = Prt = \$6412.90 \times 0.018 \times 1 \approx \$115.43$.

Exercise Set 8.5

1. a. Amount financed $= \$27{,}000 - \$5000 = \$22{,}000$

 b. Total installment price $= 60 \cdot \$410 + \$5000 = \$29{,}600$

 c. Finance charge $= \$29{,}600 - \$27{,}000 = \$2{,}600$

3. a. Amount financed $= \$1100 - 100 = \$1000$

 b. Total installment price $= 12 \cdot \$110 + \$100 = \$1420$

 c. Finance charge $= \$1420 - \$1100 = \$320$

5. In the row for 12 monthly payments, find the value $6.90. That value is in the column for 12.5%.

7. In the row for 24 monthly payments, find the value $15.80. That value is in the column for 14.5%.

9. Finance charge per $100 financed $= \dfrac{\text{Finance charge}}{\text{Amount financed}} \cdot \$100 = \dfrac{\$1279}{\$4450} \cdot \$100 \approx \$28.74$
Using table 8.5 with 48 monthly payments the APR is 13.0%.

11. a. Amount financed $= \$17{,}500 - \$500 = \$17{,}000$

 b. Total installment price $= 60 \cdot \$360.55 + \$500 = \$22{,}133$

 c. Finance charge $= \$22{,}133 - \$17{,}500 = \$4633$

 d. Finance charge per $100 financed $= \dfrac{\$4633}{\$17{,}000} \cdot \$100 \approx \$27.25$. The APR is 10.0%.

13. a. $u = \dfrac{kRV}{100+V}$, where $k = 60 - 24 = 36$, $R = \$360.55$, and $V = \$16.16$

V is found by looking up the APR from Exercise 11d in the "36-payments" row of table 8.5.

Interest saved $= u = \dfrac{36(\$360.55)(\$16.16)}{100+16.16} \approx \$1805.73$

b. Payoff amount = (Payment number 24) + (Total of remaining payments after payment 24) – (Interest saved)
$= \$360.55 + 36 \cdot \$360.55 - \$1805.73$
$= \$11,534.62$

c. $u = \dfrac{k(k+1)}{n(n+1)} \cdot F$, where $k = 36$, $n = 60$, and $F = \$4633$ (Computed in exercise 11c)

Interest saved $= u = \dfrac{36(36+1)}{60(60+1)} \cdot 4633 \approx \$1686.11$

d. Payoff amount = (Payment number 24) + (Total of remaining payments after payment 24) – (Interest saved)
$= \$360.55 + 36 \cdot \$360.55 - \$1686.11$
$= \$11,654.24$

15. a. Unpaid balance = $\$950 - \$100 = \$850$
$I = Prt = (\$850)(0.013)(1) = \$11.05$

b. Balance due June 9
$= \$850 + \$11.05 + \$85 + \$67$
$= \$1013.05$

c. Minimum monthly payment $= \dfrac{\$1013.05}{36} \approx \$28$

17. a. $I = Prt = (\$330.90)(0.015)(1) = \$4.96$

b. New Balance $= \$445.59 + \$278.06 - \$110 + \$4.96 = \$618.61$

c. Minimum monthly payment $= \dfrac{\$618.61}{10} = \$61.86$ because the new balance is over $500.

19. a. Unpaid balance = $\$3000 - \$2500 = \$500$
$I = Prt = (\$500)(0.015)(1) = \$7.50$

b. Previous balance
$I = Prt = (\$3000)(0.015)(1) = \$45$

c. Average daily balance $= \dfrac{(\$3000)(5)+(\$500)(25)}{30} \approx \$916.67$
$I = Prt = (\$916.67)(0.015)(1) = \$13.75$

31. The payments will be greater than $\dfrac{\$1400 - \$200}{30} = \$40$ because a finance charge must be paid.

Answer b is reasonable.

33. Answers will vary.

35. a. Number of days = 31

Average daily balance $= \dfrac{\$11,664.15}{30} = \$376.26$

b. Finance charge $= (0.013)(376.26) = \$4.89$

c. Balance due $= \$466.15 + \$4.89 = \$471.04$

Check Points 8.6

1. a. $PMT = \dfrac{P\left(\frac{r}{n}\right)}{1-\left(1+\frac{r}{n}\right)^{-nt}} = \dfrac{175,500\left(\frac{0.075}{12}\right)}{1-\left(1+\frac{0.075}{12}\right)^{-12\cdot15}} \approx \$1627$

b. $\$1627 \cdot 12 \cdot 15 - \$175,500 = \$117,360$

c. $\$266,220 - \$117,360 = \$148,860$

2. Interest for first month $= Prt = \$200,000 \times 0.07 \times \dfrac{1}{12} \approx \$1166.67$

Principle payment $= \$1550.00 - \$1166.67 = \$383.33$
Balance of loan $= \$200,000 - \$383.33 = \$199,616.67$

Interest for second month $= Prt = \$199,616.67 \times 0.07 \times \dfrac{1}{12} \approx \$1164.43$

Principle payment $= \$1550.00 - \$1164.43 = \$385.57$
Balance of loan $= \$199,616.67 - \$385.57 = \$199,231.10$

Payment Number	Interest Payment	Principal Payment	Balance of Loan
1	\$1166.67	\$383.33	\$199,616.67
2	\$1164.43	\$385.57	\$199,231.10

3. a. $PMT = \dfrac{P\left(\frac{r}{n}\right)}{1-\left(1+\frac{r}{n}\right)^{-nt}} = \dfrac{15,000\left(\frac{0.08}{12}\right)}{1-\left(1+\frac{0.08}{12}\right)^{-12(4)}} \approx \$367$

Total interest for loan A: $\$367 \cdot 12 \cdot 4 - \$15,000 = \$2616$

b. $PMT = \dfrac{P\left(\frac{r}{n}\right)}{1-\left(1+\frac{r}{n}\right)^{-nt}} = \dfrac{15,000\left(\frac{0.10}{12}\right)}{1-\left(1+\frac{0.10}{12}\right)^{-12(6)}} \approx \$278$

Total interest for loan A: $\$278 \cdot 12 \cdot 6 - \$15,000 = \$5016$

c. Monthly payments are less with the longer-term loan, but there is more interest with the longer-term loan.

Exercise Set 8.6

1. a. $220,000(0.20) = \$44,000$

  b. $\$220,000 - \$44,000 = \$176,000$

  c. $\$176,000(0.03) = \$5280$

  d. $PMT = \dfrac{P\left(\frac{r}{n}\right)}{1-\left(1+\frac{r}{n}\right)^{-nt}} = \dfrac{176,000\left(\frac{0.07}{12}\right)}{1-\left(1+\frac{0.07}{12}\right)^{-12(30)}} \approx \$1171$

  e. $\$1171(12)(30) - \$176,000 = \$245,560$

3. Mortgage amount: $\$100,000 - \$100,000(0.05) = \$95,000$

Payment for 20-year loan: $PMT = \dfrac{P\left(\frac{r}{n}\right)}{1-\left(1+\frac{r}{n}\right)^{-nt}} = \dfrac{95,000\left(\frac{0.08}{12}\right)}{1-\left(1+\frac{0.08}{12}\right)^{-12(20)}} \approx \$795$

Interest for 20-year loan: $\$795(12)(20) - \$100,000 = \$90,800$

Payment for 30-year loan: $PMT = \dfrac{P\left(\frac{r}{n}\right)}{1-\left(1+\frac{r}{n}\right)^{-nt}} = \dfrac{95,000\left(\frac{0.08}{12}\right)}{1-\left(1+\frac{0.08}{12}\right)^{-12(30)}} \approx \$697$

Interest for 30-year loan: $\$697(12)(30) - \$100,000 = \$150,920$
The buyer saves $\$150,920 - \$90,800 = \$60,120$

5. Payment for 30-year 8% loan: $PMT = \dfrac{P\left(\frac{r}{n}\right)}{1-\left(1+\frac{r}{n}\right)^{-nt}} = \dfrac{150,000\left(\frac{0.08}{12}\right)}{1-\left(1+\frac{0.08}{12}\right)^{-12(30)}} \approx \$1101$

Interest for 30-year loan: $\$1101(12)(30) - \$150,000 = \$246,360$

Payment for 20-year 7.5% loan: $PMT = \dfrac{P\left(\frac{r}{n}\right)}{1-\left(1+\frac{r}{n}\right)^{-nt}} = \dfrac{150,000\left(\frac{0.075}{12}\right)}{1-\left(1+\frac{0.075}{12}\right)^{-12(20)}} \approx \$1208$

Interest for 20-year loan: $\$1208(12)(20) - \$150,000 = \$139,920$
The 20-year 7.5% loan is more economical. The buyer saves $\$246,360 - 139,920 = \$106,440$

7. Payment for Mortgage A: $PMT = \dfrac{P\left(\frac{r}{n}\right)}{1-\left(1+\frac{r}{n}\right)^{-nt}} = \dfrac{120,000\left(\frac{0.07}{12}\right)}{1-\left(1+\frac{0.07}{12}\right)^{-12(30)}} \approx \$798$

Interest for Mortgage A: $\$798(12)(30) - \$120,000 = \$167,280$
Points for Mortgage A: $\$120,000(0.01) = \$1200$
Cost for Mortgage A: $\$2000 + \$1200 + \$167,280 = \$170,480$

Payment for Mortgage B: $PMT = \dfrac{P\left(\frac{r}{n}\right)}{1-\left(1+\frac{r}{n}\right)^{-nt}} = \dfrac{120,000\left(\frac{0.065}{12}\right)}{1-\left(1+\frac{0.065}{12}\right)^{-12(30)}} \approx \$758$

Interest for Mortgage B: $\$758(12)(30) - \$120,000 = \$152,880$
Points for Mortgage B: $\$120,000(0.04) = \$4800$
Cost for Mortgage B: $\$1500 + \$4800 + \$152,880 = \$159,180$
Mortgage A has the greater cost by $\$170,480 - \$159,180 = \$11,300$

9. a. $PMT = \dfrac{P\left(\frac{r}{n}\right)}{1-\left(1+\frac{r}{n}\right)^{-nt}} = \dfrac{4200\left(\frac{0.18}{12}\right)}{1-\left(1+\frac{0.18}{12}\right)^{-12(2)}} \approx \$210$

   b. $\$210(12)(2) - \$4200 = \$840$

11. a. $PMT = \dfrac{P\left(\frac{r}{n}\right)}{1-\left(1+\frac{r}{n}\right)^{-nt}} = \dfrac{4200\left(\frac{0.105}{12}\right)}{1-\left(1+\frac{0.105}{12}\right)^{-12(3)}} \approx \$137$; This payment is lower.

   b. $\$137(12)(3) - \$4200 = \$732$; This loan has less interest.

13. $PMT = \dfrac{P\left(\frac{r}{n}\right)}{1-\left(1+\frac{r}{n}\right)^{-nt}} = \dfrac{4200\left(\frac{0.18}{12}\right)}{1-\left(1+\frac{0.18}{12}\right)^{-12(1)}} \approx \$386$

Total interest: $\$386(12)(1) - \$4200 = \$432$

Additional each month: $\$386 - \$210 = \$176$

Less total interest: $\$840 - \$432 = \$408$

15. a. $PMT = \dfrac{P\left(\frac{r}{n}\right)}{1-\left(1+\frac{r}{n}\right)^{-nt}} = \dfrac{10,000\left(\frac{0.08}{12}\right)}{1-\left(1+\frac{0.08}{12}\right)^{-12(4)}} \approx \$244.13$

Total interest: $\$244.13(12)(4) - \$10,000 = \$1718.24$

   b.

Payment Number	Interest	Principal	Loan Balance
1	$10,000(0.08)\left(\frac{1}{12}\right)$ $= \$66.67$	$244.13 - 66.67$ $= \$177.46$	$10,000 - 177.46$ $= \$9822.54$
2	$9822.54(0.08)\left(\frac{1}{12}\right)$ $= \$65.48$	$244.13 - 65.48$ $= \$178.65$	$9822.54 - 178.65$ $= \$9643.89$
3	$9643.89(0.08)\left(\frac{1}{12}\right)$ $= \$64.29$	$244.13 - 64.29$ $= \$179.84$	$9643.89 - 179.84$ $= \$9464.05$

17. a. $PMT = \dfrac{P\left(\frac{r}{n}\right)}{1-\left(1+\frac{r}{n}\right)^{-nt}} = \dfrac{40,000\left(\frac{0.085}{12}\right)}{1-\left(1+\frac{0.085}{12}\right)^{-12(20)}} \approx \$347.13$

Total interest: $\$347.13(12)(20) - \$40,000 = \$43,311.20$

   b.

Payment Number	Interest	Principal	Loan Balance
1	$40,000(0.085)\left(\frac{1}{12}\right)$ $= \$283.33$	$347.13 - 283.33$ $= \$63.80$	$40,000 - 63.80$ $= \$39,936.20$
2	$39,936.20(0.085)\left(\frac{1}{12}\right)$ $= \$282.88$	$347.13 - 282.88$ $= \$64.25$	$39,936.20 - 64.25$ $= \$39,871.95$
3	$39,871.95(0.085)\left(\frac{1}{12}\right)$ $= \$282.43$	$347.13 - 282.43$ $= \$64.70$	$39,871.95 - 64.70$ $= \$39,807.25$

c. $PMT = \dfrac{P\left(\frac{r}{n}\right)}{1-\left(1+\frac{r}{n}\right)^{-nt}} = \dfrac{40{,}000\left(\frac{0.085}{12}\right)}{1-\left(1+\frac{0.085}{12}\right)^{-12(10)}} \approx \$495.94$

Amount by which the monthly payment for the 10-year loan is greater: $\$495.94 - \$347.13 = \$148.81$

Total interest for 10-year loan: $\$495.94(12)(10) - \$40{,}000 = \$19{,}512.80$

Savings from 10-year loan: $\$43{,}311.20 - \$19{,}512.80 = \$23{,}798.40$

27. $$P\left(1+\tfrac{r}{n}\right)^{nt} = \dfrac{PMT\left[\left(1+\frac{r}{n}\right)^{nt}-1\right]}{\frac{r}{n}}$$

$$P\left(\tfrac{r}{n}\right)\left(1+\tfrac{r}{n}\right)^{nt} = PMT\left[\left(1+\tfrac{r}{n}\right)^{nt}-1\right]$$

$$\dfrac{P\left(\frac{r}{n}\right)\left(1+\frac{r}{n}\right)^{nt}}{\left(1+\frac{r}{n}\right)^{nt}-1} = \dfrac{PMT\left[\left(1+\frac{r}{n}\right)^{nt}-1\right]}{\left(1+\frac{r}{n}\right)^{nt}-1}$$

$$\dfrac{P\left(\frac{r}{n}\right)\left(1+\frac{r}{n}\right)^{nt}}{\left(1+\frac{r}{n}\right)^{nt}-1} = PMT$$

$$\dfrac{\dfrac{P\left(\frac{r}{n}\right)\left(1+\frac{r}{n}\right)^{nt}}{\left(1+\frac{r}{n}\right)^{nt}}}{\dfrac{\left(1+\frac{r}{n}\right)^{nt}-1}{\left(1+\frac{r}{n}\right)^{nt}}} = PMT$$

$$\dfrac{P\left(\frac{r}{n}\right)}{\dfrac{\left(1+\frac{r}{n}\right)^{nt}}{\left(1+\frac{r}{n}\right)^{nt}}-\dfrac{1}{\left(1+\frac{r}{n}\right)^{nt}}} = PMT$$

$$\dfrac{P\left(\frac{r}{n}\right)}{1-\left(1+\frac{r}{n}\right)^{-nt}} = PMT$$

Chapter 8 Review Exercises

1. $\dfrac{4}{5} = 4 \div 5 = 0.80 = 80\%$

2. $\dfrac{1}{8} = 1 \div 8 = 0.125 = 12.5\%$

3. $\dfrac{3}{4} = 3 \div 4 = 0.75 = 75\%$

4. $0.72 = 72\%$

5. $0.0035 = 0.35\%$

6. $4.756 = 475.6\%$

7. $65\% = 0.65$

8. $99.7\% = 0.997$

9. $150\% = 1.50$

10. $3\% = 0.03$

11. $0.65\% = 0.0065$

12. $\frac{1}{4}\% = 0.25\% = 0.0025$

13. $A = PB$
     $A = 0.08 \cdot 120$
     $A = 9.6$

14. a. $\text{Tax} = 0.06(\$24) = \$1.44$

     b.   Total cost $= \$24 + \$1.44 = \$25.44$

15. a.   Amount of discount $= 0.35(\$850)$
                             $= \$297.50$

     b.   Sale price $= \$850 - \$297.50 = \$552.50$

16. Step 1. Determine the adjusted gross income.
Adj. gross income = Gross income – Adjustments
Adj. gross income = $40,000 – $2500
$$= \$37,500$$

Step 2. Determine the taxable income.
Since the total deduction of $8300 is greater than
the standard deduction of $5000, use $8300.
Taxable inc. = Adj. gross inc– (Exempt.+Deduct.)
Taxable inc. = $37,500 – ($3200 + $8300)
$$= \$26,000$$

Step 3. Determine the income tax.
Tax Computation
$$= 0.10(7300) + 0.15(26,000 - 7300)$$
$$= \$3535$$
Income tax = Tax Computation – Tax credits
Income tax = $3535 – $0
$$= \$3535$$

17. $\dfrac{45 - 40}{40} = 0.125 = 12.5\%$ increase.

18. $\dfrac{\$56.00 - \$36.40}{\$56.00} = 0.35 = 35\%$ decrease.

19. The statement is not true.
The 10% loss is $1000.
$$[0.10 \times 10,000 = 1000]$$
This leaves $9000.
The 10% rise is $900.
$$[0.10 \times 9,000 = 900]$$
Thus there is $9900 in the portfolio.
Find the percent of decrease:
$$\frac{\text{amount of decrease}}{\text{original amount}} = \frac{100}{10,000} = 0.01 = 1\%$$
The net loss of $100 is a 1% decrease from the
original.

20. $I = Prt = (\$6000)(0.03)(1) = \$180$

21. $I = Prt = (\$8400)(0.05)(6) = \$2520$

22. $I = Prt = (\$20,000)(0.08)\left(\dfrac{9}{12}\right) = \$1200$

23. $I = Prt = (\$36,000)(0.15)\left(\dfrac{60}{360}\right) = \$900$

24. a. $I = Prt = (\$3500)(0.105)\left(\dfrac{4}{12}\right)$
$$= \$122.50$$

b. Maturity value = $3500 + $122.50
$$= \$3622.50$$

25. $A = P(1 + rt)$
$A = 12,000(1 + 0.082 \times \frac{9}{12})$
$A = \$12,738$

26. $A = P(1 + rt)$
$5750 = 5000\left(1 + r(2)\right)$
$5750 = 5000 + 10,000r$
$750 = 10,000r$
$0.075 = r$
$r = 7.5\%$

27. $A = P(1 + rt)$
$16,000 = P\left(1 + (0.065)(3)\right)$
$16,000 = 1.195P$
$13,389.12 = P$
$P = \$13,389.12$

28. $A = P(1 + rt)$
$12,000 = P\left(1 + (0.073)(4)\right)$
$12,000 = 1.292P$
$9287.93 = P$
$P = \$9287.93$

29. $A = P(1 + rt)$
$1800 = 1500\left(1 + r\left(\frac{1}{2}\right)\right)$
$1800 = 1500 + 750r$
$300 = 750r$
$0.4 = r$
$r = 40\%$

30. a. $I = Prt = (1800)(0.07)\left(\frac{9}{12}\right) = \$94.50$

b. Amount received: $1800 – $94.50 = $1705.50

c.
$$I = Prt$$
$$94.50 = (1705.50)(r)\left(\tfrac{9}{12}\right)$$
$$94.50 = 1279.125r$$
$$0.0739 = r$$
$$r = 7.4\%$$

31. a.
$$A = \$7000(1 + 0.03)^5$$
$$= \$7000(1.03)^5$$
$$\approx \$8114.92$$

b. Interest $= \$8114.92 - \$7000$
$$= \$1114.92$$

32. a.
$$A = \$30,000\left(1 + \frac{0.025}{4}\right)^{4\cdot10}$$
$$= \$30,000(1.00625)^{40}$$
$$\approx \$38,490.80$$

b. Interest $= \$38,490.80 - \$30,000$
$$= \$8490.80$$

33. a.
$$A = \$2500\left(1 + \frac{0.04}{12}\right)^{12\cdot20}$$
$$= \$2500(1.003333)^{240}$$
$$\approx \$5556.46$$

b. Interest $= \$5556.46 - \$2500$
$$= \$3056.46$$

34.
$$A = P\left(1 + \tfrac{r}{n}\right)^{nt}$$
$$A = 14,000\left(1 + \tfrac{0.07}{12}\right)^{12(10)}$$
$$\approx \$28,135$$

$$A = Pe^{rt}$$
$$A = 14,000e^{0.0685(10)}$$
$$\approx \$27,773$$

The 7% compounded monthly is the better investment by $\$28,135 - \$27,773 = \$362$.

35. $P = \dfrac{100,000}{\left(1 + \dfrac{0.10}{12}\right)^{12\cdot18}} \approx \$16,653.64$

36. $P = \dfrac{75,000}{\left(1 + \dfrac{0.05}{4}\right)^{4\cdot35}} \approx \$13,175.19$

37. a.
$$A = \$2000\left(1 + \frac{0.06}{4}\right)^{4\cdot1}$$
$$= \$2000(1.015)^4$$
$$= \$2122.73$$

b.
$$A = P(1 + rt)$$
$$2122.73 = 2000\left[1 + r(1)\right]$$
$$2122.73 = 2000 + 2000r$$
$$122.73 = 2000r$$
$$0.061365 \approx r$$
$$r \approx 6.1\%$$

38. $Y = \left(1 + \dfrac{0.055}{4}\right)^4 - 1 \approx 0.0561 = 5.6\%$

5.5% compounded quarterly is equivalent to 5.6% compounded annually.

39. 6.25% compounded monthly:
$$Y = \left(1 + \frac{0.0625}{12}\right)^{12} - 1 \approx 0.0643 = 6.4\%$$
6.3% compounded annually:
$$Y = \left(1 + \frac{0.063}{1}\right)^1 - 1 \approx 0.063 = 6.3\%$$
6.25% compounded monthly is better than 6.3% compounded annually.

40. a.
$$A = \frac{P\left[(1+r)^t - 1\right]}{r}$$
$$A = \frac{520\left[(1 + 0.06)^{20} - 1\right]}{0.06}$$
$$\approx \$19,129$$

b. $\$19,129 - 20 \times \$520 = \$8729$

41. a.
$$A = \frac{P\left[\left(1 + \tfrac{r}{n}\right)^{nt} - 1\right]}{\tfrac{r}{n}}$$
$$A = \frac{100\left[\left(1 + \tfrac{0.055}{12}\right)^{12(30)} - 1\right]}{\tfrac{0.055}{12}}$$
$$\approx \$91,361$$

b. $\$91,361 - 30 \times 12 \times \$100 = \$55,361$

42. **a.** $P = \dfrac{A\left(\frac{r}{n}\right)}{\left[\left(1+\frac{r}{n}\right)^{nt}-1\right]}$

$P = \dfrac{25{,}000\left(\frac{0.0725}{4}\right)}{\left[\left(1+\frac{0.0725}{4}\right)^{4(5)}-1\right]}$

$\approx \$1049$

b. Deposits: $5 \times 4 \times \$1049 = \$20{,}980$
Interest: $\$25{,}000 - \$20{,}980 = \$4020$

43. High = $64.06, Low = $26.13

44. Dividend = $0.16(900) = $144

45. Annual return for dividends alone = 0.3%

46. Shares traded yesterday = $5458 \cdot 100$
= 545,800 shares

47. High = $61.25, Low = $59.25

48. Price at close = $61

49. Change in price = $1.75 increase

50. Annual earnings per share $\dfrac{\$61}{41} \approx \$1.49$

52. **a.** Amount financed = $16,500 − $500
= $16,000

b. Total installment price
= 60($350) + $500
= $21,500

c. Finance charge = $21,500 − $16,500
= $5000

d. Finance charge per $100 financed
$\dfrac{\$5000}{\$16{,}000} \times \$100 = \$31.25$
The APR is approximately 11.5%.

53. **a.** Interest saved
$k = 12,\ R = \$350,\ v = \$6.34$
$u = \dfrac{(12)(350)(6.34)}{100+6.34} = \$250.40$

b. Payoff amount
= $350 + (12)($350) − $250.40
= $4299.60

54. **a.** Interest saved
$k = 12,\ n = 60,\ F = \$5000$
$u = \dfrac{12(12+1)}{60(60+1)} \cdot 5000 \approx \$213.11$

b. Payoff amount
= $350 + 12($350) − $213.11 = $4336.89

55. The actuarial method saves the borrower more money.

56. **a.** Unpaid balance = $1300 - $200
= $1100
$I = Prt = (\$1100)(0.015)(1) = \$16.50$

b. Balance due
= $1100 + $380 + $120 + $140 + $16.50
= $1756.50

c. Minimum monthly payment
$= \left(\dfrac{1}{36}\right)(\$1756.50)$
= $48.79
$\approx \$49$

57. **a.** Unpaid = $3600 − $2000
= $1600
$I = Prt = (\$1600)(0.018)(1) = \$28.80$

b. Previous balance = $3600
$I = Prt = (\$3600)(0.018)(1) = \$64.80$

c. Average daily balance
$= \dfrac{(\$3600)(5)+(\$1600)(26)}{31} \approx \$1922.58$
$I = Prt = (\$1922.58)(0.018)(1) = \$34.61$

58. a. $240,000(0.20) = \$48,000$

b. $\$240,000 - \$48,000 = \$192,000$

c. $\$192,000(0.02) = \$3840$

d. $PMT = \dfrac{P\left(\frac{r}{n}\right)}{1-\left(1+\frac{r}{n}\right)^{-nt}}$

$= \dfrac{192,000\left(\frac{0.07}{12}\right)}{1-\left(1+\frac{0.07}{12}\right)^{-12(30)}}$

$\approx \$1277$

e. $\$1277(12)(30) - \$192,000 = \$267,720$

59. Payment for 30-year mortgage:

$PMT = \dfrac{P\left(\frac{r}{n}\right)}{1-\left(1+\frac{r}{n}\right)^{-nt}} = \dfrac{70,000\left(\frac{0.085}{12}\right)}{1-\left(1+\frac{0.085}{12}\right)^{-12(30)}} \approx \$538$

Interest for 30-year mortgage:
$\$538(12)(30) - \$70,000 = \$123,680$

Payment for 20-year mortgage:

$PMT = \dfrac{P\left(\frac{r}{n}\right)}{1-\left(1+\frac{r}{n}\right)^{-nt}} = \dfrac{70,000\left(\frac{0.08}{12}\right)}{1-\left(1+\frac{0.08}{12}\right)^{-12(20)}} \approx \$586$

Interest for 20-year mortgage:
$\$586(12)(20) - \$70,000 = \$70,640$

The 20-year mortgage saves
$\$123,680 - \$70,640 = \$53,040$.

An advantage of the 30-year loan is the lower monthly payment. A disadvantage of the 30-year loan is the greater total interest.
An advantage of the 20-year loan is the lower total interest. A disadvantage of the 20-year loan is the higher monthly payment.

60. a. Payment for Mortgage A:

$PMT = \dfrac{P\left(\frac{r}{n}\right)}{1-\left(1+\frac{r}{n}\right)^{-nt}} = \dfrac{100,000\left(\frac{0.085}{12}\right)}{1-\left(1+\frac{0.085}{12}\right)^{-12(30)}} \approx \$769$

Payment for Mortgage B:

$PMT = \dfrac{P\left(\frac{r}{n}\right)}{1-\left(1+\frac{r}{n}\right)^{-nt}} = \dfrac{100,000\left(\frac{0.075}{12}\right)}{1-\left(1+\frac{0.075}{12}\right)^{-12(30)}} \approx \$699$

b. Interest for Mortgage A:
$\$769(12)(30) - \$100,000 = \$176,840$
Cost for Mortgage A:
$\$0 + \$0 + \$176,840 = \$176,840$

Interest for Mortgage B:
$\$699(12)(30) - \$100,000 = \$151,640$
Points for Mortgage B:
$\$100,000(0.03) = \$3000$
Cost for Mortgage B:
$\$1300 + \$3000 + \$151,640 = \$155,940$
Mortgage A has the greater cost by
$\$176,840 - \$155,940 = \$20,900$.

61. a. Payment for Loan A:

$PMT = \dfrac{P\left(\frac{r}{n}\right)}{1-\left(1+\frac{r}{n}\right)^{-nt}} = \dfrac{100,000\left(\frac{0.072}{12}\right)}{1-\left(1+\frac{0.072}{12}\right)^{-12(3)}} \approx \$465$

Interest for Loan A:
$\$465(12)(3) - \$15,000 = \$1740$

b. Payment for Loan B:

$PMT = \dfrac{P\left(\frac{r}{n}\right)}{1-\left(1+\frac{r}{n}\right)^{-nt}} = \dfrac{100,000\left(\frac{0.081}{12}\right)}{1-\left(1+\frac{0.081}{12}\right)^{-12(5)}} \approx \$305$

Interest for Loan B:
$\$305(12)(5) - \$15,000 = \$3300$

c. The longer term has a lower monthly payment but greater total interest.

62. a. $PMT = \dfrac{P\left(\frac{r}{n}\right)}{1-\left(1+\frac{r}{n}\right)^{-nt}} = \dfrac{9312\left(\frac{0.18}{12}\right)}{1-\left(1+\frac{0.18}{12}\right)^{-12(2)}} \approx \$464.89$

b. Total interest: $\$464.89(12)(2) - \$9312 = \$1845.36$

c.

Payment Number	Interest	Principal	Loan Balance
1	$9312(0.18)\left(\frac{1}{12}\right)$ $= \$139.68$	$464.89-139.68$ $= \$325.21$	$9312-325.21$ $= \$8986.79$
2	$8986.79(0.18)\left(\frac{1}{12}\right)$ $= \$134.80$	$464.89-134.80$ $= \$330.09$	$8986.79-330.09$ $= \$8656.70$
3	$8656.70(0.18)\left(\frac{1}{12}\right)$ $= \$129.85$	$464.89-129.85$ $= \$335.04$	$8656.70-335.04$ $= \$8321.66$

Chapter 8 Test

1. a. Discount $= 0.15(\$120) = \$18$

b. Sale price $= \$120 - \$18 = \$102$

2. Step 1. Determine the adjusted gross income.
Adj. gross income = Gross income − Adjustments
Adj. gross income $= \$36,500 - \$2000$
$\qquad = \$34,500$

Step 2. Determine the taxable income.
Since the total deduction of $6000 is greater than the standard deduction of $5000, use $6000.
Taxable inc. = Adj. gross inc− (Exempt.+Deduct.)
Taxable inc. $= \$34,500 - (\$3200 + \$6000)$
$\qquad = \$25,300$

Step 3. Determine the income tax.
Tax Computation
$= 0.10(7300) + 0.15(25,300 - 7300)$
$= \$3430$
Income tax = Tax Computation − Tax credits
Income tax $= \$3430 - \$0$
$\qquad = \$3430$

3. $\dfrac{3500 - 2000}{2000} = 0.75 = 75\%$ increase

4. $A = P(1+rt)$
$A = 2400\left(1+(0.12)\left(\frac{3}{12}\right)\right)$
$A = \$2472$
The future value is $2472.
The interest earned is $72.

5. $A = P(1+rt)$
$3000 = 2000(1+r(2))$
$3000 = 2000 + 4000r$
$1000 = 4000r$
$0.25 = r$
$r = 25\%$

6. $A = P(1+rt)$
$7000 = P\left(1+(0.09)\left(\frac{6}{12}\right)\right)$
$7000 = 1.045P$
$6698.57 = P$
$P = \$6698.57$

7. $Y = \left(1+\dfrac{0.045}{4}\right)^4 - 1 \approx 0.0458 = 4.58\%$

4.5% compounded quarterly is equivalent to 4.58% compounded annually.

8. a. $A = P\left(1+\frac{r}{n}\right)^{nt}$
$A = 6000\left(1+\frac{0.065}{12}\right)^{12(5)}$
$\approx \$8297$

b. $\$8297 - \$6000 = \$2297$

9. a. $A = \dfrac{P\left[\left(1+\frac{r}{n}\right)^{nt}-1\right]}{\frac{r}{n}}$

$A = \dfrac{100\left[\left(1+\frac{0.065}{12}\right)^{12(5)}-1\right]}{\frac{0.065}{12}}$

$\approx \$7067$

b. $\$7067 - \$6000 = \$1067$

c. answers will vary

10. $P = \dfrac{A}{\left(1+\frac{r}{n}\right)^{nt}}$

$P = \dfrac{3000}{\left(1+\frac{0.095}{2}\right)^{2(4)}}$

$\approx \$2070$

11. $P = \dfrac{A\left(\frac{r}{n}\right)}{\left[\left(1+\frac{r}{n}\right)^{nt}-1\right]}$

$P = \dfrac{1,500,000\left(\frac{0.0625}{12}\right)}{\left[\left(1+\frac{0.0625}{12}\right)^{12(40)}-1\right]}$

$\approx \$704$

Interest $= \$1,500,000 - \$704(12)(40)$
$= \$1,162,080$

12. High $= \$25.75$, Low $= \$25.50$

13. Dividend $= \$2.03 \cdot 1000 = \$2030$

14. Total price paid $= 600(\$25.75) = \$15,450$
Broker's commission $= 0.025(\$15,450) = \$386.25$

15. Amount financed $= \$16,000 - \$3000$
$= \$13,000$

16. Total installment price $= 60(\$300) + \$3,000$
$= \$21,000$

17. Finance charge $= \$21,000 - \$16,000$
$= \$5000$

18. Finance charge per \$100 financed
$= \dfrac{\$5000}{\$13,000} \cdot \$100$
$= \$38.46$
The APR is approximately 13.5%.

19. Interest saved
$k = 60 - 36 = 24$, $n = 60$, $F = \$5000$
$u = \dfrac{24(24+1)}{60(60+1)} \cdot \$5000 = \$819.67$

20. Payoff amount
$= \$300 + (24)(\$300) - \$819.67$
$= \$6680.33$

21. Unpaid balance $= \$880 - \$100 = \$780$
$I = Prt = \$780\,(0.02)(1) = \$15.60$

22. Balance due
$= \$780 + \$350 + \$70 + \$120 + \$15.60$
$= \$1335.60$

23. Minimum monthly payment
$= \dfrac{1}{36}(\$1335.60)$
$= \$37$

24. Average daily balance
$= \dfrac{(\$2400)(3)+(\$900)(27)}{30} = \$1050$
$I = Prt = (\$1050)(0.016)(1) = \$16.80$

25. Down payment $= 0.10(\$120,000) = \$12,000$

26. Amount of mortgage $= \$120,000 - \$12,000$
$= \$108,000$

27. Two points $= 0.02(\$108,000) = \$2160$

28. $\dfrac{\$108,000}{\$1000} = 108$ thousands of dollars of mortgage
Monthly payment $= \$7.69 \cdot 108 = \$830.52$

29. Total cost of interest
$= 360(\$830.52) - \$108,000$
$= \$190,987.20$

30. a. $PMT = \dfrac{P\left(\frac{r}{n}\right)}{1-\left(1+\frac{r}{n}\right)^{-nt}} = \dfrac{20,000\left(\frac{0.068}{12}\right)}{1-\left(1+\frac{0.068}{12}\right)^{-12(10)}} \approx \$230$

Total interest: $\$230(12)(10) - \$20,000 = \$7600$

b.

Payment Number	Interest	Principal	Loan Balance
1	$20,000(0.068)\left(\frac{1}{12}\right)$ $= \$113.33$	$230-113.33$ $= \$116.67$	$20,000-116.67$ $= \$19,883.33$
2	$19,883.33(0.068)\left(\frac{1}{12}\right)$ $= \$112.67$	$230-112.67$ $= \$117.33$	$19,883.33-117.33$ $= \$19,766.00$

Chapter 9
Measurement

Check Points 9.1

1. a. $78 \text{ in.} = \dfrac{78 \text{ in.}}{1} \cdot \dfrac{1 \text{ ft}}{12 \text{ in.}} = 6.5 \text{ ft}$

b. $17{,}160 \text{ ft} = \dfrac{17{,}160 \text{ ft}}{1} \cdot \dfrac{1 \text{ mi}}{5280 \text{ ft}} = 3.25 \text{ mi}$

c. $3 \text{ in.} = \dfrac{3 \text{ in.}}{1} \cdot \dfrac{1 \text{ yd}}{36 \text{ in.}} = \dfrac{1}{12} \text{ yd}$

2. a. $8000 \text{ m} = 8 \text{ km}$

b. $53 \text{ m} = 53{,}000 \text{ mm}$

c. $604 \text{ cm} = 0.0604 \text{ hm}$

d. $6.72 \text{ dam} = 6720 \text{ cm}$

3. a. $8 \text{ ft} = \dfrac{8 \text{ ft}}{1} \cdot \dfrac{30.48 \text{ cm}}{1 \text{ ft}} = 243.84 \text{ cm}$

b. $20 \text{ m} = \dfrac{20 \text{ m}}{1} \cdot \dfrac{1 \text{ yd}}{0.9 \text{ m}} \approx 22.22 \text{ yd}$

c. $30 \text{ m} = 3000 \text{ cm}$
$= \dfrac{3000 \text{ cm}}{1} \cdot \dfrac{1 \text{ in.}}{2.54 \text{ cm}}$
$\approx 1181.1 \text{ in.}$

4. $\dfrac{60 \text{ km}}{\text{hr}} = \dfrac{60 \text{ km}}{\text{hr}} \cdot \dfrac{1 \text{ mi}}{1.6 \text{ km}} = 37.5 \text{ mi/hr}$

Exercise Set 9.1

1. $30 \text{ in.} = \dfrac{30 \text{ in.}}{1} \cdot \dfrac{1 \text{ ft}}{12 \text{ in.}} = 2.5 \text{ ft}$

3. $30 \text{ ft} = \dfrac{30 \text{ ft}}{1} \cdot \dfrac{12 \text{ in.}}{1 \text{ ft}} = 360 \text{ in.}$

5. $6 \text{ in.} = \dfrac{6 \text{ in.}}{1} \cdot \dfrac{1 \text{ yd}}{36 \text{ in.}} \approx 0.17 \text{ yd}$

7. $6 \text{ yd} = \dfrac{6 \text{ yd}}{1} \cdot \dfrac{36 \text{ in.}}{1 \text{ yd}} = 216 \text{ in.}$

9. $6 \text{ yd} = \dfrac{6 \text{ yd}}{1} \cdot \dfrac{3 \text{ ft}}{1 \text{ yd}} = 18 \text{ ft}$

11. $6 \text{ ft} = \dfrac{6 \text{ ft}}{1} \cdot \dfrac{1 \text{ yd}}{3 \text{ ft}} = 2 \text{ yd}$

13. $23{,}760 \text{ ft} = \dfrac{23{,}760 \text{ ft}}{1} \cdot \dfrac{1 \text{ mi}}{5280 \text{ ft}} = 4.5 \text{ mi}$

15. $0.75 \text{ mi} = \dfrac{0.75 \text{ mi}}{1} \cdot \dfrac{5280 \text{ ft}}{1 \text{ mi}} = 3960 \text{ ft}$

17. $5 \text{ m} = 500 \text{ cm}$

19. $16.3 \text{ hm} = 1630 \text{ m}$

21. $317.8 \text{ cm} = 0.03178 \text{ hm}$

23. $0.023 \text{ mm} = 0.000023 \text{ m}$

25. $2196 \text{ mm} = 21.96 \text{ dm}$

27. $14 \text{ in.} = \dfrac{14 \text{ in.}}{1} \cdot \dfrac{2.54 \text{ cm}}{1 \text{ in.}} \approx 35.56 \text{ cm}$

29. $14 \text{ cm} = \dfrac{14 \text{ cm}}{1} \cdot \dfrac{1 \text{ in.}}{2.54 \text{ cm}} \approx 5.51 \text{ in.}$

31. $265 \text{ mi} = \dfrac{265 \text{ mi}}{1} \cdot \dfrac{1.6 \text{ km}}{1 \text{ mi}} \approx 424 \text{ km}$

33. $265 \text{ km} = \dfrac{265 \text{ km}}{1} \cdot \dfrac{1 \text{ mi}}{1.6 \text{ km}} \approx 165.625 \text{ mi}$

35. $12 \text{ m} = \dfrac{12 \text{ m}}{1} \cdot \dfrac{1 \text{ yd}}{0.9 \text{ m}} \approx 13.33 \text{ yd}$

37. $14 \text{ dm} = 140 \text{ cm} = \dfrac{140 \text{ cm}}{1} \cdot \dfrac{1 \text{ in.}}{2.54 \text{ cm}} \approx 55.12 \text{ in.}$

39. $160 \text{ in.} = \dfrac{160 \text{ in.}}{1} \cdot \dfrac{2.54 \text{ cm}}{1 \text{ in.}}$
$\approx 406.4 \text{ cm}$
$= 0.4064 \text{ dam}$

41. $5 \text{ ft} = \dfrac{5 \text{ ft}}{1} \cdot \dfrac{30.48 \text{ cm}}{1 \text{ ft}} \approx 152.4 \text{ cm} \approx 1.524 \text{ m}$

43. $5 \text{ m} = 500 \text{ cm} = \dfrac{500 \text{ cm}}{1} \cdot \dfrac{1 \text{ ft}}{30.48 \text{ cm}} \approx 16.40 \text{ ft}$

45. $\dfrac{96 \text{ km}}{\text{hr}} = \dfrac{96 \text{ km}}{\text{hr}} \cdot \dfrac{1 \text{ mi}}{1.6 \text{ km}} \approx 60 \text{ mi/hr}$

47. $\dfrac{45 \text{ mi}}{\text{hr}} = \dfrac{45 \text{ mi}}{\text{hr}} \cdot \dfrac{1.6 \text{ km}}{1 \text{ mi}} \approx 72 \text{ km/hr}$

49. $5 \text{ yd} = \dfrac{5 \text{ yd}}{1} \cdot \dfrac{36 \text{ in.}}{1 \text{ yd}} \cdot \dfrac{2.54 \text{ cm}}{1 \text{ in.}} \approx 457.2 \text{ cm}$

51. $762 \text{ cm} = \dfrac{762 \text{ cm}}{1} \cdot \dfrac{1 \text{ in.}}{2.54 \text{ cm}} \cdot \dfrac{1 \text{ yd}}{36 \text{ in.}} \approx 8\dfrac{1}{3} \text{ yd}$

53. $30 \text{ mi} = \dfrac{30 \text{ mi}}{1} \cdot \dfrac{5280 \text{ ft}}{1 \text{ mi}} \cdot \dfrac{12 \text{ in.}}{1 \text{ ft}} \cdot \dfrac{2.54 \text{ cm}}{1 \text{ in.}} \cdot \dfrac{1 \text{ m}}{100 \text{ cm}} \cdot \dfrac{1 \text{ km}}{1000 \text{ m}} \approx 48.28032 \text{ km}$

55. $\dfrac{120 \text{ mi}}{\text{hr}} = \dfrac{120 \text{ mi}}{\text{hr}} \cdot \dfrac{5280 \text{ ft}}{1 \text{ mi}} \cdot \dfrac{1 \text{ hr}}{60 \text{ min.}} \cdot \dfrac{1 \text{ min.}}{60 \text{ sec}} = \dfrac{176 \text{ ft}}{1 \text{ sec}} = 176 \text{ ft/sec}$

57. meter

59. millimeter

61. meter

63. millimeter

65. millimeter

67. b.

69. a.

71. c.

73. a.

75. $2 \cdot 4 \cdot 27 \text{ m} = 216 \text{ m} = 0.216 \text{ km}$

77. 93 million miles $= \dfrac{93,000,000 \text{ mi}}{1} \cdot \dfrac{1.6 \text{ km}}{1 \text{ mi}}$
$= 148.8 \text{ million kilometers}$

87. $900 \text{ m} = 9 \text{ hm}$

89. $11,000 \text{ mm} = 11 \text{ m}$

Check Points 9.2

1. The area is 8 square units.

2. $\dfrac{35,893,799 \text{ people}}{163,696 \text{ square miles}} \approx 219.3 \text{ people per sq. mile}$

3. **a.** $1.8 \text{ acres} = \dfrac{1.8 \text{ acres}}{1} \cdot \dfrac{0.4 \text{ ha}}{1 \text{ acre}} = 0.72 \text{ ha}$

   b. $\dfrac{\$415,000}{0.72 \text{ ha}} = \$576,389 \text{ per hectare}$

4. The volume is 9 cubic units.

5. $10,000 \text{ ft}^3 = \dfrac{10,000 \text{ ft}^3}{1} \cdot \dfrac{7.48 \text{ gal}}{1 \text{ ft}^3} = 74,800 \text{ gal}$

6. $220,000 \text{ cm}^3 = \dfrac{220,000 \text{ cm}^3}{1} \cdot \dfrac{1 \text{ L}}{1000 \text{ cm}^3} = 220 \text{ L}$

Exercise Set 9.2

1. $4 \cdot 4 = 16$ square units

3. 8 square units

5. $14 \text{ cm}^2 = \dfrac{14 \text{ cm}^2}{1} \cdot \dfrac{1 \text{ in.}^2}{6.5 \text{ cm}^2} \approx 2.15 \text{ in.}^2$

7. $30 \text{ m}^2 = \dfrac{30 \text{ m}^2}{1} \cdot \dfrac{1 \text{ yd}^2}{0.8 \text{ m}^2} = 37.5 \text{ yd}^2$

9. $10.2 \text{ ha} = \dfrac{10.2 \text{ ha}}{1} \cdot \dfrac{1 \text{ acre}}{0.4 \text{ ha}} = 25.5 \text{ acres}$

11. $14 \text{ in.}^2 = \dfrac{14 \text{ in.}^2}{1} \cdot \dfrac{6.5 \text{ cm}^2}{1 \text{ in.}^2} = 91 \text{ cm}^2$

13. $2 \cdot 4 \cdot 3 = 24$ cubic units

15. $10,000 \text{ ft}^3 = \dfrac{10,000 \text{ ft}^3}{1} \cdot \dfrac{7.48 \text{ gal}}{1 \text{ ft}^3}$
$= 74,800 \text{ gal}$

17. $8 \text{ yd}^3 = \dfrac{8 \text{ yd}^3}{1} \cdot \dfrac{200 \text{ gal}}{1 \text{ yd}^3} = 1600 \text{ gal}$

19. $2079 \text{ in.}^3 = \dfrac{2079 \text{ in.}^3}{1} \cdot \dfrac{1 \text{ gal}}{231 \text{ in.}^3} = 9 \text{ gal}$

21. $2700 \text{ gal} = \dfrac{2700 \text{ gal}}{1} \cdot \dfrac{1 \text{ yd}^3}{200 \text{ gal}} = 13.5 \text{ yd}^3$

23. $45,000 \text{ cm}^3 = \dfrac{45,000 \text{ cm}^3}{1} \cdot \dfrac{1 \text{ L}}{1000 \text{ cm}^3} = 45 \text{ L}$

25. $17 \text{ cm}^3 = \dfrac{17 \text{ cm}^3}{1} \cdot \dfrac{1 \text{ L}}{1000 \text{ cm}^3} \cdot \dfrac{1 \text{ mL}}{0.001 \text{ L}} = 17 \text{ mL}$

27. $1.5 \text{ L} = \dfrac{1.5 \text{ L}}{1} \cdot \dfrac{1000 \text{ cm}^3}{1 \text{ L}} = 1500 \text{ cm}^3$

29. $150 \text{ mL} = \dfrac{150 \text{ mL}}{1} \cdot \dfrac{0.001 \text{ L}}{\text{mL}} \cdot \dfrac{1000 \text{ cm}^3}{1 \text{ L}}$
$= 150 \text{ cm}^3$

31. $12 \text{ kL} = \dfrac{12 \text{ kL}}{1} \cdot \dfrac{1000 \text{ L}}{1 \text{ kL}} \cdot \dfrac{1 \text{ dm}^3}{1 \text{ L}}$
$= 12,000 \text{ dm}^3$

33. a. Population density in 1900:
$\dfrac{75,994,575 \text{ people}}{2,969,834 \text{ square miles}}$
≈ 25.6 people per square mile
Population density in 2000:
$\dfrac{281,421,906 \text{ people}}{3,537,441 \text{ square miles}}$
≈ 79.6 people per square mile

b. $\dfrac{79.6 - 25.6}{25.6} \approx 2.109 = 210.9\%$ increase

35. $\dfrac{131,669,275 \text{ people}}{2,977,128 \text{ square miles}}$
$= \dfrac{131,669,275 \text{ people}}{2,977,128 \text{ mi}^2} \cdot \dfrac{1 \text{ mi}^2}{2.6 \text{ km}^2}$
≈ 17.0 people per square kilometer

37. a. $8 \text{ ha} = \dfrac{8 \text{ ha}}{1} \cdot \dfrac{1 \text{ acre}}{0.4 \text{ ha}} = 20 \text{ acres}$

b. $\dfrac{\$250,000}{20 \text{ acres}} = \$12,500$ per acre

39. square centimeters or square meters

41. square kilometers

43. b

45. b

47. $45,000 \text{ ft}^3 = \dfrac{45,000 \text{ ft}^3}{1} \cdot \dfrac{7.48 \text{ gal}}{1 \text{ ft}^3}$
$= 336,600 \text{ gal}$

49. $4000 \text{ cm}^3 = \dfrac{4000 \text{ cm}^3}{1} \cdot \dfrac{1 \text{ L}}{1000 \text{ cm}^3} = 4 \text{ L}$

57. $\dfrac{46,690 \text{ people}}{1000 \text{ ha}}$
$\approx \dfrac{46,690 \text{ people}}{1000 \text{ ha}} \cdot \dfrac{260 \text{ ha}}{1 \text{ square mile}}$
$\approx 12,139.4$ people per square mile

59. Answers will vary.

61. Approximately 6.5 liters. 6.5 mL is only a little more than a teaspoon and 6.5 kL is thousands of gallons.

Check Points 9.3

1. a. 4.2 dg = 420 mg

b. 620 cg = 6.2 g

2. $0.145 \text{ m}^3 = \dfrac{0.145 \text{ m}^3}{1} \cdot \dfrac{1000 \text{ kg}}{1 \text{ m}^3} = 145 \text{ kg}$

The water weighs 145 kg.

3. a. $186 \text{ lb} = \dfrac{186 \text{ lb}}{1} \cdot \dfrac{0.45 \text{ kg}}{1 \text{ lb}} = 83.7 \text{ kg}$

b. $83.7 \times 1.2 \text{ mg} = 100.44 \text{ mg dose}$

4. $F = \dfrac{9}{5} \cdot 50 + 32 = 122$

$50^\circ C = 122^\circ F$

5. $C = \dfrac{5}{9}(59 - 32) = 15$

$59^\circ F = 15^\circ C$

Exercise Set 9.3

1. 7.4 dg = 740 mg

3. 870 mg = 0.87 g

5. 8 g = 800 cg

7. 18.6 kg = 18,600 g

9. 0.018 mg = 0.000018 g

11. $0.05 \text{ m}^3 = \dfrac{0.05 \text{ m}^3}{1} \cdot \dfrac{1000 \text{ kg}}{1 \text{ m}^3} = 50 \text{ kg}$

13. $4.2 \text{ kg} = \dfrac{4.2 \text{ kg}}{1} \cdot \dfrac{1000 \text{ cm}^3}{1 \text{ kg}} = 4200 \text{ cm}^3$

15. $1100 \text{ m}^3 = 1100 \text{ t}$

17. $0.04 \text{ kL} = \dfrac{0.04 \text{ kL}}{1} \cdot \dfrac{1000 \text{ kg}}{1 \text{ kL}} \cdot \dfrac{1000 \text{ g}}{1 \text{ kg}} = 40,000 \text{ g}$

19. $36 \text{ oz} = \dfrac{36 \text{ oz}}{1} \cdot \dfrac{1 \text{ lb}}{16 \text{ oz}} = 2.25 \text{ lb}$

21. $36 \text{ oz} = \dfrac{36 \text{ oz}}{1} \cdot \dfrac{28 \text{ g}}{1 \text{ oz}} = 1008 \text{ g}$

23. $540 \text{ lb} = \dfrac{540 \text{ lb}}{1} \cdot \dfrac{0.45 \text{ kg}}{1 \text{ lb}} = 243 \text{ kg}$

25. $80 \text{ lb} = \dfrac{80 \text{ lb}}{1} \cdot \dfrac{0.45 \text{ kg}}{1 \text{ lb}} \cdot \dfrac{1000 \text{ g}}{1 \text{ kg}} = 36,000 \text{ g}$

or $80 \text{ lb} = \dfrac{80 \text{ lb}}{1} \cdot \dfrac{16 \text{ oz}}{1 \text{ lb}} \cdot \dfrac{28 \text{ g}}{1 \text{ oz}} = 35,840 \text{ g}$

27. $540 \text{ kg} = \dfrac{540 \text{ kg}}{1} \cdot \dfrac{1 \text{ lb}}{0.45 \text{ kg}} = 1200 \text{ lb}$

29. $200 \text{ t} = \dfrac{200 \text{ t}}{1} \cdot \dfrac{1 \text{ T}}{0.9 \text{ t}} \approx 222.22 \text{ T}$

31. 10° C

$F = \dfrac{9}{5} \cdot 10 + 32$

$10^\circ C = 50^\circ F$

33. 35° C

$F = \dfrac{9}{5} \cdot 35 + 32$

$35^\circ C = 95^\circ F$

35. 57° C

$F = \dfrac{9}{5} \cdot 57 + 32$

$57^\circ C = 134.6^\circ F$

37. −5° C

$F = \dfrac{9}{5}(-5) + 32$

$-5^\circ C = 23^\circ F$

39. 68° F

$C = \dfrac{5}{9}(68 - 32)$

$68^\circ F = 20^\circ C$

41. 41° F

$C = \dfrac{5}{9}(41 - 32)$

$41^\circ F = 5^\circ C$

43. 72° F

$$C=\frac{5}{9}(72-32)$$

$72°\,F \approx 22.2°\,C$

45. 23° F

$$C=\frac{5}{9}(23-32)$$

$23°\,F = -5°\,C$

47. 350° F

$$C=\frac{5}{9}(350-32)$$

$350°\,F \approx 176.7°\,C$

49. −22° F

$$C=\frac{5}{9}(-22-32)$$

$-22°\,F = -30°\,C$

51. a. $m=\dfrac{68-32}{20-0}=\dfrac{36}{20}=\dfrac{9}{5}$

This means that the Fahrenheit temperature increases by $\dfrac{9}{5}^{\circ}$ for each 1° change in Celsius temperature.

b. $y = mx + b$

$F = mC + b$

$F = \dfrac{9}{5}C + 32$

53. This statement is of the form (true ∧ true).
Thus the compound statement is true.

55. This statement is of the form (false ∨ false).
Thus the compound statement is false.

57. This statement is of the form
$[(\text{true} \wedge \text{true}) \rightarrow \text{false}]$.
Thus the compound statement is false.

59. milligram

61. gram

63. kilogram

65. kilogram

67. b

69. a

71. c

73. 720 g = 0.720 kg
14 − 0.720 = 13.28 kg

75. $85\text{ g}=\dfrac{85\text{ g}}{1}\cdot\dfrac{1\text{ oz}}{28\text{ g}} \approx 3.04\text{ oz}$

Cost = 39 + 3 · 24 = 111 cents or $1.11

77. $\dfrac{\$3.15}{3\text{ kg}} = \$1.05$ per kg for economy size

720 g = 0.72 kg

$\dfrac{\$.60}{0.72\text{ kg}} = \$.83$ per kg for regular size

It is more economical to purchase the regular size.

79. $80\text{ lb} = \dfrac{80\text{ lb}}{1}\cdot\dfrac{0.45\text{ kg}}{1\text{ lb}} = 36\text{ kg}$

$36 \times 2.5\text{ mg} = 90\text{ mg}$ dose

81. a. $\dfrac{21.5\text{ mg}}{\text{tsp}} = \dfrac{21.5\text{ mg}}{\text{tsp}}\cdot\dfrac{2\text{ tsp}}{1\text{ dose}} = 43\text{ mg/dose}$

b. $\dfrac{21.5\text{ mg}}{\text{tsp}}$

$= \dfrac{21.5\text{ mg}}{\text{tsp}}\cdot\dfrac{\text{tsp}}{5\text{ ml}}\cdot\dfrac{30\text{ ml}}{1\text{ oz}}\cdot\dfrac{4\text{ oz}}{1\text{ bottle}}$

$= 516\text{ mg/bottle}$

83. a

85. c

87. −7° C

$$F=\frac{9}{5}(-7)+32$$

$-7°\,C = 19.4°\,F$

95. a. $\dfrac{3\text{¢}}{1\text{ g}}\cdot\dfrac{1000\text{ g}}{1\text{ kg}}\cdot\dfrac{0.45\text{ kg}}{1\text{ lb}} = 1350\text{¢}$ per pound

False

b. True

c. False

d. False

Chapter 9 Review Exercises

1. $69 \text{ in.} = \dfrac{69 \text{ in.}}{1} \cdot \dfrac{1 \text{ ft}}{12 \text{ in.}} = 5.75 \text{ ft}$

2. $9 \text{ in.} = \dfrac{9 \text{ in.}}{1} \cdot \dfrac{1 \text{ yd}}{36 \text{ in.}} = 0.25 \text{ yd}$

3. $21 \text{ ft} = \dfrac{21 \text{ ft}}{1} \cdot \dfrac{1 \text{ yd}}{3 \text{ ft}} = 7 \text{ yd}$

4. $13,200 \text{ ft} = \dfrac{13,200 \text{ ft}}{1} \cdot \dfrac{1 \text{ mi}}{5280 \text{ ft}} = 2.5 \text{ mi}$

5. $22.8 \text{ m} = 2280 \text{ cm}$

6. $7 \text{ dam} = 70 \text{ m}$

7. $19.2 \text{ hm} = 1920 \text{ m}$

8. $144 \text{ cm} = 0.0144 \text{ hm}$

9. $0.5 \text{ mm} = 0.0005 \text{ m}$

10. $18 \text{ cm} = 180 \text{ mm}$

11. $23 \text{ in.} = \dfrac{23 \text{ in.}}{1} \cdot \dfrac{2.54 \text{ cm}}{1 \text{ in.}} = 58.42 \text{ cm}$

12. $19 \text{ cm} = \dfrac{19 \text{ cm}}{1} \cdot \dfrac{1 \text{ in.}}{2.54 \text{ cm}} \approx 7.48 \text{ in.}$

13. $330 \text{ mi} = \dfrac{330 \text{ mi}}{1} \cdot \dfrac{1.6 \text{ km}}{1 \text{ mi}} = 528 \text{ km}$

14. $600 \text{ km} = \dfrac{600 \text{ km}}{1} \cdot \dfrac{1 \text{ mi}}{1.6 \text{ km}} = 375 \text{ mi}$

15. $14 \text{ m} = \dfrac{14 \text{ m}}{1} \cdot \dfrac{1 \text{ yd}}{0.9 \text{ m}} \approx 15.56 \text{ yd}$

16. $12 \text{ m} = \dfrac{12 \text{ m}}{1} \cdot \dfrac{100 \text{ cm}}{1 \text{ m}} \cdot \dfrac{1 \text{ in.}}{2.54 \text{ cm}} \cdot \dfrac{1 \text{ ft}}{12 \text{ in.}} = 39.37 \text{ ft}$

17. $45 \text{ km per hour} = \dfrac{45 \text{ km}}{1 \text{ hr}} \cdot \dfrac{1 \text{ mi}}{1.6 \text{ km}}$
    $\approx 28.13 \text{ miles/hour}$

18. $60 \text{ mi per hour} = \dfrac{60 \text{ mi}}{1 \text{ hr}} \cdot \dfrac{1.6 \text{ km}}{1 \text{ mi}}$
    $= 96 \text{ km/hr}$

19. $0.024 \text{ km}; 24,000 \text{ cm}; 2400 \text{ m}$

20. $6 \cdot 800 \text{ m} = 4800 \text{ m} = 4.8 \text{ km}$

21. $3 \cdot 8 = 24$ square units

22. $\dfrac{298,923,319 \text{ people}}{3,537,441 \text{ square miles}}$
    ≈ 84.5 people per square mile
    In April 2006 the U.S. had a population density of 84.5 people per square mile.

23. $7.2 \text{ ha} = \dfrac{7.2 \text{ ha}}{1} \cdot \dfrac{1 \text{ acre}}{0.4 \text{ ha}} = 18 \text{ acres}$

24. $30 \text{ m}^2 = \dfrac{30 \text{ m}^2}{1} \cdot \dfrac{1 \text{ ft}^2}{0.09 \text{ m}^2} \approx 333.33 \text{ ft}^2$

25. $12 \text{ mi}^2 = \dfrac{12 \text{ mi}^2}{1} \cdot \dfrac{2.6 \text{ km}^2}{1 \text{ mi}^2} = 31.2 \text{ km}^2$

26. a

27. $2 \cdot 4 \cdot 3 = 24$ cubic units

28. $33,600 \text{ cubic feet} = \dfrac{33,600 \text{ ft}^3}{1} \cdot \dfrac{7.48 \text{ gal}}{1 \text{ ft}^3}$
    $= 251,328 \text{ gal}$

29. $76,000 \text{ cm}^3 = \dfrac{76,000 \text{ cm}^3}{1} \cdot \dfrac{1 \text{ L}}{1000 \text{ cm}^3} = 76 \text{ L}$

30. c

31. There are $3 \times 3 = 9$ square feet in a square yard.

32. "Cubic miles" is a unit of volume, not area.

33. $12.4 \text{ dg} = 1240 \text{ mg}$

34. $12 \text{ g} = 1200 \text{ cg}$

35. $0.012 \text{ mg} = 0.000012 \text{ g}$

36. $450 \text{ mg} = 0.00045 \text{ kg}$

37. $50 \text{ kg} = \dfrac{50 \cancel{\text{kg}}}{1} \cdot \dfrac{1000 \text{ cm}^3}{1 \cancel{\text{kg}}} = 50,000 \text{ cm}^3$

38. $4 \text{ kL} = \dfrac{4 \cancel{\text{kL}}}{1} \cdot \dfrac{1000 \cancel{\text{kg}}}{1 \cancel{\text{kL}}} \cdot \dfrac{1 \text{ dm}^3}{1 \cancel{\text{kg}}} = 4000 \text{ dm}^3$

       $4000 \text{ dm}^3 = \dfrac{4000 \cancel{\text{dm}^3}}{1} \cdot \dfrac{1000 \text{ g}}{1 \cancel{\text{dm}^3}} = 4,000,000 \text{ g}$

39. $210 \text{ lb} = \dfrac{210 \cancel{\text{lb}}}{1} \cdot \dfrac{0.45 \text{ kg}}{1 \cancel{\text{lb}}} = 94.5 \text{ kg}$

40. $392 \text{ g} = \dfrac{392 \cancel{\text{g}}}{1} \cdot \dfrac{1 \text{ oz}}{28 \cancel{\text{g}}} = 14 \text{ oz}$

41. Kilograms; Answers will vary.

42. $36 \text{ oz} = \dfrac{36 \cancel{\text{oz}}}{1} \cdot \dfrac{1 \text{ lb}}{16 \cancel{\text{oz}}} = 2.25 \text{ lb}$

43. a

44. c

45. $F = \dfrac{9}{5} \cdot 15 + 32 = 59° \text{ F}$

46. $F = \dfrac{9}{5} \cdot 100 + 32 = 212° \text{ F}$

47. $F = \dfrac{9}{5} \cdot 5 + 32 = 41° \text{ F}$

48. $F = \dfrac{9}{5} \cdot 0 + 32 = 32° \text{ F}$

49. $-F = \dfrac{9}{5}(-25) + 32 = -13° \text{ F}$

50. $C = \dfrac{5}{9}(59 - 32) = 15° \text{ C}$

51. $C = \dfrac{5}{9}(41 - 32) = 5° \text{ C}$

52. $C = \dfrac{5}{9}(212 - 32) = 100° \text{ C}$

53. $C = \dfrac{5}{9}(98.6 - 32) = 37° \text{ C}$

54. $C = \dfrac{5}{9}(0 - 32) \approx -17.8° \text{ C}$

55. $C = \dfrac{5}{9}(14 - 32) = -10° \text{ C}$

56. A decrease of $15°C$ is more than a decrease of $15°F$; Explanations will vary.

Chapter 9 Test

1. 807 mm = 0.00807 hm

2. $635 \text{ cm} = \dfrac{635 \text{ cm}}{1} \cdot \dfrac{1 \text{ in.}}{2.54 \text{ cm}} = 250 \text{ in.}$

3. $8 \cdot 600 \text{ m} = 4800 \text{ m} = 4.8 \text{ km}$

4. mm

5. cm

6. km

7. $80 \text{ miles per hour} = \dfrac{80 \text{ mi}}{1 \text{ hr}} \cdot \dfrac{1.6 \text{ km}}{1 \text{ mi}}$
   $= 128 \text{ km/hr}$

8. $1 \text{ yd}^2 = (3 \text{ ft})(3 \text{ ft}) = 9 \text{ ft}^2$
   A square yard is 9 times greater than a square foot.

9. $\dfrac{39,133,966 \text{ people}}{194,896 \text{ square miles}}$

   ≈ 200.8 people per square mile
   On average there are 200.8 people living in each square mile of Spain.

10. $18 \text{ ha} = \dfrac{18 \text{ ha}}{1} \cdot \dfrac{1 \text{ acre}}{0.4 \text{ ha}} = 45 \text{ acres}$

11. b

12. Answers will vary.
    $1 \text{ m}^3 = (10 \text{ dm})(10 \text{ dm})(10 \text{ dm}) = 1000 \text{ dm}^3$
    A cubic meter is 1000 times greater than a cubic decimeter.

13. $10,000 \text{ ft}^3 = \dfrac{10,000 \text{ ft}^3}{1} \cdot \dfrac{7.48 \text{ gal}}{1 \text{ ft}^3}$
    $= 74,800 \text{ gal}$

14. b

15. 137 g = 0.137 kg

16. $90 \text{ lb} = \dfrac{90 \text{ lb}}{1} \cdot \dfrac{0.45 \text{ kg}}{1 \text{ lb}} = 40.5 \text{ kg}$

17. kg

18. mg

19. $F = \dfrac{9}{5} \cdot 30 + 32 = 86° \text{ F}$

20. $C = \dfrac{5}{9}(176 - 32) = 80° \text{ C}$

21. d

Chapter 10
Geometry

Check Points 10.1

1. Hand moves $\frac{1}{12}$ of a rotation

$$\frac{1}{12} \cdot 360° = 30°$$

2. $90° - 19° = 71°$

3. $m\measuredangle DBC + m\measuredangle ABD = 180°$
$$x + (x + 88°) = 180°$$
$$2x + 88° = 180°$$
$$2x = 92°$$
$$x = 46°$$
Thus, $m\measuredangle DBC = 46°$ and $m\measuredangle ABD = 134°$

4. $m\measuredangle 1 = 57°$
$m\measuredangle 2 = 180° - 57° = 123°$
$m\measuredangle 3 = m\measuredangle 2 = 123°$

5. $m\measuredangle 1 = m\measuredangle 8 = 29°$
$m\measuredangle 5 = m\measuredangle 8 = 29°$
$m\measuredangle 2 = m\measuredangle 8 = 29°$
$m\measuredangle 6 = 180° - m\measuredangle 8 = 180° - 29° = 151°$
$m\measuredangle 7 = m\measuredangle 6 = 151°$
$m\measuredangle 3 = m\measuredangle 7 = 151°$
$m\measuredangle 4 = m\measuredangle 3 = 151°$

Exercise Set 10.1

1. Hand moves $\frac{5}{12}$ of a rotation

$$\frac{5}{12} \cdot 360° = 150°$$

3. Hand moves $\frac{4-1}{12} = \frac{3}{12} = \frac{1}{4}$
of a rotation
$$\frac{1}{4} \cdot 360° = 90°$$

5. $20°$ is acute.

7. $160°$ is obtuse.

9. $180°$ is straight.

11. $90° - 25° = 65°$

13. $180° - 34° = 146°$

15. Complement:
$90° - 48° = 42°$
Supplement:
$180° - 48° = 132°$

17. Complement:
$90° - 89° = 1°$
Supplement:
$180° - 89° = 91°$

19. Complement:
$90° - 37.4° = 52.6°$
Supplement:
$180° - 37.4° = 142.6°$

21. Let x = the measure of the angle's complement.
Then $x + 12°$ represents the angle.
$$x + (x + 12°) = 90°$$
$$2x + 12° = 90°$$
$$2x = 78°$$
$$x = 39°$$
$$x + 12° = 51°$$
The complements are $39°$ and $51°$.

23. Let x = the measure of the angle's supplement.
Then $3x$ represents the angle.
$$x + 3x = 180°$$
$$4x = 180°$$
$$x = 45°$$
$$3x = 135°$$
The supplements are $45°$ and $135°$.

25. $m\measuredangle 1 = 180° - 72° = 108°$
$m\measuredangle 2 = 72°$
$m\measuredangle 3 = m\measuredangle 1 = 108°$

27. $m\measuredangle 1 = 90° - 40° = 50°$
$m\measuredangle 2 = 90°$
$m\measuredangle 3 = m\measuredangle 1 = 50°$

29. $m\measuredangle 1 = 180° - 112° = 68°$
$m\measuredangle 2 = m\measuredangle 1 = 68°$
$m\measuredangle 3 = 112°$
$m\measuredangle 4 = 112°$
$m\measuredangle 5 = m\measuredangle 1 = 68°$
$m\measuredangle 6 = m\measuredangle 2 = 68°$
$m\measuredangle 7 = m\measuredangle 3 = 112°$

31. $m\angle 1 = 38°$
$m\angle 2 = 90° - 38° = 52°$
$m\angle 3 = 180° - 38° = 142°$

33. The two angles are complementary.
$(2x + 50°) + (4x + 10°) = 90°$
$6x + 60° = 90°$
$6x = 30°$
$x = 5°$
Angle 1: $2x + 50° = 2(5°) + 50° = 60°$
Angle 2: $4x + 10° = 4(5°) + 10° = 30°$

35. The two angles are equal.
$11x - 20° = 7x + 28°$
$4x = 48°$
$x = 12°$
Angle 1: $11x - 20° = 11(12°) - 20° = 112°$
Angle 2: $7x + 28° = 7(12°) + 28° = 112°$

37. $\overline{AC} \cap \overline{BD} = \overline{BC}$

39. $\overline{AC} \cup \overline{BD} = \overline{AD}$

41. $\overline{BA} \cup \overline{BC} = \overline{AD}$

43. $\overline{AD} \cap \overline{BD} = \overline{AD}$

45. $\dfrac{360°}{8} = 45°$

47. When two parallel lines are intersected by a transversal, corresponding angles have the same measure.

57. d, since $m\angle 1 = m\angle 4$ and $m\angle 4 + m\angle 5 = 90°$

Check Points 10.2

1. $m\angle A + 116° + 15° = 180°$
$m\angle A + 131° = 180°$
$m\angle A = 180° - 131°$
$m\angle A = 49°$

2. $m\angle 1 = 180° - 90° = 90°$
$m\angle 2 = 180° - 36° - m\angle 1$
$= 180° - 36° - 90°$
$= 54°$
$m\angle 3 = m\angle 2 = 54°$

$m\angle 4 = 180° - 58° - m\angle 3$
$= 180° - 58° - 54°$
$= 68°$
$m\angle 5 = 180° - m\angle 4$
$= 180° - 68°$
$= 112°$

3. $\dfrac{3}{8} = \dfrac{12}{x}$
$3 \cdot x = 8 \cdot 12$
$3x = 96$
$\dfrac{3x}{3} = \dfrac{96}{3}$
$x = 32$ in.

4. $\dfrac{h}{2} = \dfrac{56}{3.5}$
$3.5 \cdot h = 2 \cdot 56$
$3.5h = 112$
$\dfrac{3.5h}{3.5} = \dfrac{112}{3.5}$
$h = 32$ yd

5. $c^2 = a^2 + b^2$
$c^2 = 7^2 + 24^2$
$c^2 = 49 + 576$
$c^2 = 625$
$c = \sqrt{625}$
$c = 25$ ft

6. $a^2 + b^2 = c^2$
$a^2 + (50)^2 = (130)^2$
$a^2 + 2500 = 16,900$
$a^2 = 14,400$
$a = \pm 120$
-120 must be rejected.
The tower is 120 yards tall.

Exercise Set 10.2

1. $m\angle A = 180° - 46° - 67° = 67°$

3. $m\angle A = 180° - 58° - 90° = 32°$

5. $m\angle 1 = 180° - 40° - 90° = 50°$
$m\angle 2 = 180° - m\angle 1 = 180° - 50° = 130°$
$m\angle 3 = m\angle 1 = 50°$
$m\angle 4 = m\angle 2 = 130°$
$m\angle 5 = 180° - 80° - m\angle 3$
$\qquad = 180° - 80° - 50°$
$\qquad = 50°$

7. $m\angle 1 = 180° - 130° = 50°$
$m\angle 2 = m\angle 1 = 50°$
$m\angle 3 = 180° - m\angle 1 - m\angle 2$
$\qquad = 180° - 50° - 50°$
$\qquad = 80°$
$m\angle 4 = 180° - m\angle 2 = 180° - 50° = 130°$
$m\angle 5 = m\angle 4 = 130°$

9. $m\angle 1 = 55°$
$m\angle 1 + m\angle 2 = 120°$
$55° + m\angle 2 = 120°$
$\qquad m\angle 2 = 65°$
$m\angle 1 + m\angle 2 + m\angle 3 = 180°$
$55° + 65° + m\angle 3 = 180°$
$\qquad\qquad m\angle 3 = 60°$
$m\angle 4 = m\angle 2 = 65°$
$m\angle 5 = m\angle 3 = 60°$
$m\angle 6 = 120°$
$m\angle 7 = m\angle 3 = 60°$
$m\angle 8 = m\angle 7 = 60°$
$m\angle 9 = m\angle 1 = 55°$
$m\angle 10 = m\angle 9 = 55°$

11. $\dfrac{18}{9} = \dfrac{10}{x}$
$18 \cdot x = 9 \cdot 10$
$18x = 90$
$\dfrac{18x}{18} = \dfrac{90}{18}$
$x = 5$ in.

13. $\dfrac{30}{10} = \dfrac{18}{x}$
$30 \cdot x = 10 \cdot 18$
$30x = 180$
$\dfrac{30x}{30} = \dfrac{180}{30}$
$x = 6$ m

15. $\dfrac{20}{15} = \dfrac{x}{12}$
$15x = 20 \cdot 12$
$15x = 240$
$\dfrac{15x}{15} = \dfrac{240}{15}$
$x = 16$ in.

17. Let $x = \overline{CA}$
$\dfrac{CA}{EA} = \dfrac{BC}{DE}$
$\dfrac{x}{15} = \dfrac{3}{9}$
$9x = 3 \cdot 15$
$9x = 45$
$x = 5$
$\overline{CA} = 5$

19. Let $x = \overline{DA}$
$\dfrac{DA}{BA} = \dfrac{DE}{BC}$
$\dfrac{x}{3} = \dfrac{9}{3}$
$x = 9$
$\overline{DA} = 9$

21. $c^2 = 8^2 + 15^2$
$c^2 = 64 + 225$
$c^2 = 289$
$c = 17$ m

23. $c^2 = 15^2 + 36^2$
$c^2 = 225 + 1296$
$c^2 = 1521$
$c = 39$ m

25. $a^2 + 16^2 = 20^2$

$a^2 + 256 = 400$

$a^2 = 144$

$a = 12$ cm

27. The sum of the measures of the three angles of any triangle is $180°$, so $x + x + (x + 30) = 180$.

Solve for x.

$3x + 30 = 180$

$3x = 150$

$x = 50$

If $x = 50$, $x + 30 = 80$, so the three angle measures are $50°, 50°,$ and $80°$. This solution checks because $50° + 50° + 80° = 180°$.

29. Let x = the measure of the smallest angle.
Let $2x$ = the measure of the second angle.
Let $x + 20$ = the measure of the third angle.

$x + 2x + (x + 20) = 180$

$4x + 20 = 180$

$4x = 160$

$x = 40$

Measure of smallest angle is $40°$.
Measure of second angle is $2x = 80°$.
Measure of third angle is $x + 20 = 60°$.

31. If $a^2 + b^2 = c^2$, then the triangle is a right triangle.

33. $a^2 + b^2 = c^2$

$4^2 + 8^2 = 9^2$

$16 + 64 = 81$

$80 = 81$ false

This is not a right triangle.

35. Let x = height of tree.

$\dfrac{x}{5} = \dfrac{86}{6}$

$6 \cdot x = 5 \cdot 86$

$6x = 430$

$x \approx 71.7$ ft

37. Ignoring the thickness of the panel, we essentially need to find the diameter of the rectangular opening.

$a^2 + b^2 = c^2$

$4^2 + 8^2 = c^2$

$16 + 64 = c^2$

$80 = c^2$

$c = \pm\sqrt{80} = \pm 4\sqrt{5}$

Since we are looking for a length, we discard the negative solution. The solution is $4\sqrt{5} \approx 8.9$ and we conclude that a panel that is about 8.9 feet long is the longest that can be taken through the door diagonally.

39. Let x = the length of the ladder.

$x^2 + 15^2 = 20^2$

$x^2 + 225 = 400$

$x^2 = 175$

$x \approx 13.2$ ft

41.

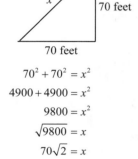

$70^2 + 70^2 = x^2$

$4900 + 4900 = x^2$

$9800 = x^2$

$\sqrt{9800} = x$

$70\sqrt{2} = x$

$x \approx 99.0$ ft

43.

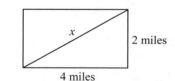

$4^2 + 2^2 = x^2$

$16 + 4 = x^2$

$20 = x^2$

$x = \pm\sqrt{20}$

$x = \pm 2\sqrt{5}$

$x \approx \pm 4.5$

We disregard -4.5 because we can't have a negative length measurement. The solution is 4.5 and we conclude that the pedestrian route is $2\sqrt{5}$ or 4.5 miles long.

45. a. $a^2 + c^2 = b^2$

$a^2 + 5^2 = 6^2$

$a^2 + 25 = 36$

$a^2 = 11$

$a = \sqrt{11}$

$a \approx 3.3$ m

Convert meters to feet:

3.3 m $= 330$ cm $= \dfrac{330 \text{ cm}}{1} \cdot \dfrac{1 \text{ ft}}{30.48 \text{ cm}} \approx 10.8$ ft

This is to high to be a realistic jump.

b. $a = \sqrt{11}$ which rounds to $a \approx 3.32$ m.

This is 2 cm more than the tests answer.
Convert centimeters to inches:

2 cm $= \dfrac{2 \text{ cm}}{1} \cdot \dfrac{1 \text{ in.}}{2.54 \text{ cm}} \approx 1$ in.

It is unlikely a carpenter would make this error.

c. 6 m $= 600$ cm $= \dfrac{600 \text{ cm}}{1} \cdot \dfrac{1 \text{ ft}}{30.48 \text{ cm}} \approx 19.7$ ft

Hardware stores do not typically carry boards of this size.

55. $m\angle PQT = 180° - 70° - 60° = 50°$

$m\angle SQR = 180° - m\angle PQT - 50°$
$= 180° - 50° - 50°$
$= 80°$

$m\angle R = 180° - m\angle SQR - 30°$
$= 180° - 80° - 30°$
$= 70°$

57.

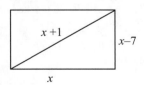

$x^2 + (x-7)^2 = (x+1)^2$

$x^2 + x^2 - 14x + 49 = x^2 + 2x + 1$

$x^2 - 16x + 48 = 0$

$(x-12)(x-4) = 0$

$x = 4$ must be rejected

$x = 12$

$x - 7 = 5$

The caret's dimensions are 5 ft by 12 ft.

Check Points 10.3

1. Note: 50 yds equals 150 ft and 30 yds equals 90 ft

$P = 2l + 2w$

$P = 2 \cdot 150$ ft $+ 2 \cdot 90$ ft $= 480$ ft

$\text{Cost} = \dfrac{480 \text{ feet}}{1} \cdot \dfrac{\$6.50}{\text{foot}} = \$3120$

2. a. Sum $= (n-2)180°$
$= (12-2)\,180°$
$= 10 \cdot 180°$
$= 1800°$

b. $m\angle A = \dfrac{1800°}{12} = 150°$

3. Each angle is $\dfrac{(n-2)\,180°}{n} = \dfrac{(8-2)\,180°}{8} = 135°$

Regular octagons can not be used to create a tessellation because $360°$ is not a multiple of $135°$.

Exercise Set 10.3

1. Quadrilateral (4 sides)

3. Pentagon (5 sides)

5. a (square), b (rhombus), d (rectangle), and e (parallelogram) all have two pairs of parallel sides.

7. a (square), d (rectangle)

9. c (trapezoid)

11. $P = 2 \cdot 3$ cm $+ 2 \cdot 12$ cm
$= 6$ cm $+ 24$ cm
$= 30$ cm

13. $P = 2 \cdot 6$ yd $+ 2 \cdot 8$ yd
$= 12$ yd $+ 16$ yd
$= 28$ yd

15. $P = 4 \cdot 250$ in. $= 1000$ in.

17. $P = 9$ ft $+ 7$ ft $+ 11$ ft $= 27$ ft

19. $P = 3 \cdot 6$ yd $= 18$ yd

21. $P = 12$ yd $+ 12$ yd $+ 9$ yd $+ 9$ yd $+ 21$ yd $+ 21$ yd
$= 84$ yd

23. First determine lengths of unknown sides.

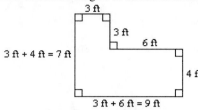

$P = 3 \text{ ft} + 3 \text{ ft} + 6 \text{ ft} + 4 \text{ ft} + 9 \text{ ft} + 7 \text{ ft} = 32 \text{ ft}$

25. Sum $= (n-2)180°$
$= (5-2)\ 180°$
$= 3 \cdot 180°$
$= 540°$

27. Sum $= (n-2)180°$
$= (4-2)180°$
$= 2 \cdot 180°$
$= 360°$

29. From Exercise 25, we know the sum of the measures of the angles of a pentagon is 540°. Since all 5 angles have the same degree measure, $m\angle A = \dfrac{540°}{5} = 108°$.
$m\angle B = 180° - 108° = 72°$

31. a. From Exercise 25, we know the sum of the measures of the angles of a pentagon is 540°.

b. $m\angle A = 540° - 70° - 150° - 90° - 90° = 140°$ and
$m\angle B = 180° - 140° = 40°$

33. a. squares, hexagons, dodecagons

b. The 3 angles that come together are $90°$, $120°$, and $150°$.

c. The tessellation is possible because $90° + 120° + 150° = 360°$.

35. a. triangles, hexagons

b. The 4 angles that come together are $60°$, $60°$, $120°$, and $120°$.

c. The tessellation is possible because $60° + 60° + 120° + 120° = 360°$.

37. Each angle is $\dfrac{(n-2)\ 180°}{n} = \dfrac{(9-2)\ 180°}{9} = 140°$
Regular nine-sided polygons can not be used to create a tessellation because $360°$ is not a multiple of $140°$.

39. Let w = the width of the field (in yards).
Let $4w$ = the length.
The perimeter of a rectangle is twice the width plus twice the length, so $2w + 2(4w) = 500$
$$2w + 8w = 500$$
$$10w = 500$$
$$w = 50$$
The width is 50 yards and the length is 4(50) = 200 yards. This checks because 2(50) + 2(200) = 500.

41. Let w = the width of a football field (in feet).
Let $w + 200$ = the length.
$$2w + 2(w + 200) = 1040$$
$$2w + 2w + 400 = 1040$$
$$4w + 400 = 1040$$
$$4w = 640$$
$$w = 160$$
The width 160 feet and the length is 160 + 200 = 360 feet. This checks because 2(160) + 2(200) = 720.

43. $x + x + (x + 5°) + (x + 5°) + 120° + 130° = (6-2)180°$
$$4x + 260° = 720°$$
$$4x = 460°$$
$$x = 115°$$
$$x + 5° = 120°$$
The angles are 115°, 115°, 120°, and 120°.

45. $\dfrac{(8-2)180°}{8} + \dfrac{(6-2)180°}{6} + \dfrac{(5-2)180°}{5} = 363°$
If the polygons were all regular polygons, the sum would be 363°. The tessellation is fake because the sum is not 360°.

47. $P = 2 \cdot 400 \text{ ft} + 2 \cdot 200 \text{ ft}$
$= 800 \text{ ft} + 400 \text{ ft}$
$= 1200 \text{ ft}$

$\text{Cost} = \dfrac{1200 \text{ ft}}{1} \cdot \dfrac{1 \text{ yd}}{3 \text{ ft}} \cdot \dfrac{\$14}{1 \text{ yd}} = \$5600$

49. Since the side of the square is 8 ft, its perimeter is 32 ft.
32 ft. is equivalent to 384 inches.
Thus, the total number of
$\text{flowers} = \dfrac{384}{8} = 48 \text{ flowers}$.

59. All sides have length a, therefore $P = 6a$.

Check Points 10.4

1. Area of large rectangle:
$$A_{large} = lw$$
$$= (13 \text{ ft} + 3 \text{ ft}) \times (3 \text{ ft} + 6 \text{ ft})$$
$$= 16 \text{ ft} \cdot 9 \text{ ft}$$
$$= 144 \text{ ft}^2$$
Area of small rectangle:
$$A_{small} = lw$$
$$= 13 \text{ ft} \cdot 6 \text{ ft}$$
$$= 78 \text{ ft}^2$$
Area of path $= 144 \text{ ft}^2 - 78 \text{ ft}^2 = 66 \text{ ft}^2$

2. First convert the linear measures in feet to linear yards.
$$18 \text{ ft} = \frac{18 \text{ ft}}{1} \cdot \frac{1 \text{ yd}}{3 \text{ ft}} = 6 \text{ yd}$$
$$21 \text{ ft} = \frac{21 \text{ ft}}{1} \cdot \frac{1 \text{ yd}}{3 \text{ ft}} = 7 \text{ yd}$$
Area of floor $= 6 \text{ yd} \cdot 7 \text{ yd} = 42 \text{ yd}^2$
$$\text{Cost of carpet} = \frac{42 \text{ yd}^2}{1} \cdot \frac{\$16}{1 \text{ yd}^2} = \$672$$

3. $A = bh$
$A = 10 \text{ in.} \cdot 6 \text{ in.} = 60 \text{ in.}^2$

4. $A = \frac{1}{2}bh$
$A = \frac{1}{2} \cdot 12 \text{ ft} \cdot 5 \text{ ft} = 30 \text{ ft}^2$

5. $A = \frac{1}{2}h(a+b)$
$$= \frac{1}{2} \cdot 7 \text{ ft} \cdot (20 \text{ ft} + 10 \text{ ft})$$
$$= \frac{1}{2} \cdot 7 \text{ ft} \cdot 30 \text{ ft}$$
$$= 105 \text{ ft}^2$$

6. $C = \pi d$
$$= \pi (10 \text{ in.})$$
$$= 10\pi \text{ in.}$$
$$\approx 31.4 \text{ in.}$$

7. Find the circumference of the semicircle:
$$C_{semicircle} = \frac{1}{2}\pi d$$
$$\approx \frac{1}{2}\pi (10 \text{ ft})$$
$$\approx 15.7 \text{ ft}$$
Length of trim $= 10 \text{ ft} + 12 \text{ ft} + 12 \text{ ft} + 15.7 \text{ ft}$
$$= 49.7 \text{ ft}$$

8. First, find the area of pizzas.

Large:	Medium:
$A = \pi r^2$	$A = \pi r^2$
$= \pi (9 \text{ in.})^2$	$= \pi (7 \text{ in.})^2$
$= 81\pi \text{ in.}^2$	$= 49\pi \text{ in.}^2$
$\approx 254 \text{ in.}^2$	$\approx 154 \text{ in.}^2$

Next, find the price per square inch.

Large:	Medium:
$\dfrac{\$20.00}{81\pi \text{ in.}^2}$	$\dfrac{\$14.00}{49\pi \text{ in.}^2}$
$\approx \dfrac{\$20.00}{254 \text{ in.}^2}$	$\approx \dfrac{\$14.00}{154 \text{ in.}^2}$
$\approx \dfrac{\$0.08}{\text{in.}^2}$	$\approx \dfrac{\$0.09}{\text{in.}^2}$

The large pizza is a better buy.

Exercise Set 10.4

1. $A = 6 \text{ m} \cdot 3 \text{ m} = 18 \text{ m}^2$

3. $A = (4 \text{ in.})^2 = 16 \text{ in.}^2$

5. $A = 50 \text{ cm} \cdot 42 \text{ cm} = 2100 \text{ cm}^2$

7. $A = \frac{1}{2} \cdot 14 \text{ in.} \cdot 8 \text{ in.} = 56 \text{ in.}^2$

9. $A = \frac{1}{2} \cdot 9.8 \text{ yd} \cdot 4.2 \text{ yd} = 20.58 \text{ yd}^2$

11. $a^2 + b^2 = c^2$
$$h^2 + 12^2 = 13^2$$
$$h^2 + 144 = 169$$
$$h^2 = 25$$
$$h = 5$$
$$A = \frac{1}{2} \cdot 12 \text{ in.} \cdot 5 \text{ in.} = 30 \text{ in.}^2$$

13. $A = \dfrac{1}{2} \cdot 7 \text{ m} \cdot (16 \text{ m} + 10 \text{ m})$

$= \dfrac{7}{2} \text{ m} \cdot (26 \text{ m})$

$= 91 \text{ m}^2$

15. $C = 2\pi \cdot 4 \text{ cm} = 8\pi \text{ cm} \approx 25.1 \text{ cm}$

$A = \pi(4 \text{ cm})^2 = 16\pi \text{ cm}^2 \approx 50.3 \text{ cm}^2$

17. $C = \pi \cdot 12 \text{ yd} = 12\pi \text{ yd} \approx 37.7 \text{ yd}$

$r = \dfrac{d}{2} = \dfrac{12 \text{ yd}}{2} = 6 \text{ yd}$

$A = \pi(6 \text{ yd})^2 = 36\pi \text{ yd}^2 \approx 113.1 \text{ yd}^2$

19. The figure breaks into a lower rectangle and an upper rectangle.

Area of lower rectangle: Area of upper rectangle:

$A = lw$ $A = lw$

$A = (12 \text{ m})(3 \text{ m})$ $A = (9 \text{ m})(4 \text{ m})$

$A = 36 \text{ m}^2$ $A = 36 \text{ m}^2$

Total area $= 36 \text{ m}^2 + 36 \text{ m}^2 = 72 \text{ m}^2$

21. The figure breaks into a lower rectangle and an upper triangle.

Area of rectangle: Area of triangle:

$A = lw$ $A = \frac{1}{2}bh$

$A = (24 \text{ m})(10 \text{ m})$ $A = \frac{1}{2}(24 \text{ m})(5 \text{ m})$

$A = 240 \text{ m}^2$ $A = 60 \text{ m}^2$

Total area $= 240 \text{ m}^2 + 60 \text{ m}^2 = 300 \text{ m}^2$

23. The figure's area can be obtained by adding the area of a square of side 10 cm, to twice the area of a circle of radius 5 cm.

Area of square: Area of circles:

$A = s^2$ $A = \pi r^2$

$A = (10 \text{ cm})^2$ $A = \pi(5 \text{ cm})^2$

$A = 100 \text{ cm}^2$ $A = 25\pi \text{ cm}^2$

Total area $= 100 \text{ cm}^2 + 25\pi \text{ cm}^2 + 25\pi \text{ cm}^2$

$= (100 + 50\pi) \text{ cm}^2$

$\approx 257.1 \text{ cm}^2$

25. Area of larger triangle:

$A = \frac{1}{2}bh$

$A = \frac{1}{2} \cdot (8 \text{ cm} + 8 \text{ cm} + 8 \text{ cm}) \cdot (12 \text{ cm} + 6 \text{ cm})$

$A = \frac{1}{2} \cdot 24 \text{ cm} \cdot 18 \text{ cm}$

$A = 216 \text{ cm}^2$

Area of smaller triangle:

$A = \frac{1}{2}bh$

$A = \frac{1}{2} \cdot 8 \text{ cm} \cdot 6 \text{ cm}$

$A = 24 \text{ cm}^2$

Shaded area $= 216 \text{ cm}^2 - 24 \text{ cm}^2 = 192 \text{ cm}^2$

27. Perimeter:

$$2\sqrt{8^2 + 15^2} + 2\sqrt{6^2 + 8^2} = 2\sqrt{289} + 2\sqrt{100}$$
$$= 2 \cdot 17 + 2 \cdot 10$$
$$= 54 \text{ ft}$$

Area:

$\frac{1}{2}(15)(8) + \frac{1}{2}(15)(8) + \frac{1}{2}(6)(8) + \frac{1}{2}(6)(8) = 168 \text{ ft}^2$

29. Let x = the width of the garden.
Then $x + 5$ = the length.

$$l \cdot w = A$$
$$(x+5)(x) = 300$$
$$x^2 + 5x = 300$$
$$x^2 + 5x - 300 = 0$$
$$(x-15)(x+20) = 0$$
$$x - 15 = 0 \quad \text{or} \quad x + 20 = 0$$
$$x = 15 \qquad \qquad x = -20$$

Discard $x = -20$ since the width cannot be negative. Then $x = 15$ and $x + 5 = 20$, so the length is 20 feet and the width is 15 feet.

31. Use the formula for the area of a triangle where x is the base and $x+1$ is the height.

$$\frac{1}{2}bh = A$$

$$\frac{1}{2}x(x+1) = 15$$

$$2\left[\frac{1}{2}x(x+1)\right] = 2\cdot15$$

$$x(x+1) = 30$$

$$x^2 + x = 30$$

$$x^2 + x - 30 = 0$$

$$(x+6)(x-5) = 0$$

$$x+6 = 0 \quad \text{or} \quad x-5 = 0$$

$$x = -6 \qquad\qquad x = 5$$

Discard $x = -6$ since the length of the base cannot be negative. Then $x = 5$ and $x+1 = 6$, so the base is 5 centimeters and the height is 6 centimeters.

33. First convert the linear measures in feet to linear yards.

$$9 \text{ ft} = \frac{9 \text{ ft}}{1} \cdot \frac{1 \text{ yd}}{3 \text{ ft}} = 3 \text{ yd}$$

$$21 \text{ ft} = \frac{21 \text{ ft}}{1} \cdot \frac{1 \text{ yd}}{3 \text{ ft}} = 7 \text{ yd}$$

Area of floor $= 3 \text{ yd} \cdot 7 \text{ yd} = 21 \text{ yd}^2$

$$\text{Cost of carpet } = \frac{21 \text{ yd}^2}{1} \cdot \frac{\$26.50}{1 \text{ yd}^2} = \$556.50$$

35. Area of tile = (Area of floor) − (Area of store) − (Area of refrigerator)
$$= (12 \text{ ft} \cdot 15 \text{ ft}) - (3 \text{ ft} \cdot 4 \text{ ft}) - (4 \text{ ft} \cdot 5 \text{ ft})$$
$$= 180 \text{ ft}^2 - 12 \text{ ft}^2 - 20 \text{ ft}^2 = 148 \text{ ft}^2$$

37. a. Area of lawn = (Area of lot) − (Area of house) − (Area of shed) − (Area of driveway)
$$= 200 \text{ ft} \cdot 500 \text{ ft} - 60 \text{ ft} \cdot 100 \text{ ft} - (20 \text{ ft})^2 - 100 \text{ ft} \cdot 20 \text{ ft}$$
$$= 100,000 \text{ ft}^2 - 6000 \text{ ft}^2 - 400 \text{ ft}^2 - 2000 \text{ ft}^2$$
$$= 91,600 \text{ ft}^2$$

$$\text{Maximum number of bags of fertilizer } = \frac{1 \text{ bag}}{4000 \text{ ft}^2} \cdot \frac{91,600 \text{ ft}^2}{1} = 22.9 \text{ bags} \rightarrow 23 \text{ bags}$$

b. Total cost of fertilizer $= \frac{\$25.00}{1 \text{ bag}} \cdot \frac{23 \text{ bags}}{2} = \$575$

39. a. Area of a front wall $=\left[20\text{ ft}\cdot40\text{ ft}\right]+\left[\frac{1}{2}\cdot40\text{ ft}\cdot10\text{ ft}\right]=1000\text{ ft}^2$

Area of a side wall $=50\text{ ft}\cdot20\text{ ft}=1000\text{ ft}^2$

Area of windows $=4\left[8\text{ ft}\cdot5\text{ ft}\right]+2\left[30\text{ ft}\cdot2\text{ ft}\right]=280\text{ ft}^2$

Area of doors $=2\left[80\text{ in.}\cdot36\text{ in.}\right]=2\left[6\frac{2}{3}\text{ ft}\cdot3\text{ ft}\right]=40\text{ ft}^2$

Area of paint = 2(Area of front wall) + 2(Area of side wall) – (Area of windows and doors)

$$=2\left(1000\text{ ft}^2\right)+2\left(1000\text{ ft}^2\right)-\left(280\text{ ft}^2+40\text{ ft}^2\right)$$
$$=2000\text{ ft}^2+2000\text{ ft}^2-320\text{ ft}^2$$
$$=3680\text{ ft}^2$$

b. Two coats will require enough paint for $2\cdot3680\text{ ft}^2=7360\text{ ft}^2$.

$$7360\text{ ft}^2=\frac{7360\text{ ft}^2}{1}\cdot\frac{1\text{ gallon}}{500\text{ ft}^2}=14.72\text{ gallons}\approx15\text{ gallons}.$$

c. $\$26.95\times15=\$404.25$ is the cost to buy the paint.

41. Area of Master Bedroom $=14\text{ ft}\cdot14\text{ ft}=196\text{ ft}^2$

Area of Bedroom #2 $=11\text{ ft}\cdot12\text{ ft}=132\text{ ft}^2$

Area of Bedroom #3 $=12\text{ ft}\cdot11\text{ ft}=132\text{ ft}^2$

Total area $=196\text{ ft}^2+132\text{ ft}^2+132\text{ ft}^2=460\text{ ft}^2$

Since there are 9 square feet in a square yard, $460\text{ ft}^2\approx51.1\text{ yd}^2$.

52 yd^2 at $\$17.95$ per square yard costs $\$933.40$.

43. Amount of fencing $=C=2\pi\cdot20\text{ m}=40\pi\text{ m}\approx125.7\text{ m}$

45. $C=2\pi\cdot30\text{ ft}\approx188.5\text{ ft}$

$$188.5\text{ ft}=\frac{188.5\text{ ft}}{1}\cdot\frac{12\text{ in.}}{1\text{ ft}}=2262\text{ in.}$$

$$\text{Number of plants}=\frac{1\text{ plant}}{6\text{ in.}}\cdot\frac{2262\text{ in.}}{1}=377\text{ plants}$$

47. First, find the area of pizzas.

Large: $A=\pi r^2=\pi(7\text{ in.})^2=49\pi\text{ in.}^2\approx153.9\text{ in.}^2$

Medium: $A=\pi r^2=\pi(3.5\text{ in.})^2=12.25\pi\text{ in.}^2\approx38.5\text{ in.}^2$

Next, find the price per square inch.

Large: $\dfrac{\$12.00}{49\pi\text{ in.}^2}\approx\dfrac{\$12.00}{153.9\text{ in.}^2}\approx\dfrac{\$0.08}{\text{in.}^2}$

Medium: $\dfrac{\$5.00}{12.25\pi\text{ in.}^2}\approx\dfrac{\$5.00}{38.5\text{ in.}^2}\approx\dfrac{\$0.13}{\text{in.}^2}$

The large pizza is a better buy.

55. Original Area $= 8 \, \text{ft} \cdot 10 \, \text{ft} = 80 \, \text{ft}^2$

New area $= 12 \, \text{ft} \cdot 15 \, \text{ft} = 180 \, \text{ft}^2$

Ratio $= \dfrac{180 \, \text{ft}^2}{80 \, \text{ft}^2} = \dfrac{9}{4}$

The cost will increase by a factor of $\dfrac{9}{4}$, or 2.25.

57. Length of pipeline $= \dfrac{16.8 \, \text{mi}}{1} \cdot \dfrac{5280 \, \text{ft}}{1 \, \text{mi}} = 88{,}704 \, \text{ft}$

Area of land $= 88{,}704 \, \text{ft} \cdot 200 \, \text{ft} = 17{,}740{,}800 \, \text{ft}^2$

Area of land in acres $= \dfrac{17{,}740{,}800 \, \text{ft}^2}{1} \cdot \dfrac{1 \, \text{acre}}{43{,}560 \, \text{ft}^2} \approx 407.2727 \, \text{acres}$

Total cost $= \dfrac{\$32}{1 \, \text{acre}} \cdot \dfrac{407.2727 \, \text{acres}}{1} = \$13{,}032.73$

Check Points 10.5

1. $V = 5 \, \text{ft} \cdot 3 \, \text{ft} \cdot 7 \, \text{ft} = 105 \, \text{ft}^3$

2. $= \dfrac{6 \, \text{ft}}{1} \cdot \dfrac{1 \, \text{yd}}{3 \, \text{ft}} = 2 \, \text{yd}$

$V = (2 \, \text{yd})^3 = 8 \, \text{yd}^3$

3. $B = (6 \, \text{ft})^2 = 36 \, \text{ft}^2$

$V = \dfrac{1}{3} \cdot 36 \, \text{ft}^2 \cdot 4 \, \text{ft}$

$= 48 \, \text{ft}^3$

4. $r = \dfrac{1}{2}(8 \, \text{cm}) = 4 \, \text{cm}$

$V = \pi(4 \, \text{in.})^2 \cdot 6 \, \text{in.} \approx 302 \, \text{in.}^3$

5. $V = \dfrac{1}{3}\pi(4 \, \text{in.})^2 \cdot 6 \, \text{in.} \approx 101 \, \text{in.}^3$

6. No, it is not enough air.

$V = \dfrac{4}{3}\pi(4.5 \, \text{in.})^3 \approx 382 \, \text{in.}^3$

7. New dimensions: $l = 16 \, \text{yd}, \ w = 10 \, \text{yd}, \ h = 6 \, \text{yd}$

$SA = 2lw + 2lh + 2wh$

$= 2 \cdot 16 \, \text{yd} \cdot 10 \, \text{yd} + 2 \cdot 16 \, \text{yd} \cdot 6 \, \text{yd} + 2 \cdot 10 \, \text{yd} \cdot 6 \, \text{yd}$

$= 320 \, \text{yd}^2 + 192 \, \text{yd}^2 + 120 \, \text{yd}^2$

$= 632 \, \text{yd}^2$

Exercise Set 10.5

1. $V = 3 \, \text{in.} \cdot 3 \, \text{in.} \cdot 4 \, \text{in.} = 36 \, \text{in.}^3$

3. $V = (4 \, \text{cm})^3 = 64 \, \text{cm}^3$

5. $B = 7 \, \text{yd} \cdot 5 \, \text{yd} = 35 \, \text{yd}^2$

$V = \dfrac{1}{3} \cdot 35 \, \text{yd}^2 \cdot 15 \, \text{yd}$

$= 175 \, \text{yd}^3$

7. $B = 4 \, \text{in.} \cdot 7 \, \text{in.} = 28 \, \text{in.}^2$

$V = \dfrac{1}{3} \cdot 28 \, \text{in.}^2 \cdot 6 \, \text{in.}$

$= 56 \, \text{in.}^3$

9. $V = \pi(5 \, \text{cm})^2 \cdot 6 \, \text{cm} = 150\pi \, \text{cm}^3 \approx 471 \, \text{cm}^3$

11. $r = \dfrac{1}{2}(24 \, \text{in.}) = 12 \, \text{in.}$

$V = \pi(12 \, \text{in.})^2 \cdot 21 \, \text{in.} = 3024\pi \, \text{in.}^3 \approx 9500 \, \text{in.}^3$

13. $V = \dfrac{1}{3}\pi(4 \, \text{m})^2 \cdot 9 \, \text{m} = 48\pi \, \text{m}^3 \approx 151 \, \text{m}^3$

15. $r = \dfrac{1}{2} \cdot 6 \text{ yd} = 3 \text{ yd}$

$V = \dfrac{1}{3}\pi(3 \text{ yd})^2 \cdot 5 \text{ yd} = 15\pi \text{ yd}^3 \approx 47 \text{ yd}^3$

17. $V = \dfrac{4}{3}\pi(6\,\text{m})^3 = 288\pi \text{ m}^3 \approx 905\,\text{m}^3$

19. $r = \dfrac{1}{2} \cdot 18 \text{ cm} = 9 \text{ cm}$

$V = \dfrac{4}{3}\pi(9 \text{ cm})^3 = 972\pi \text{ cm}^3 \approx 3054 \text{ cm}^3$

21. Surface Area $= 2(5 \text{ m} \cdot 3 \text{ m}) + 2(2 \text{ m} \cdot 3 \text{ m}) + 2(5 \text{ m} \cdot 2 \text{ m})$

$= 2 \cdot 15 \text{ m}^2 + 2 \cdot 6 \text{ m}^2 + 2 \cdot 10 \text{ m}^2$

$= 30 \text{ m}^2 + 12 \text{ m}^2 + 20 \text{ m}^2$

$= 62 \text{ m}^2$

23. Surface Area $= 6(4 \text{ ft})^2 = 96 \text{ ft}^2$

25. Volume = (volume of cone) + (volume of hemisphere)

$V = \dfrac{1}{3}\pi(6 \text{ cm})^2 \cdot 15 \text{ cm} + \dfrac{1}{2}\left[\dfrac{4}{3}\pi(6 \text{ cm})^3\right] = 324\pi \text{ cm}^3 \approx 1018 \text{ cm}^3$

27. Surface area:

$2[\overbrace{(5)(5)+(4)(3)}^{\text{front and back}}] + [\overbrace{(5)(4)+(3)(4)+(2)(4)}^{\text{left and right sides}}] + [\overbrace{(5)(4)+(4)(4)+(9)(4)}^{\text{top(s) and bottom}}] = 186 \text{ yd}^2$

Volume:

$\overbrace{(5)(5)(4)}^{\text{left part of block}} + \overbrace{(4)(4)(3)}^{\text{right part of block}} = 100 + 48 = 148 \text{ yd}^3$

29. Surface area:

$2[\overbrace{(10)(5)+\tfrac{1}{2}(4)(10+4)}^{\text{front and back}}] + \overbrace{4(15)(5)}^{\text{left, right, and 2 upper slants}} + \overbrace{(15)(4)}^{\text{top}} + \overbrace{(15)(10)}^{\text{bottom}} = 2[\overbrace{50+28}^{\text{front and back}}] + \overbrace{300}^{\text{left, right, and 2 upper slants}} + \overbrace{60}^{\text{top}} + \overbrace{150}^{\text{bottom}} = 666 \text{ yd}^2$

31. $\dfrac{\frac{4}{3}\pi 3^3}{\frac{4}{3}\pi 6^3} = \dfrac{\frac{4}{3}\cancel{\pi} 3^3}{\frac{4}{3}\cancel{\pi} 6^3} = \left(\dfrac{3}{6}\right)^3 = \left(\dfrac{1}{2}\right)^3 = \dfrac{1}{8}$

33. Smaller cylinder: $r = 3$ in, $h = 4$ in.
$V = \pi r^2 h = \pi(3)^2 \cdot 4 = 36\pi$

The volume of the smaller cylinder is $36\pi \text{ in}^3$.

Larger cylinder: r = 3(3 in) = 9 in, h = 4 in.
$V = \pi r^2 h = \pi(9)^2 \cdot 4 = 324\pi$

The volume of the larger cylinder is 324π. The ratio of the volumes of the two cylinders is $\dfrac{V_{\text{larger}}}{V_{\text{smaller}}} = \dfrac{324\pi}{36\pi} = \dfrac{9}{1}$.

So, the volume of the larger cylinder is 9 times the volume of the smaller cylinder.

35. First convert all linear measures in feet to linear yards.

$$12 \text{ ft} = \frac{12 \text{ ft}}{1} \cdot \frac{1 \text{ yd}}{3 \text{ ft}} = 4 \text{ yd}$$

$$9 \text{ ft} = \frac{9 \text{ ft}}{1} \cdot \frac{1 \text{ yd}}{3 \text{ ft}} = 3 \text{ yd}$$

$$6 \text{ ft} = \frac{6 \text{ ft}}{1} \cdot \frac{1 \text{ yd}}{3 \text{ ft}} = 2 \text{ yd}$$

Total dirt $= 4 \text{ yd} \cdot 3 \text{ yd} \cdot 2 \text{ yd} = 24 \text{ yd}^3$

$$\text{Total cost} = \frac{24 \text{ yd}^3}{1} \cdot \frac{1 \text{ truck}}{6 \text{ yd}^3} \cdot \frac{\$10}{1 \text{ truck}} = \$40$$

37. Volume of house $= 1400 \text{ ft}^2 \cdot 9 \text{ ft} = 12,600 \text{ ft}^3$
No. This furnace will not be adequate.

39. a. First convert linear measures in feet to linear yards.

$$756 \text{ ft} = \frac{756 \text{ ft}}{1} \cdot \frac{1 \text{ yd}}{3 \text{ ft}} = 252 \text{ yd}$$

$$480 \text{ ft} = \frac{480 \text{ ft}}{1} \cdot \frac{1 \text{ yd}}{3 \text{ ft}} = 160 \text{ yd}$$

$$B = (252 \text{ yd})^2 = 63,504 \text{ yd}^2$$

$$V = \frac{1}{3} \cdot 63,504 \text{ yd}^2 \cdot 160 \text{ yd}$$

$$= 3,386,880 \text{ yd}^3$$

b. $\dfrac{1 \text{ block}}{1.5 \text{ yd}^3} \cdot \dfrac{3,386,880 \text{ yd}^3}{1} = 2,257,920 \text{ blocks}$

41. Volume of tank $= \pi(3 \text{ ft})^2 \cdot \dfrac{7}{3} \text{ ft} \approx 66 \text{ ft}^3$

Yes. The volume of the tank is less than 67 cubic feet.

43. Volume of pool (in cubic feet) $= \pi(12 \text{ ft})^2 \cdot 4 \text{ ft} = 576\pi \text{ ft}^3 \approx 1809.6 \text{ ft}^3$

Volume of pool (in gallons) $= 1809.6 \text{ ft}^3 = \dfrac{1809.6 \text{ ft}^3}{1} \cdot \dfrac{7.48 \text{ gallons}}{1 \text{ ft}^3} \approx 13,536 \text{ gallons}$

Cost to fill the pool $= \$2 \cdot 13.535 \approx \$27$

47. New volume $= \dfrac{4}{3}\pi(2r)^3 = \dfrac{4}{3}\pi \cdot 8r^3 = 8\left(\dfrac{4}{3}\pi r^3\right)$

The volume is multiplied by 8.

49. Volume of darkly shaded region $=$ (Volume of rectangular solid) $-$ (Volume of pyramid)

$$= 6 \text{ cm} \cdot 6 \text{ cm} \cdot 7 \text{ cm} - \frac{1}{3}(6 \text{ cm})^2 \cdot 7 \text{ cm}$$

$$= 168 \text{ cm}^3$$

51. Surface area = (Areas of 3 rectangles) + (Area of 2 triangles)

$$= (5 \text{ cm} \cdot 6 \text{ cm} + 4 \text{ cm} \cdot 6 \text{ cm} + 3 \text{ cm} \cdot 6 \text{ cm}) + 2\left(\frac{1}{2} \cdot 3 \text{ cm} \cdot 4 \text{ cm}\right)$$

$$= 72 \text{ cm}^2 + 12 \text{ cm}^2$$

$$= 84 \text{ cm}^2$$

Check Points 10.6

1. Begin by finding the measure of the hypotenuse c using the Pythagorean Theorem.

$$c^2 = a^2 + b^2 = 3^2 + 4^2 = 25$$

$$c = \sqrt{25} = 5$$

$$\sin A = \frac{3}{5}$$

$$\cos A = \frac{4}{5}$$

$$\tan A = \frac{3}{4}$$

2. $\tan A = \dfrac{a}{b}$

$$\tan 62° = \frac{a}{140}$$

$$a = 140 \tan 62° \approx 263 \text{ cm}$$

3. $\cos A = \dfrac{b}{c}$

$$\cos 62° = \frac{140}{c}$$

$$c \cos 62° = 140$$

$$c = \frac{140}{\cos 62°}$$

$$c \approx 298 \text{ cm}$$

4. Let a = the height of the tower.

$$\tan 85.4° = \frac{a}{80}$$

$$a = 80 \tan 85.4° \approx 994 \text{ ft}$$

5. $\tan A = \dfrac{14}{10}$

$$A = \tan^{-1}\left(\frac{14}{10}\right) \approx 54°$$

Exercise Set 10.6

1. $\sin A = \dfrac{3}{5}$

$$\cos A = \frac{4}{5}$$

$$\tan A = \frac{3}{4}$$

3. First find the length of missing side.

$$a^2 = 29^2 - 21^2 = 400$$

$$a = 20$$

$$\sin A = \frac{20}{29}$$

$$\cos A = \frac{21}{29}$$

$$\tan A = \frac{20}{21}$$

5. First find the length of missing side.

$$b^2 = 26^2 - 10^2 = 576$$

$$b = 24$$

$$\sin A = \frac{10}{26} = \frac{5}{13}$$

$$\cos A = \frac{24}{26} = \frac{12}{13}$$

$$\tan A = \frac{10}{24} = \frac{5}{12}$$

7. First find the length of missing side.

$$a^2 = 35^2 - 21^2 = 784$$

$$a = 28$$

$$\sin A = \frac{28}{35} = \frac{4}{5}$$

$$\cos A = \frac{21}{35} = \frac{3}{5}$$

$$\tan A = \frac{28}{21} = \frac{4}{3}$$

9. $\tan A = \dfrac{a}{b}$

$\tan 37° = \dfrac{a}{250}$

$a = 250 \tan 37° \approx 188$ cm

11. $\cos 34° = \dfrac{b}{220}$

$b = 220 \cos 34° \approx 182$ in.

13. $\sin 34° = \dfrac{a}{13}$

$a = 13 \sin 34° \approx 7$ m

15. $\tan 33° = \dfrac{14}{b}$

$b = \dfrac{14}{\tan 33°} \approx 22$ yd

17. $\sin 30° = \dfrac{20}{c}$

$c = \dfrac{20}{\sin 30°} = 40$ m

19. $m\angle B = 90° - 40° = 50°$

Side a: $\tan 40° = \dfrac{a}{22}$

$a = 22 \tan 40° \approx 18$ yd

Side c: $\cos 40° = \dfrac{22}{c}$

$c = \dfrac{22}{\cos 40°} \approx 29$ yd

$m\angle B = 50°,\ a \approx 18$ yd, $c \approx 28$ yd

21. $m\angle B = 90° - 52° = 38°$

Side a: $\sin 52° = \dfrac{a}{54}$

$a = 54 \sin 52° \approx 43$ cm

Side b: $\cos 52° = \dfrac{b}{54}$

$b = 54 \cos 52° \approx 33$ cm

$m\angle B = 38°,\ a \approx 43$ cm, $b \approx 33$ cm

23. $\sin A = \dfrac{30}{50}$

$A = \sin^{-1}\left(\dfrac{30}{50}\right) \approx 37°$

25. $\cos A = \dfrac{15}{17}$

$A = \cos^{-1}\left(\dfrac{15}{17}\right) \approx 28°$

27. $x = 500 \tan 40° + 500 \tan 25°$

$x \approx 653$

29. $x = 600 \tan 28° - 600 \tan 25°$

$x \approx 39$

31. $x = \dfrac{300}{\tan 34°} - \dfrac{300}{\tan 64°}$

$x \approx 298$

33. $x = \dfrac{400 \tan 40° \tan 20°}{\tan 40° - \tan 20°}$

$x \approx 257$

35. $\tan 40° = \dfrac{a}{630}$

$a = 630 \tan 40° \approx 529$ yd

37. $\sin 10° = \dfrac{500}{c}$

$c = \dfrac{500}{\sin 10°} \approx 2879$ ft

39. Let h = the height of the tower.

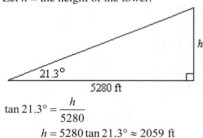

$\tan 21.3° = \dfrac{h}{5280}$

$h = 5280 \tan 21.3° \approx 2059$ ft

41. Let x = the distance.

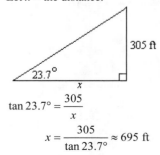

$\tan 23.7° = \dfrac{305}{x}$

$x = \dfrac{305}{\tan 23.7°} \approx 695$ ft

43. $\tan x = \dfrac{125}{172}$

$x = \tan^{-1}\left(\dfrac{125}{172}\right) \approx 36°$

45. $m\measuredangle P = 36°$

$\tan 36° = \dfrac{1000}{d}$

$d = \dfrac{1000}{\tan 36°} \approx 1376$ ft

47. Let A = the angle of elevation.

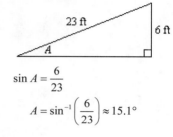

$\sin A = \dfrac{6}{23}$

$A = \sin^{-1}\left(\dfrac{6}{23}\right) \approx 15.1°$

57. The sine and cosine of an acute angle cannot be greater than or equal to 1 because they are each the ratio of a leg of a right triangle to the hypotenuse. The hypotenuse of a right triangle is always the longest side; this results in a value less than 1.

59.

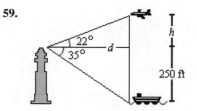

a. $\tan 35° = \dfrac{250}{d}$

$d = \dfrac{250}{\tan 35°} \approx 357$ ft

b. $\tan 22° = \dfrac{h}{d} = \dfrac{h}{357}$

$h = 357 \tan 22° \approx 144$ ft

Height of plane = 250 ft + 144 ft = 394 ft.

Check Points 10.7

1. Answers will vary. Possible answer:

The upper left and lower right vertices are odd.
The lower left and upper right vertices are even.
One possible tracing:
Start at the upper left, trace around the square, then trace down the diagonal.

Exercise Set 10.7

1. a. A and C are even vertices.
   B and D are odd vertices.
   Because this graph has two odd vertices, by Euler's second rule, it is traversable.

   b. Sample path: D, A, B, D, C, B

3. a. C, D, E are even vertices.
   A and B are odd vertices.
   Because this graph has two odd vertices, by Euler's second rule, it is traversable.

   b. Sample path: A, D, C, B, D, E, A, B

5. A, B, D, E are odd vertices.
   Because this graph has more than two odd vertices, by Euler's third rule, it is not traversable.

7.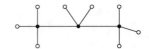

9. No, the graph is not traversable because there are more than 2 odd vertices.

11. 2 doors connect room *C* to the outside. This is shown in the graph by connecting 2 edges from *C* to *E*.

13. Sample path: *B, E, A, B, D, C, A, E, C, E, D*

15. 2

17. 4

19. Answers will vary.

21. The sum of the angles of such a quadrilateral is greater than 360°.

23. Yes

Chapter 10 Review Exercises

1. $\angle 3$

2. $\angle 5$

3. $\angle 4$ and $\angle 6$

4. $\angle 1$ and $\angle 6$

5. $\angle 1$ and $\angle 4$

6. $\angle 2$

7. $\angle 5$

8. $180° - 115° = 65°$

9. $90° - 41° = 49°$

10. Measure of complement $= 90° - 73° = 17°$

11. Measure of supplement $= 180° - 46° = 134°$

12. $m\angle 1 = 180° - 70° = 110°$
    $m\angle 2 = 70°$
    $m\angle 3 = m\angle 1 = 110°$

13. $m\angle 1 = 180° - 42° = 138°$
    $m\angle 2 = 42°$
    $m\angle 3 = m\angle 1 = 138°$
    $m\angle 4 = m\angle 1 = 138°$
    $m\angle 5 = m\angle 2 = 42°$
    $m\angle 6 = 42°$
    $m\angle 7 = m\angle 3 = 138°$

14. $m\angle A = 180° - 60° - 48° = 72°$

15. $m\angle A = 90° - 39° = 51°$

16. $m\angle 1 = 180° - 50° - 40° = 90°$
    $m\angle 2 = 180° - 90° = 90°$
    $m\angle 3 = 180° - 40° = 140°$
    $m\angle 4 = 40°$
    $m\angle 5 = m\angle 3 = 140°$

17. $m\angle 2 = 180° - 115° = 65°$
    $\angle 1$ is in a triangle with angles of $65°$ and $35°$.
    Thus, $m\angle 1 = 180° - 65° - 35° = 80°$
    $m\angle 3 = 115°$
    $m\angle 4 = m\angle 1 = 80°$
    $m\angle 5 = 180° - 80° = 100°$
    $m\angle 6 = m\angle 1 = 80°$

18. $\dfrac{8}{4} = \dfrac{10}{x}$
    $8x = 40$
    $x = 5$ ft

19. $\dfrac{9}{x} = \dfrac{7+5}{5}$
    $\dfrac{9}{x} = \dfrac{12}{5}$
    $12x = 45$
    $x = \dfrac{45}{12} = 3.75$ ft

20. $c^2 = 8^2 + 6^2$
$c^2 = 64 + 36$
$c^2 = 100$
$c = 10 \text{ ft}$

21. $c^2 = 6^2 + 4^2$
$c^2 = 36 + 16$
$c^2 = 52$
$c \approx 7.2 \text{ in.}$

22. $b^2 = 15^2 - 11^2$
$b^2 = 225 - 121$
$b^2 = 104$
$b \approx 10.2 \text{ cm}$

23. $\dfrac{x}{5} = \dfrac{9+6}{6}$
$\dfrac{x}{5} = \dfrac{15}{6}$
$6x = 75$
$x = 12.5 \text{ ft}$

24. $a^2 = 25^2 - 20^2$
$a^2 = 625 - 400$
$a^2 = 225$
$a = 15 \text{ ft}$

25. $b^2 = 13^2 + 5^2$
$b^2 = 169 - 25$
$b^2 = 144$
$b = 12 \text{ yd}$

26. Rectangle, square

27. Rhombus, square

28. Parallelogram, rhombus, trapezoid

29. $P = 2 \,(6 \text{ cm}) + 2(9 \text{ cm})$
$= 12 \text{ cm} + 18 \text{ cm}$
$= 30 \text{ cm}$

30. $P = 2 \cdot 1000 \text{ yd} + 2 \cdot 1240 \text{ yd}$
$= 2000 \text{ yd} + 2480 \text{ yd}$
$= 4480 \text{ yd}$

31. First find the lengths of missing sides.

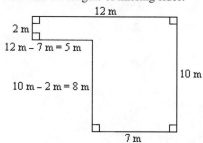

$P = 12 \text{ m} + 10 \text{ m} + 7 \text{ m} + 8 \text{ m} + 5 \text{ m} + 2 \text{ m}$
$= 44 \text{ m}$

32. Sum $= (n - 2) \, 180°$
$= (12 - 2) \, 180°$
$= 10 \cdot 180°$
$= 1800°$

33. Sum $= (n - 2) \, 180°$
$= (8 - 2) \, 180°$
$= 6 \cdot 180°$
$= 1080°$

34. Sum of measures of angles $= (n - 2)180°$
$= (8 - 2)180°$
$= 6 \cdot 180°$
$= 1080°$

$m\angle 1 = \dfrac{1080°}{8} = 135°$
$m\angle 2 = 180° - 135° = 45°$

35. Amount of baseboard
$=$ Perimeter of room $-$ Lengths of doorways
$= 2 \cdot 35 \text{ ft} + 2 \cdot 15 \text{ ft} - 4 \cdot 3 \text{ ft}$
$= 70 \text{ ft} + 30 \text{ ft} - 12 \text{ ft}$
$= 88 \text{ ft}$
$\text{Cost} = \dfrac{\$1.50}{1\,\cancel{\text{ft}}} \cdot \dfrac{88\,\cancel{\text{ft}}}{1} = \$132$

36. a. triangles, hexagons

b. The 5 angles that come together are
$60°$, $60°$, $60°$, $60°$, and $120°$.

c. The tessellation is possible because
$60° + 60° + 60° + 60° + 120° = 360°$.

37. Each angle is $\dfrac{(n-2)\,180°}{n} = \dfrac{(6-2)\,180°}{6} = 120°$
Regular hexagons can be used to create a
tessellation because $360°$ is a multiple of $120°$.

38. $A = 5 \text{ ft} \cdot 6.5 \text{ ft} = 32.5 \text{ ft}^2$

39. $A = 5 \text{ m} \cdot 4 \text{ m} = 20 \text{ m}^2$

40. $A = \frac{1}{2} \cdot 20 \text{ cm} \cdot 5 \text{ cm} = 50 \text{ cm}^2$

41. $A = \frac{1}{2} \cdot 10 \text{ yd} \cdot (22 \text{ yd} + 5 \text{ yd})$

$\qquad = \frac{1}{2} \cdot 10 \text{ yd} \cdot (27 \text{ yd})$

$\qquad = 135 \text{ yd}^2$

42. $C = \pi \cdot 20 \text{ m} = 20\pi \text{ m} \approx 62.8 \text{ m}$

$\qquad r = \frac{1}{2} d = \frac{1}{2} \cdot 20 \text{ m} = 10 \text{ m}$

$\qquad A = \pi r^2 = \pi (10 \text{ m})^2 = 100\pi \text{ m}^2 \approx 314.2 \text{ m}^2$

43. Area = (Area of square) + (Area of triangle)

$\qquad = (12 \text{ in.})^2 + \frac{1}{2} \cdot 12 \text{ in.} \cdot 8 \text{ in.}$

$\qquad = 144 \text{ in.}^2 + 48 \text{ in.}^2$

$\qquad = 192 \text{ in.}^2$

44. Area = (Area of top rectangle)

$\qquad\quad$ + (Area of bottom rectangle)

$\qquad = 8 \text{ m} \cdot 2 \text{ m} + 6 \text{ m} \cdot 2 \text{ m}$

$\qquad = 16 \text{ m}^2 + 12 \text{ m}^2$

$\qquad = 28 \text{ m}^2$

45. First convert linear measurements in feet to linear yards.

$\qquad 15 \text{ ft} = \frac{15 \text{ ft}}{1} \cdot \frac{1 \text{ yd}}{3 \text{ ft}} = 5 \text{ yd}$

$\qquad 21 \text{ ft} = \frac{21 \text{ ft}}{1} \cdot \frac{1 \text{ yd}}{3 \text{ ft}} = 7 \text{ yd}$

$\qquad$ Area = 5 yd $\cdot$ 7 yd = 35 yd^2

$\qquad$ Cost $= \frac{\$22.50}{1 \text{ yd}^2} \cdot \frac{35 \text{ yd}^2}{1} = \$787.50$

46. Area of floor = 40 ft $\cdot$ 50 ft = 2000 ft^2

$\qquad$ Area of each tile = $(2 \text{ ft})^2 = 4 \text{ ft}^2$

$\qquad$ Number of tiles $= \frac{2000 \text{ ft}^2}{4 \text{ ft}^2} = 500$ tiles

$\qquad$ Cost $= \frac{\$13}{10 \text{ tiles}} \cdot \frac{500 \text{ tiles}}{1} = \$650$

47. $C = \pi d = \pi \cdot 10 \text{ yd} = 10\pi \text{ yd} \approx 31 \text{ yd}$

48. $V = 5 \text{ cm} \cdot 3 \text{ cm} \cdot 4 \text{ cm} = 60 \text{ cm}^3$

49. V = (Volume of rectangular solid)

$\qquad\quad$ + (Volume of Pyramid)

$\qquad = 8 \text{ m} \cdot 9 \text{ m} \cdot 10 \text{ m} + \frac{1}{3}(8 \text{ m} \cdot 9 \text{ m}) \, 10 \text{ m}$

$\qquad \approx 720 \text{ m}^3 + 240 \text{ m}^3 = 960 \text{ m}^3$

50. $V = \pi (4 \text{ yd})^2 \cdot 8 \text{ yd} = 128\pi \text{ yd}^3 \approx 402 \text{ yd}^3$

51. $V = \frac{1}{3} \pi (40 \text{ in.})^2 \cdot 28 \text{ in.}$

$\qquad = \frac{44,800}{3} \pi \text{ in.}^3 \approx 46,914 \text{ in.}^3$

52. $V = \frac{4}{3} \pi (6 \text{ m})^3 = 288\pi \text{ m}^3 \approx 905 \text{ m}^3$

53. Surface area

$\qquad = 2(5 \text{ m})(3 \text{ m}) + 2(3 \text{ m})(6 \text{ m}) + 2(5 \text{ m})(6 \text{ m})$

$\qquad = 30 \text{ m}^2 + 36 \text{ m}^2 + 60 \text{ m}^2$

$\qquad = 126 \text{ m}^2$

54. Volume of one box = 8 m $\cdot$ 4 m $\cdot$ 3 m = 96 m^3

$\qquad$ Volume of 50 boxes = 50 $\cdot$ 96 m^3 = 4800 m^3

55. $V = \frac{1}{3}(145 \text{ m})^2 \cdot 93 \text{ m} = 651,775 \text{ m}^3$

56. First convert linear measures in feet to linear yards.

$\qquad 27 \text{ ft} = \frac{27 \text{ ft}}{1} \cdot \frac{1 \text{ yd}}{3 \text{ ft}} = 9 \text{ yd}$

$\qquad 4 \text{ ft} = \frac{4 \text{ ft}}{1} \cdot \frac{1 \text{ yd}}{3 \text{ ft}} = \frac{4}{3} \text{ yd}$

$\qquad 6 \text{ in.} = \frac{6 \text{ in.}}{1} \cdot \frac{1 \text{ yd}}{36 \text{ in.}} = \frac{1}{6} \text{ yd}$

$\qquad$ Volume $= 9 \text{ yd} \cdot \frac{4}{3} \text{ yd} \cdot \frac{1}{6} \text{ yd} = 2 \text{ yd}^3$

$\qquad$ Cost $= \frac{\$40}{1 \text{ yd}^3} \cdot \frac{2 \text{ yd}^3}{1} = \$80$

57. First compute length of hypotenuse
$$c^2 = 12^2 + 9^2 = 144 + 81 = 225$$
$$c = 15$$
$$\sin A = \frac{9}{15} = \frac{3}{5}$$
$$\cos A = \frac{12}{15} = \frac{4}{5}$$
$$\tan A = \frac{9}{12} = \frac{3}{4}$$

58. $\tan 23° = \dfrac{a}{100}$
$$a = 100 \tan 23° \approx 42 \text{ mm}$$

59. $\sin 61° = \dfrac{20}{c}$
$$c = \frac{20}{\sin 61°} \approx 23 \text{ cm}$$

60. $\sin 48° = \dfrac{a}{50}$
$$a = 50 \sin 48° \approx 37 \text{ in.}$$

61. $\sin A = \dfrac{17}{20}$
$$A = \sin^{-1}\left(\frac{17}{20}\right) \approx 58°$$

62. $\dfrac{1}{2}\text{ mi} = \dfrac{0.5\text{ mi}}{1} \cdot \dfrac{5280\text{ ft}}{1\text{ mi}} = 2640\text{ ft}$
$$\sin 17° = \frac{h}{2640}$$
$$h = 2640 \sin 17° \approx 772 \text{ ft}$$

63. $\tan 32° = \dfrac{d}{50}$
$$d = 50 \tan 32° \approx 31 \text{ m}$$

64.

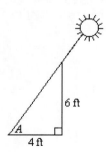

$$\tan A = \frac{6}{4}$$
$$A = \tan^{-1}\left(\frac{6}{4}\right) \approx 56°$$

65. The graph is not traversable because there are more than two odd vertices.

66. All vertices have even degrees, so the graph is traversable. Possible path: *A, B, C, D, A, B, C, D, A*

67. 0

68. 2

69. 1

70. 2

Chapter 10 Test

1. Measure of complement = 90° − 54° = 36°
Measure of supplement = 180° − 54° = 126°

2. $m\angle 1 = 133°$ because alternate exterior angles are equal.

3. $m\angle 1 = 180° - 40° - 70° = 70°$

4. First find measures of other angles of triangle.

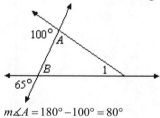

$$m\angle A = 180° - 100° = 80°$$
$$m\angle B = 65°$$
$$m\angle 1 = 180° - 80° - 65° = 35°$$

5. $\dfrac{x}{8} = \dfrac{4}{10}$
$$10x = 4 \cdot 8$$
$$10x = 32$$
$$x = \frac{32}{10} = 3.2 \text{ in.}$$

6. $b^2 = 26^2 - 24^2$
$$b^2 = 676 - 576$$
$$b^2 = 100$$
$$b = 10 \text{ ft}$$

7. Sum $= (n-2)\,180°$
$= (10-2)\,180°$
$= 8 \cdot 180°$
$= 1440°$

8. First find lengths of missing sides.

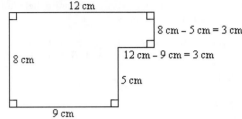

$P = 12 \text{ cm} + 3 \text{ cm} + 3 \text{ cm} + 5 \text{ cm} + 9 \text{ cm} + 8 \text{ cm}$
$= 40 \text{ cm}$

9. d

10. a. triangles, squares

   b. The 5 angles that come together are
   $60°$, $60°$, $60°$, $90°$, and $90°$.

   c. The tessellation is possible because
   $60° + 60° + 60° + 90° + 90° = 360°$.

11. $A = \dfrac{1}{2} bh$

$A = \dfrac{1}{2} \cdot 47 \text{ m} \cdot 22 \text{ m} \ = 517 \text{ m}^2$

12. $A = \dfrac{1}{2} \cdot 15 \text{ in.}(40 \text{ in.} + 30 \text{ in.})$

$= \dfrac{1}{2} \cdot 15 \text{ in.}(70 \text{ in.})$

$= 525 \text{ in.}^2$

13. a. $a^2 + b^2 = c^2$
$a^2 + 5^2 = 13^2$
$a^2 + 25 = 169$
$a^2 = 144$
$a = 12 \text{ cm}$

   b. $P = 5 \text{ cm} + 12 \text{ cm} + 13 \text{ cm} = 30 \text{ cm}$

   c. $A = \frac{1}{2} bh = \frac{1}{2} \cdot 12 \text{ cm} \cdot 5 \text{ cm} = 30 \text{ cm}^2$

14. $C = \pi d = \pi \cdot 40 \text{ m} = 40\pi \text{ m} \approx 125.7 \text{ m}$
$A = \pi r^2 = \pi (20 \text{ m})^2 = 400\pi \text{ m}^2 \approx 1256.6 \text{ m}^2$

15. Area of floor $8 \text{ ft} \cdot 6 \text{ ft} = 48 \text{ ft}^2$
Convert inches to feet:

$8 \text{ in.} = \dfrac{8 \text{ in.}}{1} \cdot \dfrac{1 \text{ ft}}{12 \text{ in.}} = \dfrac{2}{3} \text{ ft}$

Area of one tile

$= \left(\dfrac{2}{3} \text{ ft} \right)^2 = \dfrac{4}{9} \text{ ft}^2$

Number of tiles

$= \dfrac{48 \text{ ft}^2}{\frac{4}{9} \text{ ft}^2} = 108 \text{ tiles}$

16. $V = 3 \text{ ft} \cdot 2 \text{ ft} \cdot 3 \text{ ft} = 18 \text{ ft}^3$

17. $V = \dfrac{1}{3}(4 \text{ m} \cdot 3 \text{ m})\, 4 \text{ m} = 16 \text{ m}^3$

18. $V = \pi(5 \text{ cm})^2 \cdot 7 \text{ cm} = 175\pi \text{ cm}^3 \approx 550 \text{ cm}^3$

19. $\sin 28° = \dfrac{40}{c}$

$c = \dfrac{40}{\sin 28°} \approx 85 \text{ cm}$

20.

$\tan 34° = \dfrac{h}{104}$
$h = 104 \tan 34° \approx 70 \text{ ft}$

21. The graph is traversable because there are two odd
vertices (B and E). Sample path: $BCAECDE$

22. Answers will vary.

Chapter 11
Counting Methods and Probability Theory

Check Points 11.1

1. Multiply the number of choices for each of the two courses of the meal:

   <u>Appetizers</u> : <u>Main Courses:</u>

      10   ×   15   = 150

2. Multiply the number of choices for each of the two courses:

   <u>Psychology</u> : <u>Social Science:</u>

      10   ×   4   = 40

3. Multiply the number of choices for each of the three decisions:

   <u>Size</u> : <u>Crust</u> : <u>Topping:</u>

     2 × 3 × 5   = 30

4. Multiply the number of choices for each of the five options:

   <u>Color:</u> <u>A/C:</u> <u>Electric/Gas:</u> <u>Onboard Computer:</u> <u>Global Positioning System:</u>

     10 × 2 ×   2   ×   2   ×   2   = 160

5. Multiply the number of choices for each of the six questions:

   <u>Question #1:</u> <u>Question #2:</u> <u>Question #3:</u> <u>Question #4:</u> <u>Question #5:</u> <u>Question #6:</u>

      3   ×   3   ×   3   ×   3   ×   3   ×   3   = 3^6 = 729

6. Multiply the number of choices for each of the five digits:

   $\overbrace{}^{1-9}$ $\overbrace{}^{0-9}$ $\overbrace{}^{0-9}$ $\overbrace{}^{0-9}$ $\overbrace{}^{0-9}$

   <u>Digit 1:</u> <u>Digit 2:</u> <u>Digit 3:</u> <u>Digit 4:</u> <u>Digit 5:</u>

      9   ×   10   ×   10   ×   10   ×   10   = 90,000

Exercise Set 11.1

1. $8 \cdot 10 = 80$

3. $3 \cdot 4 = 12$

5. $3 \cdot 2 = 6$

7. Multiply the number of choices for each of the three decisions:

   <u>Drink:</u>   <u>Size:</u>   <u>Flavor:</u>

     2 × 4 × 5   = 40

9. Multiply the number of choices for each of the four menu categories:

   <u>Main Course:</u> <u>Vegetables:</u> <u>Beverages:</u> <u>Desserts:</u>

      4   ×   3   ×   4   ×   3   = 144

   This includes, for example, an order of ham and peas with tea and cake.
   This also includes an order of beef and peas with milk and pie.

11. Multiply the number of choices for each of the three categories:
Gender: Age: Payment method:

  2   ×   2   ×   2   = 8

13. Multiply the number of choices for each of the five options:
Color: A/C: Transmission: Windows: CD Player:

  6   ×   2   ×   2   ×   2   ×   2   = 96

15. Multiply the number of choices for each of the five questions:
Question 1:   Question 2:   Question 3:   Question 4:   Question 5:

  3   ×   3   ×   3   ×   3   ×   3   = 243

17. Multiply the number of choices for each of the three digits:
Digit 1:   Digit 2:   Digit 3:

  8   ×   2   ×   9   = 144

19. Multiply the number of choices for each of the letters and digits:
Letter 1:   Letter 2:   Digit 1:   Digit 2:   Digit 3:

  26   ×   26   ×   10   ×   10   ×   10   = 676,000

21. This situation involves making choices with seven groups of items. Each stock is a group, and each group has three choices. Multiply choices: $3 \times 3 \times 3 \times 3 \times 3 \times 3 \times 3 = 3^7 = 2187$

27. Multiply the number of choices for each of the four groups of items:
Bun: Sauce: Lettuce: Tomatoes:

  12   ×   30   ×   4   ×   3   = 4320

Total time $= 10 \times 4320 = 43,200$ minutes, which is $43,200 \div 60 = 720$ hours.

Check Points 11.2

1. There is only one choice each for the first and last performers. This leaves two choices for the second performer and then one for the third.

      U2   Bruce Springsteen, Aerosmith   Rolling Stones
1st Performer: 2nd Performer: 3rd Performer: 4th Performer:

  1   ×   2   ×   1   ×   1 $= 2$

2. The number of choices decreases by 1 each time a book is selected.
1st Book: 2nd Book: 3rd Book: 4th Book: 5th Book:

  5   ×   4   ×   3   ×   2   ×   1 $= 120$

3. a. $\dfrac{9!}{6!} = \dfrac{9 \cdot 8 \cdot 7 \cdot 6!}{6!} = \dfrac{9 \cdot 8 \cdot 7 \cdot \cancel{6!}}{\cancel{6!}} = 9 \cdot 8 \cdot 7 = 504$

b. $\dfrac{16!}{11!} = \dfrac{16 \cdot 15 \cdot 14 \cdot 13 \cdot 12 \cdot 11!}{11!} = \dfrac{16 \cdot 15 \cdot 14 \cdot 13 \cdot 12 \cdot \cancel{11!}}{\cancel{11!}} = 16 \cdot 15 \cdot 14 \cdot 13 \cdot 12 = 524,160$

c. $\dfrac{100!}{99!} = \dfrac{100 \cdot 99!}{99!} = \dfrac{100 \cdot \cancel{99!}}{\cancel{99!}} = 100$

4. $_7P_4 = \dfrac{7!}{(7-4)!} = \dfrac{7!}{3!} = \dfrac{7 \cdot 6 \cdot 5 \cdot 4 \cdot 3!}{3!} = \dfrac{7 \cdot 6 \cdot 5 \cdot 4 \cdot \cancel{3!}}{\cancel{3!}} = 7 \cdot 6 \cdot 5 \cdot 4 = 840$

5. $_9P_5 = \dfrac{9!}{(9-5)!} = \dfrac{9!}{4!} = \dfrac{9 \cdot 8 \cdot 7 \cdot 6 \cdot 5 \cdot 4!}{4!} = \dfrac{9 \cdot 8 \cdot 7 \cdot 6 \cdot 5 \cdot \cancel{4!}}{\cancel{4!}} = 9 \cdot 8 \cdot 7 \cdot 6 \cdot 5 = 15,120$

6. There a 7 letters with 2 O's and 3 S's. Thus, $\dfrac{n!}{p!q!} = \dfrac{7!}{2!3!} = \dfrac{7 \cdot 6 \cdot 5 \cdot 4 \cdot \cancel{3!}}{2 \cdot 1 \cdot \cancel{3!}} = 420$

Exercise Set 11.2

1. The number of choices decreases by 1 each time a performer is selected.

1st Performer :	2nd Performer:	3rd Performer:	4th Performer:	5th Performer:	6th Performer:
6 $\times$	5 $\times$	4 $\times$	3 $\times$	2 $\times$	1 $= 720$

3. The number of choices decreases by 1 each time a sentence is selected.

1st Sentence:	2nd Sentence:	3rd Sentence:	4th Sentence:	5th Sentence:
5 $\times$	4 $\times$	3 $\times$	2 $\times$	1 $= 120$

5. There is only one choice for the 6th performer. The number of choices decreases by 1 each time a performer is selected.

1st Performer :	2nd Performer:	3rd Performer:	4th Performer:	5th Performer:	6th Performer:
5 $\times$	4 $\times$	3 $\times$	2 $\times$	1 $\times$	1 $= 120$

7. The number of choices decreases by 1 each time a book is selected.

1st Book:	2nd:	3rd Book:	4th:	5th Book:	6th:	7th Book:	8th:	9th Book:
9 $\times$	8 $\times$	7 $\times$	6 $\times$	5 $\times$	4 $\times$	3 $\times$	2 $\times$	1 $= 362,880$

9. There is only one choice each for the first and last sentences. For the other values, the number of choices decreases by 1 each time a sentence is selected.

1st Sentence:	2nd Sentence:	3rd Sentence:	4th Sentence:	5th Sentence:
1 $\times$	3 $\times$	2 $\times$	1 $\times$	1 $= 6$

11. There are two choices for the first movie and one for the second. There is only one choice for the last movie. This leaves two choices for the third movie and one for the fourth.

	G rated		Other two movies		NC-17 Rated	
1st Movie:	2nd Movie:	3rd Movie:	4th Movie:	5th Movie:		
2 $\times$	1 $\times$	2 $\times$	1 $\times$	1 $= 4$		

13. $\dfrac{9!}{6!} = \dfrac{9 \cdot 8 \cdot 7 \cdot 6!}{6!} = 9 \cdot 8 \cdot 7 = 504$

15. $\dfrac{29!}{25!} = \dfrac{29 \cdot 28 \cdot 27 \cdot 26 \cdot 25!}{25!}$
$= 29 \cdot 28 \cdot 27 \cdot 26$
$= 570,024$

17. $\dfrac{19!}{11!} = \dfrac{19 \cdot 18 \cdot 17 \cdot 16 \cdot 15 \cdot 14 \cdot 13 \cdot 12 \cdot 11!}{11!}$
$= 19 \cdot 18 \cdot 17 \cdot 16 \cdot 15 \cdot 14 \cdot 13 \cdot 12$
$= 3,047,466,240$

19. $\dfrac{600!}{599!} = \dfrac{600 \cdot 599!}{599!} = 600$

21. $\dfrac{104!}{102!} = \dfrac{104 \cdot 103 \cdot 102!}{102!} = 104 \cdot 103 = 10,712$

23. $7! - 3! = 5040 - 6 = 5034$

25. $(7-3)! = 4! = 4 \cdot 3 \cdot 2 \cdot 1 = 24$

27. $\left(\dfrac{12}{4}\right)! = 3! = 3 \cdot 2 \cdot 1 = 6$

29. $\dfrac{7!}{(7-2)!} = \dfrac{7!}{5!} = \dfrac{7 \cdot 6 \cdot 5!}{5!} = 7 \cdot 6 = 42$

31. $\dfrac{13!}{(13-3)!} = \dfrac{13!}{10!}$
$= \dfrac{13 \cdot 12 \cdot 11 \cdot 10!}{10!}$
$= 13 \cdot 12 \cdot 11$
$= 1716$

33. $_9P_4 = \dfrac{9!}{(9-4)!}$
$= \dfrac{9!}{5!}$
$= \dfrac{9 \cdot 8 \cdot 7 \cdot 6 \cdot 5!}{5!}$
$= 9 \cdot 8 \cdot 7 \cdot 6$
$= 3024$

35. $_8P_5 = \dfrac{8!}{(8-5)!}$
$= \dfrac{8!}{3!}$
$= \dfrac{8 \cdot 7 \cdot 6 \cdot 5 \cdot 4 \cdot 3!}{3!}$
$= 8 \cdot 7 \cdot 6 \cdot 5 \cdot 4$
$= 6720$

37. $_6P_6 = \dfrac{6!}{(6-6)!} = \dfrac{6!}{0!} = \dfrac{6 \cdot 5 \cdot 4 \cdot 3 \cdot 2 \cdot 1}{1} = 720$

39. $_8P_0 = \dfrac{8!}{(8-0)!} = \dfrac{8!}{8!} = 1$

41. $_{10}P_3 = \dfrac{10!}{(10-3)!}$
$= \dfrac{10!}{7!}$
$= \dfrac{10 \cdot 9 \cdot 8 \cdot 7!}{7!}$
$= 10 \cdot 9 \cdot 8$
$= 720$

43. $_{13}P_7 = \dfrac{13!}{(13-7)!}$
$= \dfrac{13!}{6!}$
$= \dfrac{13 \cdot 12 \cdot 11 \cdot 10 \cdot 9 \cdot 8 \cdot 7 \cdot 6!}{6!}$
$= 13 \cdot 12 \cdot 11 \cdot 10 \cdot 9 \cdot 8 \cdot 7$
$= 8,648,640$

45. $_6P_3 = \dfrac{6!}{(6-3)!}$
$= \dfrac{6!}{3!}$
$= \dfrac{6 \cdot 5 \cdot 4 \cdot 3!}{3!}$
$= 6 \cdot 5 \cdot 4$
$= 120$

47. $_9P_5 = \dfrac{9!}{(9-5)!}$
$= \dfrac{9!}{4!}$
$= \dfrac{9 \cdot 8 \cdot 7 \cdot 6 \cdot 5 \cdot 4!}{4!}$
$= 9 \cdot 8 \cdot 7 \cdot 6 \cdot 5$
$= 15,120$

49. $\dfrac{n!}{p!q!} = \dfrac{6!}{2!2!} = \dfrac{6 \cdot 5 \cdot 4 \cdot 3 \cdot 2 \cdot 1}{2 \cdot 1 \cdot 2 \cdot 1} = 180$

51. $\dfrac{n!}{p!q!r!s!} = \dfrac{11!}{3!2!2!2!}$
$= \dfrac{11 \cdot 10 \cdot 9 \cdot 8 \cdot 7 \cdot 6 \cdot 5 \cdot 4 \cdot 3!}{3! \cdot 2 \cdot 1 \cdot 2 \cdot 1 \cdot 2 \cdot 1}$
$= 831,600$

53. $\dfrac{n!}{p!q!} = \dfrac{7!}{4!2!} = \dfrac{7 \cdot 6 \cdot 5 \cdot 4!}{4! \cdot 2 \cdot 1} = 105$

55. $\dfrac{n!}{p!q!} = \dfrac{8!}{4!3!} = \dfrac{8 \cdot 7 \cdot 6 \cdot 5 \cdot 4!}{4! \cdot 3 \cdot 2 \cdot 1} = 280$

63. Because the letter B is repeated in the word BABE, the number of permutations is given by
$\dfrac{n!}{p!} = \dfrac{4!}{2!} = \dfrac{4 \cdot 3 \cdot 2 \cdot 1}{2 \cdot 1} = 12$

65. Multiply the number of ways to select the two first place horses by the number of orders in which the remaining four horses can finish.

$$_6C_2 \times _4P_4 = 15 \times 24 = 360$$

67. There are 5! ways to arrange the women, and 5! ways to arrange the men. The total number of arrangements is found by multiplying these values: (5!)(5!) = 120 · 120 = 14,400

69. $$_nP_{n-2} = \frac{n!}{(n-(n-2))!} = \frac{n!}{(n-n+2)!} = \frac{n!}{2!} = \frac{n(n-1)(n-2) \times \cdots \times 3 \times 2 \times 1}{2} = n(n-1)(n-2) \times \cdots \times 3$$

Check Points 11.3

1. **a.** The order in which you select the DVDs does not matter. This problem involves combinations.

   b. Order matters. This problem involves permutations.

2. $$_7C_3 = \frac{7!}{(7-3)!3!} = \frac{7!}{4!3!} = \frac{7 \cdot 6 \cdot 5 \cdot 4!}{4! \cdot 3 \cdot 2 \cdot 1} = \frac{7 \cdot 6 \cdot 5 \cdot \cancel{4!}}{\cancel{4!} \cdot 3 \cdot 2 \cdot 1} = \frac{7 \cdot 6 \cdot 5}{3 \cdot 2 \cdot 1} = 35$$

3. $$_{16}C_4 = \frac{16!}{(16-4)!4!} = \frac{16!}{12!4!} = \frac{16 \cdot 15 \cdot 14 \cdot 13 \cdot 12!}{12! \cdot 4 \cdot 3 \cdot 2 \cdot 1} = \frac{16 \cdot 15 \cdot 14 \cdot 13 \cdot \cancel{12!}}{\cancel{12!} \cdot 4 \cdot 3 \cdot 2 \cdot 1} = \frac{16 \cdot 15 \cdot 14 \cdot 13}{4 \cdot 3 \cdot 2 \cdot 1} = 1820$$

4. Choose the Democrats: $$_{50}C_3 = \frac{50!}{(50-3)!3!} = \frac{50!}{47!3!} = \frac{50 \cdot 49 \cdot 48 \cdot 47!}{47! \cdot 3 \cdot 2 \cdot 1} = \frac{50 \cdot 49 \cdot 48 \cdot \cancel{47!}}{\cancel{47!} \cdot 3 \cdot 2 \cdot 1} = \frac{50 \cdot 49 \cdot 48}{3 \cdot 2 \cdot 1} = 19,600$$

   Choose the Republicans: $$_{49}C_2 = \frac{49!}{(49-2)!2!} = \frac{49!}{47!2!} = \frac{49 \cdot 48 \cdot 47!}{47! \cdot 2 \cdot 1} = \frac{49 \cdot 48 \cdot \cancel{47!}}{\cancel{47!} \cdot 2 \cdot 1} = \frac{49 \cdot 48}{2 \cdot 1} = 1176$$

   Multiply the choices: $19,600 \times 1176 = 23,049,600$

Exercise Set 11.3

1. Order does not matter. This problem involves combinations.

3. Order matters. This problem involves permutations.

5. $$_6C_5 = \frac{6!}{(6-5)!5!} = \frac{6!}{1!5!} = \frac{6 \cdot 5!}{1 \cdot 5!} = 6$$

7. $$_9C_5 = \frac{9!}{(9-5)!5!} = \frac{9!}{4!5!} = \frac{9 \cdot 8 \cdot 7 \cdot 6 \cdot 5!}{4 \cdot 3 \cdot 2 \cdot 1 \cdot 5!} = 126$$

9. $$_{11}C_4 = \frac{11!}{(11-4)!4!} = \frac{11!}{7!4!} = \frac{11 \cdot 10 \cdot 9 \cdot 8 \cdot 7!}{7! \cdot 4 \cdot 3 \cdot 2 \cdot 1} = 330$$

11. $$_8C_1 = \frac{8!}{(8-1)!1!} = \frac{8!}{7!1!} = \frac{8 \cdot 7!}{7!1} = 8$$

13. $$_7C_7 = \frac{7!}{(7-7)!7!} = \frac{7!}{0!7!} = 1$$

15. $\displaystyle {}_{30}C_3 = \frac{30!}{(30-3)!3!} = \frac{30!}{27!3!} = \frac{30\cdot 29\cdot 28\cdot 27!}{27!\cdot 3\cdot 2\cdot 1} = 4060$

17. $\displaystyle {}_5C_0 = \frac{5!}{(5-0)!0!} = \frac{5!}{5!0!} = 1$

19. $\displaystyle \frac{{}_7C_3}{{}_5C_4} = \frac{\frac{7!}{(7-3)!3!}}{\frac{5!}{(5-4)!4!}} = \frac{\frac{7!}{4!3!}}{\frac{5!}{1!4!}} = \frac{\frac{7\cdot6\cdot5\cdot4!}{4!\cdot3\cdot2\cdot1}}{\frac{5\cdot4!}{1\cdot4!}} = \frac{35}{5} = 7$

21. $\displaystyle \frac{{}_7P_3}{3!} - {}_7C_3 = \frac{\frac{7!}{(7-3)!}}{3!} - \frac{7!}{(7-3)!3!} = \frac{\frac{7!}{4!}}{3!} - \frac{7!}{4!3!} = \frac{7!}{4!3!} - \frac{7!}{4!3!} = 0$

23. $\displaystyle 1 - \frac{{}_3P_2}{{}_4P_3} = 1 - \frac{\frac{3!}{(3-2)!}}{\frac{4!}{(4-3)!}} = 1 - \frac{\frac{3!}{1!}}{\frac{4!}{1!}} = 1 - \frac{3!}{4!} = 1 - \frac{3!}{4\cdot3!} = 1 - \frac{1}{4} = \frac{3}{4}$

25. $\displaystyle \frac{{}_7C_3}{{}_5C_4} - \frac{98!}{96!} = \frac{\frac{7!}{(7-3)!3!}}{\frac{5!}{(5-4)!4!}} - \frac{98\cdot97\cdot96!}{96!} = \frac{\frac{7!}{4!3!}}{\frac{5!}{1!4!}} - 95067 = \frac{\frac{7\cdot6\cdot5\cdot4!}{4!3\cdot2\cdot1}}{\frac{5\cdot4!}{1!4!}} - 9506 = \frac{35}{5} - 9506 = 7 - 9506 = -9499$

27. $\displaystyle \frac{{}_4C_2\cdot{}_6C_1}{{}_{18}C_3} = \frac{\frac{4!}{(4-2)!2!}\cdot\frac{6!}{(6-1)!1!}}{\frac{18!}{(18-3)!3!}} = \frac{\frac{4!}{2!2!}\cdot\frac{6!}{5!1!}}{\frac{18!}{15!3!}} = \frac{\frac{4\cdot3\cdot2!}{2!2\cdot1}\cdot\frac{6\cdot5!}{5!1!}}{\frac{18\cdot17\cdot16\cdot15!}{15!3\cdot2\cdot1}} = \frac{36}{816} = \frac{3}{68}$

29. $\displaystyle {}_6C_3 = \frac{6!}{(6-3)!3!} = \frac{6!}{3!3!} = \frac{6\cdot5\cdot4\cdot3!}{3!3\cdot2\cdot1} = 20$

31. $\displaystyle {}_{12}C_4 = \frac{12!}{(12-4)!4!} = \frac{12!}{8!4!} = \frac{12\cdot11\cdot10\cdot9\cdot8!}{8!4\cdot3\cdot2\cdot1} = 495$

33. $\displaystyle {}_{17}C_8 = \frac{17!}{(17-8)!8!} = \frac{17!}{9!8!} = \frac{17\cdot16\cdot15\cdot14\cdot13\cdot12\cdot11\cdot10\cdot9!}{9!8\cdot7\cdot6\cdot5\cdot4\cdot3\cdot2\cdot1} = 24,310$

35. $\displaystyle {}_{53}C_6 = \frac{53!}{(53-6)!6!} = \frac{53!}{47!6!} = \frac{53\cdot52\cdot51\cdot50\cdot49\cdot48\cdot47!}{47!6\cdot5\cdot4\cdot3\cdot2\cdot1} = 22,957,480$

37. $\displaystyle {}_6P_4 = \frac{6!}{2!} = 6\cdot5\cdot4\cdot3 = 360$ ways

39. $\displaystyle {}_{13}C_6 = \frac{13!}{7!6!} = \frac{13\cdot12\cdot11\cdot10\cdot9\cdot8}{6\cdot5\cdot4\cdot3\cdot2\cdot1}$
$= 1716$ ways

41. $_{20}C_3 = \dfrac{20!}{17!3!} = \dfrac{20 \cdot 19 \cdot 18}{3 \cdot 2 \cdot 1} = 1140$ ways

43. $_7P_4 = \dfrac{7!}{3!} = 840$ passwords

45. $_{15}P_3 = \dfrac{15!}{12!} = 15 \cdot 14 \cdot 13 = 2730$ cones

47. Choose the men: $_7C_4 = \dfrac{7!}{(7-4)!4!} = \dfrac{7!}{3!4!} = \dfrac{7 \cdot 6 \cdot 5 \cdot 4!}{3 \cdot 2 \cdot 1 \cdot 4!} = 35$

Choose the women: $_7C_5 = \dfrac{7!}{(7-5)!5!} = \dfrac{7!}{2!5!} = \dfrac{7 \cdot 6 \cdot 5!}{2 \cdot 1 \cdot 5!} = 21$

Multiply the choices: $35 \cdot 21 = 735$

49. Choose the Republicans: $_{55}C_4 = \dfrac{55!}{(55-4)!4!} = \dfrac{55!}{51!4!} = \dfrac{55 \cdot 54 \cdot 53 \cdot 52 \cdot 51!}{51! \cdot 4 \cdot 3 \cdot 2 \cdot 1} = 341,055$

Choose the Democrats: $_{44}C_3 = \dfrac{44!}{(44-3)!3!} = \dfrac{44!}{41!3!} = \dfrac{44 \cdot 43 \cdot 42 \cdot 41!}{3 \cdot 2 \cdot 1 \cdot 41!} = 13,244$

Multiply the choices: $341,055 \times 13,244 = 4,516,932,420$

51. $_6P_6 = \dfrac{6!}{(6-6)!} = 6 \cdot 5 \cdot 4 \cdot 3 \cdot 2 \cdot 1 = 720$ ways

53. $_6C_3 = \dfrac{6!}{(6-3)!3!} = \dfrac{6!}{3!3!} = \dfrac{6 \cdot 5 \cdot 4 \cdot 3!}{3 \cdot 2 \cdot 1 \cdot 3!} = 20$ ways

55. $_4P_4 = \dfrac{4!}{(4-4)!} = 4 \cdot 3 \cdot 2 \cdot 1 = 24$ ways

57. $2 \cdot_4 C_2 = 2 \cdot \dfrac{4!}{(4-2)!2!} = 2 \cdot \dfrac{4!}{2!2!} = 2 \cdot \dfrac{4 \cdot 3 \cdot 2!}{2 \cdot 1 \cdot 2!} = 12$ ways

63. Selections for 6/53 lottery: $_{53}C_6 = \dfrac{53!}{(53-6)!6!} = \dfrac{53!}{47!6!} = \dfrac{53 \cdot 52 \cdot 51 \cdot 50 \cdot 49 \cdot 48 \cdot 47!}{47!6 \cdot 5 \cdot 4 \cdot 3 \cdot 2 \cdot 1} = 22,957,480$

Selections for 5/36 lottery: $_{36}C_5 = \dfrac{36!}{(36-5)!5!} = \dfrac{36!}{31!5!} = \dfrac{36 \cdot 35 \cdot 34 \cdot 33 \cdot 32 \cdot 31!}{31!5 \cdot 4 \cdot 3 \cdot 2 \cdot 1} = 376,992$

The 5/36 lottery is easier to win because there are fewer possible selections.

65. For a group of 20 people:

$_{20}C_2 = \dfrac{20!}{(20-2)!2!} = \dfrac{20!}{18!2!} = \dfrac{20 \cdot 19 \cdot 18!}{18!2 \cdot 1} = 190$ handshakes

Time $= 3 \times 190 = 570$ seconds, which gives $570 \div 60 = 9.5$ minutes.

For a group of 40 people:

$_{40}C_2 = \dfrac{40!}{(40-2)!2!} = \dfrac{40!}{38!2!} = \dfrac{40 \cdot 39 \cdot 38!}{38!2 \cdot 1} = 780$ handshakes

Time $= 3 \times 780 = 2340$ seconds, which gives $2340 \div 60 = 39$ minutes.

Check Points 11.4

1. **a.** The event of getting a 2 can occur in one way.
$$P(2) = \frac{\text{number of ways a 2 can occur}}{\text{total number of possible outcomes}} = \frac{1}{6}$$

   b. The event of getting a number less than 4 can occur in three ways: 1, 2, 3.
$$P(\text{less than } 4) = \frac{\text{number of ways a number less than 4 can occur}}{\text{total number of possible outcomes}} = \frac{3}{6} = \frac{1}{2}$$

   c. The event of getting a number greater than 7 cannot occur.
$$P(\text{greater than } 7) = \frac{\text{number of ways a number greater than 7 can occur}}{\text{total number of possible outcomes}} = \frac{0}{6} = 0$$
   The probability of an event that cannot occur is 0.

   d. The event of getting a number less than 7 can occur in six ways: 1, 2, 3, 4, 5, 6.
$$P(\text{less than } 7) = \frac{\text{number of ways a number less than 7 can occur}}{\text{total number of possible outcomes}} = \frac{6}{6} = 1$$
   The probability of any certain event is 1.

2. **a.** $P(\text{ace}) = \dfrac{\text{number of ways a ace can occur}}{\text{total number of possibilities}} = \dfrac{4}{52} = \dfrac{1}{13}$

   b. $P(\text{red card}) = \dfrac{\text{number of ways a red card can occur}}{\text{total number of possible outcomes}} = \dfrac{26}{52} = \dfrac{1}{2}$

   c. $P(\text{red king}) = \dfrac{\text{number of ways a red king can occur}}{\text{total number of possible outcomes}} = \dfrac{2}{52} = \dfrac{1}{26}$

3. The table shows the four equally likely outcomes. The Cc and cC children will be carriers who are not actually sick.
$$P(\text{carrier, not sick}) = P(Cc) = \frac{\text{number of ways } Cc \text{ or } cC \text{ can occur}}{\text{total number of possible outcomes}} = \frac{2}{4} = \frac{1}{2}$$

4. **a.** $P(\text{never married}) = \dfrac{\text{number of persons never married}}{\text{total number of U.S. adults}} = \dfrac{51.9}{212.5} \approx 0.24$

   b. $P(\text{male}) = \dfrac{\text{number of males}}{\text{total number of U.S. adults}} = \dfrac{102.4}{212.5} \approx 0.48$

Exercise Set 11.4

1. $P(4) = \dfrac{\text{number of ways a 4 can occur}}{\text{total number of possible outcomes}} = \dfrac{1}{6}$

3. $P(\text{odd number}) = \dfrac{\text{number of ways an odd number can occur}}{\text{total number of possible outcomes}} = \dfrac{3}{6} = \dfrac{1}{2}$

5. $P(\text{less than } 3) = \dfrac{\text{number of ways a number less than 3 can occur}}{\text{total number of possible outcomes}} = \dfrac{2}{6} = \dfrac{1}{3}$

7. $P(\text{less than } 7) = \dfrac{\text{number of ways a number less than 7 can occur}}{\text{total number of possible outcomes}} = \dfrac{6}{6} = 1$

9. $P(\text{greater than } 7) = \dfrac{\text{number of ways a number greater than 7 can occur}}{\text{total number of possible outcomes}} = \dfrac{0}{6} = 0$

11. $P(\text{queen}) = \dfrac{\text{number of ways a queen can occur}}{\text{total number of possibilities}} = \dfrac{4}{52} = \dfrac{1}{13}$

13. $P(\text{club}) = \dfrac{\text{number of ways a club can occur}}{\text{total number of possibilities}} = \dfrac{13}{52} = \dfrac{1}{4}$

15. $P(\text{picture card}) = \dfrac{\text{number of ways a picture card can occur}}{\text{total number of possibilities}} = \dfrac{12}{52} = \dfrac{3}{13}$

17. $P(\text{queen of spades}) = \dfrac{\text{number of ways a queen of spades can occur}}{\text{total number of possibilities}} = \dfrac{1}{52}$

19. $P(\text{diamond and spade}) = \dfrac{\text{number of ways a diamond and a spade can occur}}{\text{total number of possibilities}} = \dfrac{0}{52} = 0$

21. $P(\text{two heads}) = \dfrac{\text{number of ways two heads can occur}}{\text{total number of possibilities}} = \dfrac{1}{4}$

23. $P(\text{same on each toss}) = \dfrac{\text{number of ways the same outcome on each toss can occur}}{\text{total number of possibilities}} = \dfrac{2}{4} = \dfrac{1}{2}$

25. $P(\text{head on second toss}) = \dfrac{\text{number of ways a head on the second toss can occur}}{\text{total number of possibilities}} = \dfrac{2}{4} = \dfrac{1}{2}$

27. $P(\text{exactly one female child}) = \dfrac{\text{number of ways exactly one female child can occur}}{\text{total number of possibilities}} = \dfrac{3}{8}$

29. $P(\text{exactly two male children}) = \dfrac{\text{number of ways exactly two male children can occur}}{\text{total number of possibilities}} = \dfrac{3}{8}$

31. $P(\text{at least one male child}) = \dfrac{\text{number of ways at least one male child can occur}}{\text{total number of possiblities}} = \dfrac{7}{8}$

33. $P(\text{four male children}) = \dfrac{\text{number of ways four male children can occur}}{\text{total number of possibilities}} = \dfrac{0}{8} = 0$

35. $P(\text{two even numbers}) = \dfrac{\text{number of ways two even numbers can occur}}{\text{total number of possibilities}} = \dfrac{9}{36} = \dfrac{1}{4}$

37. $P(\text{two numbers whose sum is } 5) = \dfrac{\text{number of ways two numbers whose sum is 5 can occur}}{\text{total number of possibilities}} = \dfrac{4}{36} = \dfrac{1}{9}$

39. $P(\text{two numbers whose sum exceeds 12})$

$$= \frac{\text{number of ways two numbers whose sum exceeds 12 can occur}}{\text{total number of possibilities}} = \frac{0}{36} = 0$$

41. $P(\text{red region}) = \dfrac{\text{number of ways a red region can occur}}{\text{total number of possibilities}} = \dfrac{3}{10}$

43. $P(\text{blue region}) = \dfrac{\text{number of ways a blue region can occur}}{\text{total number of possibilities}} = \dfrac{2}{10} = \dfrac{1}{5}$

45. $P(\text{region that is red or blue}) = \dfrac{\text{number of ways a region that is red or blue can occur}}{\text{total number of possibilities}} = \dfrac{5}{10} = \dfrac{1}{2}$

47. $P(\text{region that is red and blue}) = \dfrac{\text{number of ways a region that is red and blue can occur}}{\text{total number of possibilities}} = \dfrac{0}{10} = 0$

49. $P(\text{sickle cell anemia}) = \dfrac{\text{number of ways sickle cell anemia can occur}}{\text{total number of possibilities}} = \dfrac{1}{4}$

51. $P(\text{healthy}) = \dfrac{\text{number of ways a healthy child can occur}}{\text{total number of possibilities}} = \dfrac{1}{4}$

53. $P(\text{sickle cell trait}) = \dfrac{\text{number of ways sickle cell trait can occur}}{\text{total number of possibilities}} = \dfrac{2}{4} = \dfrac{1}{2}$

55. $P(\text{male}) = \dfrac{\text{number of males}}{\text{total number of Americans living alone}} = \dfrac{12.5}{29.3} \approx 0.43$

57. $P(25-34 \text{ age range}) = \dfrac{\text{number in } 25-34 \text{ age range}}{\text{total number of Americans living alone}} = \dfrac{3.8}{29.3} \approx 0.13$

59. $P(\text{woman in } 15-24 \text{ age range}) = \dfrac{\text{number of women in } 15-24 \text{ age range}}{\text{total number of Americans living alone}} = \dfrac{0.8}{29.3} \approx 0.03$

Table For #61–66	Moved to Same State	Moved to Different State	Moved to Different Country	Total
Owner	11.7	2.8	0.3	14.8
Renter	18.7	4.5	1.0	24.2
Total	30.4	7.3	1.3	39.0

61. $P(\text{owner}) = \dfrac{\text{number of owners}}{\text{total number of Americans who moved in 2004}} = \dfrac{14.8}{39.0} \approx 0.38$

63. $P(\text{moved within state}) = \dfrac{\text{number that moved within state}}{\text{total number of Americans who moved in 2004}} = \dfrac{30.4}{39.0} \approx 0.78$

65. $P(\text{renter who moved to different state}) = \dfrac{\text{number of renters who moved to a different state}}{\text{total number of Americans who moved in 2004}} = \dfrac{4.5}{39.0} \approx 0.12$

75. The area of the target is $(12 \text{ in.})^2 = 144 \text{ in.}^2$

The area of the yellow region is $(9 \text{ in.})^2 - (6 \text{ in.})^2 + (3 \text{ in.})^2 = 54 \text{ in.}^2$

The probability that the dart hits a yellow region is $\dfrac{54 \text{ in.}^2}{144 \text{ in.}^2} = 0.375$

Check Points 11.5

1. total number of permutations $= 5! = 5 \cdot 4 \cdot 3 \cdot 2 \cdot 1 = 120$
number of arrangements with U2 first, the Rolling Stones fourth, and the Beatles last.

U2	Bruce Springsteen, Aerosmith		Rolling Stones	Beatles
1st:	2nd:	3rd:	4th:	5th:
1 ×	2 ×	1 ×	1 ×	1 = 2

$P(\text{U2 first, Rolling Stones fourth, and the Beatles last}) = \dfrac{2}{120} = \dfrac{1}{60}$

2. Number of LOTTO selections: $_{49}C_6 = \dfrac{49!}{(49-6)!6!} = \dfrac{49!}{43!6!} = \dfrac{49 \cdot 48 \cdot 47 \cdot 46 \cdot 45 \cdot 44 \cdot 43!}{43!6 \cdot 5 \cdot 4 \cdot 3 \cdot 2 \cdot 1} = 13{,}983{,}816$

$P(\text{winning}) = \dfrac{\text{one LOTTO ticket}}{\text{total number of LOTTO combinations}} = \dfrac{1}{13{,}983{,}816} \approx 0.0000000715$

3. total number of combinations: $_{10}C_3 = \dfrac{10!}{(10-3)!3!} = \dfrac{10!}{7!3!} = \dfrac{10 \cdot 9 \cdot 8 \cdot 7!}{7!3 \cdot 2 \cdot 1} = 120$

a. total number of combinations of 3 men: $_6C_3 = \dfrac{6!}{(6-3)!3!} = \dfrac{6!}{3!3!} = \dfrac{6 \cdot 5 \cdot 4 \cdot 3!}{3 \cdot 2 \cdot 1 \cdot 3!} = 20$

$P(3 \text{ men}) = \dfrac{\text{number of combinations with 3 men}}{\text{total number of combinations}} = \dfrac{20}{120} = \dfrac{1}{6}$

b. Select 2 out of 6 men: $_6C_2 = \dfrac{6!}{(6-2)!2!} = \dfrac{6!}{4!2!} = \dfrac{6 \cdot 5 \cdot 4!}{4!2 \cdot 1} = 15$

Select 1 out of 4 women: $_4C_1 = \dfrac{4!}{(4-1)!1!} = \dfrac{4!}{3!1!} = \dfrac{4 \cdot 3!}{3!} = \dfrac{4 \cdot 3!}{3!} = 4$

total number of combinations of 2 men and 1 woman: $15 \times 4 = 60$

$P(2 \text{ men, 1 woman}) = \dfrac{\text{number of combinations with 2 men, 1 woman}}{\text{total number of combinations}} = \dfrac{60}{120} = \dfrac{1}{2}$

Exercise Set 11.5

1. **a.** $5! = 5 \cdot 4 \cdot 3 \cdot 2 \cdot 1 = 120$

   b.
   $$\underbrace{\text{Martha}}_{} \quad \overbrace{\text{Lee, Nancy, Paul}}^{} \quad \underbrace{\text{Armando}}_{}$$
   $$\underline{\text{1st:}} \quad \underline{\text{2nd:}} \quad \underline{\text{3rd:}} \quad \underline{\text{4th:}} \quad \underline{\text{5th:}}$$
   $$1 \times 3 \times 2 \times 1 \times 1 = 6$$

   c. $P(\text{Martha first and Armando last}) = \dfrac{6}{120} = \dfrac{1}{20}$

3. **a.** total number of permutations $= 6! = 6 \cdot 5 \cdot 4 \cdot 3 \cdot 2 \cdot 1 = 720$
   number of permutations with E first $= 1 \cdot 5 \cdot 4 \cdot 3 \cdot 2 \cdot 1 = 120$
   $$P(\text{E first}) = \frac{\text{number of permutations with E first}}{\text{total number of permutations}} = \frac{120}{720} = \frac{1}{6}$$

   b. number of permutations with C fifth and B last $= 4 \cdot 3 \cdot 2 \cdot 1 \cdot 1 \cdot 1 = 24$
   $$P(\text{C fifth and B last}) = \frac{\text{number of permutations with C fifth and B last}}{\text{total number of permutations}} = \frac{24}{720} = \frac{1}{30}$$

   c. $P(\text{D, E, C, A, B, F}) = \dfrac{\text{number of permutations with order D, E, C, A, B, F}}{\text{total number of permutations}} = \dfrac{1}{720}$

   d. number of permutations with A or B first $= 2 \cdot 5 \cdot 4 \cdot 3 \cdot 2 \cdot 1 = 240$
   $$P(\text{A or B first}) = \frac{\text{number of permutations with A or B first}}{\text{total number of permutations}} = \frac{240}{720} = \frac{1}{3}$$

5. **a.** $_9C_3 = \dfrac{9!}{(9-3)!3!} = \dfrac{9!}{6!3!} = \dfrac{9 \cdot 8 \cdot 7 \cdot 6!}{6!3 \cdot 2 \cdot 1} = 84$

   b. $_5C_3 = \dfrac{5!}{(5-3)!3!} = \dfrac{5!}{2!3!} = \dfrac{5 \cdot 4 \cdot 3!}{2 \cdot 1 \cdot 3!} = 10$

   c. $P(\text{all women}) = \dfrac{\text{number of ways to select 3 women}}{\text{total number of possible combinations}} = \dfrac{10}{84} = \dfrac{5}{42}$

7. $_{51}C_6 = \dfrac{51!}{(51-6)!6!} = \dfrac{51!}{45!6!} = \dfrac{51 \cdot 50 \cdot 49 \cdot 48 \cdot 47 \cdot 46 \cdot 45!}{45!6 \cdot 5 \cdot 4 \cdot 3 \cdot 2 \cdot 1} = 18,009,460$

   $$P(\text{winning}) = \frac{\text{number of ways of winning}}{\text{total number of possible combinations}} = \frac{1}{18,009,460} \approx 0.0000000555$$

   If 100 different tickets are purchased, $P(\text{winning}) = \dfrac{100}{18,009,460} \approx 0.00000555$

9. **a.** $_{25}C_6 = \dfrac{25!}{(25-6)!6!} = \dfrac{25!}{19!6!} = \dfrac{25 \cdot 24 \cdot 23 \cdot 22 \cdot 21 \cdot 20 \cdot 19!}{19!6 \cdot 5 \cdot 4 \cdot 3 \cdot 2 \cdot 1} = 177,100$

   $$P(\text{all are defective}) = \frac{\text{number of ways to choose 6 defective transistors}}{\text{total number of possible combinations}} = \frac{1}{177,100} \approx 0.00000565$$

b. $_{19}C_6 = \dfrac{19!}{(19-6)!6!} = \dfrac{19!}{13!6!} = \dfrac{19 \cdot 18 \cdot 17 \cdot 16 \cdot 15 \cdot 14 \cdot 13!}{13! 6 \cdot 5 \cdot 4 \cdot 3 \cdot 2 \cdot 1} = 27,132$

$P(\text{none are defective}) = \dfrac{\text{number of ways to choose 6 good transistors}}{\text{total number of possible permutations}} = \dfrac{27,132}{177,100} = \dfrac{969}{6325} \approx 0.153$

11. total number of possible combinations: $_{10}C_3 = \dfrac{10!}{(10-3)!3!} = \dfrac{10!}{7!3!} = \dfrac{10 \cdot 9 \cdot 8 \cdot 7!}{7! 3 \cdot 2 \cdot 1} = 120$

number of ways to select one Democrat: $_6C_1 = \dfrac{6!}{(6-1)!1!} = \dfrac{6!}{5!1!} = \dfrac{6 \cdot 5!}{5!1} = 6$

number of ways to select two Republicans: $_4C_2 = \dfrac{4!}{(4-2)!2!} = \dfrac{4!}{2!2!} = \dfrac{4 \cdot 3 \cdot 2!}{2!2 \cdot 1} = 6$

number of ways to select one Democrat and two Republicans: $_6C_1 \cdot {}_4C_2 = 6 \cdot 6 = 36$

$P(\text{one Democrat and two Republicans}) = \dfrac{36}{120} = \dfrac{3}{10} = 0.3$

13. a. $_{52}C_5 = \dfrac{52!}{(52-5)!5!} = \dfrac{52!}{47!5!} = \dfrac{52 \cdot 51 \cdot 50 \cdot 49 \cdot 48 \cdot 47!}{47! 5 \cdot 4 \cdot 3 \cdot 2 \cdot 1} = 2,598,960$

b. $_{13}C_5 = \dfrac{13!}{(13-5)!5!} = \dfrac{13!}{8!5!} = \dfrac{13 \cdot 12 \cdot 11 \cdot 10 \cdot 9 \cdot 8!}{8! 5 \cdot 4 \cdot 3 \cdot 2 \cdot 1} = 1287$

c. $P(\text{diamond flush}) = \dfrac{\text{number of possible 5-card diamond flushes}}{\text{total number of possible combinations}} = \dfrac{1287}{2,598,960} \approx 0.000495$

15. total number of possible combinations: $_{52}C_3 = \dfrac{52!}{(52-3)!3!} = \dfrac{52!}{49!3!} = \dfrac{52 \cdot 51 \cdot 50 \cdot 49!}{49! 3 \cdot 2 \cdot 1} = 22,100$

number of ways to select 3 picture cards: $_{12}C_3 = \dfrac{12!}{(12-3)!3!} = \dfrac{12!}{9!3!} = \dfrac{12 \cdot 11 \cdot 10 \cdot 9!}{9! 3 \cdot 2 \cdot 1} = 220$

$P(\text{3 picture cards}) = \dfrac{220}{22,100} = \dfrac{11}{1105} \approx 0.00995$

17. total number of possible combinations: $_{52}C_4 = \dfrac{52!}{(52-4)!4!} = \dfrac{52!}{48!4!} = \dfrac{52 \cdot 51 \cdot 50 \cdot 49 \cdot 48!}{48! 4 \cdot 3 \cdot 2 \cdot 1} = 270,725$

number of ways to select 2 queens: $_4C_2 = \dfrac{4!}{(4-2)!2!} = \dfrac{4!}{2!2!} = \dfrac{4 \cdot 3 \cdot 2!}{2!2 \cdot 1} = 6$

number of ways to select 2 kings: $_4C_2 = 6$

number of ways to select 2 queens and 2 kings: $_4C_2 \cdot {}_4C_2 = 6 \cdot 6 = 36$

$P(\text{2 queens and 2 kings}) = \dfrac{36}{270,725} \approx 0.000133$

23. Refer to solution 7: $_{51}C_6 = 18,009,460$

$P(\text{winning}) = \dfrac{\text{number of ways of winning } (x)}{\text{total number of possible combinations}} = \dfrac{x}{18,009,460} = \dfrac{1}{2}$, therefore $x = 9,004,730$.

At \$1 per ticket, a person must spend \$9,004,730 to have a probability of winning of $\dfrac{1}{2}$.

25. total number of possible combinations: $\dfrac{\text{Digit 1:}}{5} \times \dfrac{\text{Digit 2:}}{4} \times \dfrac{\text{Digit 3:}}{3} = 60$

number of even numbers greater than 500: $\dfrac{\overbrace{5}^{}}{\text{Digit 1:}}\; \dfrac{\overbrace{1,\,3,\text{ and }2\text{ or }4}^{}}{\text{Digit 2:}}\; \dfrac{\overbrace{2\text{ or }4}^{}}{\text{Digit 3:}}$

$1 \times 3 \times 2 = 6$

$P(\text{even and greater than 500}) = \dfrac{\text{number of even numbers greater than 500}}{\text{total number of possible combinations}} = \dfrac{6}{60} = \dfrac{1}{10}$

Check Points 11.6

1. $P(\text{not a diamond}) = 1 - P(\text{diamond}) = 1 - \dfrac{13}{52} = \dfrac{39}{52} = \dfrac{3}{4}$

2. a. $P(\text{not } 50-59) = 1 - P(50-59) = 1 - \dfrac{31}{191} = \dfrac{160}{191}$

b. $P(\text{at least 20 years old}) = 1 - P(\text{less than 20 years}) = 1 - \dfrac{9}{191} = \dfrac{182}{191}$

3. $P(4 \text{ or } 5) = P(4) + P(5) = \dfrac{1}{6} + \dfrac{1}{6} = \dfrac{2}{6} = \dfrac{1}{3}$

4. $P(\text{math or psychology}) = P(\text{math}) + P(\text{psychology}) - P(\text{math and psychology}) = \dfrac{23}{50} + \dfrac{11}{50} - \dfrac{7}{50} = \dfrac{27}{50}$

5. $P(\text{odd or less than 5}) = P(\text{odd}) + P(\text{less than 5}) - P(\text{odd and less than 5}) = \dfrac{4}{8} + \dfrac{4}{8} - \dfrac{2}{8} = \dfrac{6}{8} = \dfrac{3}{4}$

6. a. These events are not mutually exclusive.
$P(\text{at least \$100,000 or was not audited})$
$= P(\text{at least \$100,000}) + P(\text{was not audited}) - P(\text{at least \$100,000 and was not audited})$
$= \dfrac{10,927,511}{120,851,273} + \dfrac{120,035,962}{120,851,273} - \dfrac{10,775,542}{120,851,273} = \dfrac{120,187,931}{120,851,273} \approx 0.99$

b. These events are mutually exclusive.
$P(\text{less than \$25,000 or between \$50,000 and \$99,999, inclusive})$
$= P(\text{less than \$25,000}) + P(\text{between \$50,000 and \$99,999, inclusive})$
$= \dfrac{53,207,268}{120,851,273} + \dfrac{25,616,486}{120,851,273} = \dfrac{78,823,754}{120,851,273} \approx 0.65$

7. There are 2 red queens. Number of favorable outcomes = 2, Number of unfavorable outcomes = 50

a. Odds in favor of getting a red queen are 2 to 50 or 2:50 which reduces to 1:25.

b. Odds against getting a red queen are 50 to 2 or 50:2 which reduces to 25:1.

8. number of unfavorable outcomes = 995, number of favorable outcomes = 5
Odds against winning the scholarship are 995 to 5 or 995:5 which reduces to 199:1.

9. number of unfavorable outcomes = 15, number of favorable outcomes = 1
Odds in favor of the horse winning the race are 1 to 15

$$P(\text{the horse wins race}) = \frac{1}{1+15} = \frac{1}{16} = 0.0625 \text{ or } 6.3\%.$$

Exercise Set 11.6

1. $P(\text{not an ace}) = 1 - P(\text{ace}) = 1 - \frac{4}{52} = \frac{48}{52} = \frac{12}{13}$

3. $P(\text{not a heart}) = 1 - P(\text{heart}) = 1 - \frac{13}{52} = \frac{39}{52} = \frac{3}{4}$

5. $P(\text{not a picture card}) = 1 - P(\text{picture card}) = 1 - \frac{12}{52} = \frac{40}{52} = \frac{10}{13}$

7. $P(\text{not a straight flush}) = 1 - P(\text{straight flush}) = 1 - \frac{36}{2,598,960} = \frac{2,598,924}{2,598,960} \approx 0.999986$

9. $P(\text{not a full house}) = 1 - P(\text{full house}) = 1 - \frac{3744}{2,598,960} = \frac{2,595,216}{2,598,960} \approx 0.998559$

11. **a.** 0.10 (read from graph)

    b. $1.00 - 0.10 = 0.90$

13. $P(\text{not } \$50,000 - \$74,999) = 1 - P(\$50,000 - \$74,999) = 1 - \frac{20}{112} = \frac{92}{112} = \frac{23}{28}$

15. $P(\text{less than } \$100,000) = 1 - P(\$100,000 \text{ or more}) = 1 - \frac{17}{112} = \frac{95}{112}$

17. $P(2 \text{ or } 3) = P(2) + P(3) = \frac{4}{52} + \frac{4}{52} = \frac{8}{52} = \frac{2}{13}$

19. $P(\text{red 2 or black 3}) = P(\text{red 2}) + P(\text{black 3}) = \frac{2}{52} + \frac{2}{52} = \frac{4}{52} = \frac{1}{13}$

21. $P(2 \text{ of hearts or 3 of spades}) = P(2 \text{ of hearts}) + P(3 \text{ of spades}) = \frac{1}{52} + \frac{1}{52} = \frac{2}{52} = \frac{1}{26}$

23. $P(\text{professor or instructor}) = P(\text{professor}) + P(\text{instructor}) = \frac{8}{44} + \frac{10}{44} = \frac{18}{44} = \frac{9}{22}$

25. $P(\text{even or less than 5}) = P(\text{even}) + P(\text{less than 5}) - P(\text{even and less than 5}) = \frac{3}{6} + \frac{4}{6} - \frac{2}{6} = \frac{5}{6}$

27. $P(7 \text{ or red}) = P(7) + P(\text{red}) - P(\text{red 7}) = \frac{4}{52} + \frac{26}{52} - \frac{2}{52} = \frac{28}{52} = \frac{7}{13}$

29. $P(\text{heart or picture card}) = P(\text{heart}) + P(\text{picture card}) - P(\text{heart and picture card}) = \dfrac{13}{52} + \dfrac{12}{52} - \dfrac{3}{52} = \dfrac{22}{52} = \dfrac{11}{26}$

31. $P(\text{odd or less than 6}) = P(\text{odd}) + P(\text{less than 6}) - P(\text{odd and less than 6}) = \dfrac{4}{8} + \dfrac{5}{8} - \dfrac{3}{8} = \dfrac{6}{8} = \dfrac{3}{4}$

33. $P(\text{even or greater than 5}) = P(\text{even}) + P(\text{greater than 5}) - P(\text{even and greater than 5}) = \dfrac{4}{8} + \dfrac{3}{8} - \dfrac{2}{8} = \dfrac{5}{8}$

35. $P(\text{professor or male}) = P(\text{professor}) + P(\text{male}) - P(\text{male professor}) = \dfrac{19}{40} + \dfrac{22}{40} - \dfrac{8}{40} = \dfrac{33}{40}$

37. $P(\text{teach. assist. or female}) = P(\text{teach. assist.}) + P(\text{female}) - P(\text{female teach. assist.}) = \dfrac{21}{40} + \dfrac{18}{40} - \dfrac{7}{40} = \dfrac{32}{40} = \dfrac{4}{5}$

39. $P(\text{Democrat or business major}) = P(\text{Democrat}) + P(\text{business major}) - P(\text{Democrat and business major})$

$= \dfrac{29}{50} + \dfrac{11}{50} - \dfrac{5}{50} = \dfrac{35}{50} = \dfrac{7}{10}$

41. $P(\text{not completed 4 years or more}) = 1 - P(\text{completed 4 years or more}) = 1 - \dfrac{51}{187} = \dfrac{136}{187} = \dfrac{8}{11}$

43. $P(\text{completed 4 years of high school only or less than four years of college})$

$= P(\text{completed 4 years of high school only}) + P(\text{less than four years of college})$

$= \dfrac{60}{187} + \dfrac{48}{187} = \dfrac{108}{187}$

45. $P(\text{completed 4 years of high school only or is a man})$

$= P(\text{completed 4 years of high school only}) + P(\text{male}) - P(\text{completed 4 years of high school only and is a man})$

$= \dfrac{60}{187} + \dfrac{90}{187} - \dfrac{28}{187} = \dfrac{122}{187}$

47. The number that meets the characteristic is 51. The number that does not meet the characteristic is $187 - 51 = 136$.
    Odds in favor: 51 to 136 which reduces to 3 to 8
    Odds against: 136 to 51 which reduces to 8 to 3

49. $P(\text{not in the Army}) = 1 - P(\text{in the Army}) = 1 - \dfrac{420 + 80}{1430} = \dfrac{930}{1430} = \dfrac{93}{143}$

51. $P(\text{in the Navy or a man}) = P(\text{in the Navy}) + P(\text{a man}) - P(\text{in the Navy and a man})$

$= \dfrac{320 + 60}{1430} + \dfrac{300 + 420 + 170 + 320}{1430} - \dfrac{320}{1430} = \dfrac{1270}{1430} = \dfrac{127}{143}$

53. $P(\text{in the Air Force or the Marines}) = P(\text{in the Air Force}) + P(\text{in the Marines}) = \dfrac{300 + 70}{1430} + \dfrac{170 + 10}{1430} = \dfrac{550}{1430} = \dfrac{5}{13}$

55. The number that meets the characteristic is $320 + 60 = 380$.
    The number that does not meet the characteristic is $1430 - 380 = 1050$.
    Odds in favor: 380 to 1050 which reduce 38 to 105
    Odds against: 1050 to 380 which reduce 105 to 38

57. The number that meets the characteristic is 10. The number that does not meet the characteristic is $1430 - 10 = 1420$.
Odds in favor: 10 to 1420 which reduce 1 to 142
Odds against: 1420 to 10 which reduce 142 to 1

59. The number that meets the characteristic is $300 + 420 + 170 + 320 = 1210$.
The number that does not meet the characteristic is $1430 - 1210 = 220$.
Odds in favor: 1210 to 220 which reduce 11 to 2
Odds against: 220 to 1210 which reduce 2 to 11

61. number of favorable outcomes = 4, number of unfavorable outcomes = 2
Odds in favor of getting a number greater than 2 are 4:2, or 2:1.

63. number of unfavorable outcomes = 2, number of favorable outcomes = 4
Odds against getting a number greater than 2 or 2:4, or 1:2.

65. number of favorable outcomes = 9, number of unfavorable outcomes $= 100 - 9 = 91$

    a. Odds in favor of a child in a one-parent household having a parent who is a college graduate are 9:91.

    b. Odds against a child in a one-parent household having a parent who is a college graduate are 91:9.

67. number of favorable outcomes = 13, number of unfavorable outcomes = 39
Odds in favor of a heart are 13:39, or 1:3.

69. number of favorable outcomes = 26, number of unfavorable outcomes = 26
Odds in favor of a red card are 26:26, or 1:1.

71. number of unfavorable outcomes = 48, number of favorable outcomes = 4
Odds against a 9 are 48:4, or 12:1.

73. number of unfavorable outcomes = 50, number of favorable outcomes = 2
Odds against a black king are 50:2, or 25:1.

75. number of unfavorable outcomes = 47, number of favorable outcomes = 5
Odds against a spade greater than 3 and less than 9 are 47:5.

77. number of unfavorable outcomes = 980, number of favorable outcomes = 20
Odds against winning are 980:20, or 49:1.

79. The number that meets the characteristic is 18. The number that does not meet the characteristic is $38 - 18 = 20$.
Odds in favor: 18 to 20 which reduce 9 to 10

81. The number that meets the characteristic is 10. The number that does not meet the characteristic is $38 - 10 = 28$.
Odds against: 28 to 10 which reduce 14 to 5

83. The number that meets the characteristic is $18 + 10 = 28$.
The number that does not meet the characteristic is $38 - 28 = 10$.
Odds in favor: 28 to 10 which reduce 14 to 5

85. The number that meets the characteristic is $10 + 10 = 20$.
The number that does not meet the characteristic is $38 - 20 = 18$.
Odds against: 18 to 20 which reduce 9 to 10

87. $P(\text{winning}) = \dfrac{3}{3+4} = \dfrac{3}{7}$

89. $P(\text{miss free throw}) = \dfrac{4}{21+4} = \dfrac{4}{25} = 0.16 = 16\%$

   In 100 free throws, on average he missed 16, so he made $100 - 16 = 84$.

91. $P(\text{contracting an airborn illness}) = \dfrac{1}{1+999} = \dfrac{1}{1000}$

101. $P(\text{driving intoxicated or driving accident})$

   $=P(\text{driving intoxicated}) + P(\text{driving accident}) - P(\text{driving accident while intoxicated})$

   Substitute the three given probabilities and solve for the unknown probability:
   $$0.35 = 0.32 + 0.09 - P(\text{driving accident while intoxicated})$$
   $P(\text{driving accident while intoxicated}) = 0.32 + 0.09 - 0.35$
   $P(\text{driving accident while intoxicated}) = 0.06$

Check Points 11.7

1. $P(\text{green and green}) = P(\text{green}) \cdot P(\text{green}) = \dfrac{2}{38} \cdot \dfrac{2}{38} = \dfrac{1}{19} \cdot \dfrac{1}{19} = \dfrac{1}{361} \approx 0.00277$

2. $P(\text{4 boys in a row}) = P(\text{boy and boy and boy and boy}) = P(\text{boy}) \cdot P(\text{boy}) \cdot P(\text{boy}) \cdot P(\text{boy}) = \dfrac{1}{2} \cdot \dfrac{1}{2} \cdot \dfrac{1}{2} \cdot \dfrac{1}{2} = \dfrac{1}{16}$

3. a. $P(\text{hit four years in a row}) = P(\text{hit}) \cdot P(\text{hit}) \cdot P(\text{hit}) \cdot P(\text{hit}) = \dfrac{5}{19} \cdot \dfrac{5}{19} \cdot \dfrac{5}{19} \cdot \dfrac{5}{19} = \dfrac{625}{130,321} \approx 0.005$

b. Note: $P(\text{not hit in any single year}) = 1 - P(\text{hit in any single year}) = 1 - \dfrac{5}{19} = \dfrac{14}{19}$,. Therefore,

   $P(\text{not hit in next four years})$
   $= P(\text{not hit}) \cdot P(\text{not hit}) \cdot P(\text{not hit}) \cdot P(\text{not hit}) = \dfrac{14}{19} \cdot \dfrac{14}{19} \cdot \dfrac{14}{19} \cdot \dfrac{14}{19} = \dfrac{38,416}{130,321} \approx 0.295$

c. $P(\text{hit at least once in next four years}) = 1 - P(\text{not hit in next four years}) = 1 - \frac{38,416}{130,321} = \frac{91,905}{130,321} \approx 0.705$

4. $P(\text{2 kings}) = P(\text{king}) \cdot P(\text{king given the first card was a king}) = \dfrac{4}{52} \cdot \dfrac{3}{51} = \dfrac{1}{13} \cdot \dfrac{1}{17} = \dfrac{1}{221} \approx 0.00452$

5. $P(\text{3 hearts}) = P(\text{heart}) \cdot P(\text{heart given the first card was a heart}) \cdot P(\text{heart given the first two cards were hearts})$

   $= \dfrac{13}{52} \cdot \dfrac{12}{51} \cdot \dfrac{11}{50} = \dfrac{1}{4} \cdot \dfrac{4}{17} \cdot \dfrac{11}{50} = \dfrac{1}{1} \cdot \dfrac{1}{17} \cdot \dfrac{11}{50} = \dfrac{11}{850} \approx 0.0129$

6. $P(\text{heart}|\text{red}) = \dfrac{13}{26} = \dfrac{1}{2}$

7. a. $P(\text{positive mammogram}|\text{breast cancer}) = \dfrac{720}{720+80} = \dfrac{720}{800} = \dfrac{9}{10} \approx 0.9$

b. $P(\text{breast cancer}|\text{positive mammogram}) = \dfrac{720}{720+6944} = \dfrac{720}{7664} = \dfrac{45}{479} \approx 0.094$

Exercise Set 11.7

1. $P(\text{green and then red}) = P(\text{green}) \cdot P(\text{red}) = \dfrac{2}{6} \cdot \dfrac{3}{6} = \dfrac{1}{3} \cdot \dfrac{1}{2} = \dfrac{1}{6}$

3. $P(\text{yellow and then yellow}) = P(\text{yellow}) \cdot P(\text{yellow}) = \dfrac{1}{6} \cdot \dfrac{1}{6} = \dfrac{1}{36}$

5. $P(\text{color other than red each time}) = P(\text{not red}) \cdot P(\text{not red}) = \dfrac{3}{6} \cdot \dfrac{3}{6} = \dfrac{1}{2} \cdot \dfrac{1}{2} = \dfrac{1}{4}$

7. $P(\text{green and then red and then yellow}) = P(\text{green}) \cdot P(\text{red}) \cdot P(\text{yellow}) = \dfrac{2}{6} \cdot \dfrac{3}{6} \cdot \dfrac{1}{6} = \dfrac{1}{3} \cdot \dfrac{1}{2} \cdot \dfrac{1}{6} = \dfrac{1}{36}$

9. $P(\text{red every time}) = P(\text{red}) \cdot P(\text{red}) \cdot P(\text{red}) = \dfrac{3}{6} \cdot \dfrac{3}{6} \cdot \dfrac{3}{6} = \dfrac{1}{2} \cdot \dfrac{1}{2} \cdot \dfrac{1}{2} = \dfrac{1}{8}$

11. $P(2 \text{ and then } 3) = P(2) \cdot P(3) = \dfrac{1}{6} \cdot \dfrac{1}{6} = \dfrac{1}{36}$

13. $P(\text{even and then greater than 2}) = P(\text{even}) \cdot P(\text{greater than 2}) = \dfrac{3}{6} \cdot \dfrac{4}{6} = \dfrac{1}{2} \cdot \dfrac{2}{3} = \dfrac{1}{3}$

15. $P(\text{picture card and then heart}) = P(\text{picture card}) \cdot P(\text{heart}) = \dfrac{12}{52} \cdot \dfrac{13}{52} = \dfrac{3}{13} \cdot \dfrac{1}{4} = \dfrac{3}{52}$

17. $P(2 \text{ kings}) = P(\text{king}) \cdot P(\text{king}) = \dfrac{4}{52} \cdot \dfrac{4}{52} = \dfrac{1}{13} \cdot \dfrac{1}{13} = \dfrac{1}{169}$

19. $P(\text{red each time}) = P(\text{red}) \cdot P(\text{red}) = \dfrac{26}{52} \cdot \dfrac{26}{52} = \dfrac{1}{2} \cdot \dfrac{1}{2} = \dfrac{1}{4}$

21. $P(\text{all heads}) = P(\text{heads}) \cdot P(\text{heads}) \cdot P(\text{heads}) \cdot P(\text{heads}) \cdot P(\text{heads}) \cdot P(\text{heads}) = \dfrac{1}{2} \cdot \dfrac{1}{2} \cdot \dfrac{1}{2} \cdot \dfrac{1}{2} \cdot \dfrac{1}{2} \cdot \dfrac{1}{2} = \dfrac{1}{64}$

23. $P(\text{head and number greater than 4}) = P(\text{head}) \cdot P(\text{number greater than 4}) = \dfrac{1}{2} \cdot \dfrac{2}{6} = \dfrac{1}{6}$

25. **a.** $P(\text{hit two years in a row}) = P(\text{hit}) \cdot P(\text{hit}) = \dfrac{1}{16} \cdot \dfrac{1}{16} = \dfrac{1}{256} \approx 0.00391$

 b. $P(\text{Hit three consecutive years}) = P(\text{hit}) \cdot P(\text{hit}) \cdot P(\text{hit}) = \dfrac{1}{16} \cdot \dfrac{1}{16} \cdot \dfrac{1}{16} = \dfrac{1}{4096} \approx 0.000244$

 c. $P(\text{not hit in next ten years}) = [P(\text{not hit})]^{10} = \left(1 - \dfrac{1}{16}\right)^{10} = \left(\dfrac{15}{16}\right)^{10} \approx 0.524$

 d. $P(\text{hit at least once in next ten years}) = 1 - P(\text{not hit in next ten years}) \approx 1 - 0.524 \approx 0.476$

27. $P(\text{both suffer from depression - from general population}) = P(\text{depression}) \cdot P(\text{depression}) = 0.12 \cdot 0.12 = 0.0144$

29. P(all three suffer from frequent hangovers - from population of smokers)

$= P$(frequent hangovers) $\cdot P$(frequent hangovers) $\cdot P$(frequent hangovers) $= 0.20 \cdot 0.20 \cdot 0.20 = 0.008$

31. P(at least one of three suffers from anxiety/panic disorder - from population of smokers)

$= 1 - \left[1 - P(\text{anxiety/panic disorder})\right] \cdot \left[1 - P(\text{anxiety/panic disorder})\right] \cdot \left[1 - P(\text{anxiety/panic disorder})\right]$

$= 1 - [1 - 0.19] \cdot [1 - 0.19] \cdot [1 - 0.19]$

$= 1 - [0.81] \cdot [0.81] \cdot [0.81]$

$\approx 1 - 0.5314$

$= 0.4686$

33. P(solid and solid) $= P$(solid) $\cdot P$(solid given first was solid) $= \dfrac{15}{30} \cdot \dfrac{14}{29} = \dfrac{1}{2} \cdot \dfrac{14}{29} = \dfrac{7}{29}$

35. P(coconut then caramel) $= P$(coconut) $\cdot P$(caramel given first was coconut) $= \dfrac{5}{30} \cdot \dfrac{10}{29} = \dfrac{1}{6} \cdot \dfrac{10}{29} = \dfrac{5}{87}$

37. P(two Democrats) $= P$(Democrat) $\cdot P$(Democrat given first was Democrat) $= \dfrac{5}{15} \cdot \dfrac{4}{14} = \dfrac{1}{3} \cdot \dfrac{2}{7} = \dfrac{2}{21}$

39. P(Independent then Republican) $= P$(Independent) $\cdot P$(Republican given first was Independent)

$= \dfrac{4}{15} \cdot \dfrac{6}{14} = \dfrac{4}{15} \cdot \dfrac{3}{7} = \dfrac{4}{35}$

41. P(no Independents) $= P$(not Independent) $\cdot P$(not Independent given first was not Independent)

$= \dfrac{11}{15} \cdot \dfrac{10}{14} = \dfrac{11}{15} \cdot \dfrac{5}{7} = \dfrac{11}{21}$

43. P(three cans of apple juice)

$= P(\text{apple juice}) \cdot P\left(\begin{array}{c}\text{apple juice given} \\ \text{first was apple juice}\end{array}\right) \cdot P\left(\begin{array}{c}\text{apple juice given first} \\ \text{two were apple juice}\end{array}\right) = \dfrac{6}{20} \cdot \dfrac{5}{19} \cdot \dfrac{4}{18} = \dfrac{1}{57}$

45. P(grape juice then orange juice then mango juice)

$= P(\text{grape juice}) \cdot P\left(\begin{array}{c}\text{orange juice given} \\ \text{first was grape juice}\end{array}\right) \cdot P\left(\begin{array}{c}\text{mango juice given first was grape juice} \\ \text{and second was orange juice}\end{array}\right) = \dfrac{8}{20} \cdot \dfrac{4}{19} \cdot \dfrac{2}{18} = \dfrac{8}{855}$

47. P(no grape juice)

$= P(\text{not grape juice}) \cdot P\left(\begin{array}{c}\text{not grape juice given} \\ \text{first was not grape juice}\end{array}\right) \cdot P\left(\begin{array}{c}\text{not grape juice given first} \\ \text{two were not grape juice}\end{array}\right) = \dfrac{12}{20} \cdot \dfrac{11}{19} \cdot \dfrac{10}{18} = \dfrac{11}{57}$

49. $P\left(3 \mid \text{red}\right) = \dfrac{1}{5}$

51. $P\left(\text{even} \mid \text{yellow}\right) = \dfrac{2}{3}$

53. $P\left(\text{red} \mid \text{odd}\right) = \dfrac{3}{4}$

55. $P\left(\text{red}|\text{at least 5}\right) = \dfrac{3}{4}$

57. $P\left(\text{surviving}|\text{wore seat belt}\right) = \dfrac{412,368}{412,878} = \dfrac{68,728}{68,813} \approx 0.999$

59. $P\left(\text{wore seat belt}|\text{driver survived}\right) = \dfrac{412,368}{574,895} \approx 0.717$

61. $P(\text{not divorced}) = 1 - P(\text{divorced}) = 1 - \dfrac{21.7}{212.5} \approx 0.898$

63. $P(\text{widowed or divorced}) = P(\text{widowed}) + P(\text{divorced}) = \dfrac{14.0}{212.5} + \dfrac{21.7}{212.5} = \dfrac{35.7}{212.5} \approx 0.168$

65. $P(\text{male or is divorced}) = P(\text{male}) + P(\text{divorced}) - P(\text{male and is divorced}) = \dfrac{102.4}{212.5} + \dfrac{21.7}{212.5} - \dfrac{9.0}{212.5} = \dfrac{115.1}{212.5} \approx 0.542$

67. $P\left(\text{male}|\text{divorced}\right) = \dfrac{9.0}{21.7} \approx 0.415$

69. $P\left(\text{widowed}|\text{woman}\right) = \dfrac{11.3}{110.1} \approx 0.103$

71. $P\left(\text{never married or married}|\text{man}\right) = \dfrac{28.6}{102.4} + \dfrac{62.1}{102.4} = \dfrac{90.7}{102.4} \approx 0.886$

81. $P(\text{no one hospitalized}) = [P(\text{not hospitalized})]^5 = (0.9)(0.9)(0.9)(0.9)(0.9) = (0.9)^5 \approx 0.59049 \approx 59.0\%$

83. a. The first person can have any of 365 birthdays. To not match, the second person can then have any of the remaining 364 birthdays.

b. $P(\text{three different birthdays}) = \dfrac{365}{365} \cdot \dfrac{364}{365} \cdot \dfrac{363}{365} \approx 0.992$

c. $P(\text{at least two have same birthday}) = 1 - P(\text{three different birthdays}) = 1 - 0.992 = 0.008$

d. $P(\text{20 different birthdays})$
$= \dfrac{365 \cdot 364 \cdot 363 \cdot 362 \cdot 361 \cdot 360 \cdot 359 \cdot 358 \cdot 357 \cdot 356 \cdot 355 \cdot 354 \cdot 353 \cdot 352 \cdot 351 \cdot 350 \cdot 349 \cdot 348 \cdot 347 \cdot 346}{365 \cdot 365 \cdot 365 \cdot 365 \cdot 365 \cdot 365 \cdot 365 \cdot 365 \cdot 365 \cdot 365 \cdot 365 \cdot 365 \cdot 365 \cdot 365 \cdot 365 \cdot 365 \cdot 365 \cdot 365 \cdot 365 \cdot 365} \approx 0.589$
$P(\text{at least two have same birthday}) = 1 - P(\text{20 different birthdays}) = 1 - 0.589 = 0.411$

e. 23 people (determine by trial-and-error using method shown in part d)

Check Points 11.8

1. $E = 1 \cdot \dfrac{1}{4} + 2 \cdot \dfrac{1}{4} + 3 \cdot \dfrac{1}{4} + 4 \cdot \dfrac{1}{4} = \dfrac{1+2+3+4}{4} = \dfrac{10}{4} = 2.5$

2. $E = 0 \cdot \dfrac{1}{16} + 1 \cdot \dfrac{4}{16} + 2 \cdot \dfrac{6}{16} + 3 \cdot \dfrac{4}{16} + 4 \cdot \dfrac{1}{16} = \dfrac{0+4+12+12+4}{16} = \dfrac{32}{16} = 2$

3. **a.** $E = \$0(0.01) + \$2000(0.15) + \$4000(0.08) + \$6000(0.05) + \$8000(0.01) + \$10{,}000(0.70) = \$8000$
    This means that in the long run, the average cost of a claim is expected to be $8000.

    b. An average premium charge of \$8000 would cause the company to neither lose nor gain money.

4. $E = (1)\left(\dfrac{1}{5}\right) + \left(-\dfrac{1}{4}\right)\left(\dfrac{4}{5}\right) = \dfrac{1}{5} + \left(-\dfrac{1}{5}\right) = 0$
    Since the expected value is 0, there is nothing to gain or lose on average by guessing.

5. Values of gain or loss:
    Grand Prize: $\$1000 - \$2 = \$998$, Consolation Prize: $\$50 - \$2 = \$48$, Nothing: $\$0 - \$2 = -\$2$
    $E = (-\$2)\left(\dfrac{997}{1000}\right) + (\$48)\left(\dfrac{2}{1000}\right) + (\$998)\left(\dfrac{1}{1000}\right) = \dfrac{-\$1994 + \$96 + \$998}{1000} = -\dfrac{\$900}{1000} = -\$0.90$
    The expected value for one ticket is $-\$0.90$. This means that in the long run a player can expect to lose \$0.90 for each ticket bought. Buying five tickets will make your likelihood of winning five times greater, however there is no advantage to this strategy because the *cost* of five tickets is also five times greater than one ticket.

6. $E = (\$2.20)\left(\dfrac{20}{80}\right) + (-\$1.00)\left(\dfrac{60}{80}\right) = \dfrac{\$44 - \$60}{80} = \dfrac{-\$16}{80} = -\$0.20$
    This means that in the long run a player can expect to lose an average of \$0.20 for each \$1 bet.

Exercise Set 11.8

1. $E = 1 \cdot \dfrac{1}{2} + 2 \cdot \dfrac{1}{4} + 3 \cdot \dfrac{1}{4} = 1.75$

3. **a.** $E = \$0(0.65) + \$50{,}000(0.20) + \$100{,}000(0.10) + \$150{,}000(0.03) + \$200{,}000(0.01) + \$250{,}000(0.01) = \$29{,}000$
    This means that in the long run the average cost of a claim is \$29,000.

    b. \$29,000

    c. \$29,050

5. $E = -\$10{,}000(0.9) + \$90{,}000(0.1) = \$0$. This means on the average there will be no gain or loss.

7. $E = -\$99{,}999\left(\dfrac{27}{10{,}000{,}000}\right) + \$1\left(\dfrac{9{,}999{,}973}{10{,}000{,}000}\right) = \$0.73$

9. Probabilities after eliminating one possible answer: Guess Correctly: $\dfrac{1}{4}$, Guess Incorrectly: $\dfrac{3}{4}$
    $E = (1)\left(\dfrac{1}{4}\right) + \left(-\dfrac{1}{4}\right)\left(\dfrac{3}{4}\right) = \dfrac{1}{4} + \left(-\dfrac{3}{16}\right) = \dfrac{1}{16}$ expected points on a guess if one answer is eliminated.
    Yes, it is advantageous to guess after eliminating one possible answer.

11. First mall: $E = \$300,000\left(\dfrac{1}{2}\right) - \$100,000\left(\dfrac{1}{2}\right) = \$100,000$

Second mall: $E = \$200,000\left(\dfrac{3}{4}\right) - \$60,000\left(\dfrac{1}{4}\right) = \$135,000$

Choose the second mall.

13. a. $E = \$700,000(0.2) + \$0(0.8) = \$140,000$

   b. No

15. $E = \$4\left(\dfrac{1}{6}\right) - \$1\left(\dfrac{5}{6}\right) = -\$\dfrac{1}{6} \approx -\$0.17$. This means an expected loss of approximately $0.17 per game.

17. $E = \$1\left(\dfrac{18}{38}\right) - \$1\left(\dfrac{20}{38}\right) \approx -\$0.053$. This means an expected loss of approximately $0.053 per $1.00 bet.

19. $E = \$499\left(\dfrac{1}{1000}\right) - \$1\left(\dfrac{999}{1000}\right) = -\$0.50$. This means an expected loss of $0.50 per $1.00 bet.

27. Let x = the charge for the policy. Note, the expected value, $E = \$60$.
$$\$60 = (x - \$200,000)(0.0005) + (x)(0.9995)$$
$$\$60 = 0.0005x - \$100 + 0.9995x$$
$$\$160 = x$$
The insurance company should charge $160 for the policy.

Chapter 11 Review Exercises

1. Use the Fundamental Counting Principle with two groups of items. $20 \cdot 40 = 800$

2. Use the Fundamental Counting Principle with two groups of items. $4 \cdot 5 = 20$

3. Use the Fundamental Counting Principle with two groups of items. $100 \cdot 99 = 9900$

4. Use the Fundamental Counting Principle with three groups of items. $5 \cdot 5 \cdot 5 = 125$

5. Use the Fundamental Counting Principle with five groups of items. $3 \cdot 3 \cdot 3 \cdot 3 \cdot 3 = 243$

6. Use the Fundamental Counting Principle with four groups of items. $5 \cdot 2 \cdot 2 \cdot 3 = 60$

7. Use the Fundamental Counting Principle with six groups of items. $6 \cdot 5 \cdot 4 \cdot 3 \cdot 2 \cdot 1 = 720$

8. Use the Fundamental Counting Principle with five groups of items. $5 \cdot 4 \cdot 3 \cdot 2 \cdot 1 = 120$

9. Use the Fundamental Counting Principle with seven groups of items. $1 \cdot 5 \cdot 4 \cdot 3 \cdot 2 \cdot 1 \cdot 1 = 120$

10. $\dfrac{16!}{14!} = \dfrac{16 \cdot 15 \cdot 14!}{14!} = 240$

11. $\dfrac{800!}{799!} = \dfrac{800 \cdot 799!}{799!} = 800$

12. $5! - 3! = 5 \cdot 4 \cdot 3 \cdot 2 \cdot 1 - 3 \cdot 2 \cdot 1 = 120 - 6 = 114$

13. $\dfrac{11!}{(11-3)!} = \dfrac{11!}{8!} = \dfrac{11 \cdot 10 \cdot 9 \cdot 8!}{8!} = 990$

14. $_{10}P_6 = \dfrac{10!}{(10-6)!} = \dfrac{10!}{4!} = \dfrac{10\cdot9\cdot8\cdot7\cdot6\cdot5\cdot4!}{4!} = 151,200$

15. $_{100}P_2 = \dfrac{100!}{(100-2)!} = \dfrac{100!}{98!} = \dfrac{100\cdot99\cdot98!}{98!} = 9900$

16. $_{15}P_4 = \dfrac{15!}{(15-4)!} = \dfrac{15!}{11!} = \dfrac{15\cdot14\cdot13\cdot12\cdot11!}{11!} = 32,760$

17. $_{20}P_5 = \dfrac{20!}{(20-5)!} = \dfrac{20!}{15!} = \dfrac{20\cdot19\cdot18\cdot17\cdot16\cdot15!}{15!} = 1,860,480$

18. $\dfrac{n!}{p!q!} = \dfrac{7!}{3!2!} = \dfrac{7\cdot6\cdot5\cdot4\cdot3!}{3!\cdot2\cdot1} = 420$

19. $\dfrac{n!}{p!q!} = \dfrac{6!}{3!2!} = \dfrac{6\cdot5\cdot4\cdot3!}{3!\cdot2\cdot1} = 60$

20. Order does not matter. This problem involves combinations.

21. Order matters. This problem involves permutations.

22. Order does not matter. This problem involves combinations.

23. $_{11}C_7 = \dfrac{11!}{(11-7)!7!} = \dfrac{11!}{4!7!} = \dfrac{11\cdot10\cdot9\cdot8\cdot7!}{4\cdot3\cdot2\cdot1\cdot7!} = 330$

24. $_{14}C_5 = \dfrac{14!}{(14-5)!5!} = \dfrac{14!}{9!5!} = \dfrac{14\cdot13\cdot12\cdot11\cdot10\cdot9!}{9!\cdot5\cdot4\cdot3\cdot2\cdot1} = 2002$

25. $_{10}C_4 = \dfrac{10!}{(10-4)!4!} = \dfrac{10!}{6!4!} = \dfrac{10\cdot9\cdot8\cdot7\cdot6!}{6!4\cdot3\cdot2\cdot1} = 210$

26. $_{13}C_5 = \dfrac{13!}{(13-5)!5!} = \dfrac{13!}{8!5!} = \dfrac{13\cdot12\cdot11\cdot10\cdot9\cdot8!}{8!5\cdot4\cdot3\cdot2\cdot1} = 1287$

27. $_{20}C_3 = \dfrac{20!}{(20-3)!3!} = \dfrac{20!}{17!3!} = \dfrac{20\cdot19\cdot18\cdot17!}{17!3\cdot2\cdot1} = 1140$

28. Choose the Republicans: $_{12}C_5 = \dfrac{12!}{(12-5)!5!} = \dfrac{12!}{7!5!} = \dfrac{12\cdot11\cdot10\cdot9\cdot8\cdot7!}{7!5\cdot4\cdot3\cdot2\cdot1} = 792$

Choose the Democrats: $_8C_4 = \dfrac{8!}{(8-4)!4!} = \dfrac{8!}{4!4!} = \dfrac{8\cdot7\cdot6\cdot5\cdot4!}{4!4\cdot3\cdot2\cdot1} = 70$

Multiply the choices: $792 \cdot 70 = 55,440$

29. $P(6) = \dfrac{\text{number of ways a 6 can occur}}{\text{total number of possible outcomes}} = \dfrac{1}{6}$

30. $P(\text{less than } 5) = \dfrac{\text{number of ways a number less than 5 can occur}}{\text{total number of possible outcomes}} = \dfrac{4}{6} = \dfrac{2}{3}$

31. $P(\text{less than } 7) = \dfrac{\text{number of ways a number less than 7 can occur}}{\text{total number of possible outcomes}} = \dfrac{6}{6} = 1$

32. $P(\text{greater than } 6) = \dfrac{\text{number of ways a number greater than 6 can occur}}{\text{total number of possible outcomes}} = \dfrac{0}{6} = 0$

33. $P(5) = \dfrac{\text{number of ways a 5 can occur}}{\text{total number of possible outcomes}} = \dfrac{4}{52} = \dfrac{1}{13}$

34. $P(\text{picture card}) = \dfrac{\text{number of ways a picture card can occur}}{\text{total number of possible outcomes}} = \dfrac{12}{52} = \dfrac{3}{13}$

35. $P(\text{greater than 4 and less than 8}) = \dfrac{\text{number of ways a card greater than 4 and less than 8 can occur}}{\text{total number of possible outcomes}} = \dfrac{12}{52} = \dfrac{3}{13}$

36. $P(\text{4 of diamonds}) = \dfrac{\text{number of ways a 4 of diamonds can occur}}{\text{total number of possible outcomes}} = \dfrac{1}{52}$

37. $P(\text{red ace}) = \dfrac{\text{number of ways a red ace can occur}}{\text{total number of possible outcomes}} = \dfrac{2}{52} = \dfrac{1}{26}$

38. $P(\text{chocolate}) = \dfrac{\text{number of ways a chocolate can occur}}{\text{total number of possible outcomes}} = \dfrac{15}{30} = \dfrac{1}{2}$

39. $P(\text{caramel}) = \dfrac{\text{number of ways a caramel can occur}}{\text{total number of possible outcomes}} = \dfrac{10}{30} = \dfrac{1}{3}$

40. $P(\text{peppermint}) = \dfrac{\text{number of ways a peppermint can occur}}{\text{total number of possible outcomes}} = \dfrac{5}{30} = \dfrac{1}{6}$

41. a. $P(\text{carrier without the disease}) = \dfrac{\text{number of ways to be a carrier without the disease}}{\text{total number of possible outcomes}} = \dfrac{2}{4} = \dfrac{1}{2}$

     b. $P(\text{disease}) = \dfrac{\text{number of ways to have the disease}}{\text{total number of possible outcomes}} = \dfrac{0}{4} = 0$

42. $P(\text{employed}) = \dfrac{139.2}{223.4} \approx 0.623$

43. $P(\text{female}) = \dfrac{115.7}{223.4} \approx 0.518$

44. $P(\text{unemployed male}) = \dfrac{33.2}{223.4} \approx 0.149$

45. number of ways to visit in order D, B, A, C = 1
   total number of possible permutations = $4 \cdot 3 \cdot 2 \cdot 1 = 24$

   $$P(\text{D, B, A, C}) = \frac{1}{24}$$

46. number of permutations with C last = $5 \cdot 4 \cdot 3 \cdot 2 \cdot 1 \cdot 1 = 120$
   total number of possible permutations = $6 \cdot 5 \cdot 4 \cdot 3 \cdot 2 \cdot 1 = 720$

   $$P(\text{C last}) = \frac{120}{720} = \frac{1}{6}$$

47. number of permutations with B first and A last = $1 \cdot 4 \cdot 3 \cdot 2 \cdot 1 \cdot 1 = 24$
   total number of possible permutations = $6 \cdot 5 \cdot 4 \cdot 3 \cdot 2 \cdot 1 = 720$

   $$P(\text{B first and A last}) = \frac{24}{720} = \frac{1}{30}$$

48. number of permutations in order F, E, A, D, C, B = 1
   total number of possible permutations = $6 \cdot 5 \cdot 4 \cdot 3 \cdot 2 \cdot 1 = 720$

   $$P(\text{F, E, A, D, C, B}) = \frac{1}{720}$$

49. number of permutations with A or C first = $2 \cdot 5 \cdot 4 \cdot 3 \cdot 2 \cdot 1 = 240$
   total number of possible permutations = $6 \cdot 5 \cdot 4 \cdot 3 \cdot 2 \cdot 1 = 720$

   $$P(\text{A or C first}) = \frac{240}{720} = \frac{1}{3}$$

50. a. number of ways to win = 1
   total number of possible combinations:

   $$_{20}C_5 = \frac{20!}{(20-5)!5!} = \frac{20!}{15!5!} = \frac{20 \cdot 19 \cdot 18 \cdot 17 \cdot 16 \cdot 15!}{15!5 \cdot 4 \cdot 3 \cdot 2 \cdot 1} = 15,504$$

   $$P(\text{winning with one ticket}) = \frac{1}{15,504} \approx 0.0000645$$

   b. number of ways to win = 100

   $$P(\text{winning with 100 different tickets}) = \frac{100}{15,504} \approx 0.00645$$

51. a. number of ways to select 4 Democrats: $_6C_4 = \dfrac{6!}{(6-4)!4!} = \dfrac{6!}{2!4!} = \dfrac{6 \cdot 5 \cdot 4!}{2 \cdot 1 \cdot 4!} = 15$

   total number of possible combinations: $_{10}C_4 = \dfrac{10!}{(10-4)!4!} = \dfrac{10!}{6!4!} = \dfrac{10 \cdot 9 \cdot 8 \cdot 7 \cdot 6!}{6!4 \cdot 3 \cdot 2 \cdot 1} = 210$

   $$P(\text{all Democrats}) = \frac{15}{210} = \frac{1}{14}$$

   b. number of ways to select 2 Democrats: $_6C_2 = \dfrac{6!}{(6-2)!2!} = \dfrac{6!}{4!2!} = \dfrac{6 \cdot 5 \cdot 4!}{4!2 \cdot 1} = 15$

   number of ways to select 2 Republicans: $_4C_2 = \dfrac{4!}{(4-2)!2!} = \dfrac{4!}{2!2!} = \dfrac{4 \cdot 3 \cdot 2!}{2!2 \cdot 1} = 6$

   number of ways to select 2 Democrats and 2 Republicans = $15 \cdot 6 = 90$

   $$P(\text{2 Democrats and 2 Republicans}) = \frac{90}{210} = \frac{3}{7}$$

52. number of ways to get 2 picture cards: $_6C_2 = \dfrac{6!}{(6-2)!2!} = \dfrac{6!}{4!2!} = \dfrac{6 \cdot 5 \cdot 4!}{4!2 \cdot 1} = 15$

number of ways to get one non-picture card = 20
number of ways to get 2 picture cards and one non-picture card = $15 \cdot 20 = 300$

total number of possible combinations: $_{26}C_3 = \dfrac{26!}{(26-3)!3!} = \dfrac{26!}{23!3!} = \dfrac{26 \cdot 25 \cdot 24 \cdot 23!}{23!3 \cdot 2 \cdot 1} = 2600$

$P(\text{2 picture cards}) = \dfrac{300}{2600} = \dfrac{3}{26}$

53. $P(\text{not a 5}) = 1 - P(5) = 1 - \dfrac{1}{6} = \dfrac{5}{6}$

54. $P(\text{not less than 4}) = 1 - P(\text{less than 4}) = 1 - \dfrac{3}{6} = 1 - \dfrac{1}{2} = \dfrac{1}{2}$

55. $P(\text{3 or 5}) = P(3) + P(5) = \dfrac{1}{6} + \dfrac{1}{6} = \dfrac{2}{6} = \dfrac{1}{3}$

56. $P(\text{less than 3 or greater than 4}) = P(\text{less than 3}) + P(\text{greater than 4}) = \dfrac{2}{6} + \dfrac{2}{6} = \dfrac{1}{3} + \dfrac{1}{3} = \dfrac{2}{3}$

57. $P(\text{less than 5 or greater than 2}) = P(\text{less than 5}) + P(\text{greater than 2}) - P(\text{less than 5 and greater than 2})$

$$= \dfrac{4}{6} + \dfrac{4}{6} - \dfrac{2}{6} = 1$$

58. $P(\text{not a picture card}) = 1 - P(\text{picture card}) = 1 - \dfrac{12}{52} = 1 - \dfrac{3}{13} = \dfrac{10}{13}$

59. $P(\text{not a diamond}) = 1 - P(\text{diamond}) = 1 - \dfrac{13}{52} = 1 - \dfrac{1}{4} = \dfrac{3}{4}$

60. $P(\text{ace or king}) = P(\text{ace}) + P(\text{king}) = \dfrac{4}{52} + \dfrac{4}{52} = \dfrac{1}{13} + \dfrac{1}{13} = \dfrac{2}{13}$

61. $P(\text{black 6 or red 7}) = P(\text{black 6}) + P(\text{red 7}) = \dfrac{2}{52} + \dfrac{2}{52} = \dfrac{1}{26} + \dfrac{1}{26} = \dfrac{2}{26} = \dfrac{1}{13}$

62. $P(\text{queen or red card}) = P(\text{queen}) + P(\text{red card}) - P(\text{red queen}) = \dfrac{4}{52} + \dfrac{26}{52} - \dfrac{2}{52} = \dfrac{28}{52} = \dfrac{7}{13}$

63. $P(\text{club or picture card}) = P(\text{club}) + P(\text{picture card}) - P(\text{club and picture card}) = \dfrac{13}{52} + \dfrac{12}{52} - \dfrac{3}{52} = \dfrac{22}{52} = \dfrac{11}{26}$

64. $P(\text{not 4}) = 1 - P(4) = 1 - \dfrac{1}{6} = \dfrac{5}{6}$

65. $P(\text{not yellow}) = 1 - P(\text{yellow}) = 1 - \dfrac{1}{6} = \dfrac{5}{6}$

66. $P(\text{not red}) = 1 - P(\text{red}) = 1 - \dfrac{3}{6} = 1 - \dfrac{1}{2} = \dfrac{1}{2}$

67. $P(\text{red or yellow}) = P(\text{red}) + P(\text{yellow}) = \dfrac{3}{6} + \dfrac{1}{6} = \dfrac{4}{6} = \dfrac{2}{3}$

68. $P(\text{red or even}) = P(\text{red}) + P(\text{even}) - P(\text{red and even}) = \dfrac{3}{6} + \dfrac{3}{6} - \dfrac{0}{6} = 1$

69. $P(\text{red or greater than 3}) = P(\text{red}) + P(\text{greater than 3}) - P(\text{red and greater than 3}) = \dfrac{3}{6} + \dfrac{3}{6} - \dfrac{1}{6} = \dfrac{5}{6}$

70. $P(\text{African American or male}) = P(\text{African American}) + P(\text{male}) - P(\text{African American male})$
$$= \dfrac{50+20}{200} + \dfrac{50+90}{200} - \dfrac{50}{200} = \dfrac{160}{200} = \dfrac{4}{5}$$

71. $P(\text{female or white}) = P(\text{female}) + P(\text{white}) - P(\text{white female}) = \dfrac{20+40}{200} + \dfrac{90+40}{200} - \dfrac{40}{200} = \dfrac{150}{200} = \dfrac{3}{4}$

72. $P(\text{public college}) = \dfrac{252}{350} = \dfrac{18}{25}$

73. $P(\text{not from high-income family}) = 1 - P(\text{from high-income family}) = 1 - \dfrac{50}{350} = \dfrac{350}{350} - \dfrac{50}{350} = \dfrac{300}{350} = \dfrac{6}{7}$

74. $P(\text{from middle-income family or high-income family}) = \dfrac{160+50}{350} = \dfrac{210}{350} = \dfrac{3}{5}$

75. $P(\text{attended private college or is from a high income family})$
$= P(\text{private college}) + P(\text{high income family}) - P(\text{attended private college and is from a high income family})$
$$= \dfrac{98}{350} + \dfrac{50}{350} - \dfrac{28}{350} = \dfrac{120}{350} = \dfrac{12}{35}$$

76. number of favorable outcomes = 4, number of unfavorable outcomes = 48
Odds in favor of getting a queen are 4:48, or 1:12. Odds against getting a queen are 12:1.

77. number of favorable outcomes = 20, number of unfavorable outcomes = 1980
Odds against winning are 1980: 20, or 99:1.

78. $P(\text{win}) = \dfrac{3}{3+1} = \dfrac{3}{4}$

79. $P(\text{yellow then red}) = P(\text{yellow}) \cdot P(\text{red}) = \dfrac{2}{6} \cdot \dfrac{4}{6} = \dfrac{1}{3} \cdot \dfrac{2}{3} = \dfrac{2}{9}$

80. $P(1 \text{ then } 3) = P(1) \cdot P(3) = \dfrac{1}{6} \cdot \dfrac{1}{6} = \dfrac{1}{36}$

81. $P(\text{yellow both times}) = P(\text{yellow}) \cdot P(\text{yellow}) = \dfrac{2}{6} \cdot \dfrac{2}{6} = \dfrac{1}{3} \cdot \dfrac{1}{3} = \dfrac{1}{9}$

82. $P(\text{yellow then 4 then odd}) = P(\text{yellow}) \cdot P(4) \cdot P(\text{odd}) = \dfrac{2}{6} \cdot \dfrac{1}{6} \cdot \dfrac{3}{6} = \dfrac{1}{3} \cdot \dfrac{1}{6} \cdot \dfrac{1}{2} = \dfrac{1}{36}$

83. $P(\text{red every time}) = P(\text{red}) \cdot P(\text{red}) \cdot P(\text{red}) = \dfrac{4}{6} \cdot \dfrac{4}{6} \cdot \dfrac{4}{6} = \dfrac{2}{3} \cdot \dfrac{2}{3} \cdot \dfrac{2}{3} = \dfrac{8}{27}$

84. $P(\text{five boys in a row}) = P(\text{boy}) \cdot P(\text{boy}) \cdot P(\text{boy}) \cdot P(\text{boy}) \cdot P(\text{boy}) = \dfrac{1}{2} \cdot \dfrac{1}{2} \cdot \dfrac{1}{2} \cdot \dfrac{1}{2} \cdot \dfrac{1}{2} = \dfrac{1}{2^5} = \dfrac{1}{32}$

85. **a.** $P(\text{flood two years in a row}) = P(\text{flood}) \cdot P(\text{flood}) = (0.2)(0.2) = 0.04$

    b. $P(\text{flood for three consecutive years}) = P(\text{flood}) \cdot P(\text{flood}) \cdot P(\text{flood}) = (0.2)(0.2)(0.2) = 0.008$

    c. $P(\text{no flooding for four consecutive years}) = [1 - P(\text{flood})]^4 = (1 - 0.2)^4 = (0.8)^4 = 0.4096$

    d. $P(\text{flood at least once in next four years}) = 1 - P(\text{no flooding for four consecutive years})$
$$= 1 - 0.4096 = 0.5904$$

86. $P(\text{music major then psychology major}) = P(\text{music major}) \cdot P\left(\begin{array}{c}\text{psychology major given}\\\text{first was music major}\end{array}\right) = \dfrac{2}{9} \cdot \dfrac{4}{8} = \dfrac{2}{9} \cdot \dfrac{1}{2} = \dfrac{1}{9}$

87. $P(\text{two business majors}) = P(\text{bus. major}) \cdot P(\text{bus. major given first was bus. major}) = \dfrac{3}{9} \cdot \dfrac{2}{8} = \dfrac{1}{3} \cdot \dfrac{1}{4} = \dfrac{1}{12}$

88. $P(\text{solid then two cherry})$
$$= P(\text{solid}) \cdot P\left(\begin{array}{c}\text{cherry given}\\\text{first was solid}\end{array}\right) \cdot P\left(\begin{array}{c}\text{cherry given first was solid}\\\text{and second was cherry}\end{array}\right) = \dfrac{30}{50} \cdot \dfrac{5}{49} \cdot \dfrac{4}{48} = \dfrac{3}{5} \cdot \dfrac{5}{49} \cdot \dfrac{1}{12} = \dfrac{1}{196}$$

89. $P\left(5 \mid \text{odd}\right) = \dfrac{1}{3}$

90. $P\left(\text{vowel} \mid \text{precedes the letter k}\right) = \dfrac{3}{10}$

91. **a.** $P\left(\text{odd} \mid \text{red}\right) = \dfrac{2}{4} = \dfrac{1}{2}$

    b. $P\left(\text{yellow} \mid \text{at least 3}\right) = \dfrac{2}{7}$

92. $P(\text{does not have TB}) = \dfrac{11 + 124}{9 + 1 + 11 + 124} = \dfrac{135}{145} = \dfrac{27}{29}$

93. $P(\text{tests positive}) = \dfrac{9 + 11}{9 + 1 + 11 + 124} = \dfrac{20}{145} = \dfrac{4}{29}$

94. $P(\text{does not have TB or tests positive})$
$$= P(\text{does not have TB}) + P(\text{tests positive}) - P(\text{does not have TB and tests positive})$$
$$= \dfrac{11 + 124}{145} + \dfrac{9 + 11}{145} - \dfrac{11}{145} = \dfrac{144}{145}$$

95. $P(\text{does not have TB}|\text{positive test}) = \dfrac{11}{9+11} = \dfrac{11}{20}$

96. $P(\text{tests positive}|\text{does not have TB}) = \dfrac{11}{11+124} = \dfrac{11}{135}$

97. $P(\text{has TB}|\text{negative Test}) = \dfrac{1}{1+124} = \dfrac{1}{125}$

98. $P(\text{two people with TB}) = P(\text{TB}) \cdot P(\text{TB}|\text{first person selected has TB}) = \dfrac{10}{145} \cdot \dfrac{9}{144} = \dfrac{1}{232}$

99. $P(\text{two people with positive tests}) = P(\text{positive test}) \cdot P(\text{positive test}|\text{first person has positive test}) = \dfrac{20}{145} \cdot \dfrac{19}{144} = \dfrac{19}{1044}$

100. $P(\text{male}) = \dfrac{26,098}{30,242} = \dfrac{13,049}{15,121} \approx 0.863$

101. $P(\text{age } 25-44) = \dfrac{11,586}{30,242} = \dfrac{5793}{15,121} \approx 0.383$

102. $P(\text{less than } 75) = 1 - P(\text{greater than or equal to } 75) = 1 - \dfrac{2424}{30,242} = \dfrac{27,818}{30,242} = \dfrac{13909}{15,121} \approx 0.920$

103. $P(\text{age } 20-24 \text{ or } 25-44) = P(\text{age } 20-24) + P(\text{age } 25-44) = \dfrac{4306}{30,242} + \dfrac{11,586}{30,242} = \dfrac{15,892}{30,242} = \dfrac{7946}{15,121} \approx 0.525$

104. $P(\text{female or younger than } 5) = P(\text{female}) + P(\text{younger than } 5) - P(\text{female and younger than } 5)$
$$= \dfrac{4144}{30,242} + \dfrac{71}{30,242} - \dfrac{29}{30,242} = \dfrac{4186}{30,242} = \dfrac{2093}{15,121} \approx 0.138$$

105. $P(\text{age } 20-24|\text{male}) = \dfrac{3887}{26,098} \approx 0.149$

106. $P(\text{male}|\text{at least } 75) = \dfrac{2225}{2424} \approx 0.918$

107. $E = 1 \cdot \dfrac{1}{4} + 2 \cdot \dfrac{1}{8} + 3 \cdot \dfrac{1}{8} + 4 \cdot \dfrac{1}{4} + 5 \cdot \dfrac{1}{4} = 3.125$

108. a. $E = \$0(0.9999995) + (-\$1,000,000)(0.0000005) = -\$.50$
  The insurance company spends an average of \$0.50 per person insured.

  b. charge $\$9.50 - (-\$0.50) = \$10.00$

109. $E = \$27,000\left(\dfrac{1}{4}\right) + (-\$3000)\left(\dfrac{3}{4}\right) = \$4500$. The expected gain is \$4500 per bid.

110. $E = \$1\left(\dfrac{2}{4}\right) + \$1\left(\dfrac{1}{4}\right) + (-\$4)\left(\dfrac{1}{4}\right) = -\$0.25$. The expected loss is \$0.25 per game.

Chapter 11 Test

1. Use the Fundamental Counting Principle with five groups of items. $10 \cdot 2 \cdot 2 \cdot 2 \cdot 3 = 240$

2. Use the Fundamental Counting Principle with four groups of items. $4 \cdot 3 \cdot 2 \cdot 1 = 24$

3. Use the Fundamental Counting Principle with seven groups of items. $1 \cdot 6 \cdot 5 \cdot 4 \cdot 3 \cdot 2 \cdot 1 = 720$

4. $_{11}P_3 = \dfrac{11!}{(11-3)!} = \dfrac{11!}{8!} = \dfrac{11 \cdot 10 \cdot 9 \cdot 8!}{8!} = 990$

5. $_{10}C_4 = \dfrac{10!}{(10-4)!4!} = \dfrac{10!}{6!4!} = \dfrac{10 \cdot 9 \cdot 8 \cdot 7 \cdot 6!}{6!4 \cdot 3 \cdot 2 \cdot 1} = 210$

6. $\dfrac{n!}{p!q!} = \dfrac{7!}{3!2!} = \dfrac{7 \cdot 6 \cdot 5 \cdot 4 \cdot \cancel{3!}}{\cancel{3!} \cdot 2 \cdot 1} = 420$

7. $P(\text{freshman}) = \dfrac{12}{50} = \dfrac{6}{25}$

8. $P(\text{not a sophomore}) = 1 - P(\text{sophomore}) = 1 - \dfrac{16}{50} = 1 - \dfrac{8}{25} = \dfrac{17}{25}$

9. $P(\text{junior or senior}) = P(\text{junior}) + P(\text{senior}) = \dfrac{20}{50} + \dfrac{2}{50} = \dfrac{22}{50} = \dfrac{11}{25}$

10. $P(\text{greater than 4 and less than 10}) = \dfrac{20}{52} = \dfrac{5}{13}$

11. $P(C \text{ first}, A \text{ next-to-last}, E \text{ last})$

   $= P(C) \cdot P(A \text{ given } C \text{ was first}) \cdot P(E \text{ given } C \text{ was first and } A \text{ was next-to-last}) = \dfrac{1}{7} \cdot \dfrac{1}{6} \cdot \dfrac{1}{5} = \dfrac{1}{210}$

12. total number of possible combinations: $_{15}C_6 = \dfrac{15!}{(15-6)!6!} = \dfrac{15!}{9!6!} = \dfrac{15 \cdot 14 \cdot 13 \cdot 12 \cdot 11 \cdot 10 \cdot 9!}{9!6 \cdot 5 \cdot 4 \cdot 3 \cdot 2 \cdot 1} = 5005$

   $P(\text{winning with 50 tickets}) = \dfrac{50}{5005} = \dfrac{10}{1001} \approx 0.00999$

13. $P(\text{red or blue}) = P(\text{red}) + P(\text{blue}) = \dfrac{2}{8} + \dfrac{2}{8} = \dfrac{4}{8} = \dfrac{1}{2}$

14. $P(\text{red then blue}) = P(\text{red}) \cdot P(\text{blue}) = \dfrac{2}{8} \cdot \dfrac{2}{8} = \dfrac{1}{4} \cdot \dfrac{1}{4} = \dfrac{1}{16}$

15. $P(\text{flooding for three consecutive years}) = P(\text{flood}) \cdot P(\text{flood}) \cdot P(\text{flood}) = \dfrac{1}{20} \cdot \dfrac{1}{20} \cdot \dfrac{1}{20} = \dfrac{1}{8000}$

16. $P(\text{black or picture card}) = P(\text{black}) + P(\text{picture card}) - P(\text{black picture card}) = \dfrac{26}{52} + \dfrac{12}{52} - \dfrac{6}{52} = \dfrac{32}{52} = \dfrac{8}{13}$

17. $P(\text{freshman or female}) = P(\text{freshman}) + P(\text{female}) - P(\text{female freshman}) = \dfrac{10+15}{50} + \dfrac{15+5}{50} - \dfrac{15}{50} = \dfrac{30}{50} = \dfrac{3}{5}$

18. $P(\text{both red}) = P(\text{red}) \cdot P(\text{red given first ball was red}) = \dfrac{5}{20} \cdot \dfrac{4}{19} = \dfrac{1}{4} \cdot \dfrac{4}{19} = \dfrac{1}{19}$

19. $P(\text{all correct}) = P(\text{correct}) \cdot P(\text{correct}) \cdot P(\text{correct}) \cdot P(\text{correct}) = \dfrac{1}{4} \cdot \dfrac{1}{4} \cdot \dfrac{1}{4} \cdot \dfrac{1}{4} = \left(\dfrac{1}{4}\right)^4 = \dfrac{1}{256}$

20. number of favorable outcomes = 20, number of unfavorable outcomes = 15
Odds against being a man are 15:20, or 3:4.

21. a. Odds in favor are 4:1.   **b.** $P(\text{win}) = \dfrac{4}{1+4} = \dfrac{4}{5}$

22. $P(\text{not brown eyes}) = \dfrac{18+10+20+12}{22+18+10+18+20+12} = \dfrac{60}{100} = \dfrac{3}{5}$

23. $P(\text{brown eyes or blue eyes}) = \dfrac{22+18+18+20}{22+18+10+18+20+12} = \dfrac{78}{100} = \dfrac{39}{50}$

24. $P(\text{female or green eyes}) = P(\text{female}) + P(\text{green eyes}) - P(\text{female and green eyes})$

$$= \dfrac{18+20+12}{100} + \dfrac{10+12}{100} - \dfrac{12}{100}$$
$$= \dfrac{50}{100} + \dfrac{22}{100} - \dfrac{12}{100}$$
$$= \dfrac{60}{100}$$
$$= \dfrac{3}{5}$$

25. $P(\text{male}|\text{blue eyes}) = \dfrac{18}{18+20} = \dfrac{18}{38} = \dfrac{9}{19}$

26. $P(\text{two people with green eyes}) = P(\text{green eyes}) \cdot P(\text{green eyes}|\text{first person has green eyes}) = \dfrac{22}{100} \cdot \dfrac{21}{99} = \dfrac{7}{150}$

27. $E = \$65{,}000(0.2) + (-\$15{,}000)(0.8) = \$1000.$ This means the expected gain is $1000 for this bid.

28. $E = (-\$19) \cdot \dfrac{10}{20} + (-\$18) \cdot \dfrac{5}{20} + (-\$15) \cdot \dfrac{3}{20} + (-\$10) \cdot \dfrac{1}{20} + (\$80) \cdot \dfrac{1}{20}$

$$= \dfrac{-\$190 - \$90 - \$45 - \$10 + \$80}{20} = \dfrac{-\$255}{20} = -\$12.75$$

This expected value of $-\$12.75$ means that a player will lose an average of $12.75 per play in the long run.

Chapter 12
Statistics

Check Points 12.1

1. **a.** The population is the set containing all the of the city's homeless people.

   b. This is not a good idea. This sample of people currently in a shelter is more likely to hold opinions that favor required residence in city shelters than the population of all the city's homeless.

2. The sampling technique described in Check Point 1b does not produce a random sample because homeless people who do not go to shelters have no chance of being selected for the survey. In this instance, an appropriate method would be to randomly select neighborhoods of the city and then randomly survey homeless people within the selected neighborhood.

3.

Grade	Number of students
A	3
B	5
C	9
D	2
F	1
	20

4.

Exam Scores (class)	Tally	Number of students (frequency)
40 – 49		1
50 – 59	卌	5
60 – 69	|||	4
70 – 79	卌 卌 卌	15
80 – 89	卌	5
90 – 99	卌		7
		37

5.

Stems	Leaves
4	1
5	8 2 8 0 7
6	8 2 9 9
7	3 5 9 9 7 5 5 3 3 6 7 1 7 1 5
8	7 3 9 9 1
9	4 6 9 7 5 8 0

Exercise Set 12.1

1. **a.** The *population* is all American men ages 18 and older. The *sample* is the group of 1014 that were randomly selected.

    b. The variable measured was health. The data was qualitative.

3. c

5. A stress rating of 7 was reported by 31 students.

7. Totaling the frequency column shows that 151 students were involved in the study.

9.

Time Spent on Homework (in hours)	Number of students
15	4
16	5
17	6
18	5
19	4
20	2
21	2
22	0
23	0
24	2
	30

11. The lower class limits are 0, 5, 10, 15, 20, 25, 30, 35, 40, and 45.

13. The class width is 5, the difference between successive lower limits.

15. $4 + 3 + 3 + 3 = 13$. Thus, 13 students had at least 30 social interactions.

17. The 5 – 9 class.

19.

Age	Frequency
41–45	2
46–50	8
51–55	15
56–60	9
61–65	7
66–70	2
	43

21. Histogram for Stress Rating:

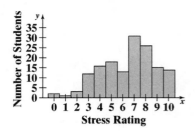

Frequency Polygon for Stress Rating:

23. Histogram for Height: Frequency Polygon for Height:

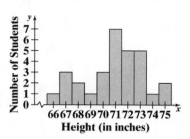   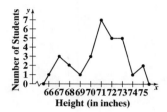

25. b

27.

Stems	Leaves
2	8 8 9 5
3	8 7 0 1 2 7 6 4 0 5
4	8 2 2 1 4 5 4 6 2 0 8 2 7 9
5	9 4 1 9 1 0
6	3 2 3 6 6 3

The greatest number of college professors are in their 40s.

29. The bars on the horizontal axis are evenly spaced, yet the time intervals that they represent vary greatly. This may give the misleading impression of linear growth.

31. The sectors representing these six countries use up 100% of the pie graph, yet the percentages for these six countries total only 57%. This may give the misleading impression that the U.S. has about 50% of the world's computer use.

33. It is not clear whether the lengths of the bars (arms) represent the percentage or the median loan amount. Regardless, the lengths of the bars are not proportional to either set of data.

45.

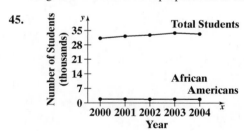

Check Points 12.2

1. a. $\dfrac{10+20+30+40+50}{5} = \dfrac{150}{5} = 30$

   b. $\dfrac{3+10+10+10+117}{5} = \dfrac{150}{5} = 30$

2.

x	f	xf
30	3	$30 \cdot 3 = 90$
33	4	$33 \cdot 4 = 132$
40	4	$40 \cdot 4 = 160$
50	1	$50 \cdot 1 = 50$
	12	$\sum xf = 432$

$$\text{Mean} = \frac{\sum xf}{n} = \frac{432}{12} = 36$$

3. a. First arrange the data items from smallest to largest: 25, 28, <u>35</u>, 40, 42
The number of data items is odd, so the median is the middle number. The median is 35.

  b. First arrange the data items from smallest to largest: 61, 72, <u>79</u>, <u>85</u>, 87, 93
The number of data items is even, so the median is the mean of the two middle data items.
The median is $\dfrac{79+85}{2} = \dfrac{164}{2} = 82$.

4. The data items are arranged from smallest to largest with $n = 19$, which gives $\dfrac{n+1}{2} = \dfrac{19+1}{2} = \dfrac{20}{2} = 10$

The median is in the 10th position, which means the median is 5.

5. The data items are arranged from smallest to largest with $n = 14$, which gives $\dfrac{n+1}{2} = \dfrac{14+1}{2} = \dfrac{15}{2} = 7.5$ position

The median is the mean of the data items in positions 7 and 8.
Thus, the median is $\dfrac{157+168}{2} = \dfrac{325}{2} = 162.5$.

6. The total frequency is $1+1+1+3+1+2+2+2+1+2+1+1 = 18$, therefore $n = 18$

The median's position is $\dfrac{n+1}{2} = \dfrac{18+1}{2} = \dfrac{19}{2} = 9.5$.

Therefore, the median is the mean of the data items in positions 9 and 10.
Counting through the frequency row identifies that the 9th data item is 54 and the 10th data item is 55.
Thus, the median is $\dfrac{54+55}{2} = \dfrac{109}{2} = 54.5$.

7. a. $\text{Mean} = \dfrac{417.4+46.9+37.1+35.0+32.8+27.2+20.8+19.2+19.1+13.9}{10} = \$66.94$ billion

  b. Position of mean: $\dfrac{n+1}{2} = \dfrac{10+1}{2} = 5.5$ position
The median is the mean of the data items in positions 5 and 6.
Thus, the median is $\dfrac{27.2+32.8}{2} = \$30$ billion.

  c. The mean is so much greater than the median because one data item ($417.4) is much greater than the others.

8. The number 8 occurs more often than any other. The mode is 8.

9. $\text{Midrange} = \dfrac{70,000+179,000}{2} = \$124,500$

10. a. Mean $= \dfrac{173+191+182+190+172+147+146+138+175+136+179+153+107+195+135+140+138}{17}$

$= \dfrac{2697}{17} = 158.6$ calories

b. Order the data items: 107, 135, 136, 138, 138, 140, 146, 147, <u>153</u>, 172, 173, 175, 179, 182, 190, 191, 195
The number of data items is odd, so the median is the middle number. The median is 153 calories.

c. The number 138 occurs more often than any other. The mode is 138 calories.

d. Midrange $= \dfrac{107+195}{2} = \dfrac{302}{2} = 151$ calories

Exercise Set 12.2

1. $\dfrac{7+4+3+2+8+5+1+3}{8} = \dfrac{33}{8} = 4.125$

3. $\dfrac{91+95+99+97+93+95}{6} = \dfrac{570}{6} = 95$

5. $\dfrac{100+40+70+40+60}{5} = \dfrac{310}{5} = 62$

7. $\dfrac{1.6+3.8+5.0+2.7+4.2+4.2+3.2+4.7+3.6+2.5+2.5}{11} = \dfrac{38}{11} \approx 3.45$

9.

x	f	xf
1	1	$1 \cdot 1 = 1$
2	3	$2 \cdot 3 = 6$
3	4	$3 \cdot 4 = 12$
4	4	$4 \cdot 4 = 16$
5	6	$5 \cdot 6 = 30$
6	5	$6 \cdot 5 = 30$
7	3	$7 \cdot 3 = 21$
8	2	$8 \cdot 2 = 16$
	28	$\sum xf = 132$

Mean $= \dfrac{\sum xf}{n} = \dfrac{132}{28} \approx 4.71$

11.

x	f	xf
1	1	$1 \cdot 1 = 1$
2	1	$2 \cdot 1 = 2$
3	2	$3 \cdot 2 = 6$
4	5	$4 \cdot 5 = 20$
5	7	$5 \cdot 7 = 35$
6	9	$6 \cdot 9 = 54$
7	8	$7 \cdot 8 = 56$
8	6	$8 \cdot 6 = 48$
9	4	$9 \cdot 4 = 36$
10	3	$10 \cdot 3 = 30$
	46	$\sum xf = 288$

Mean $= \dfrac{\sum xf}{n} = \dfrac{288}{46} \approx 6.26$

13. First arrange the data items from smallest to largest: 1, 2, 3, 3, 4, 5, 7, 8
The number of data items is even, so the median is the mean of the two middle data items. The median is 3.5.

15. First arrange the data items from smallest to largest: 91, 93, 95, 95, 97, 99
The number of data items is even, so the median is the mean of the two middle data items.
$$\text{Median} = \frac{95+95}{2} = 95$$

17. First arrange the data items from smallest to largest: 40, 40, 60, 70, 100
The number of data items is odd, so the median is the middle number. The median is 60.

19. First arrange the data items from smallest to largest: 1.6, 2.5, 2.5, 2.7, 3.2, 3.6, 3.8, 4.2, 4.2, 4.7, 5.0
The number of data items is odd, so the median is the middle number. The median is 3.6.

21. $n = 28$
$$\frac{n+1}{2} = \frac{28+1}{2} = \frac{29}{2} = 14.5$$
The median is in the 14.5 position, which means the median is the mean of the data items in positions 14 and 15.
Counting down the frequency column, the 14th and 15th data items are both 5.
$$\text{Median} = \frac{5+5}{2} = 5$$

23. $n = 46$
$$\frac{n+1}{2} = \frac{46+1}{2} = 23.5$$
The median is in the 23.5 position, which means the median is the mean of the data items in positions 23 and 24.
Counting down the frequency column, the 23rd and 24th data items are both 6.
$$\text{Median} = \frac{6+6}{2} = 6$$

25. The mode is 3.

27. The mode is 95.

29. The mode is 40.

31. The modes are 2.5 and 4.2 (bimodal).

33. The mode is 5.

35. The mode is 6.

37. lowest data value = 1, highest data value = 8
$$\text{Midrange} = \frac{1+8}{2} = 4.5$$

39. lowest data value = 91, highest data value = 99
$$\text{Midrange} = \frac{91+99}{2} = 95$$

41. lowest data value = 40, highest data value = 100
$$\text{Midrange} = \frac{40+100}{2} = 70$$

43. lowest data value = 1.6, highest data value = 5.0

$$\text{Midrange} = \frac{1.6 + 5.0}{2} = 3.3$$

45. $\text{Midrange} = \dfrac{1 + 8}{2} = 4.5$

47. $\text{Midrange} = \dfrac{1 + 10}{2} = 5.5$

49.

x	f	xf
10	1	10
20	2	40
30	4	120
40	2	80
50	1	50
	10	$\sum xf = 300$

$$\text{Mean} = \frac{\sum xf}{n} = \frac{300}{10} = 30$$

The median is the mean of the 5th and 6th data items. Since these items are both 30, the median is 30. The mode is 30 (it has the highest frequency).

$$\text{Midrange} = \frac{10 + 50}{2} = 30$$

51.

x	f	xf
10	2	20
11	2	22
12	3	36
13	4	52
14	1	14
15	2	30
	14	$\sum xf = 174$

$$\text{Mean} = \frac{\sum xf}{n} = \frac{174}{14} \approx 12.4$$

The median is the mean of the 7th and 8th data items. $\text{Median} = \dfrac{12 + 13}{2} = 12.5$

The mode is 13 (it has the highest frequency).

$$\text{Midrange} = \frac{10 + 15}{2} = 12.5$$

53. The data items are 21, 24, 25, 30, 31, 31, 33, 42, 45

$$\text{Mean} = \frac{21 + 24 + 25 + 30 + 31 + 31 + 33 + 42 + 45}{9} = \frac{282}{9} \approx 31.3$$

The median is the 5th data item, or 31.
The mode is 31.

$$\text{Midrange} = \frac{21 + 45}{2} = 33$$

55. The data items are 52, 56, 59, 68, 69, 71, 71, 74, 87.

   a.   Mean $= \dfrac{52+56+59+68+69+71+71+74+87}{9} = \dfrac{607}{9} \approx 67.4$

   b.   The number of data items is odd, so the median is the middle number. The median is 69.

   c.   The number 71 occurs more often than any other. The mode is 71.

   d.   Midrange $= \dfrac{52+87}{2} = \dfrac{139}{2} = 69.5$

57. The data items are 0, 4, 8, 12, 14, 17, 19, 19, 23, 32, 34, 48.

   a.   Mean $= \dfrac{0+4+8+12+14+17+19+19+23+32+34+48}{12} = \dfrac{230}{12} \approx 19.2$

   b.   The median is the mean of the 6th and 7th data items.   Median $= \dfrac{17+19}{2} = 18$

   c.   The mode is 19.

   d.   Midrange $= \dfrac{0+48}{2} = 24$

59.

x	f	xf
2	12	$2 \cdot 12 = 24$
7	16	$7 \cdot 16 = 112$
12	16	$12 \cdot 16 = 192$
17	16	$17 \cdot 16 = 272$
22	10	$22 \cdot 10 = 220$
27	11	$27 \cdot 11 = 297$
32	4	$32 \cdot 4 = 128$
37	3	$37 \cdot 3 = 111$
42	3	$42 \cdot 3 = 126$
47	3	$47 \cdot 3 = 141$
	94	$\sum xf = 1623$

   a.   Mean $= \dfrac{\sum xf}{n} = \dfrac{1623}{94} \approx 17.27$

   b.   The median is 17 because the 47th and 48th data items both are 17.

   c.   The modes are 7, 12, and 17.

   d.   Midrange $= \dfrac{47+2}{2} = \dfrac{49}{2} = 24.5$

61. $n = 40$, $\dfrac{n+1}{2} = \dfrac{40+1}{2} = \dfrac{41}{2} = 20.5$

The median is in the 20.5 position, which means the median is the mean of the data items in positions 20 and 21.

Median $= \dfrac{175+175}{2} = 175$ lb

63. Midrange $= \dfrac{150+205}{2} = 177.5$ lb

65. Find the weighted mean by treating the number of credits as the "frequency."

Course	Grade	Value (x)	Credits (f)	xf
Sociology	A	4	3	$4 \cdot 3 = 12$
Biology	C	2	3.5	$2 \cdot 3.5 = 7$
Music	B	3	1	$3 \cdot 1 = 3$
Math	B	3	4	$3 \cdot 4 = 12$
English	C	2	3	$2 \cdot 3 = 6$
			14.5	$\sum xf = 40$

$$\text{Mean} = \frac{\sum xf}{n} = \frac{40}{14.5} \approx 2.76$$

77. All 30 students had the same grade.

Check Points 12.3

1. Range $= 11 - 2 = 9$

2. Mean $= \dfrac{2+4+7+11}{4} = \dfrac{24}{4} = 6$

Data item	Deviation: Data item – mean
2	$2 - 6 = -4$
4	$4 - 6 = -2$
7	$7 - 6 = 1$
11	$11 - 6 = 5$

3. Mean $= \dfrac{2+4+7+11}{4} = \dfrac{24}{4} = 6$

Data item	Deviation: Data item – mean	$(\text{Deviation})^2$: $(\text{Data item–mean})^2$
2	$2 - 6 = -4$	$(-4)^2 = 16$
4	$4 - 6 = -2$	$(-2)^2 = 4$
7	$7 - 6 = 1$	$1^2 = 1$
11	$11 - 6 = 5$	$5^2 = 25$
		$\sum (\text{data item–mean})^2 = 46$

$$\text{Standard deviation} = \sqrt{\frac{46}{4-1}} = \sqrt{\frac{46}{3}} \approx 3.92$$

4. *Sample A*

$$\text{Mean} = \frac{73 + 75 + 77 + 79 + 81 + 83}{6} = \frac{468}{6} = 78$$

Data item	Deviation: Data item − mean	(Deviation)^2 : (Data item−mean)^2
73	$73 - 78 = -5$	$(-5)^2 = 25$
75	$75 - 78 = -3$	$(-3)^2 = 9$
77	$77 - 78 = -1$	$(-1)^2 = 1$
79	$79 - 78 = 1$	$1^2 = 1$
81	$81 - 78 = 3$	$3^2 = 9$
83	$83 - 78 = 5$	$5^2 = 25$

$$\sum (\text{data item−mean})^2 = 70$$

$$\text{Standard deviation} = \sqrt{\frac{70}{6-1}} = \sqrt{\frac{70}{5}} \approx 3.74$$

Sample B

$$\text{Mean} = \frac{40 + 44 + 92 + 94 + 98 + 100}{6} = \frac{468}{6} = 78$$

Data item	Deviation: Data item − mean	(Deviation)^2 : (Data item−mean)^2
40	$40 - 78 = -38$	$(-38)^2 = 1444$
44	$44 - 78 = -34$	$(-34)^2 = 1156$
92	$92 - 78 = 14$	$14^2 = 196$
94	$94 - 78 = 16$	$16^2 = 256$
98	$98 - 78 = 20$	$20^2 = 400$
100	$100 - 78 = 22$	$22^2 = 484$

$$\sum (\text{data item−mean})^2 = 3936$$

$$\text{Standard deviation} = \sqrt{\frac{3936}{6-1}} = \sqrt{\frac{3936}{5}} \approx 28.06$$

5. a. Stocks had a greater return on investment.

b. Stocks have the greater risk. The high standard deviation indicates that stocks are more likely to lose money.

Exercise Set 12.3

1. Range $= 5 - 1 = 4$

3. Range $= 15 - 7 = 8$

5. Range $= 5 - 3 = 2$

7. **a.**

Data item	Deviation: Data item – mean
3	$3 - 12 = -9$
5	$5 - 12 = -7$
7	$7 - 12 = -5$
12	$12 - 12 = 0$
18	$18 - 12 = 6$
27	$27 - 12 = 15$

   b. $-9 - 7 - 5 + 0 + 6 + 15 = 0$

9. **a.**

Data item	Deviation: Data item – mean
29	$29 - 49 = -20$
38	$38 - 49 = -11$
48	$48 - 49 = -1$
49	$49 - 49 = 0$
53	$53 - 49 = 4$
77	$77 - 49 = 28$

   b. $-20 - 11 - 1 + 0 + 4 + 28 = 0$

11. **a.** Mean $= \dfrac{85 + 95 + 90 + 85 + 100}{5} = 91$

   b.

Data item	Deviation: Data item – mean
85	$85 - 91 = -6$
95	$95 - 91 = 4$
90	$90 - 91 = -1$
85	$85 - 91 = -6$
100	$100 - 91 = 9$

   c. $-6 + 4 - 1 - 6 + 9 = 0$

13. **a.** Mean $= \dfrac{146 + 153 + 155 + 160 + 161}{5} = 155$

   b.

Data item	Deviation: Data item – mean
146	$146 - 155 = -9$
153	$153 - 155 = -2$
155	$155 - 155 = 0$
160	$160 - 155 = 5$
161	$161 - 155 = 6$

   c. $-9 - 2 + 0 + 5 + 6 = 0$

15. a. $\text{Mean} = \dfrac{2.25 + 3.50 + 2.75 + 3.10 + 1.90}{5} = 2.70$

b.

Data item	Deviation: Data item – mean
2.25	$2.25 - 2.70 = -0.45$
3.50	$3.50 - 2.70 = 0.80$
2.75	$2.75 - 2.70 = 0.05$
3.10	$3.10 - 2.70 = 0.40$
1.90	$1.90 - 2.70 = -0.80$

c. $-0.45 + 0.80 + 0.05 + 0.40 - 0.80 = 0$

17. $\text{Mean} = \dfrac{1 + 2 + 3 + 4 + 5}{5} = 3$

Data item	Deviation: Data item – mean	$(\text{Deviation})^2$: $(\text{Data item–mean})^2$
1	$1 - 3 = -2$	$(-2)^2 = 4$
2	$2 - 3 = -1$	$(-1)^2 = 1$
3	$3 - 3 = 0$	$0^2 = 0$
4	$4 - 3 = 1$	$1^2 = 1$
5	$5 - 3 = 2$	$2^2 = 4$
		$\sum(\text{data item–mean})^2 = 10$

$\text{Standard deviation} = \sqrt{\dfrac{10}{5-1}} = \sqrt{\dfrac{10}{4}} \approx 1.58$

19. $\text{Mean} = \dfrac{7 + 9 + 9 + 15}{4} = 10$

Data item	Deviation: Data item – mean	$(\text{Deviation})^2$: $(\text{Data item–mean})^2$
7	$7 - 10 = -3$	$(-3)^2 = 9$
9	$9 - 10 = -1$	$(-1)^2 = 1$
9	$9 - 10 = -1$	$(-1)^2 = 1$
15	$15 - 10 = 5$	$5^2 = 25$
		$\sum(\text{data item–mean})^2 = 36$

$\text{Standard deviation} = \sqrt{\dfrac{36}{4-1}} = \sqrt{\dfrac{36}{3}} \approx 3.46$

21. Mean $= \dfrac{3+3+4+4+5+5}{6} = 4$

Data item	Deviation: Data item − mean	$(\text{Deviation})^2$: $(\text{Data item–mean})^2$
3	$3 - 4 = -1$	$(-1)^2 = 1$
3	$3 - 4 = -1$	$(-1)^2 = 1$
4	$4 - 4 = 0$	$0^2 = 0$
4	$4 - 4 = 0$	$0^2 = 0$
5	$5 - 4 = 1$	$1^2 = 1$
5	$5 - 4 = 1$	$1^2 = 1$

$$\sum (\text{data item–mean})^2 = 4$$

Standard deviation $= \sqrt{\dfrac{4}{6-1}} = \sqrt{\dfrac{4}{5}} \approx 0.89$

23. Mean $= \dfrac{1+1+1+4+7+7+7}{7} = 4$

Data item	Deviation: Data item − mean	$(\text{Deviation})^2$: $(\text{Data item–mean})^2$
1	$1 - 4 = -3$	$(-3)^2 = 9$
1	$1 - 4 = -3$	$(-3)^2 = 9$
1	$1 - 4 = -3$	$(-3)^2 = 9$
4	$4 - 4 = 0$	$0^2 = 0$
7	$7 - 4 = 3$	$3^2 = 9$
7	$7 - 4 = 3$	$3^2 = 9$
7	$7 - 4 = 3$	$3^2 = 9$

$$\sum (\text{data item–mean})^2 = 54$$

Standard deviation $= \sqrt{\dfrac{54}{7-1}} = \sqrt{\dfrac{54}{6}} = 3$

25. Mean $= \dfrac{9+5+9+5+9+5+9+5}{8} = 7$

Data item	Deviation: Data item − mean	(Deviation)^2 : (Data item−mean)^2
9	$9 - 7 = 2$	$2^2 = 4$
5	$5 - 7 = -2$	$(-2)^2 = 4$
9	$9 - 7 = 2$	$2^2 = 4$
5	$5 - 7 = -2$	$(-2)^2 = 4$
9	$9 - 7 = 2$	$2^2 = 4$
5	$5 - 7 = -2$	$(-2)^2 = 4$
9	$9 - 7 = 2$	$2^2 = 4$
5	$5 - 7 = -2$	$(-2)^2 = 4$

$$\sum (\text{data item–mean})^2 = 32$$

$$\text{Standard deviation} = \sqrt{\frac{32}{8-1}} = \sqrt{\frac{32}{7}} \approx 2.14$$

27. *Sample A*

Mean $= \dfrac{6+8+10+12+14+16+18}{7} = 12$

Range $= 18 - 6 = 12$

Data item	Deviation: Data item − mean	(Deviation)^2 : (Data item−mean)^2
6	$6 - 12 = -6$	$(-6)^2 = 36$
8	$8 - 12 = -4$	$(-4)^2 = 16$
10	$10 - 12 = -2$	$(-2)^2 = 4$
12	$12 - 12 = 0$	$0^2 = 0$
14	$14 - 12 = 2$	$2^2 = 4$
16	$16 - 12 = 4$	$4^2 = 16$
18	$18 - 12 = 6$	$6^2 = 36$

$$\sum (\text{data item–mean})^2 = 112$$

$$\text{Standard deviation} = \sqrt{\frac{112}{7-1}} = \sqrt{\frac{112}{6}} \approx 4.32$$

Sample B

Mean $= \dfrac{6+7+8+12+16+17+18}{7} = 12$

Range $= 18 - 6 = 12$

Data item	Deviation: Data item – mean	(Deviation)^2 : (Data item–mean)^2
6	$6 - 12 = -6$	$(-6)^2 = 36$
7	$7 - 12 = -5$	$(-5)^2 = 25$
8	$8 - 12 = -4$	$(-4)^2 = 16$
12	$12 - 12 = 0$	$0^2 = 0$
16	$16 - 12 = 4$	$4^2 = 16$
17	$17 - 12 = 5$	$5^2 = 25$
18	$18 - 12 = 6$	$6^2 = 36$
		$\sum (\text{data item–mean})^2 = 154$

Standard deviation $= \sqrt{\dfrac{154}{7-1}} = \sqrt{\dfrac{154}{6}} \approx 5.07$

Sample C

Mean $= \dfrac{6+6+6+12+18+18+18}{7} = 12$

Range $= 18 - 6 = 12$

Data item	Deviation: Data item – mean	(Deviation)^2 : (Data item–mean)^2
6	$6 - 12 = -6$	$(-6)^2 = 36$
6	$6 - 12 = -6$	$(-6)^2 = 36$
6	$6 - 12 = -6$	$(-6)^2 = 36$
12	$12 - 12 = 0$	$0^2 = 0$
18	$18 - 12 = 6$	$6^2 = 36$
18	$18 - 12 = 6$	$6^2 = 36$
18	$18 - 12 = 6$	$6^2 = 36$
		$\sum (\text{data item–mean})^2 = 216$

Standard deviation $= \sqrt{\dfrac{216}{7-1}} = \sqrt{\dfrac{216}{6}} = 6$

The samples have the same mean and range, but different standard deviations.

29. Mean $= \dfrac{9+9+9+9+9+9+9}{7} = \dfrac{63}{7} = 9$

Data item	Deviation: Data item − mean	$(\text{Deviation})^2$: $(\text{Data item–mean})^2$
9	$9 - 9 = 0$	$(0)^2 = 0$
9	$9 - 9 = 0$	$(0)^2 = 0$
9	$9 - 9 = 0$	$(0)^2 = 0$
9	$9 - 9 = 0$	$(0)^2 = 0$
9	$9 - 9 = 0$	$(0)^2 = 0$
9	$9 - 9 = 0$	$(0)^2 = 0$
9	$9 - 9 = 0$	$(0)^2 = 0$
		$\sum (\text{data item} - \text{mean})^2 = 0$

Standard deviation $= \sqrt{\dfrac{0}{7-1}} = \sqrt{\dfrac{0}{6}} = 0$

31. Mean $= \dfrac{8+8+8+9+10+10+10}{7} = \dfrac{63}{7} = 9$

Data item	Deviation: Data item − mean	$(\text{Deviation})^2$: $(\text{Data item–mean})^2$
8	$8 - 9 = -1$	$(-1)^2 = 1$
8	$8 - 9 = -1$	$(-1)^2 = 1$
8	$8 - 9 = -1$	$(-1)^2 = 1$
9	$9 - 9 = 0$	$(0)^2 = 0$
10	$10 - 9 = 1$	$(-1)^2 = 1$
10	$10 - 9 = 1$	$(-1)^2 = 1$
10	$10 - 9 = 1$	$(-1)^2 = 1$
		$\sum (\text{data item} - \text{mean})^2 = 6$

Standard deviation $= \sqrt{\dfrac{6}{7-1}} = \sqrt{\dfrac{6}{6}} = 1$

33. Mean $= \dfrac{5+10+15+20+25}{5} = \dfrac{75}{5} = 15$

Data item	Deviation: Data item − mean	(Deviation)^2 : (Data item−mean)^2
5	$5 - 15 = -10$	$(-10)^2 = 100$
10	$10 - 15 = -5$	$(-5)^2 = 25$
15	$15 - 15 = 0$	$(0)^2 = 0$
20	$20 - 15 = 5$	$(5)^2 = 25$
25	$25 - 15 = 10$	$(10)^2 = 100$
		$\sum (\text{data item} - \text{mean})^2 = 250$

Standard deviation $= \sqrt{\dfrac{250}{5-1}} = \sqrt{\dfrac{250}{4}} \approx 7.91$

35. Mean $= \dfrac{17+18+18+18+19+19+20+20+21+22}{10} = \dfrac{192}{10} = 19.2$

Data item	Deviation: Data item − mean	(Deviation)^2 : (Data item−mean)^2
17	$17 - 19.2 = -2.2$	$(-2.2)^2 = 4.84$
18	$18 - 19.2 = -2.2$	$(-1.2)^2 = 1.44$
18	$18 - 19.2 = -2.2$	$(-1.2)^2 = 1.44$
18	$18 - 19.2 = -2.2$	$(-1.2)^2 = 1.44$
19	$19 - 19.2 = -0.2$	$(-0.2)^2 = 0.04$
19	$19 - 19.2 = -0.2$	$(-0.2)^2 = 0.04$
20	$20 - 19.2 = 0.8$	$(0.8)^2 = 0.64$
20	$20 - 19.2 = 0.8$	$(0.8)^2 = 0.64$
21	$21 - 19.2 = 1.8$	$(1.8)^2 = 3.24$
22	$22 - 19.2 = 2.8$	$(2.8)^2 = 7.84$
		$\sum (\text{data item} - \text{mean})^2 = 21.6$

Standard deviation $= \sqrt{\dfrac{21.6}{10-1}} = \sqrt{\dfrac{21.6}{9}} \approx 1.55$

37. a. Best Actor has the greater mean. This can be seen without calculating by observing that most of the ages for Best Actor are in the upper 30s and lower 40s while all of the ages for Best Actress are in the upper 20s and lower 30s.

    b.   Mean (Best Actor) $= \dfrac{270}{7} \approx 38.57$;   Mean (Best Actress) $= \dfrac{215}{7} \approx 30.71$

    c.   Best Actor has the greater standard deviation. This can be seen without calculating by observing that the ages for Best Actor have a greater spread as compared to the ages for Best Actress.

    d.   Standard deviation (Best Actor) $= \sqrt{\dfrac{193.7143}{7-1}} = \sqrt{\dfrac{193.7143}{6}} \approx 5.68$

            Standard deviation (Best Actress) $= \sqrt{\dfrac{85.4287}{7-1}} = \sqrt{\dfrac{85.4287}{6}} \approx 3.77$

47. a is true

51 Original data:

    Mean $= \dfrac{0+1+3+4+4+6}{6} = \dfrac{18}{6} = 3$

    Standard deviation $= \sqrt{\dfrac{24}{6-1}} = \sqrt{\dfrac{24}{5}} \approx 2.19$

    Adjusted data:

    Mean $= \dfrac{2+3+5+6+6+8}{6} = \dfrac{30}{6} = 5$

    Standard deviation $= \sqrt{\dfrac{24}{6-1}} = \sqrt{\dfrac{24}{5}} \approx 2.19$

    Adding 2 to each data item raises the mean by 2, but does not affect the standard deviation.

Check Points 12.4

1. a.   Height $=$ mean $+ 3 \cdot$ standard deviation

                 $= 65 + 3 \cdot 3.5 = 75.5$ in.

    b.   Height $=$ mean $- 2 \cdot$ standard deviation

                 $= 65 - 2 \cdot 3.5 = 58$ in.

2. a.   The 68-95-99.7 Rule states that approximately 95% of the data items fall within 2 standard deviations of the mean. The figure shows that 95% of male adults have heights between 62 inches and 78 inches.

    b.   The 68-95-99.7 Rule states that approximately 95% of the data items fall within 2 standard deviations of the mean. Since the mean is 70 inches, the figure shows that half of the 95%, or 47.5% of male adults have heights between 70 inches and 78 inches.

    c.   The 68-95-99.7 Rule states that approximately 68% of the data items fall within 1 standard deviation of the mean, thus 32% of the data falls outside this range. Half of the 32%, or 16% of male adults will have heights above 74 inches.

3. a. $z_{342} = \dfrac{\text{data item} - \text{mean}}{\text{standard deviation}} = \dfrac{342 - 336}{3} = \dfrac{6}{3} = 2$

b. $z_{336} = \dfrac{\text{data item} - \text{mean}}{\text{standard deviation}} = \dfrac{336 - 336}{3} = \dfrac{0}{3} = 0$

c. $z_{333} = \dfrac{\text{data item} - \text{mean}}{\text{standard deviation}} = \dfrac{333 - 336}{3} = \dfrac{-3}{3} = -1$

4. Find the z-score for each test taken.

SAT: $z_{550} = \dfrac{\text{data item} - \text{mean}}{\text{standard deviation}} = \dfrac{550 - 500}{100} = \dfrac{50}{100} = 0.5$

ACT: $z_{24} = \dfrac{\text{data item} - \text{mean}}{\text{standard deviation}} = \dfrac{24 - 18}{6} = \dfrac{6}{6} = 1$

You scored better on the ACT test because the score is 1 standard deviation above the mean. The SAT score is only half a standard deviation above the mean.

5. a. Score $= \text{mean} - 2.25 \cdot \text{standard deviation} = 100 - 2.25(16) = 64$

b. Score $= \text{mean} + 1.75 \cdot \text{standard deviation} = 100 + 1.75(16) = 128$

6. This means that 75% of the scores on the SAT are less than this student's score.

7. $z_{83.60} = \dfrac{\text{data item} - \text{mean}}{\text{standard deviation}} = \dfrac{83.60 - 62}{18} = 1.2$

A z-score of 1.2 corresponds to a percentile of 88.49. Thus, 88.49% of plans have charges less than \$83.60.

8. $z_{69.9} = \dfrac{\text{data item} - \text{mean}}{\text{standard deviation}} = \dfrac{69.9 - 65}{3.5} = 1.4$

A z-score of 1.4 corresponds to a percentile of 91.93. Thus, 100% − 91.92% = 8.08% of women have heights greater than 69.9 inches.

9. $z_{11} = \dfrac{\text{data item} - \text{mean}}{\text{standard deviation}} = \dfrac{11 - 14}{2.5} = -1.2$ which corresponds to a percentile of 11.51.

$z_{18} = \dfrac{\text{data item} - \text{mean}}{\text{standard deviation}} = \dfrac{18 - 14}{2.5} = 1.6$ which corresponds to a percentile of 94.52.

Thus, 94.52% − 11.51% = 83.01% of refrigerators have lives between 11 and 18 years.

10. a. Margin of error $= \pm \dfrac{1}{\sqrt{n}} = \pm \dfrac{1}{\sqrt{485}} \approx \pm 0.045 = \pm 4.5\%$

b. 54% − 4.5% = 49.5% and 54% + 4.5% = 58.5%
There is a 95% probability that the true population percentage lies between 49.5% and 58.5%.
We can be 95% confident that between 49.5% and 58.5% of American adults support physician-assisted suicide.

c. The percentage of American adults who support physician-assisted suicide may be less than 50%.

Exercise Set 12.4

1. Score $= 100 + 1 \cdot 20 = 100 + 20 = 120$

3. Score $= 100 + 3 \cdot 20 = 100 + 60 = 160$

5. Score $= 100 + 2.5(20) = 100 + 50 = 150$

7. Score $= 100 - 2 \cdot 20 = 100 - 40 = 60$

9. Score $= 100 - 0.5(20) = 100 - 10 = 90$

11. $16,500 is 1 standard deviation below the mean and $17,500 is 1 standard deviation above the mean. The Rule and the figure indicate that 68% of the buyers paid between $16,500 and $17,500.

13. $17,500 is 1 standard deviation above the mean. 68% of the buyers paid between $16,500 and $17,500. Because of symmetry, the percent that paid between $17,000 and $17,500 is $\frac{1}{2}(68\%) = 34\%$.

15. $16,000 is 2 standard deviations below the mean. 95% of the buyers paid between $16,000 and $18,000. Because of symmetry, the percent that paid between $16,000 and $17,000 is $\frac{1}{2}(95\%) = 47.5\%$.

17. $15,500 is 3 standard deviations below the mean. 99.7% of the buyers paid between $15,500 and $18,500. Because of symmetry, the percent that paid between $15,500 and $17,000 is $\frac{1}{2}(99.7\%) = 49.85\%$.

19. $17,500 is 1 standard deviation above the mean. Since 68% of the data items fall within 1 standard deviation of the mean, $100\% - 68\% = 32\%$ fall farther than 1 standard deviation from the mean. Because of symmetry, the percent that paid more than $17,500 is $\frac{1}{2}(32\%) = 16\%$.

21. $16,000 is 2 standard deviations below the mean. Since 95% of the data items fall within 2 standard deviations of the mean, $100\% - 95\% = 5\%$ fall farther than 2 standard deviations from the mean. Because of symmetry, the percent that paid less than $16,000 is $\frac{1}{2}(5\%) = 2.5\%$.

23. The 68-95-99.7 Rule states that approximately 95% of the data items fall within 2 standard deviations of the mean. 95% of people will have IQs between 68 and 132.

25. The 68-95-99.7 Rule states that approximately 95% of the data items fall within 2 standard deviations of the mean. Half of the 95%, or 47.5% of people will have IQs between 68 and 100.

27. The 68-95-99.7 Rule states that approximately 68% of the data items fall within 1 standard deviation of the mean. Thus, $100\% - 68\% = 32\%$ will fall outside this range. Half of the 32%, or 16% of people will have IQs above 116.

29. The 68-95-99.7 Rule states that approximately 95% of the data items fall within 2 standard deviations of the mean. Thus, $100\% - 95\% = 5\%$ will fall outside this range. Half of the 5%, or 2.5% of people will have IQs below 68.

31. The 68-95-99.7 Rule states that approximately 99.7% of the data items fall within 3 standard deviations of the mean. Thus, $100\% - 99.7\% = 0.3\%$ will fall outside this range. Half of the 0.3%, or 0.15% of people will have IQs above 148.

33. $z_{68} = \dfrac{68-60}{8} = \dfrac{8}{8} = 1$

35. $z_{84} = \dfrac{84-60}{8} = \dfrac{24}{8} = 3$

37. $z_{64} = \dfrac{64-60}{8} = \dfrac{4}{8} = 0.5$

39. $z_{74} = \dfrac{74-60}{8} = \dfrac{14}{8} = 1.75$

41. $z_{60} = \dfrac{60-60}{8} = \dfrac{0}{8} = 0$

43. $z_{52} = \dfrac{52-60}{8} = \dfrac{-8}{8} = -1$

45. $z_{48} = \dfrac{48-60}{8} = \dfrac{-12}{8} = -1.5$

47. $z_{34} = \dfrac{34-60}{8} = \dfrac{-26}{8} = -3.25$

49. $z = \dfrac{\text{data item} - \text{mean}}{\text{standard deviation}} = \dfrac{43-12.4}{20.4} = 1.5$

51. $z = \dfrac{\text{data item} - \text{mean}}{\text{standard deviation}} = \dfrac{58.3-12.4}{20.4} = 2.25$

53. $z = \dfrac{\text{data item} - \text{mean}}{\text{standard deviation}} = \dfrac{-13.1 - 12.4}{20.4} = -1.25$

55. $z = \dfrac{\text{data item} - \text{mean}}{\text{standard deviation}} = \dfrac{-18.2 - 12.4}{20.4} = -1.5$

57. z-score of 128 on the Stanford-Binet:

$z = \dfrac{\text{data item} - \text{mean}}{\text{standard deviation}} = \dfrac{128 - 100}{16} = 1.75$

z-score of 127 on the Wechsler:

$z = \dfrac{\text{data item} - \text{mean}}{\text{standard deviation}} = \dfrac{127 - 100}{15} = 1.8$

The person who scores 127 on the Wechsler has the higher IQ.

59. $2 \cdot 50 = 100$
The data item is 100 units above the mean.
$400 + 100 = 500$

61. $1.5(50) = 75$
The data item is 75 units above the mean.
$400 + 75 = 475$

63. $-3 \cdot 50 = -150$
The data item is 150 units below the mean.
$400 - 150 = 250$

65. $-2.5(50) = -125$
The data item is 125 units below the mean.
$400 - 125 = 275$

67. a. 72.57%

b. $100\% - 72.57\% = 27.43\%$

69. a. 88.49%

b. $100\% - 88.49\% = 11.51\%$

71. a. 24.20%

b. $100\% - 24.20\% = 75.8\%$

73. a. 11.51%

b. $100\% - 11.51\% = 88.49\%$

75. $z = 0.2 \rightarrow 57.93\%$
$z = 1.4 \rightarrow 91.92\%$
$91.92\% - 57.93\% = 33.99\%$

77. $z = 1 \rightarrow 84.13\%$
$z = 3 \rightarrow 99.87\%$
$99.87\% - 84.13\% = 15.74\%$

79. $z = -1.5 \rightarrow 6.68\%$
$z = 1.5 \rightarrow 93.32\%$
$93.32\% - 6.68\% = 86.64\%$

81. $z = -2 \rightarrow 2.28\%$
$z = -0.5 \rightarrow 30.85\%$
$30.85\% - 2.28\% = 28.57\%$

83. $z_{142} = \dfrac{\text{data item} - \text{mean}}{\text{standard deviation}} = \dfrac{142 - 121}{15} = 1.4$

A z-score of 1.4 corresponds to a percentile of 91.92. Thus, 91.92% of people have blood pressure below 142

85. $z_{130} = \dfrac{\text{data item} - \text{mean}}{\text{standard deviation}} = \dfrac{130 - 121}{15} = 0.6$

A z-score of 0.6 corresponds to a percentile of 72.57. Thus, $100\% - 72.97\% = 27.43\%$ of people have blood pressure above 130.

87. $z_{103} = \dfrac{\text{data item} - \text{mean}}{\text{standard deviation}} = \dfrac{103 - 121}{15} = -1.2$

A z-score of -1.2 corresponds to a percentile of 11.51. Thus, $100\% - 11.51\% = 88.49\%$ of people have blood pressure above 103.

89. $z_{142} = \dfrac{\text{data item} - \text{mean}}{\text{standard deviation}} = \dfrac{142 - 121}{15} = 1.4$

A z-score of 1.4 corresponds to a percentile of 91.92.

$z_{154} = \dfrac{\text{data item} - \text{mean}}{\text{standard deviation}} = \dfrac{154 - 121}{15} = 2.2$

A z-score of 2.2 corresponds to a percentile of 98.61. Thus, $98.61\% - 91.92\% = 6.69\%$ of people have blood pressure between 142 and 154.

91. $z_{112} = \dfrac{\text{data item} - \text{mean}}{\text{standard deviation}} = \dfrac{112 - 121}{15} = -0.6$

A z-score of -0.6 corresponds to a percentile of 27.43.

$z_{130} = \dfrac{\text{data item} - \text{mean}}{\text{standard deviation}} = \dfrac{130 - 121}{15} = 0.6$

A z-score of 0.6 corresponds to a percentile of 72.57. Thus, $72.57\% - 27.43\% = 45.14\%$ of people have blood pressure between 112 and 130.

93. $z_{25.8} = \dfrac{25.8 - 22.5}{2.2} = 1.5$

$z = 1.5 \rightarrow 93.32\%$
$100\% - 93.32\% = 6.68\%$ weigh more than 25.8 pounds.

95. $z_{19.2} = \dfrac{19.2 - 22.5}{2.2} = -1.5$

$z = -1.5 \rightarrow 6.68\%$

$z_{21.4} = \dfrac{21.4 - 22.5}{2.2} = -0.5$

$z = -0.5 \rightarrow 30.85\%$
30.85% − 6.68% = 24.17% weigh between 19.2 and 21.4 pounds.

97. a. margin of error $= \pm \dfrac{1}{\sqrt{2272}}$

$\approx \pm 0.021$

$\approx \pm 2.1\%$

b. 18% − 2.1% = 15.9%
18% + 2.1% = 20.1%
We can be 95% confident that between 15.9% and 20.1% of parents in the population consider a doctor as the dream job for their child.

99. a. margin of error $= \pm \dfrac{1}{\sqrt{4000}}$

$\approx \pm 0.016$

$\approx \pm 1.6\%$

b. 60.2% − 1.6% = 58.6%
60.2% + 1.6% = 61.8%
We can be 95% confident that between 58.6% and 61.8% of all TV households watched the final episode of *M*A*S*H*.

101. new margin of error $= \pm \dfrac{1}{\sqrt{5000}}$

$\approx \pm 0.014$

$\approx \pm 1.4\%$
improvement = 1.6% − 1.4% = 0.2%

103. a. The graph is skewed to the right.

b.

x	f	xf
1	3	$1 \cdot 3 = 3$
2	8	$2 \cdot 8 = 16$
3	9	$3 \cdot 9 = 27$
4	2	$4 \cdot 2 = 8$
5	8	$5 \cdot 8 = 40$
6	9	$6 \cdot 9 = 54$
7	5	$7 \cdot 5 = 35$
8	2	$8 \cdot 2 = 16$
9	2	$9 \cdot 2 = 18$
10	1	$10 \cdot 1 = 10$
13	1	$13 \cdot 1 = 13$
44	1	$44 \cdot 1 = 44$
	51	$\sum xf = 284$

$\text{Mean} = \dfrac{\sum xf}{n} = \dfrac{284}{51} \approx 5.6$

The mean rate is 5.6 murders per 100,000 residents.

c. The median is in the 26th position. The median rate is 5 murders per 100,000 residents.

d. Yes, these rates are consistent with the graph. The mean is greater than the median, which is expected with a distribution that is skewed to the right.

e. $z_{44} = \dfrac{44 - 5.6}{6.1} = 6.3$

Yes, this is unusually high. For a normal distribution, almost 100% of the z-scores are between −3 and 3

119. A z-score of 1.3 has 90.32% of the data items below it, and 9.68% above it. So find the score corresponding to $z = 1.3$.
500 + 1.3(100) = 630
The cutoff score is 630.

Check Points 12.5

1. 0.51 would indicate a moderate correlation between the two.

2.

x	y	xy	x^2	y^2
2.5	211	527.5	6.25	44,521
3.9	167	651.3	15.21	27,889
2.9	131	379.9	8.41	17,161
2.4	191	458.4	5.76	36,481
2.9	220	638	8.41	48,400
0.8	297	237.6	0.64	88,209
9.1	71	646.1	82.81	5041
0.8	211	168.8	0.64	44,521
0.7	300	210	0.49	90,000
7.9	107	845.3	62.41	11,449
1.8	167	300.6	3.24	27,889
1.9	266	505.4	3.61	70,756
0.8	227	181.6	0.64	51,529
6.5	86	559	42.25	7396
1.6	207	331.2	2.56	42,849
5.8	115	667	33.64	13,225
1.3	285	370.5	1.69	81,225
1.2	199	238.8	1.44	39,601
2.7	172	464.4	7.29	29,584

$$\sum x = 57.5 \quad \sum y = 3630 \quad \sum xy = 8381.4 \quad \sum x^2 = 287.39 \quad \sum y^2 = 777,726$$

$$\left(\sum x\right)^2 = (57.5)^2 = 3306.25 \text{ and } \left(\sum y\right)^2 = (3630)^2 = 13,176,900$$

$$r = \frac{19(8381.4) - (57.5)(3630)}{\sqrt{19(287.39) - 3306.25}\sqrt{19(777,726) - 13,176,900}} = \frac{-49,478.4}{\sqrt{2154.16}\sqrt{1599894}} \approx -0.84$$

This value for r is fairly close to -1 and indicates a strong negative correlation. This means the more a person drinks, the less likely the person is to die from heart disease.

3. $$m = \frac{19(8381.4) - (57.5)(3630)}{19(287.39) - 3306.25} = \frac{-49,478.4}{2154.16} \approx -22.97$$

$$b = \frac{3630 - (-22.97)(57.5)}{19} = \frac{4950.775}{19} \approx 260.56$$

The equation of the regression line is $y = -22.97x + 260.56$.

The predicted heart disease death rate in a country where adults average 10 liters of alcohol per person per year can be found by substituting 10 for x.

$$y = -22.97x + 260.56$$
$$= -22.97(10) + 260.56$$
$$= 30.86$$

4. Yes, $|r| = 0.84$. Since $0.84 > 0.456$ and 0.575 (using table 12.16), we may conclude that a correlation does exist.

Exercise Set 12.5

1. There appears to be a positive correlation.

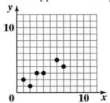

3. There appears to be a negative correlation.

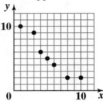

5. There appears to be a positive correlation.

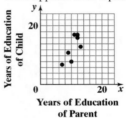

7. There appears to be a positive correlation.

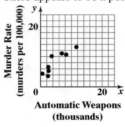

9. False; the correlation is negative.

11. True

13. True

15. False; see for example, Syria and Vietnam.

17. True

19. False; Generally speaking, as per capita income rises, the percentage of people who call themselves "happy" rises.

21. True

23. False; The lowest the lowest level was reported by the Ukraine, yet several countries have lower per capita income (points to the left).

25. False; The correlation is positive, but not that strong..

27. a

29. d

31.

x	y	xy	x^2	y^2
1	2	2	1	4
6	5	30	36	25
4	3	12	16	9
3	3	9	9	9
7	4	28	49	16
2	1	2	4	1

$\sum x = 23 \qquad \sum y = 18 \qquad \sum xy = 83 \qquad \sum x^2 = 115 \qquad \sum y^2 = 64$

$\left(\sum x\right)^2 = (23)^2 = 529$ and $\left(\sum y\right)^2 = (18)^2 = 324$

$r = \dfrac{6(83) - (23)(18)}{\sqrt{6(115) - 529}\sqrt{6(64) - (324)}}$

$= \dfrac{84}{\sqrt{161}\sqrt{60}}$

≈ 0.855

33.

x	y	xy	x^2	y^2
8	2	16	64	4
6	4	24	36	16
1	10	10	1	100
5	5	25	25	25
4	6	24	16	36
10	2	20	100	4
3	9	27	9	81

$\sum x = 37 \qquad \sum y = 38 \qquad \sum xy = 146 \qquad \sum x^2 = 251 \qquad \sum y^2 = 266$

$\left(\sum x\right)^2 = (37)^2 = 1369$ and $\left(\sum y\right)^2 = (38)^2 = 1444$

$r = \dfrac{7(146) - (37)(38)}{\sqrt{7(251) - 1369}\sqrt{7(266) - 1444}}$

$= \dfrac{-384}{\sqrt{388}\sqrt{418}}$

≈ -0.954

35. a.

x	y	xy	x^2	y^2
13	13	169	169	169
9	11	99	81	121
7	7	49	49	49
12	16	192	144	256
12	17	204	144	289
10	8	80	100	64
11	17	187	121	289

$\sum x = 74 \qquad \sum y = 89 \qquad \sum xy = 980 \qquad \sum x^2 = 808 \qquad \sum y^2 = 1237$

$\left(\sum x\right) = (74)^2 = 5476$ and $\left(\sum y\right)^2 = (89)^2 = 7921$

$r = \dfrac{7(980) - (74)(89)}{\sqrt{7(808) - 5476}\sqrt{7(1237) - 7921}}$

$= \dfrac{274}{\sqrt{180}\sqrt{738}}$

≈ 0.75

b. $m = \dfrac{7(980) - (74)(89)}{7(808) - 5476} = \dfrac{274}{180} \approx 1.52$

$b = \dfrac{89 - 1.52(74)}{7} = \dfrac{-23.48}{7} \approx -3.38$

$y = 1.52x - 3.38$

c. $x = 16$

$y = 1.52(16) - 3.38 = 20.98$

21 years

37. a.

x	y	xy	x^2	y^2
11.6	13.1	151.96	134.56	171.61
8.3	10.6	87.98	68.89	112.36
6.9	11.5	79.35	47.61	132.25
3.6	10.1	36.36	12.96	102.01
2.6	5.3	13.78	6.76	28.09
2.5	6.6	16.50	6.25	43.56
2.4	3.6	8.64	5.76	12.96
0.6	4.4	2.64	0.36	19.36

$\sum x = 38.5 \qquad \sum y = 65.2 \qquad \sum xy = 397.21 \qquad \sum x^2 = 283.15 \qquad \sum y^2 = 622.2$

$\left(\sum x\right)^2 = (38.5)^2 = 1482.25$ and $\left(\sum y\right)^2 = (65.2)^2 = 4251.04$

$r = \dfrac{8(397.21) - (38.5)(65.2)}{\sqrt{8(283.15) - 1482.25}\sqrt{8(622.2) - 4251.04}}$

$= \dfrac{667.48}{\sqrt{782.95}\sqrt{726.56}}$

≈ 0.885

b. $m = \dfrac{8(397.21)-(38.5)(65.2)}{8(283.15)-1482.25} = \dfrac{667.48}{782.95} \approx 0.8525 \approx 0.85$

$b = \dfrac{65.2 - 0.8525(38.5)}{8} = \dfrac{32.378}{8} \approx 4.05$

$y = 0.85x + 4.05$

c. $x = 14$
$y = 0.85(14) + 4.05 = 15.95$
16 murders per 100,000 people

39. $|r| = 0.5$

Since $0.5 > 0.444$, conclude that a correlation does exist.

41. $|r| = 0.5$

Since $0.5 < 0.576$, conclude that a correlation does not exist.

43. $|r| = 0.351$

Since $0.351 > 0.232$, conclude that a correlation does exist.

45. $|r| = 0.37$

Since $0.37 < 0.444$, conclude that a correlation does not exist.

Chapter 12 Review Exercises

1. The population is American women ages 25 through 60.
The sample is the 1511 women randomly selected from the population.

2. The variable measured is how often the woman entertains guests for dinner.

3. a

4.

Time Spent on Homework (in hours)	Number of students
6	1
7	3
8	3
9	2
10	1
	10

5.

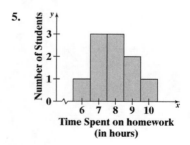

6.

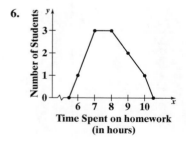

7.

Grades	Number of students
0–39	19
40–49	8
50–59	6
60–69	6
70–79	5
80–89	3
90–100	3
	50

8.

Stems	Leaves
1	3 4 1 3 7 8
2	4 9 6 9 2 7
3	4 9 6 5 1 1 1
4	4 0 2 7 9 1 2 5
5	7 9 6 4 0 1
6	3 3 7 0 8 9
7	2 3 4 0 5
8	7 1 6
9	5 1 0

9. The size of the barrels are not scaled proportionally in terms of the data they represent.

10. $\text{Mean} = \dfrac{84+90+95+89+98}{5}$

$= \dfrac{456}{5}$

$= 91.2$

11. $\text{Mean} = \dfrac{33+27+9+10+6+7+11+23+27}{9}$

$= \dfrac{153}{9}$

$= 17$

12. $\text{Mean} = \dfrac{1\cdot2+2\cdot4+3\cdot3+4\cdot1}{10}$

$= \dfrac{2+8+9+4}{10}$

$= \dfrac{23}{10}$

$= 2.3$

13. First arrange the data items from smallest to largest.
6, 7, 9, 10, $\underline{11}$, 23, 27, 27, 33
There is an odd number of data items, so the median is the middle number. The median is 11.

14. First arrange the data items from smallest to largest.
16, 22, $\underline{28}$, 28, 34
There is an odd number of data items, so the median is the middle number. The median is 28.

15. The median is the value in the
$\dfrac{n+1}{2} = \dfrac{10+1}{2} = \dfrac{11}{2} = 5.5$ position, which means the
median is the mean of the 5th and 6th values. The 5th and 6th values are both 2, therefore the median is 2.

16. The number 27 occurs most frequently, so the mode is 27.

17. Bimodal; 585 and 587 each occur twice.

18. The number 2 occurs most frequently, so the mode is 2.

19. lowest data value = 84, highest data value = 98
$\text{Midrange} = \dfrac{84+98}{2} = \dfrac{182}{2} = 91$

20. lowest data value = 6, highest data value = 33
$\text{Midrange} = \dfrac{6+33}{2} = \dfrac{39}{2} = 19.5$

21. lowest data value = 1, highest data value = 4
$\text{Midrange} = \dfrac{1+4}{2} = \dfrac{5}{2} = 2.5$

24. a.

Age at first inauguration	Number of Presidents
42	1
43	1
44	0
45	0
46	2
47	1
48	1
49	2
50	1
51	5
52	2
53	0
54	5
55	4
56	3
57	4
58	1
59	0
60	1
61	3
62	1
63	0
64	2
65	1
66	0
67	0
68	1
69	1
	43

b.

$$\text{Mean} = \frac{\left(\begin{array}{c} 42 \cdot 1 + 43 \cdot 1 + 46 \cdot 2 + 47 \cdot 1 + 48 \cdot 1 + 49 \cdot 2 + 50 \cdot 1 + 51 \cdot 5 + 52 \cdot 2 + 54 \cdot 5 + 55 \cdot 4 \\ + 56 \cdot 3 + 57 \cdot 4 + 58 \cdot 1 + 60 \cdot 1 + 61 \cdot 3 + 62 \cdot 1 + 64 \cdot 2 + 65 \cdot 1 + 68 \cdot 1 + 69 \cdot 1 \end{array}\right)}{43} = \frac{2358}{43} \approx 54.84 \text{ years}$$

The median is the value in the $\frac{n+1}{2} = \frac{43+1}{2} = \frac{44}{2} = 22$ position, which means the median is 22nd value.

Median = 55 years

The model ages are 51 and 54 years (bimodal).

Midrange = $\frac{42+69}{2} = 55.5$ years

25. Range = 34 − 16 = 18

26. Range = 783 − 219 = 564

27. a.

Data item	Deviation: Data item – mean
29	$29 - 35 = -6$
9	$9 - 35 = -26$
8	$8 - 35 = -27$
22	$22 - 35 = -13$
46	$46 - 35 = 11$
51	$51 - 35 = 16$
48	$48 - 35 = 13$
42	$42 - 35 = 7$
53	$53 - 35 = 18$
42	$42 - 35 = 7$

b. $-6 - 26 - 27 - 13 + 11 + 16 + 13 + 7 + 18 + 7 = 0$

28. a. $\text{Mean} = \dfrac{36 + 26 + 24 + 90 + 74}{5} = \dfrac{250}{5} = 50$

b.

Data item	Deviation: Data item – mean
36	$36 - 50 = -14$
26	$26 - 50 = -24$
24	$24 - 50 = -26$
90	$90 - 50 = 40$
74	$74 - 50 = 24$

c. $-14 - 24 - 26 + 40 + 24 = 0$

29. $\text{Mean} = \dfrac{3 + 3 + 5 + 8 + 10 + 13}{6} = \dfrac{42}{6} = 7$

Data item	Deviation: Data item – mean	$(\text{Deviation})^2$: $(\text{Data item–mean})^2$
3	$3 - 7 = -4$	$(-4)^2 = 16$
3	$3 - 7 = -4$	$(-4)^2 = 16$
5	$5 - 7 = -2$	$(-2)^2 = 4$
8	$8 - 7 = 1$	$1^2 = 1$
10	$10 - 7 = 3$	$3^2 = 9$
13	$13 - 7 = 6$	$6^2 = 36$

$$\sum (\text{data item–mean})^2 = 82$$

$$\text{Standard deviation} = \sqrt{\frac{82}{6-1}} = \sqrt{\frac{82}{5}} \approx 4.05$$

30. Mean = $\dfrac{20+27+23+26+28+32+33+35}{8} = \dfrac{224}{8} = 28$

Data item	Deviation: Data item − mean	(Deviation)^2 : (Data item−mean)^2
20	$20 - 28 = -8$	$(-8)^2 = 64$
27	$27 - 28 = -1$	$(-1)^2 = 1$
23	$23 - 28 = -5$	$(-5)^2 = 25$
26	$26 - 28 = -2$	$(-2)^2 = 4$
28	$28 - 28 = 0$	$0^2 = 0$
32	$32 - 28 = 4$	$4^2 = 16$
33	$33 - 28 = 5$	$5^2 = 25$
35	$35 - 28 = 7$	$7^2 = 49$

$\sum (\text{data item−mean})^2 = 184$

Standard deviation = $\sqrt{\dfrac{184}{8-1}} = \sqrt{\dfrac{184}{7}} \approx 5.13$

31. Mean = $\dfrac{10+30+37+40+43+44+45+69+86+86}{10} = \dfrac{490}{10} = 49$

Range = $86 - 10 = 76$

Data item	Deviation: Data item − mean	(Deviation)^2 : (Data item−mean)^2
10	$10 - 49 = -39$	$(-39)^2 = 1521$
30	$30 - 49 = -19$	$(-19)^2 = 361$
37	$37 - 49 = -12$	$(-12)^2 = 144$
40	$40 - 49 = -9$	$(-9)^2 = 81$
43	$43 - 49 = -6$	$(-6)^2 = 36$
44	$44 - 49 = -5$	$(-5)^2 = 25$
45	$45 - 49 = -4$	$(-4)^2 = 16$
69	$69 - 49 = 20$	$20^2 = 400$
86	$86 - 49 = 37$	$37^2 = 1369$
86	$86 - 49 = 37$	$37^2 = 1369$

$\sum (\text{data item−mean})^2 = 5322$

Standard deviation = $\sqrt{\dfrac{5322}{10-1}} = \sqrt{\dfrac{5322}{9}} \approx 24.32$

32. Set A:

$$\text{Mean} = \frac{80 + 80 + 80 + 80}{4} = \frac{320}{4} = 80$$

Data item	Deviation: Data item − mean	(Deviation)^2 : (Data item−mean)^2
80	$80 - 80 = 0$	$0^2 = 0$
80	$80 - 80 = 0$	$0^2 = 0$
80	$80 - 80 = 0$	$0^2 = 0$
80	$80 - 80 = 0$	$0^2 = 0$
		$\sum(\text{data item−mean})^2 = 0$

$$\text{Standard deviation} = \sqrt{\frac{0}{4-1}} = \sqrt{\frac{0}{3}} = 0$$

Set B:

$$\text{Mean} = \frac{70 + 70 + 90 + 90}{4} = \frac{320}{4} = 80$$

Data item	Deviation: Data item − mean	(Deviation)^2 : (Data item−mean)^2
70	$70 - 80 = -10$	$(-10)^2 = 100$
70	$70 - 80 = -10$	$(-10)^2 = 100$
90	$90 - 80 = 10$	$10^2 = 100$
90	$90 - 80 = 10$	$10^2 = 100$
		$\sum(\text{data item−mean})^2 = 400$

$$\text{Standard deviation} = \sqrt{\frac{400}{4-1}} = \sqrt{\frac{400}{3}} \approx 11.55$$

Written descriptions of the similarities and differences between the two sets of data will vary.

33. Answers will vary.

34. $70 + 2 \cdot 8 = 70 + 16 = 86$

35. $70 + 3.5(8) = 70 + 28 = 98$

36. $70 - 1.25(8) = 70 - 10 = 60$

37. 64 is one standard deviation below the mean and 72 is one standard deviation above the mean, so 68% of the people in the retirement community are between 64 and 72 years old.

38. 60 is two standard deviations below the mean and 76 is two standard deviations above the mean, so 95% of the people in the retirement community are between 60 and 76 years old.

39. 68 is the mean and 72 is one standard deviation above the mean, so half of 68%, or 34% of the people in the retirement community are between 68 and 72 years old.

40. 56 is three standard deviations below the mean and 80 is three standard deviations above the mean, so 99.7% of the people in the retirement community are between 56 and 80 years old.

41. 72 is one standard deviation above the mean, so 16% of the people in the retirement community are over 72 years old. (Note: 100% − 68% = 32%, half of 32% is 16%).

42. 72 is one standard deviation above the mean, so 84% of the people in the retirement community are under 72 years old. (Note: Question #41 showed that 16% is above 72, 100% − 16% = 84%)

43. 76 is two standard deviations above the mean, so 2.5% of the people in the retirement community are over 76 years old. (Note: 100% − 95% = 5%, half of 5% is 2.5%).

44. $z_{50} = \dfrac{50-50}{5} = \dfrac{0}{5} = 0$

45. $z_{60} = \dfrac{60-50}{5} = \dfrac{10}{5} = 2$

46. $z_{58} = \dfrac{58-50}{5} = \dfrac{8}{5} = 1.6$

47. $z_{35} = \dfrac{35-50}{5} = \dfrac{-15}{5} = -3$

48. $z_{44} = \dfrac{44-50}{5} = \dfrac{-6}{5} = -1.2$

49. vocabulary test: $z_{60} = \dfrac{60-50}{5} = \dfrac{10}{5} = 2$

grammar test: $z_{80} = \dfrac{80-72}{6} = \dfrac{8}{6} \approx 1.3$

The student scored better on the vocabulary test because it has a higher z-score.

50. $1.5(4000) = 6000$
$32{,}000 + 6000 = 38{,}000$ miles

51. $2.25(4000) = 9000$
$32{,}000 + 9000 = 41{,}000$ miles

52. $-2.5(4000) = -10{,}000$
$32{,}000 - 10{,}000 = 22{,}000$ miles

53. $z_{221} = \dfrac{221-200}{15} = \dfrac{21}{15} = 1.4$
$z = 1.4 \rightarrow 91.92\%$
91.92% have cholesterol less than 221.

54. $z_{173} = \dfrac{173-200}{15} = \dfrac{-27}{15} = -1.8$
$z = -1.8 \rightarrow 3.59\%$
$100\% - 3.59\% = 96.41\%$ have cholesterol greater than 173.

55. $z_{173} = \dfrac{173-200}{15} = \dfrac{-27}{15} = -1.8$
and $z = -1.8 \rightarrow 3.59\%$
$z_{221} = \dfrac{221-200}{15} = \dfrac{21}{15} = 1.4$
and $z = 1.4 \rightarrow 91.92\%$
$91.92\% - 3.59\% = 88.33\%$ have cholesterol between 173 and 221.

56. $z_{164} = \dfrac{164-200}{15} = \dfrac{-36}{15} = -2.4$
and $z = -2.4 \rightarrow 0.82\%$
$z_{182} = \dfrac{182-200}{15} = \dfrac{-18}{15} = -1.2$
and $z = -1.2 \rightarrow 11.51\%$
$11.51\% - 0.82\% = 10.69\%$ have cholesterol between 164 and 182.

57. 75%

58. $100\% - 86\% = 14\%$

59. $86\% - 75\% = 11\%$

60. a. Margin of error $= \pm\dfrac{1}{\sqrt{n}} = \pm\dfrac{1}{\sqrt{2041}}$
$\approx \pm0.022$
$= \pm2.2\%$

b. $15\% - 2.2\% = 12.8\%$ and $15\% + 2.2\% = 17.2\%$
There is a 95% probability that the true population percentage lies between 12.8% and 17.2%. We can be 95% confident that between 12.8% and 17.2% of executives at American companies have careers not related to their college degree.

61. a. The graph is skewed to the right.

b.

x	f	xf
1	36	$1 \cdot 36 = 36$
2	34	$2 \cdot 34 = 68$
3	18	$3 \cdot 18 = 54$
4	9	$4 \cdot 9 = 36$
5	2	$5 \cdot 2 = 10$
6	1	$6 \cdot 1 = 6$
	100	$\sum xf = 210$

$$\text{Mean} = \frac{\sum xf}{n} = \frac{210}{100} = 2.1 \text{ syllables}$$

The median is the mean of the 50th and 51st positions. Since these data items are both 2, the median is 2 syllables.
The mode is 1 syllable.

c. Yes, these measures of central tendency are consistent with the graph. The mean is greater than the median, which is expected with a distribution that is skewed to the right.

62. There appears to be a positive correlation.

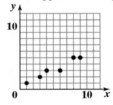

63. There appears to be a negative correlation.

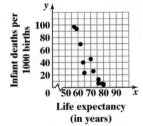

64. False; the correlation is only moderate.

65. True

66. False

67. False; data points that are vertically aligned dispute this statement.

68. True

69. False; there is a moderate negative correlation.

70. True

71. c

72. a.

x	y	xy	x^2	y^2
1	1	1	1	1
3	2	6	9	4
4	3	12	16	9
6	3	18	36	9
8	5	40	64	25
9	5	45	81	25
$\sum x = 31$	$\sum y = 19$	$\sum xy = 122$	$\sum x^2 = 207$	$\sum y^2 = 73$

$$\left(\sum x\right)^2 = (31)^2 = 961 \text{ and } \left(\sum y\right)^2 = (19)^2 = 361$$

$$r = \frac{6(122) - (31)(19)}{\sqrt{6(207) - 961}\sqrt{6(73) - 361}} = \frac{143}{\sqrt{281}\sqrt{77}} \approx 0.972$$

b.
$$m = \frac{6(122) - (31)(19)}{6(207) - 961} = \frac{143}{281} \approx 0.509$$

$$b = \frac{19 - (0.509)(31)}{6} = \frac{3.221}{6} \approx 0.537$$

$$y = 0.509x + 0.537$$

73. a.

x	y	xy	x^2	y^2
22	26	1	1	1
32	32	6	9	4
42	34	12	16	9
52	39	18	36	9
62	44	45	81	25

$$\sum x = 210 \qquad \sum y = 175 \qquad \sum xy = 7780 \qquad \sum x^2 = 9820 \qquad \sum y^2 = 6313$$

$$\left(\sum x\right)^2 = (210)^2 = 44,100 \text{ and } \left(\sum y\right)^2 = (175)^2 = 30,625$$

$$r = \frac{5(7780) - (210)(175)}{\sqrt{5(9820) - 44,100}\sqrt{5(6313) - 30,625}} = \frac{2150}{\sqrt{5000}\sqrt{940}} \approx 0.99$$

b.   There is a correlation.

Chapter 12 Test

1. d

2.

Score	Frequency
3	1
4	2
5	3
6	2
7	2
8	3
9	2
10	1
	16

3.

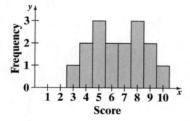

4.

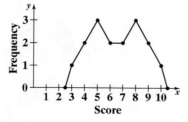

5.

Class	Frequency
40–49	3
50–59	6
60–69	6
70–79	7
80–89	6
90–99	2
	30

6.

Stems	Leaves
4	1 8 6
5	9 1 0 5 0 0
6	2 3 7 0 1 1
7	9 3 1 5 8 9 1
8	8 9 9 1 3 0
9	0 3

7. The roofline gives the impression that the percentage of home schooled students grew at the same rate each year between the years shown. This may be misleading if the growth rate was not constant from year to year.

8. Mean $= \dfrac{3+6+2+1+7+3}{6} = \dfrac{22}{6} \approx 3.67$

9. First arrange the numbers from smallest to largest.
1, 2, 3, 3, 6, 7
There is an even number of data items, so the median is the mean of the middle two data values.

Median $= \dfrac{3+3}{2} = \dfrac{6}{2} = 3$

10. lowest data value $= 1$
highest data value $= 7$

Midrange $= \dfrac{1+7}{2} = \dfrac{8}{2} = 4$

11.

Data item	Deviation: Data item – mean	$(\text{Deviation})^2$: $(\text{Data item–mean})^2$
3	$3 - 3.7 = -0.7$	$(-0.7)^2 = 0.49$
6	$6 - 3.7 = 2.3$	$(2.3)^2 = 5.29$
2	$2 - 3.7 = -1.7$	$(-1.7)^2 = 2.89$
1	$1 - 3.7 = -2.7$	$(-2.7)^2 = 7.29$
7	$7 - 3.7 = 3.3$	$(3.3)^2 = 10.89$
3	$3 - 3.7 = -0.7$	$(-0.7)^2 = 0.49$

$\sum (\text{data item–mean})^2 = 27.34$

Standard deviation $= \sqrt{\dfrac{27.34}{6-1}} = \sqrt{\dfrac{27.34}{5}} \approx 2.34$

12. Mean $= \dfrac{1 \cdot 3 + 2 \cdot 5 + 3 \cdot 2 + 4 \cdot 2}{12}$

$\qquad = \dfrac{3 + 10 + 6 + 8}{12}$

$\qquad = \dfrac{27}{12}$

$\qquad = 2.25$

13. The median is in the $\dfrac{n+1}{2} = \dfrac{12+1}{2} = \dfrac{13}{2} = 6.5$ position, which means the median is the mean of the values in the 6th and 7th positions.

$\qquad$ Median $= \dfrac{2+2}{2} = \dfrac{4}{2} = 2$

14. Mode $= 2$

15. Answers will vary.

16. $7 + 1(5.3) = 12.3$

$\qquad$ 68% of the data values are within 1 standard deviation of the mean. Because of symmetry, $\dfrac{1}{2}(68\%) = 34\%$ of college freshmen study between 7 and 12.3 hours per week.

17. $7 + 2(5.3) = 17.6$

$\qquad$ 95% of the data values are within 2 standard deviations of the mean. $100\% - 95\% = 5\%$ of the values are farther than 2 standard deviations from the mean. Because of symmetry, $\dfrac{1}{2}(5\%) = 2.5\%$ of college freshmen study more than 17.6 hours per week.

18. student: $z_{120} = \dfrac{120 - 100}{10} = \dfrac{20}{10} = 2$

$\qquad$ professor: $z_{128} = \dfrac{128 - 100}{15} = \dfrac{28}{15} \approx 1.9$

$\qquad$ The student scored better, because the student's z-score is higher.

19. $z_{88} = \dfrac{88 - 74}{10} = \dfrac{14}{10} = 1.4$

$\qquad z = 1.4 \rightarrow 91.92\%$

$\qquad 100\% - 91.92\% = 8.08\%$ of the scores are above 88.

20. $49\% - 8\% = 41\%$

21. a.   margin of error $= \pm \dfrac{1}{\sqrt{n}}$

$= \pm \dfrac{1}{\sqrt{100}}$

$= \pm 0.1$

$= \pm 10\%$

b.   We can be 95% confident that between 50% and 70% of all students are very satisfied with their professors.

22.   There appears to be a strong negative correlation.

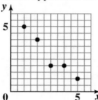

23.   False; Though the data shows that there is a <u>correlation</u>, it does not prove <u>causation</u>.

24.   False

25.   True

26.   Answers will vary.

Chapter 13
Mathematical Systems

Check Points 13.1

1. $O + O = E$. This means that the sum of two odd numbers is an even number.

2. No, the set is not closed under addition. Example: 2 + 2 = 4; 4 is not an element of the set.

3. Yes, the natural numbers are closed under multiplication.

4. No, the natural numbers are not closed under division. Example: 2 divided by 3 is not a natural number.

5. We must show that $(1 \oplus 3) \oplus 2 = 1 \oplus (3 \oplus 2)$.
$$(1 \oplus 3) \oplus 2 = 1 \oplus (3 \oplus 2)$$
$$0 \oplus 2 = 1 \oplus 1$$
$$2 = 2$$

6. The identity element is g, because it does not change anything.

7. The inverse is -12. (because $12 + (-12) = 0$)

8. The inverse is $\dfrac{1}{12}$. (because $12 \cdot \dfrac{1}{12} = 1$)

9. a. The identity element is k, because it does not change anything.

b. The inverse of j is l because $j \circ l = k$
The inverse of k is k because $k \circ k = k$
The inverse of l is j because $l \circ j = k$
The inverse of m is m because $m \circ m = k$

Exercise Set 13.1

1. $\{e, a, b, c\}$

3. $a \circ b = c$

5. $b \circ c = a$

7. $e \circ e = e$

9. $b \circ e = b$

11. $a \circ a = e$

13. $(c \circ c) \circ a = e \circ a = a$

15. No. For example, $4 + 1 = 5$, and 5 is not in the set.

17. Yes. The answer to any possible combination of two elements of the set is an element that is in the set.

19. No. For example, $1 - 2 = -1$, and -1 is not in the set.

21. No, the system is not closed under $*$. $b * a = c$, and c is not in the set.

23. $2 \oplus 4 = 1$
$4 \oplus 2 = 1$
So $2 \oplus 4 = 4 \oplus 2$

25. $4 \oplus 1 = 0$
$1 \oplus 4 = 0$
So $4 \oplus 1 = 1 \oplus 4$

27. The Commutative Property; since table entries are mirror images of each other across the main diagonal, $\oplus$ is commutative.

29. $(4 \oplus 3) \oplus 2 = 2 \oplus 2 = 4$
$4 \oplus (3 \oplus 2) = 4 \oplus 0 = 4$
$(4 \oplus 3) \oplus 2 = 4 \oplus (3 \oplus 2)$

31. $(4 \oplus 4) \oplus 2 = 3 \oplus 2 = 0$
$4 \oplus (4 \oplus 2) = 4 \oplus 1 = 0$
$(4 \oplus 4) \oplus 2 = 4 \oplus (4 \oplus 2)$

33. a. $(b \circ c) \circ b = c \circ b = b$

b. $b \circ (c \circ b) = b \circ b = a$

c. No. $(b \circ c) \circ b \neq b \circ (c \circ b)$
When the grouping changed, the answer changed.

35. Answers will vary. Sample answer:
$(5 - 2) - 1 = 3 - 1 = 2$ but $5 - (2 - 1) = 5 - 1 = 4$

37. $29 + (-29) = 0$
-29 is the inverse of 29 under the operation of addition.

39. $29 \cdot \dfrac{1}{29} = 1$

$\dfrac{1}{29}$ is the inverse of 29 under the operation of multiplication.

41. $a \circ a = a$

43. $a \circ c = c$

45. $a \circ e = e$

47. $c \circ a = c$

49. $e \circ a = e$

51. a

53. d

55. b

57. 0 is the identity element.

59. $1 \oplus 4 = 0$
4 is the inverse for 1.

61. $3 \oplus 2 = 0$
2 is the inverse for 3.

63. **a.** d is the identity element.

b.

element	inverse
a	none
b	none
c	none
d	d

65. **a.** $e_\triangle(c \square d) = (e_\triangle c)\square(e_\triangle d)$
$e_\triangle a = d \square c$
$a = a$

b. distributive property

67. $c_\triangle\big[c\square(c_\triangle c)\big] = c_\triangle\big[c \square e\big]$
$= c_\triangle b$
$= c$

69. $x \square d = e$ is true if $x = b$.

71. $x_\triangle(e \square c) = d$
$x_\triangle b = d$ is true if $x = d$.

73. The mathematical system shows all possible starting and ending positions for the four-way switch. The 25 entries in the table represent the final position of the switch for each starting position (shown in the left column) and number of clockwise turns (shown across the top).

75–79.

$\times$	E	O
E	E	E
O	E	O

75. Yes. The answer to any possible combination of two elements of the set is an element that is in the set.

77. $(O \times E) \times O = E$ and $O \times (E \times O) = E$

79. E does not have an inverse.

89. **a.** $\begin{bmatrix} 2 & 3 \\ 4 & 7 \end{bmatrix} \times \begin{bmatrix} 0 & 1 \\ 5 & 6 \end{bmatrix}$

$= \begin{bmatrix} 2\cdot0+3\cdot5 & 2\cdot1+3\cdot6 \\ 4\cdot0+7\cdot5 & 4\cdot1+7\cdot6 \end{bmatrix}$

$= \begin{bmatrix} 0+15 & 2+18 \\ 0+35 & 4+42 \end{bmatrix}$

$= \begin{bmatrix} 15 & 20 \\ 35 & 46 \end{bmatrix}$

b. $\begin{bmatrix} 0 & 1 \\ 5 & 6 \end{bmatrix} \times \begin{bmatrix} 2 & 3 \\ 4 & 7 \end{bmatrix}$

$= \begin{bmatrix} 0\cdot2+1\cdot4 & 0\cdot3+1\cdot7 \\ 5\cdot2+6\cdot4 & 5\cdot3+6\cdot7 \end{bmatrix}$

$= \begin{bmatrix} 0+4 & 0+7 \\ 10+24 & 15+42 \end{bmatrix}$

$= \begin{bmatrix} 4 & 7 \\ 34 & 57 \end{bmatrix}$

c. Matrix multiplication is not commutative.

Check Points 13.2

1. a. 1. The set is closed under the binary operation because the entries in the body of the table are all elements of the set.

2. Associative Property: For example,
$$(O \circ E) \circ O = O \circ (E \circ O)$$
$$O \circ O = O \circ O$$
$$E = E$$
and
$$(E \circ O) \circ O = E \circ (O \circ O)$$
$$O \circ O = E \circ E$$
$$E = E$$

3. E is the identity element.

4.

element	inverse
E	E
O	O

Each element has an inverse.
Since the system meets the four requirements, the system is a group.

b. The Commutative Property holds for this group (as can be seen by the symmetry along the diagonal from the upper left to lower right). Therefore, this system is a commutative group.

2. a. $(8+5)+11 = 8+(5+11)$
$$1+11 = 8+4$$
$$0 = 0$$

b. Locating 9 on the left and 4 across the top indicates that $9+4 = 1$.
Locating 4 on the left and 9 across the top indicates that $4+9 = 1$.

3. a. true; $61 \equiv 5 \pmod 7$ because
$61 \div 7 = 8$, remainder 5.

b. true; $36 \equiv 0 \pmod 6$ because
$36 \div 6 = 6$, remainder 0.

c. false; $57 \equiv 2 \pmod{11}$ because
$57 \div 11 = 5$, remainder 2 (not 3).

4. a. $(1+3)(\bmod 5) \equiv 4(\bmod 5)$

b. $(5+4)(\bmod 7) \equiv 9(\bmod 7) \equiv 2(\bmod 7)$

c. $(8+10)(\bmod 13) \equiv 18(\bmod 13) \equiv 5(\bmod 13)$

5. $97 \equiv 6(\bmod 7)$ thus, the desired day of the week is 6 days past Wednesday, or Tuesday.

Exercise Set 13.2

1. 8-fold rotational symmetry.

3. 18-fold rotational symmetry.

5. For any 2 elements in the set, the result is also in the set.

7. $(r \circ t) \circ q = r \circ (t \circ q)$
$$p \circ q = r \circ r$$
$$e = e$$

9. e is the identity element.

11. $p \circ q = e$
q is the inverse of p.

13. $r \circ r = e$
r is the inverse of r.

15. $t \circ t = e$
t is the inverse of t.

17. $r \circ p = t$

19. This mathematical system is not commutative.

21. Most elements do not have an inverse. For example, no natural number will satisfy the expression $2 \times ? = 1$.

23. a.

+	0	1	2	3	4	5
0	0	1	2	3	4	5
1	1	2	3	4	5	0
2	2	3	4	5	0	1
3	3	4	5	0	1	2
4	4	5	0	1	2	3
5	5	0	1	2	3	4

b. 1. The set is closed under the operation of clock addition because the entries in the body of the table are all elements of the set.

2. Associative Property: For example,
$(2+3)+4 = 2+(3+4)$
$5+4 = 2+1$
$3 = 3$
and
$(3+4)+0 = 3+(4+0)$
$1+0 = 3+4$
$1 = 1$

3. 0 is the identity element.

4.

element	inverse
0	0
1	5
2	4
3	3
4	2
5	1

Each element has an inverse.

5. The table is symmetric, so the Commutative Property holds. Therefore, this system is a commutative group.

25. $7 \equiv 2 \pmod 5$
$7 \div 5 = 1$, remainder 2
True

27. $41 \equiv 6 \pmod 7$
$41 \div 7 = 5$, remainder 6
True

29. $84 \equiv 1 \pmod 7$
$84 \div 7 = 12$, remainder 0
False
A true statement is $84 \equiv 0 \pmod 7$

31. $23 \equiv 2 \pmod 4$
$23 \div 4 = 5$, remainder 3
False
A true statement is $23 \equiv 3 \pmod 4$

33. $55 \equiv 0 \pmod{11}$
$55 \div 11 = 5$, remainder 0
True

35. $(3+2)\pmod 6$
$3+2 = 5,\ 5 < 6$
$3+2 \equiv 5 \pmod 6$

37. $(4+5)\pmod 6$
$4+5 = 9,\ 9 > 6$
$9 \div 6 = 1$, remainder 3
$4+5 \equiv 3 \pmod 6$

39. $(6+5)\pmod 7$
$6+5 = 11,\ 11 > 7$
$11 \div 7 = 1$, remainder 4
$6+5 \equiv 4 \pmod 7$

41. $(49+49)\pmod 5$
$49+49 = 98,\ 98 > 50$
$98 \div 50 = 1$, remainder 48
$49+49 \equiv 48 \pmod{50}$

43. $(5 \times 7)\pmod{12} \equiv 35 \pmod{12}$
$\equiv 11 \pmod{12}$

45. a. $[3 \times (4+5)]\pmod 7 \equiv 27 \pmod 7$
$\equiv 6 \pmod 7$

b. $[(3 \times 4) + (3 \times 5)]\pmod 7 \equiv 27 \pmod 7$
$\equiv 6 \pmod 7$

c. distributive property

47. $3x \equiv 1 \pmod 5$ is true for $x = 2$.
$3(2) \equiv 1 \pmod 5$
$6 \equiv 1 \pmod 5$

49. $3x \equiv 3 \pmod 6$ is true for $x = 1, 3$, and 5.

$3(1) \equiv 3 \pmod 6$
$3 \equiv 3 \pmod 6$

$3(3) \equiv 3 \pmod 6$
$9 \equiv 3 \pmod 6$

$3(5) \equiv 3 \pmod 6$
$15 \equiv 3 \pmod 6$

51. $4x \equiv 5 \pmod 8$ is false for all values of x.
No replacements exist.

53. $(1200 + 0600) \pmod{2400}$
$1200 + 0600 = 1800, \; 1800 < 2400$
$1200 + 0600 \equiv 1800 \pmod{2400}$

55. $(0830 + 1550) \pmod{2400}$
$0830 + 1550 = 2380,$
$23 \text{ hr } 80 \text{ min } \equiv 24 \text{ hr } 20 \text{ min}$
$2420 \equiv 0020 \pmod{2400}$

57. $67 \div 7 = 9$, remainder 4
$67 \equiv 4 \pmod 7$
Thus, the desired day of the week is 4 days past Wednesday, or Sunday.

59. *Beam me up*

Code:	9	12	8	20	7	20	12	7	1	23
Add 20:	29	32	28	40	27	40	32	27	21	43
Mod 27:	2	5	1	13	0	13	5	0	21	16
Letter:	B	E	A	M	_	M	E	_	U	P

61. $0 + 2(1) + 3(3) + 4(2) + 5(1) + 6(9) + 7(1) + 8(4) + 9(1) = 126$
Since $126 \equiv 5 \pmod{11}$ and the last digit is 5, the number is a valid ISBN.

63. $2 + 2(4) + 3(1) + 4(9) + 5(7) + 6(1) + 7(3) + 8(2) + 9(9) = 208$
Since $208 \equiv 10 \pmod{11}$ and the last digit is 0, the number is a not a valid ISBN.

65. $0 + 2(1) + 3(3) + 4(2) + 5(2) + 6(5) + 7(9) + 8(8) + 9(0) = 186$
Since $186 \equiv 10 \pmod{11}$ and the last digit is X, the number is a valid ISBN.

67. $\left[3(2+2+9+1+3+4) + (5+9+8+7+9+9) \right] = 110$
Since $110 \equiv 0 \pmod{10}$ the number is a valid universal product number.

69. $\left[3(0+4+0+1+5+9) + (6+2+0+1+8+7) \right] = 81$
Since $81 \equiv 1 \pmod{10}$ the number is not a valid universal product number.

79. Example involving associative property:
$$(2 \circ 3) \circ 5 \neq 2 \circ (3 \circ 5)$$
$$(2 - 3 + 2 \cdot 3) \circ 5 \neq 2 \circ (3 - 5 + 3 \cdot 5)$$
$$5 \circ 5 \neq 2 \circ 13$$
$$5 - 5 + 5 \cdot 5 \neq 2 - 13 + 2 \cdot 13$$
$$25 \neq 15$$
Since the associative property is not satisfied, the mathematical system is not a group.

81. $99,999,999 \div 24 = 4,166,666$, remainder 15
$99,999,999 = 15 \pmod{24}$
It will be 15 hours past 5:00 P.M., or 8:00 A.M.

Chapter 13 Review Exercises

1. $\{e, c, f, r\}$

2. Yes. Any possible combination of two elements of the set is an element of the set.

3. $c \circ f = r$

4. $r \circ r = f$

5. $e \circ c = c$

6. $c \circ r = e$
   $r \circ c = e$

7. $f \circ r = c$
   $r \circ f = c$

8. $f \circ e = f$
   $e \circ f = f$

9. The Commutative Property

10. Since the table entries are symmetric about the main diagonal, the operation is commutative.

11. $(c \circ r) \circ f = e \circ f = f$
    $c \circ (r \circ f) = c \circ c = f$

12. $(r \circ e) \circ c = r \circ c = e$
    $r \circ (e \circ c) = r \circ c = e$

13. The Associative Property

14. e is the identity element.

15. $e \circ e = e$
    e is the inverse for e.

16. $c \circ r = e$
    r is the inverse for c.

17. $f \circ f = e$
    f is the inverse for f.

18. $r \circ c = e$
    c is the inverse for r.

19. No. $1 + 1 = 2$, and 2 is not in the set.

20. Yes. Any possible combination of two elements of the set is an element of the set.

21. No. For example, $1 \div 2 = \dfrac{1}{2}$, and $\dfrac{1}{2}$ is not in the set.

22. $123 + (-123) = 0$
    -123 is the additive inverse of 123.

23. $123 \cdot \dfrac{1}{123} = 1$
    $\dfrac{1}{123}$ is the multiplicative inverse of 123.

24. a.

$\circ$	0	1	2	3	4
0	0	1	2	3	4
1	1	1	2	3	4
2	2	2	2	3	4
3	3	3	3	3	4
4	4	4	4	4	4

   b. Zero is the identity element.

   c. No. There is no element in the set such that $2 \circ ? = 0$.

25. 3-fold rotational symmetry

26. 18-fold rotational symmetry

27. a.

$+$	0	1	2	3	4
0	0	1	2	3	4
1	1	2	3	4	0
2	2	3	4	0	1
3	3	4	0	1	2
4	4	0	1	2	3

   b. 1. The set is closed under the operation of clock addition because the entries in the body of the table are all elements of the set.

      2. Associative Property: For example,
      $$(1+2)+3 = 1+(2+3)$$
      $$3+3 = 1+0$$
      $$1 = 1$$

      3. 0 is the identity element.

4.

element	inverse
0	0
1	4
2	3
3	2
4	1

Each element has an inverse.

5. The table is symmetric, so the Commutative Property holds. Therefore, this system is a commutative group.

28. $17 \equiv 2 \pmod 8$

$17 \div 8 = 2,$ remainder 1
False
A true statement is $17 \equiv 1 \pmod 8$.

29. $37 \equiv 3 \pmod 5$

$37 \div 5 = 7,$ remainder 2
False
A true statement is $37 \equiv 2 \pmod 5$.

30. $60 \equiv 0 \pmod{10}$

$60 \div 10 = 6,$ remainder 0
True

31. $(4+3) \pmod 6$

$4+3 = 7, 7 > 6$
$7 \div 6 = 1,$ remainder 1
$4+3 \equiv 1 \pmod 6$

32. $(7+7) \pmod 8$

$7+7 = 14, 14 > 8$
$14 \div 8 = 1,$ remainder 6
$7+7 \equiv 6 \pmod 8$

33. $(4+3) \pmod 9$

$4+3 = 7, 7 < 9$
$4+3 = 7 \pmod 9$

34. $(3+18) \pmod{20}$

$3+18 = 21, 21 > 20$
$21 \div 20 = 1,$ remainder 1
$3+18 \equiv 1 \pmod{20}$

Chapter 13 Test

1. Yes. Any possible combination of two elements of the set is an element of the set.

2. $z \circ y = x$

$y \circ z = x$
This illustrates the Commutative Property.

3. $(x \circ z) \circ z = z \circ z = y$

$x \circ (z \circ z) = x \circ y = y$
This illustrates the Associative Property.

4. x is the identity element.

5.

element	inverse
x	x
y	z
z	y

6. No. For example, $1 + 1 = 2$, and 2 is not in the set.

7. $5 \cdot \dfrac{1}{5} = 1$

$\dfrac{1}{5}$ is the multiplicative inverse of 5.

8. 6-fold rotational symmetry; Since it takes 6 equal turns to restore the design to its original position and each of these turns is a design that is identical to the original, the design has 6-fold rotational symmetry.

9.

+	0	1	2	3
0	0	1	2	3
1	1	2	3	0
2	2	3	0	1
3	3	0	1	2

10. 1. The set is closed under the operation of clock addition because the entries in the body of the table are all elements of the set.

  2. Associative Property:
      For example,
      $(1+2)+3 = 1+(2+3)$
      $3+3 = 1+1$
      $2 = 2$

  3. 0 is the identity element.

  4.

element	inverse
0	0
1	3
2	2
3	1

      Each element has an inverse.

  5. The table is symmetric, so the Commutative Property holds. Therefore, this system is a commutative group.

11. $39 \equiv 3 \ (\text{mod} \, 6)$

  $39 \div 6 = 6$, remainder 3

  True

12. $14 \equiv 2 \ (\text{mod} \, 7)$

  $14 \div 7 = 2$, remainder 0

  False

  A true statement is $14 \equiv 0 \ (\text{mod} \, 7)$

13. $(9+1) \ (\text{mod} \, 11)$

  $9+1 = 10, \ 10 < 11$

  $9+1 \equiv 10 \ (\text{mod} \, 11)$

14. $(9+6) \ (\text{mod} \, 10)$

  $9+6 = 15, \ 15 > 10$

  $15 \div 10 = 1$, remainder 5

  $9+6 \equiv 5 \ (\text{mod} \, 10)$

Chapter 14
Voting and Apportionment

Check Points 14.1

1. **a.** We find the number of people who voted in the election by adding the numbers in the row labeled Number of Votes: 2100 + 1305 + 765 + 40 = 4210. Thus, 4210 people voted in the election.

   b. We find how many people selected the candidates in the order B, S, A, C by referring to the fourth column of letters in the preference table. Above this column is the number 40. Thus, 40 people voted in the order B, S, A, C.

   c. We find the number of people who selected S as their first choice by reading across the row that says First Choice: 2100 + 765 = 2865. Thus, 2865 students selected S (Samir) as their first choice for student body president.

2. The candidate with the most first-place votes is the winner. When using Table 14.2, it is only necessary to look at the row which indicates the number of first-place votes. This indicates that A (Antonio) gets 130 first-place votes, C (Carmen) gets 150 first-place votes, and D (Donna) gets 120 + 100 = 220 first-place votes. Thus Donna is declared the winner using the plurality method.

3. Because there are four candidates, a first-place vote is worth 4 points, a second-place vote is worth 3 points, a third-place vote is worth 2 points, and a fourth-place vote is worth 1 point. We show the points produced by the votes in the preference table.

Number of Votes	130	120	100	150
First Choice: 4 points	A: $130 \times 4 = 520$ pts	D: $120 \times 4 = 480$ pts	D: $100 \times 4 = 400$ pts	C: $150 \times 4 = 600$ pts
Second Choice: 3 points	B: $130 \times 3 = 390$ pts	B: $120 \times 3 = 360$ pts	B: $100 \times 3 = 300$ pts	B: $150 \times 3 = 450$ pts
Third Choice: 2 points	C: $130 \times 2 = 260$ pts	C: $120 \times 2 = 240$ pts	A: $100 \times 2 = 200$ pts	A: $150 \times 2 = 300$ pts
Fourth Choice: 1 point	D: $130 \times 1 = 130$ pts	A: $120 \times 1 = 120$ pts	C: $100 \times 1 = 100$ pts	D: $150 \times 1 = 150$ pts

Now we read down each column and total the points for each candidate separately.

A gets 520 + 120 + 200 + 300 = 1140 points
B gets 390 + 360 + 300 + 450 = 1500 points
C gets 260 + 240 + 100 + 600 = 1200 points
D gets 130 + 480 + 400 + 150 = 1160 points

Because B (Bob) has received the most points, he is the winner and the new mayor of Smallville.

4. There are 130 + 120 + 100 + 150, or 500, people voting. In order to receive a majority, a candidate must receive more than 50% of the votes, meaning more than 250 votes. The number of first-place votes for each candidate is
   A (Antonio) = 130   B (Bob) = 0   C (Carmen) = 150   D (Donna) = 220

We see that no candidate receives a majority of first-place votes. Because Bob received the fewest first-place votes, he is eliminated in the next round. We construct a new preference table in which B is removed. Each candidate below B moves up one place, while the positions of candidates above B remain unchanged.

Number of Votes	130	120	100	150
First Choice	A	D	D	C
Second Choice	C	C	A	A
Third Choice	D	A	C	D

The number of first-place votes for each candidate is now A (Antonio) = 130; C (Carmen) = 150; D (Donna) = 220

No candidate receives a majority of first-place votes. Because Antonio received the fewest first-place votes, he is eliminated in the next round.

Number of Votes	130	120	100	150
First Choice	C	D	D	C
Second Choice	D	C	C	D

The number of first-place votes for each candidate is now C (Carmen) = 280; D (Donna) = 220

Because Carmen has received the majority of first-place votes, she is the winner and the new mayor of Smallville.

5.

A vs. B

130	120	100	150
A	D	D	C
B	B	B	B
C	C	A	A
D	A	C	D

130 voters prefer A to B.
120 + 100 + 150 = 370 voters prefer B to A.

Conclusion: B wins this comparison and gets one point.

A vs. C

130	120	100	150
A	D	D	C
B	B	B	B
C	C	A	A
D	A	C	D

130 + 100 = 230 voters prefer A to C.
120 + 150 = 270 voters prefer C to A.

Conclusion: C wins this comparison and gets one point.

A vs. D

130	120	100	150
A	D	D	C
B	B	B	B
C	C	A	A
D	A	C	D

130 + 150 = 280 voters prefer A to D.
120 + 100 = 220 voters prefer D to A.

Conclusion: A wins this comparison and gets one point.

B vs. C

130	120	100	150
A	D	D	C
B	B	B	B
C	C	A	A
D	A	C	D

130 + 120 + 100 = 350 voters prefer B to C.
150 voters prefer C to B.

Conclusion: B wins this comparison and gets one point.

352

<table>
<tr><th colspan="5" align="center">B vs. D</th></tr>
<tr><td></td><td>130</td><td>120</td><td>100</td><td>150</td></tr>
<tr><td></td><td>A</td><td>**D**</td><td>**D**</td><td>C</td></tr>
<tr><td></td><td>**B**</td><td>*B*</td><td>*B*</td><td>**B**</td></tr>
<tr><td></td><td>C</td><td>C</td><td>A</td><td>A</td></tr>
<tr><td></td><td>*D*</td><td>A</td><td>C</td><td>D</td></tr>
</table>

130 + 150 = 280 voters prefer B to D.
120 + 100 = 220 voters prefer D to B.

Conclusion: B wins this comparison and gets one point.

<table>
<tr><th colspan="5" align="center">C vs. D</th></tr>
<tr><td></td><td>130</td><td>120</td><td>100</td><td>150</td></tr>
<tr><td></td><td>A</td><td>**D**</td><td>**D**</td><td>C</td></tr>
<tr><td></td><td>B</td><td>B</td><td>B</td><td>B</td></tr>
<tr><td></td><td>**C**</td><td>*C*</td><td>A</td><td>A</td></tr>
<tr><td></td><td>*D*</td><td>A</td><td>*C*</td><td>D</td></tr>
</table>

130 + 150 = 280 voters prefer C to D.
120 + 100 = 220 voters prefer D to C.

Conclusion: C wins this comparison and gets one point.

We now use each of the six conclusions and add points for the six comparisons.
   A gets 1 point.
   B gets 1 + 1 + 1 = 3 points.
   C gets 1 + 1 = 2 points.

After all comparisons have been made, the candidate receiving the most points is B (Bob). He is the winner and the new mayor of Smallville.

Exercise Set 14.1

1.

Number of Votes	7	5	4
First Choice	A	B	C
Second Choice	B	C	B
Third Choice	C	A	A

3.

Number of Votes	5	1	4	2
First Choice	A	B	C	C
Second Choice	B	D	B	B
Third Choice	C	C	D	A
Fourth Choice	D	A	A	D

5. a. 14 + 8 + 3 + 1 = 26

   b. 8

   c. 14 + 8 = 22

   d. 3

7. "Musical" received 12 first-place votes, "comedy" received 10 first-place votes, and "drama" received 8 first-place votes, so the type of play selected is a musical.

9. Darwin received 30 first-place votes, Einstein received 22 first-place votes, Freud received 20 first-place votes, and Hawking received 14 first-place votes, so the professor declared chair is Darwin.

11.

Number of Votes	10	6	6	4	2	2
First Choice: 3 points	M: $10 \times 3 = 30$	C: $6 \times 3 = 18$	D: $6 \times 3 = 18$	C: $4 \times 3 = 12$	D: $2 \times 3 = 6$	M: $2 \times 3 = 6$
Second Choice: 2 points	C: $10 \times 2 = 20$	M: $6 \times 2 = 12$	C: $6 \times 2 = 12$	D: $4 \times 2 = 8$	M: $2 \times 2 = 4$	D: $2 \times 2 = 4$
Third Choice: 1 point	D: $10 \times 1 = 10$	D: $6 \times 1 = 6$	M: $6 \times 1 = 6$	M: $4 \times 1 = 4$	C: $2 \times 1 = 2$	C: $2 \times 1 = 2$

C gets $20 + 18 + 12 + 12 + 2 + 2 = 66$ points.
D gets $10 + 6 + 18 + 8 + 6 + 4 = 52$ points.
M gets $30 + 12 + 6 + 4 + 4 + 6 = 62$ points.

C (Comedy) receives the most points, and is selected.

13.

Number of Votes	30	22	20	12	2
First Choice: 4 points	D: $30 \times 4 = 120$	E: $22 \times 4 = 88$	F: $20 \times 4 = 80$	H: $12 \times 4 = 48$	H: $2 \times 4 = 8$
Second Choice: 3 points	H: $30 \times 3 = 90$	F: $22 \times 3 = 66$	E: $20 \times 3 = 60$	E: $12 \times 3 = 36$	F: $2 \times 3 = 6$
Third Choice: 2 points	F: $30 \times 2 = 60$	H: $22 \times 2 = 44$	H: $20 \times 2 = 40$	F: $12 \times 2 = 24$	D: $2 \times 2 = 4$
Fourth Choice: 1 point	E: $30 \times 1 = 30$	D: $22 \times 1 = 22$	D: $20 \times 1 = 20$	D: $12 \times 1 = 12$	E: $2 \times 1 = 2$

D gets $120 + 22 + 20 + 12 + 4 = 178$ points.
E gets $30 + 88 + 60 + 36 + 2 = 216$ points.
F gets $60 + 66 + 80 + 24 + 6 = 236$ points.
H gets $90 + 44 + 40 + 48 + 8 = 230$ points.

F (Freud) receives the most points and is declared the new division chair.

15. There are 30 people voting, so the winner needs more than 15 votes for a majority.
The number of first-place votes for each candidate is

C (Comedy) = 10   D (Drama) = 8   M (Musical) = 12

No candidate has a majority. Drama received the fewest first-place votes, so we eliminate it in the next round.

Number of Votes	10	6	6	4	2	2
First Choice	M	C	C	C	M	M
Second Choice	C	M	M	M	C	C

The number of first-place votes for each candidate is now

C (Comedy) = 16   M (Musical) = 14

C (Comedy) has 16 votes, which is a majority, so "Comedy" is selected.

17. There are 86 people voting, so the winner needs more than 43 votes for a majority. The number of first-place votes for each candidate is

D (Darwin) = 30   E (Einstein) = 22
F (Freud) = 20   H (Hawking) = 14

No candidate has a majority. Hawking received the fewest first-place votes, so we eliminate him in the next round.

Number of Votes	30	22	20	12	2
First Choice	D	E	F	E	F
Second Choice	F	F	E	F	D
Third Choice	E	D	D	D	E

The number of first-place votes for each candidate is now

D (Darwin) = 30 E (Einstein) = 34 F (Freud) = 22

No candidate has a majority. Freud received the fewest first-place votes, so we eliminate him in the next round:

Number of Votes	30	22	20	12	2
First Choice	D	E	E	E	D
Second Choice	E	D	D	D	E

The number of first-place votes for each candidate is now

D (Darwin) = 32 E (Einstein) = 54

E (Einstein) has 54 votes, which is a majority, so Einstein is declared the new division chair.

19. With $n = 5$, there are $\dfrac{5(5-1)}{2} = 10$ comparisons.

21. With $n = 8$, there are $\dfrac{8(8-1)}{2} = 28$ comparisons.

23.

10	6	6	4	2	2
M	C	D	C	D	M
C	M	C	D	M	D
D	D	M	M	C	C

C vs. D
$10 + 6 + 4 = 20$ voters prefer C to D.
$6 + 2 + 2 = 10$ voters prefer D to C.
C wins this comparison and gets one point.

D vs. M
$6 + 4 + 2 = 12$ voters prefer D to M.
$10 + 6 + 2 = 18$ voters prefer M to D.
M wins this comparison and gets one point.

C vs. M
$6 + 6 + 4 = 16$ voters prefer C to M.
$10 + 2 + 2 = 14$ voters prefer M to C.
C wins this comparison and gets one point

Adding points for the three comparisons:
C gets $1 + 1 = 2$ points.
D gets 0 points.
M gets 1 point.

C (Comedy) receives the most points, so a comedy is selected.

25.

30	22	20	12	2
D	E	F	H	H
H	F	E	E	F
F	H	H	F	D
E	D	D	D	E

D vs. E
30 + 2 = 32 voters prefer D to E.
22 + 20 + 12 = 54 voters prefer E to D.
E wins the comparison and gets one point

D vs. F
30 voters prefer D to F.
22 + 20 + 12 + 2 = 56 voters prefer F to D.
F wins this comparison and gets one point.

D vs. H
30 voters prefer D to H.
22 + 20 + 12 + 2 = 56 voters prefer H to D.
H wins this comparison and gets one point.

E vs. F
22 + 12 = 34 voters prefer E to F.
30 + 20 + 2 = 52 voters prefer F to E.
F wins this comparison and gets one point

E vs. H
22 + 20 = 42 voters prefer E to H.
30 + 12 + 2 = 44 voters prefer H to E.
H wins this comparison and gets one point.

F vs. H
22 + 20 = 42 voters prefer F to H.
30 + 12 + 2 = 44 voters prefer H to F.
H wins this comparison and gets one point.

Adding points for the six comparisons:
D gets 0 points.
E gets 1 point.
F gets 1 + 1 = 2 points.
H gets 1 + 1 + 1 = 3 points.

H (Hawking) receives the most points, so Hawking is declared the new division chair.

27. A received 34 first-place votes, B received 30 first-place votes, C received 6 first-place votes, and D received 2 first-place votes, so A is the winner.

29. There are 72 people voting, so the winner needs more than 36 votes for a majority. The number of first-place votes for each candidate is: A = 34;   B = 30;   C = 6;   D = 2

No candidate has a majority. D received the fewest first-place votes, so we eliminate it in the next round.

Number of Voters	34	30	6	2
First Choice	A	B	C	B
Second Choice	B	C	B	C
Third Choice	C	A	A	A

The number of first-place votes for each candidate is now   A = 34;   B = 32;   C = 6

No candidate has a majority. C received the fewest first-place votes, so we eliminate it in the next round.

Number of Voters	34	30	6	2
First Choice	A	B	B	B
Second Choice	B	A	A	A

The number of first-place votes for each candidate is now   A = 34;   B = 38

B has 38 votes, which is a majority, so B is selected.

31.

Number of Votes	5	5	4	3	3	2
First choice: 5 points	C: $5 \times 5 = 25$	S: $5 \times 5 = 25$	C: $4 \times 5 = 20$	W: $3 \times 5 = 15$	W: $3 \times 5 = 15$	P: $2 \times 5 = 10$
Second choice: 4 points	R: $5 \times 4 = 20$	R: $5 \times 4 = 20$	P: $4 \times 4 = 16$	P: $3 \times 4 = 12$	R: $3 \times 4 = 12$	S: $2 \times 4 = 8$
Third choice: 3 points	P: $5 \times 3 = 15$	W: $5 \times 3 = 15$	R: $4 \times 3 = 12$	R: $3 \times 3 = 9$	S: $3 \times 3 = 9$	C: $2 \times 3 = 6$
Fourth choice: 2 points	W: $5 \times 2 = 10$	P: $5 \times 2 = 10$	S: $4 \times 2 = 8$	S: $3 \times 2 = 6$	C: $3 \times 2 = 6$	R: $2 \times 2 = 4$
Fifth choice: 1 point	S: $5 \times 1 = 5$	C: $5 \times 1 = 5$	W: $4 \times 1 = 4$	C: $3 \times 1 = 3$	P: $3 \times 1 = 3$	W: $2 \times 1 = 2$

C gets $25 + 5 + 20 + 3 + 6 + 6 = 65$ points.
P gets $15 + 10 + 16 + 12 + 3 + 10 = 66$ points.
R gets $20 + 20 + 12 + 9 + 12 + 4 = 77$ points.
S gets $5 + 25 + 8 + 6 + 9 + 8 = 61$ points.
W gets $10 + 15 + 4 + 15 + 15 + 2 = 61$ points.

R (Rent) receives the most points and is selected.

33.

5	5	4	3	3	2
C	S	C	W	W	P
R	R	P	P	R	S
P	W	R	R	S	C
W	P	S	S	C	R
S	C	W	C	P	W

C vs. P
5 + 4 + 3 = 12 voters prefer C to P.
5 + 3 + 2 = 10 voters prefer P to C.
C wins this comparison and gets one point.

C vs. R
5 + 4 + 2 = 11 voters prefer C to R.
5 + 3 + 3 = 11 voters prefer R to C.
C and R are tied. Each gets $\frac{1}{2}$ point.

C vs. S
5 + 4 = 9 voters prefer C to S.
5 + 3 + 3 + 2 = 13 voters prefer S to C.
S wins this comparison and gets one point.

C vs. W
5 + 4 + 2 = 11 voters prefer C to W.
5 + 3 + 3 = 11 voters prefer W to C.
C and W are tied. Each gets $\frac{1}{2}$ point.

P vs. R
4 + 3 + 2 = 9 voters prefer P to R.
5 + 5 + 3 = 13 voters prefer R to P.
R wins this comparison and gets one point.

P vs. S
5 + 4 + 3 + 2 = 14 voters prefer P to S.
5 + 3 = 8 voters prefer S to P.
P wins this comparison and gets one point.

P vs. W
5 + 4 + 2 = 11 voters prefer P to W.
5 + 3 + 3 = 11 voters prefer W to P.
P and W are tied. Each gets $\frac{1}{2}$ point.

R vs. S
5 + 4 + 3 + 3 = 15 voters prefer R to S.
5 + 2 = 7 voters prefer S to R.
R wins this comparison and gets one point.

R vs. W
5 + 5 + 4 + 2 = 16 voters prefer R to W.
3 + 3 = 6 voters prefer W to R.
R wins this comparison and gets one point.

S vs. W
5 + 4 + 2 = 11 voters prefer S to W.
5 + 3 + 3 = 11 voters prefer W to S.
S and W are tied. Each gets $\frac{1}{2}$ point.

Adding points for 10 comparisons:

C gets $1 + \frac{1}{2} + \frac{1}{2} = 2$ points.

P gets $1 + \frac{1}{2} = 1\frac{1}{2}$ points.

R gets $\frac{1}{2} + 1 + 1 + 1 = 3\frac{1}{2}$ points.

S gets $1 + \frac{1}{2} = 1\frac{1}{2}$ points.

W gets $\frac{1}{2} + \frac{1}{2} + \frac{1}{2} = 1\frac{1}{2}$ points.

R (Rent) receives the most points, so Rent is the winner.

35. a.

Number of Votes	5	5	3	3	3	2
First Choice: 5 points	A: 5 × 5 = 25	C: 5 × 5 = 25	D: 3 × 5 = 15	A: 3 × 5 = 15	B: 3 × 5 = 15	D: 2 × 5 = 10
Second Choice: 4 points	B: 5 × 4 = 20	E: 5 × 4 = 20	C: 3 × 4 = 12	D: 3 × 4 = 12	E: 3 × 4 = 12	C: 2 × 4 = 8
Third Choice: 3 points	C: 5 × 3 = 15	D: 5 × 3 = 15	B: 3 × 3 = 9	B: 3 × 3 = 9	A: 3 × 3 = 9	B: 2 × 3 = 6
Fourth Choice: 2 points	D: 5 × 2 = 10	A: 5 × 2 = 10	E: 3 × 2 = 6	C: 3 × 2 = 6	C: 3 × 2 = 6	A: 2 × 2 = 4
Fifth Choice: 1 point	E: 5 × 1 = 5	B: 5 × 1 = 5	A: 3 × 1 = 3	E: 3 × 1 = 3	D: 3 × 1 = 3	E: 2 × 1 = 2

A gets 25 + 10 + 3 + 15 + 9 + 4 = 66 points.
B gets 20 + 5 + 9 + 9 + 15 + 6 = 64 points.
C gets 15 + 25 + 12 + 6 + 6 + 8 = 72 points.
D gets 10 + 15 + 15 + 12 + 3 + 10 = 65 points.
E gets 5 + 20 + 6 + 3 + 12 + 2 = 48 points.

C receives the most points and is the winner.

b.

Number of Votes	5	5	3	3	3	2
First Choice: 4 points	A: 5 × 4 = 20	C: 5 × 4 = 20	D: 3 × 4 = 12	A: 3 × 4 = 12	B: 3 × 4 = 12	D: 2 × 4 = 8
Second Choice: 3 points	B: 5 × 3 = 15	D: 5 × 3 = 15	C: 3 × 3 = 9	D: 3 × 3 = 9	A: 3 × 3 = 9	C: 2 × 3 = 6
Third Choice: 2 points	C: 5 × 2 = 10	A: 5 × 2 = 10	B: 3 × 2 = 6	B: 3 × 2 = 6	C: 3 × 2 = 6	B: 2 × 2 = 4
Fourth Choice: 1 points	D: 5 × 1 = 5	B: 5 × 1 = 5	A: 3 × 1 = 3	C: 3 × 1 = 3	D: 3 × 1 = 3	A: 2 × 1 = 2

A gets 20 + 10 + 3 + 12 + 9 + 2 = 56 points.
B gets 15 + 5 + 6 + 6 + 12 + 4 = 48 points.
C gets 10 + 20 + 9 + 3 + 6 + 6 = 54 points.
D gets 5 + 15 + 12 + 9 + 3 + 8 = 52 points.

A receives the most points and is the winner.

37. First use the plurality method: C receives 12,000 first-place votes, and A receives 12,000 first-place votes. This results in a tie, so we use the Borda count method.

Number of Votes	12,000	7500	4500
First Choice: 3 points	C: 12,000 × 3 = 36,000	A: 7500 × 3 = 22,500	A: 4500 × 3 = 13,500
Second Choice: 2 points	B: 12,000 × 2 = 24,000	B: 7500 × 2 = 15,000	C: 4500 × 3 = 9000
Third Choice: 1 points	A: 12,000 × 1 = 12,000	C: 7500 × 1 = 7500	B: 4500 × 1 = 4500

A gets 12,000 + 22,500 + 13,500 = 48,000 points.
B gets 24,000 + 15,000 + 4500 = 43,500 points.
C gets 36,000 + 7500 + 9000 = 52,500 points.
C receives the most points and is the winner.

49. b

Check Points 14.2

1. **a.** There are 14 first-place votes. A candidate with more than half of these receives a majority. The first-choice row shows that candidate A received 8 first-place votes. Thus, candidate A has a majority of first-place votes.

    b. Using the Borda count method with four candidates, a first-place vote is worth 4 points, a second-place vote is worth 3 points, a third-place vote is worth 2 points, and a fourth-place vote is worth 1 point.

Number of Votes	6	4	2	2
First Choice: 4 points	A: $6 \times 4 = 24$ pts	B: $4 \times 4 = 16$ pts	B: $2 \times 4 = 8$ pts	A: $2 \times 4 = 8$ pts
Second Choice: 3 points	B: $6 \times 3 = 18$ pts	C: $4 \times 3 = 12$ pts	D: $2 \times 3 = 6$ pts	B: $2 \times 3 = 6$ pts
Third Choice: 2 points	C: $6 \times 2 = 12$ pts	D: $4 \times 2 = 8$ pts	C: $2 \times 2 = 4$ pts	D: $2 \times 2 = 4$ pts
Fourth Choice: 1 point	D: $6 \times 1 = 6$ pts	A: $4 \times 1 = 4$ pts	A: $2 \times 1 = 2$ pts	C: $2 \times 1 = 2$ pts

Now we read down the columns and total the points for each candidate.
A gets $24 + 4 + 2 + 8 = 38$ points.
B gets $18 + 16 + 8 + 6 = 48$ points.
C gets $12 + 12 + 4 + 2 = 30$ points.
D gets $6 + 8 + 6 + 4 = 24$ points.

Because candidate B has received the most points, candidate B is declared the new principal using the Borda count method.

2. **a.** We begin by comparing A and B. A is favored over B in column 1, giving A 3 votes. B is favored over A in columns 2 and 3, giving B $2 + 2$, or 4, votes. Thus, B is favored when compared to A.

    Now we compare B to C. B is favored over C in columns 1 and 2, giving B $3 + 2$, or 5, votes. C is favored over B in column 3, giving C 2 votes. Thus, B is favored when compared to C.

    We see that B is favored over both A and C using a head-to-head comparison.

    b. Using the plurality method, the brand with the most first-place votes is the winner. In the row indicating first choice, A received 3 votes, B received 2 votes, and C received 2 votes. A wins using the plurality method.

3. **a.** There are 120 people voting. No candidate initially receives more than 60 votes. Because C receives the fewest first-place votes, C is eliminated in the next round. The new preference table is

Number of Votes	42	34	28	16
First Choice	A	A	B	B
Second Choice	B	B	A	A

Because A has received a majority of first-place votes, A is the winner of the straw poll.

    b. No candidate initially receives more than 60 votes. Because B receives the fewest first-place votes, B is eliminated in the next round. The new preference table is

Number of Votes	54	34	28	4
First Choice	A	C	C	A
Second Choice	C	A	A	C

Because C has received a majority of first-place votes, C is the winner of the second election.

    c. A won the first election. A then gained additional support with the 12 voters who changed their ballots to make A their first choice. A lost the second election. This violates the monotonicity criterion.

4.   a.   Because there are 4 candidates, $n = 4$ and the number of comparisons we must make is
$\dfrac{n(n-1)}{2} = \dfrac{4(4-1)}{2} = \dfrac{4 \cdot 3}{2} = \dfrac{12}{2} = 6$.

The following table shows the results of these 6 comparisons.

Comparison	Vote Results	Conclusion
A vs. B	270 voters prefer A to B. 90 voters prefer B to A.	A wins and gets 1 point.
A vs. C	270 voters prefer A to C. 90 voters prefer C to A.	A wins and gets 1 point.
A vs. D	150 voters prefer A to D. 210 voters prefer D to A.	D wins and gets 1 point.
B vs. C	180 voters prefer B to C. 180 voters prefer C to B.	B and C tie. Each gets $\frac{1}{2}$ point.
B vs. D	240 voters prefer B to D. 120 voters prefer D to B.	B wins and gets 1 point.
C vs. D	240 voters prefer C to D. 120 voters prefer D to C.	C wins and gets 1 point.

Thus A gets 2 points, B gets $1\frac{1}{2}$ points, C gets $1\frac{1}{2}$ points, and D gets 1 point. Therefore A is the winner.

b.   After B and C withdraw, there is a new preference table:

Number of Votes	150	90	90	30
First Choice	A	D	D	D
Second Choice	D	A	A	A

Using the pairwise comparison test with 2 candidates, there is only one comparison to make namely A vs. D.

150 voters prefer A to D, and 210 voters prefer D to A. D gets 1 point, A gets 0 points, and D wins the election.

c.   The first election count produced A as the winner. The removal of B and C from the ballots produced D as the winner. This violates the irrelevant alternatives criterion.

Exercise Set 14.2

1.   a.   D has 300 first-place votes, which is more than half of the 570 total votes, so D has a majority of first-place votes.

b.

Number of Votes	300	120	90	60
First Choice: 4 points	D: $300 \times 4 = 1200$	C: $120 \times 4 = 480$	C: $90 \times 4 = 360$	A: $60 \times 4 = 240$
Second Choice: 3 points	A: $300 \times 3 = 900$	A: $120 \times 3 = 360$	A: $90 \times 3 = 270$	D: $60 \times 3 = 180$
Third Choice: 2 points	B: $300 \times 2 = 600$	B: $120 \times 2 = 240$	D: $90 \times 2 = 180$	B: $60 \times 2 = 120$
Fourth Choice: 1 point	C: $300 \times 1 = 300$	D: $120 \times 1 = 120$	B: $90 \times 1 = 90$	C: $60 \times 1 = 60$

A gets $900 + 360 + 270 + 240 = 1770$ points.
B gets $600 + 240 + 90 + 120 = 1050$ points.
C gets $300 + 480 + 360 + 60 = 1200$ points.
D gets $1200 + 120 + 180 + 180 = 1680$ points.

A receives the most points, so A is the chosen design.

c.   No. D receives a majority of first-place votes, but A is chosen by the Borda count method.

3. a. A is favored over R in columns 1 and 3, giving A 12 + 4, or 16, votes. R is favored over A in columns 2 and 4, giving R 9 + 4, or 13, votes. Thus, A is favored when compared to R.

A is favored over V in columns 1 and 4, giving A 12 + 4, or 16, votes. V is favored over A in columns 2 and 3, giving V 9 + 4, or 13, votes. Thus, A is favored when compared to V.

We see that A is favored over the other two cities using a head-to-head comparison.

b. A gets 12 first-place votes, V gets 13 first-place votes, and R gets 4 first-place votes, so V wins using the plurality method.

c. No. A wins the head-to-head comparison, but V wins the election.

5. a. A is favored over B in columns 1 and 4, giving A 120 + 30, or 150, votes. B is favored over A in columns 2, 3, and 5, giving B 60 + 30 + 30, or 120 votes. Thus, A is favored when compared to B.

A is favored over C in columns 1 and 3, giving A 120 + 30, or 150 votes. C is favored over A in columns 2, 4, and 5, giving C 60 + 30 + 30, or 120, votes. Thus, A is favored when compared to C.

We see that A is favored over the other two options using a head-to-head comparison.

b.

Number of Votes	120	60	30	30	30
First Choice: 3 points	A: $120 \times 3 = 360$	C: $60 \times 3 = 180$	B: $30 \times 3 = 90$	C: $30 \times 3 = 90$	B: $30 \times 3 = 90$
Second Choice: 2 points	C: $120 \times 2 = 240$	B: $60 \times 2 = 120$	A: $30 \times 2 = 60$	A: $30 \times 2 = 60$	C: $30 \times 2 = 60$
Third Choice: 1 point	B: $120 \times 1 = 120$	A: $60 \times 1 = 60$	C: $30 \times 1 = 30$	B: $30 \times 1 = 30$	A: $30 \times 1 = 30$

A gets $360 + 60 + 60 + 60 + 30 = 570$ points.
B gets $120 + 120 + 90 + 30 + 90 = 450$ points.
C gets $240 + 180 + 30 + 90 + 60 = 600$ points.

C receives the most points, so C is the winner.

c. No. A wins the head-to-head comparison, but C wins the election.

7. a. There are 29 people voting. No one receives the 15 first-place votes needed for a majority. B receives the fewest first-place votes and is eliminated in the next round.

Number of Votes	18	11
First Choice	C	A
Second Choice	A	C

C receives the majority of first-place votes, so C is the winner.

b. With the voting change, a new preference table results.

Number of Votes	14	8	7
First Choice	C	B	A
Second Choice	A	C	B
Third Choice	B	A	C

No one receives a majority of first-place votes. A receives the fewest first-place votes, and is eliminated in the next round.

Number of Votes	14	15
First Choice	C	B
Second Choice	B	C

B receives the majority of first-place votes, so B is the winner.

c. No. C wins the straw vote, and the only change increases the number of first-place votes for C, but B wins the election.

9. a. There are 3 candidates, so $n = 3$ and the number of comparisons we must make is $\frac{n(n-1)}{2} = \frac{3(2)}{2} = 3$.

Comparison	Vote Results	Conclusion
H vs. L	10 voters prefer H to L. 13 voters prefer L to H.	L wins and gets one point.
H vs. S	10 voters prefer H to S. 13 voters prefer S to H.	S wins and gets one point.
L vs. S	8 voters prefer L to S. 15 voters prefer S to L.	S wins and gets one point.

Thus, L gets 1 point and S gets 2 points. Therefore, S is the winner when candidates H and L are included.

b. New preference table:

Number of Votes	15	8
First Choice	S	L
Second Choice	L	S

With only two candidates, we can only make one comparison. We see that S wins, defeating L by 15 votes to 8 votes. Thus S gets 1 point, L gets 0 points, and S is the winner.

c. Yes. S wins whether or not H withdraws.

11. a.

Number of Votes	20	16	10	4
First Choice: 4 points	D: $20 \times 4 = 80$	C: $16 \times 4 = 64$	C: $10 \times 4 = 40$	A: $4 \times 4 = 16$
Second Choice: 3 points	A: $20 \times 3 = 60$	A: $16 \times 3 = 48$	B: $10 \times 3 = 30$	B: $4 \times 3 = 12$
Third Choice: 2 points	B: $20 \times 2 = 40$	B: $16 \times 2 = 32$	D: $10 \times 2 = 20$	D: $4 \times 2 = 8$
Fourth Choice: 1 point	C: $20 \times 1 = 20$	D: $16 \times 1 = 16$	A: $10 \times 1 = 10$	C: $4 \times 1 = 4$

A gets $60 + 48 + 10 + 16 = 134$ points.
B gets $40 + 32 + 30 + 12 = 114$ points.
C gets $20 + 64 + 40 + 4 = 128$ points.
D gets $80 + 16 + 20 + 8 = 124$ points.

A receives the most points, so A is the winner.

b. No. A has only 4 first-place votes, out of 50 total votes. C has 26 first-place votes, which is a majority, but A wins the election.

13. a. There are 70 people voting. No one receives the 36 first-place votes needed for a majority. B receives the fewest first-place votes and is eliminated in the next round.

Number of Votes	24	20	10	8	8
First Choice	D	C	A	A	C
Second Choice	A	A	D	C	D
Third Choice	C	D	C	D	A

No one receives a majority of first-place votes. A receives the fewest first-place votes and is eliminated in the next round.

Number of Votes	34	36
First Choice	D	C
Second Choice	C	D

C receives 36 first-place votes, which is a majority, so C is the winner.

b. No. When compared individually to B, A wins with 60 votes to 10. Compared with C, A wins with 42 votes to 28. Compared with D, A wins with 38 votes to 32. So A is favored in all head-to-head contests but C wins the election.

15. a.

Number of Votes	14	8	4
First Choice: 4 points	A: $14 \times 4 = 56$	B: $8 \times 4 = 32$	D: $4 \times 4 = 16$
Second Choice: 3 points	B: $14 \times 3 = 42$	D: $8 \times 3 = 24$	A: $4 \times 3 = 12$
Third Choice: 2 points	C: $14 \times 2 = 28$	C: $8 \times 2 = 16$	C: $4 \times 2 = 8$
Fourth Choice: 1 point	D: $14 \times 1 = 14$	A: $8 \times 1 = 8$	B: $4 \times 1 = 4$

A gets $56 + 8 + 12 = 76$ points.
B gets $42 + 32 + 4 = 78$ points.
C gets $28 + 16 + 8 = 52$ points.
D gets $14 + 24 + 16 = 54$ points.

B receives the most points, so B is the winner.

b. No. A receives the majority of first-place votes, but B wins the election.

c. No. A wins all head-to-head comparisons, but B wins the election.

d. Using the Borda count method with C removed:

Number of Votes	14	8	4
First Choice: 3 points	A: $14 \times 3 = 42$	B: $8 \times 3 = 24$	D: $4 \times 3 = 12$
Second Choice: 2 points	B: $14 \times 2 = 28$	D: $8 \times 2 = 16$	A: $4 \times 2 = 8$
Third Choice: 1 point	D: $14 \times 1 = 14$	A: $8 \times 1 = 8$	B: $4 \times 1 = 4$

A gets $42 + 8 + 8 = 58$ points.
B gets $28 + 24 + 4 = 56$ points.
D gets $14 + 16 + 12 = 42$ points.

A receives the most points, and wins the election.

The irrelevant alternatives criterion is not satisfied. Candidate C's dropping out changed the outcome of the election.

17. a.

Number of Votes	16	14	12	4	2
First Choice: 5 points	A: $16 \times 5 = 80$	D: $14 \times 5 = 70$	D: $12 \times 5 = 60$	C: $4 \times 5 = 20$	E: $2 \times 5 = 10$
Second Choice: 4 points	B: $16 \times 4 = 64$	B: $14 \times 4 = 56$	B: $12 \times 4 = 48$	A: $4 \times 4 = 16$	A: $2 \times 4 = 8$
Third Choice: 3 points	C: $16 \times 3 = 48$	A: $14 \times 3 = 42$	E: $12 \times 3 = 36$	B: $4 \times 3 = 12$	D: $2 \times 3 = 6$
Fourth Choice: 2 points	D: $16 \times 2 = 32$	C: $14 \times 2 = 28$	C: $12 \times 2 = 24$	D: $4 \times 2 = 8$	B: $2 \times 2 = 4$
Fifth Choice: 1 point	E: $16 \times 1 = 16$	E: $14 \times 1 = 14$	A: $12 \times 1 = 12$	E: $4 \times 1 = 4$	C: $2 \times 1 = 2$

A gets $80 + 42 + 12 + 16 + 8 = 158$ points.
B gets $64 + 56 + 48 + 12 + 4 = 184$ points.
C gets $48 + 28 + 24 + 20 + 2 = 122$ points.
D gets $32 + 70 + 60 + 8 + 6 = 176$ points.
E gets $16 + 14 + 36 + 4 + 10 = 80$ points.

B receives the most points, so B is the winner.

b. No. D gets a majority of first-place votes, but B wins the election.

c. No. D wins all head-to-head comparisons, but B wins the election.

19. a. A receives the most first-place votes, and is the winner.

b. Yes. A has a majority of the first-place votes, and wins.

c. Yes. A wins in comparisons to B and C.

d. New preference table:

Number of Votes	7	3	2
First Choice	A	B	A
Second Choice	B	C	C
Third Choice	C	A	B

A has the majority of first-place votes, and wins using the plurality method.

e. Yes. A still receives the most first-place votes, and wins.

f. No. The fact that all four criteria are satisfied in a particular case does not mean that the method used always satisfies all four criteria.

Check Points 14.3

1. **a.** Standard divisor $= \dfrac{\text{total population}}{\text{number of allocated items}} = \dfrac{10,000}{200} = 50$

   b. Standard quota for state A $= \dfrac{\text{population of state A}}{\text{standard divisor}} = \dfrac{1112}{50} = 22.24$

   Standard quota for state B $= \dfrac{\text{population of state B}}{\text{standard divisor}} = \dfrac{1118}{50} = 22.36$

   Standard quota for state C $= \dfrac{\text{population of state C}}{\text{standard divisor}} = \dfrac{1320}{50} = 26.4$

   Standard quota for state D $= \dfrac{\text{population of state D}}{\text{standard divisor}} = \dfrac{1515}{50} = 30.3$

   Standard quota for state E $= \dfrac{\text{population of state E}}{\text{standard divisor}} = \dfrac{4935}{50} = 98.7$

Table 14.27   **Population of Amador by State**

State	A	B	C	D	E	Total
Population (in thousands)	1112	1118	1320	1515	4935	10,000
Standard quota	22.24	22.36	26.4	30.3	98.7	200

2.

State	Population (in thousands)	Standard Quota	Lower Quota	Fractional Part	Surplus	Final Apportionment
A	1112	22.24	22	0.24		22
B	1118	22.36	22	0.36		22
C	1320	26.4	26	0.4 (next largest)	1	27
D	1515	30.3	30	0.3		30
E	4935	98.7	98	0.7 (largest)	1	99
Total	10,000	200	198			200

3.

State	Population (in thousands)	Modified Quota (using $d = 49.3$)	Modified Lower Quota	Final Apportionment
A	1112	22.56	22	22
B	1118	22.68	22	22
C	1320	26.77	26	26
D	1515	30.73	30	30
E	4935	100.10	100	100
Total	10,000		200	200

4.

State	Population (in thousands)	Modified Quota (using $d = 50.5$)	Modified Upper Quota
A	1112	22.02	23
B	1118	22.14	23
C	1320	26.14	27
D	1515	30	30
E	4935	97.72	98
Total	10,000		201

This sum should be 200, not 201.

State	Population (in thousands)	Modified Quota (using $d = 50.6$)	Modified Upper Quota	Final Apportionment
A	1112	21.98	22	22
B	1118	22.09	23	23
C	1320	26.09	27	27
D	1515	29.94	30	30
E	4935	97.53	98	98
Total	10,000		200	200

5.

State	Population (in thousands)	Modified Quota (using $d = 49.8$)	Modified Rounded Quota
A	1112	22.33	22
B	1118	22.45	22
C	1320	26.51	27
D	1515	30.42	30
E	4935	99.10	99
Total	10,000		200

Exercise Set 14.3

1. a. Standard divisor $= \frac{1600}{80} = 20$. There are 20,000 people for each seat in congress.

b–c.

State	A	B	C	D
Standard quota	$\frac{138}{20} = 6.9$	$\frac{266}{20} = 13.3$	$\frac{534}{20} = 26.7$	$\frac{662}{20} = 33.1$
Lower quota	6	13	26	33
Upper Quota	7	14	27	34

3.

State	Population (in thousands)	Standard Quota	Lower Quota	Fractional Part	Surplus	Final Apportionment
A	138	6.9	6	0.9	1	7
B	266	13.3	13	0.3		13
C	534	26.7	26	0.7	1	27
D	662	33.1	33	0.1		33
Total	1600	80	78			80

5.	School	Enrollment	Standard Quota	Lower Quota	Fractional Part	Surplus	Final Apportionment
	Humanities	1050	30.26	30	0.26		30
	Social Science	1410	40.63	40	0.63	1	41
	Engineering	1830	52.74	52	0.74	1	53
	Business	2540	73.20	73	0.20		73
	Education	3580	103.17	103	0.17		103
	Total	10,410	300	298			300

We use $\frac{10,410}{300} = 34.7$ as the standard divisor.

7.	State	Population	Modified Quota ($d = 32,920$)	Modified Lower Quota	Final Apportionment
	A	126,316	3.84	3	3
	B	196,492	5.97	5	5
	C	425,264	12.92	12	12
	D	526,664	15.998	15	15
	E	725,264	22.03	22	22
	Total	2,000,000		57	57

9. There are 15,000 patients. The standard divisor is $\frac{15,000}{150}$, or 100. Try a modified divisor of 98.

Clinic	Average Weekly Patient Load	Modified Quota	Modified Lower Quota	Final Apportionment
A	1714	17.49	17	17
B	5460	55.71	55	55
C	2440	24.90	24	24
D	5386	54.96	54	54
Total	15,000		150	150

11.	Precinct	Crimes	Modified Quota ($d = 16$)	Modified Upper Quota	Final Apportionment
	A	446	27.88	28	28
	B	526	32.88	33	33
	C	835	52.19	53	53
	D	227	14.19	15	15
	E	338	21.13	22	22
	F	456	28.5	29	29
	Total	2828		180	180

13. There is a total of \$2025 to be invested. The standard divisor is $\frac{2025}{30}$, or 67.5. Try a modified divisor of 72.

Person	Amount	Modified Quota	Modified Upper Quota	Final Apportionment
A	795	11.04	12	12
B	705	9.79	10	10
C	525	7.29	8	8
Total	2025		30	30

15.

Course	Enrollment	Modified Quota ($d = 29.6$)	Modified Rounded Quota	Final Apportionment
Introductory Algebra	130	4.39	4	4
Intermediate Algebra	282	9.53	10	10
Liberal Arts Math	188	6.35	6	6
Total	600		20	20

17. The total number of passengers is 11,060. The standard divisor is $\frac{11,060}{200}$ or 55.3. Try a modified divisor of 55.5.

Route	Average Number of Passengers	Modified Quota	Modified Rounded Quota	Final Apportionment
A	1087	19.59	20	20
B	1323	23.84	24	24
C	1592	28.68	29	29
D	1596	28.76	29	29
E	5462	98.41	98	98
Total	11,060		200	200

19. The total number of patients is 2000. The standard divisor is $\frac{2000}{250}$, or 8. Use Hamilton's method.

Shift	Average Number of Patients	Standard Quota	Lower Quota	Fractional Part	Surplus	Final Apportionment
A	453	56.625	56	0 .625	1	57
B	650	81.25	81	0.25		81
C	547	68.375	68	0.375		68
D	350	43.75	43	0.75	1	44
Total	2000	250	248			250

21. Try a modified divisor of 8.06. Use Adams' method.

Shift	Average Number of Patients	Modified Quota	Modified Upper Quota	Final Apportionment
A	453	56.20	57	57
B	650	80.65	81	81
C	547	67.87	68	68
D	350	43.42	44	44
Total	2000		250	250

23. The total population is 3,615,920. The standard divisor is $\frac{3,615,920}{105}$, or 34,437.333. Use Hamilton's method.

State	Population	Standard Quota	Lower Quota	Fractional Part	Surplus	Final Apportionment
Connecticut	236,841	6.88	6	0.88	1	7
Delaware	55,540	1.61	1	0.61	1	2
Georgia	70,835	2.06	2	0.06		2
Kentucky	68,705	1.995	1	0.995	1	2
Maryland	278,514	8.09	8	0.09		8
Massachusetts	475,327	13.80	13	0.80	1	14
New Hampshire	141,822	4.12	4	0.12		4
New Jersey	179,570	5.21	5	0.21		5
New York	331,589	9.63	9	0.63	1	10
North Carolina	353,523	10.27	10	0.27		10
Pennsylvania	432,879	12.57	12	0.57	1	13
Rhode Island	68,446	1.99	1	0.99	1	2
South Carolina	206,236	5.99	5	0.99	1	6
Vermont	85,533	2.48	2	0.48		2
Virginia	630,560	18.31	18	0.31		18
Total	3,615,920	105.005	97			105

25. Use Adams' method with $d = 36,100$.

State	Population	Modified Quota	Modified Upper Quota	Final Apportionment
Connecticut	236,841	6.56	7	7
Delaware	55,540	1.54	2	2
Georgia	70,835	1.96	2	2
Kentucky	68,705	1.90	2	2
Maryland	278,514	7.72	8	8
Massachusetts	475,327	13.17	14	14
New Hampshire	141,822	3.93	4	4
New Jersey	179,570	4.97	5	5
New York	331,589	9.19	10	10
North Carolina	353,523	9.79	10	10
Pennsylvania	432,879	11.99	12	12
Rhode Island	68,446	1.90	2	2
South Carolina	206,236	5.71	6	6
Vermont	85,533	2.37	3	3
Virginia	630,560	17.47	18	18
Total	3,615,920		105	105

Check Points 14.4

1. We begin with 99 seats in the Congress.

First we compute the standard divisor: Standard divisor = $\dfrac{\text{total population}}{\text{number of allocated items}} = \dfrac{20,000}{99} = 202.02$

Using this value, make a table showing apportionment using Hamilton's method.

State	Population	Standard Quota	Lower Quota	Fractional Part	Surplus Seats	Final Apportionment
A	2060	10.20	10	0.20		10
B	2080	10.30	10	0.30	1	11
C	7730	38.26	38	0.26		38
D	8130	40.24	40	0.24		40
Total	20,000	99	98			99

Now let's see what happens with 100 seats in Congress.

First we compute the standard divisor: Standard divisor $= \dfrac{\text{total population}}{\text{number of allocated items}} = \dfrac{20,000}{100} = 200$.

Using this value, make a table showing apportionment using Hamilton's method.

State	Population	Standard Quota	Lower Quota	Fractional Part	Surplus Seats	Final Apportionment
A	2060	10.3	10	0.3		10
B	2080	10.4	10	0.4		10
C	7730	38.65	38	0.65	1	39
D	8130	40.65	40	0.65	1	41
Total	20,000	100	98			100

The final apportionments are summarized in the following table.

State	Apportionment with 99 seats	Apportionment with 100 seats
A	10	10
B	11	10
C	38	39
D	40	41

When the number of seats increased from 99 to 100, B's apportionment decreased from 11 to 10.

2. a. We use Hamilton's method to find the apportionment for each state with its original population. First we compute the standard divisor.

Standard divisor $= \dfrac{\text{total population}}{\text{number of allocated items}} = \dfrac{200,000}{100} = 2000$

Using this value, we show the apportionment in the following table.

State	Original Population	Standard Quota	Lower Quota	Fractional Part	Surplus Seats	Final Apportionment
A	19,110	9.56	9	0.56	1	10
B	39,090	19.55	19	0.55		19
C	141,800	70.9	70	0.9	1	71
Total	200,000	100.01	98			100

b. The fraction for percent increase is the amount of increase divided by the original amount. The percent increase in the population of each state is determined as follows.

State A: $\dfrac{19,302-19,110}{19,110} = \dfrac{192}{19,110} \approx 0.01005 = 1.005\%$

State B: $\dfrac{39,480-39,090}{39,090} = \dfrac{390}{39,090} \approx 0.00998 = 0.998\%$

State A is increasing at a rate of 1.005%. This is faster than State B, which is increasing at a rate of 0.998%.

c. We use Hamilton's method to find the apportionment for each state with its new population. First we compute the standard divisor.

$$\text{Standard divisor} = \dfrac{\text{total population}}{\text{number of allocated items}} = \dfrac{200,582}{100} = 2005.82$$

Using this value, we show the apportionment in the following table.

State	New Population	Standard Quota	Lower Quota	Fractional Part	Surplus Seats	Final Apportionment
A	19,302	9.62	9	0.62		9
B	39,480	19.68	19	0.68	1	20
C	141,800	70.69	70	0.69	1	71
Total	200,582	99.99	98			100

The final apportionments are summarized in the following table.

State	Growth Rate	Original Apportionment	New Apportionment
A	1.005%	10	9
B	0.998%	19	20
C	0%	71	71

State A loses a seat to State B, even though the population of State A is increasing at a faster rate. This is an example of the population paradox.

3. a. We use Hamilton's method to find the apportionment for each school. First we compute the standard divisor.

$$\text{Standard divisor} = \frac{\text{total population}}{\text{number of allocated items}} = \frac{12,000}{100} = 120$$

Using this value, we show the apportionment in the following table.

School	Enrollment	Standard Quota	Lower Quota	Fractional Part	Surplus	Final Apportionment
East High	2574	21.45	21	0.45		21
West High	9426	78.55	78	0.55	1	79
Total	12,000	100	99			100

b. Again we use Hamilton's method.

$$\text{Standard divisor} = \frac{\text{total population}}{\text{number of allocated items}} = \frac{12,750}{106} = 120.28$$

Using this value, we show the apportionment in the following table

School	Enrollment	Standard Quota	Lower Quota	Fractional Part	Surplus	Final Apportionment
East High	2574	21.40	21	0.40	1	22
West High	9426	78.37	78	0.37		78
North High	750	6.24	6	0.24		6
Total	12,750	106.01	105			106

West High has lost a counselor to East High.

Exercise Set 14.4

1. a. The standard divisor is $\frac{1800}{30}$, or 60.

Course	Enrollment	Standard Quota	Lower Quota	Fractional Part	Surplus	Final Apportionment
College Algebra	978	16.30	16	0.30		16
Statistics	500	8.33	8	0.33		8
Liberal Arts Math	322	5.37	5	0.37	1	6
Total	1800	30	29			30

b. The standard divisor is $\frac{1800}{31}$, or 58.06.

Course	Enrollment	Standard Quota	Lower Quota	Fractional Part	Surplus	Final Apportionment
College Algebra	978	16.84	16	0.84	1	17
Statistics	500	8.61	8	0.61	1	9
Liberal Arts Math	322	5.55	5	0.55		5
Total	1800	31	29			31

Liberal Arts Math loses a teaching assistant when the total number of teaching assistants is raised from 30 to 31. This is an example of the Alabama paradox.

3. Standard divisor with 40 seats: $\frac{20,000}{40} = 500$. Use Hamilton's method.

State	Population	Standard Quota	Lower Quota	Fractional Part	Surplus	Final Apportionment
A	680	1.36	1	0.36	1	2
B	9150	18.30	18	0.30		18
C	10,170	20.34	20	0.34		20
Total	20,000	40	39			40

Standard divisor with 41 seats: $\frac{20,000}{41} = 487.8$. Use Hamilton's method.

State	Population	Standard Quota	Lower Quota	Fractional Part	Surplus	Final Apportionment
A	680	1.39	1	0.39		1
B	9150	18.76	18	0.76	1	19
C	10,170	20.85	20	0.85	1	21
Total	20,000	41	39			41

State A loses a seat when the total number of seats increases from 40 to 41.

5. a. Standard divisor: $\frac{3760}{24} = 156.7$. Use Hamilton's method.

State	Original Population	Standard Quota	Lower Quota	Fractional Part	Surplus	Final Apportionment
A	530	3.38	3	0.38	1	4
B	990	6.32	6	0.32		6
C	2240	14.30	14	0.30		14
Total	3760	24	23			24

b. Percent increase for state A: $\dfrac{680 - 530}{530} \approx 0.283 = 28.3\%$

Percent increase for state B: $\dfrac{1250 - 990}{990} \approx 0.263 = 26.3\%$

Percent increase for state C: $\dfrac{2570 - 2240}{2240} \approx 0.147 = 14.7\%$

c. Standard divisor: $\frac{4500}{24} = 187.5$. Use Hamilton's method.

State	New Population	Standard Quota	Lower Quota	Fractional Part	Surplus	Final Apportionment
A	680	3.63	3	0.63		3
B	1250	6.67	6	0.67	1	7
C	2570	13.71	13	0.71	1	14
Total	4500	24.01	22			24

A loses a seat while B gains, even though A has a faster increasing population. The population paradox does occur.

7. Original standard divisor: $\frac{8880}{40} = 222$

District	Original Population	Standard Quota	Lower Quota	Fractional Part	Surplus	Final Apportionment
A	1188	5.35	5	0.35		5
B	1424	6.41	6	0.41		6
C	2538	11.43	11	0.43	1	12
D	3730	16.80	16	0.80	1	17
Total	8880	39.99	38			40

New standard divisor: $\frac{9000}{40} = 225$

District	New Population	Standard Quota	Lower Quota	Fractional Part	Surplus	Final Apportionment
A	1188	5.28	5	0.28		5
B	1420	6.311	6	0.311	1	7
C	2544	11.307	11	0.307		11
D	3848	17.10	17	0.10		17
Total	9000	39.998	39			40

Percent increase by state:

A: 0% (no change)

B: $\frac{1420 - 1424}{1424} \approx -0.0028 = -0.28\%$

C: $\frac{2544 - 2538}{2538} \approx 0.0024 = 0.24\%$

D: $\frac{3848 - 3730}{3730} \approx 0.032 = 3.2\%$

C loses a truck to B even though C increased in population faster than B. This shows the population paradox occurs.

9. a. Standard divisor: $\frac{10,000}{100} = 100$

Branch	Employees	Standard Quota	Lower Quota	Fractional Part	Surplus	Final Apportionment
A	1045	10.45	10	0.45		10
B	8955	89.55	89	0.55	1	90
Total	10,000	100	99			100

b. New standard divisor: $\dfrac{10{,}525}{105} = 100.238$

Branch	Employees	Standard Quota	Lower Quota	Fractional Part	Surplus	Final Apportionment
A	1045	10.43	10	0.43	1	11
B	8955	89.34	89	0.34		89
C	525	5.24	5	0.24		5
Total	10,525	105.01	104			105

Branch B loses a promotion when branch C is added. This means the new-states paradox has occurred.

11. a. Standard divisor: $\dfrac{9450 + 90{,}550}{100} = 1000$

State	Population	Standard Quota	Lower Quota	Fractional Part	Surplus	Final Apportionment
A	9450	9.45	9	0.45		9
B	90,550	90.55	90	0.55	1	91
Total	100,000	100	99			100

b. New standard divisor: $\dfrac{100{,}000 + 10{,}400}{110} = 1003.64$

State	Population	Standard Quota	Lower Quota	Fractional Part	Surplus	Final Apportionment
A	9450	9.42	9	0.42	1	10
B	90,550	90.22	90	0.22		90
C	10,400	10.36	10	0.36		10
Total	110,400	110	109			110

State B loses a seat when state C is added.

13. a.

State	Population	Modified Quota	Modified Lower Quota	Final Apportionment
A	99,000	6.39	6	6
B	214,000	13.81	13	13
C	487,000	31.42	31	31
Total	800,000		50	50

b.

State	Population	Modified Quota	Modified Lower Quota	Final Apportionment
A	99,000	6.39	6	6
B	214,000	13.81	13	13
C	487,000	31.42	31	37
D	116,000	7.48	7	7
Total	916,000		57	57

The new-states paradox does not occur. As long as the modified divisor, d, remains the same, adding a new state cannot change the number of seats held by existing states.

Chapter 14 Review Exercises

1.

Number of Votes	4	3	3	2
First Choice	A	B	C	C
Second Choice	B	D	B	B
Third Choice	C	C	D	A
Fourth Choice	D	A	A	D

2. $9 + 5 + 4 + 2 + 2 + 1 = 23$

3. 4

4. $9 + 5 + 2 = 16$

5. $9 + 5 = 14$

6. M receives 12 first-choice votes, compared to 10 for C and 2 for D, so M (Musical) is selected.

7.

Number of Votes	10	8	4	2
First Choice: 3 points	C: $10 \times 3 = 30$	M: $8 \times 3 = 24$	M: $4 \times 3 = 12$	D: $2 \times 3 = 6$
Second Choice: 2 points	D: $10 \times 2 = 20$	C: $8 \times 2 = 16$	D: $4 \times 2 = 8$	M: $2 \times 2 = 4$
Third Choice: 1 point	M: $10 \times 1 = 10$	D: $8 \times 1 = 8$	C: $4 \times 1 = 4$	C: $2 \times 1 = 2$

C gets $30 + 16 + 4 + 2 = 52$ points.
D gets $20 + 8 + 8 + 6 = 42$ points.
M gets $10 + 24 + 12 + 4 = 50$ points.

C (Comedy) gets the most points and is chosen.

8. There are 24 voters, so 13 votes are needed for a majority. None of the candidates has 13 first-place votes. D has the fewest first-place votes and is eliminated in the next round.

Number of Votes	10	14
First Choice	C	M
Second Choice	M	C

M (Musical) has 14 first-place votes, a majority, so a musical is selected.

9. There are 3 choices so we make $\frac{3(3-1)}{2} = 3$ comparisons.

Comparison	Vote Results	Conclusion
C vs. D	18 voters prefer C to D. 6 voters prefer D to C.	C wins and gets 1 point.
C vs. M	10 voters prefer C to M. 14 voters prefer M to C.	M wins and gets 1 point.
D vs. M	12 voters prefer D to M. 12 voters prefer M to D.	D and M tie. Each gets $\frac{1}{2}$ point.

C gets 1 point, D gets $\frac{1}{2}$ point, and M gets $1\frac{1}{2}$ points. So M (Musical) wins, and is selected.

10. A receives 40 first-place votes, compared to 30 for B, 6 for C, and 2 for D. So A wins.

11.

Number of Votes	40	30	6	2
First Choice: 4 points	A: $40 \times 4 = 160$	B: $30 \times 4 = 120$	C: $6 \times 4 = 24$	D: $2 \times 4 = 8$
Second Choice: 3 points	B: $40 \times 3 = 120$	C: $30 \times 3 = 90$	D: $6 \times 3 = 18$	B: $2 \times 3 = 6$
Third Choice: 2 points	C: $40 \times 2 = 80$	D: $30 \times 2 = 60$	B: $6 \times 2 = 12$	C: $2 \times 2 = 4$
Fourth Choice: 1 point	D: $40 \times 1 = 40$	A: $30 \times 1 = 30$	A: $6 \times 1 = 6$	A: $2 \times 1 = 2$

A gets $160 + 30 + 6 + 2 = 198$ points.
B gets $120 + 120 + 12 + 6 = 258$ points.
C gets $80 + 90 + 24 + 4 = 198$ points.
D gets $40 + 60 + 18 + 8 = 126$ points.

B receives the most points, and wins.

12. There are 78 voters, so 40 first-place votes are needed for a majority. A has 40 first-place votes, and wins.

13. There are 4 candidates, so $\frac{4(4-1)}{2} = 6$ comparisons are needed.

Comparison	Vote Results	Conclusion
A vs. B	40 voters prefer A to B. 38 voters prefer B to A.	A wins and gets 1 point.
A vs. C	40 voters prefer A to C. 38 voters prefer C to A.	A wins and gets 1 point.
A vs. D	40 voters prefer A to D. 38 voters prefer D to A.	A wins and gets 1 point.
B vs. C	72 voters prefer B to C. 6 voters prefer C to B.	B wins and gets 1 point.
B vs. D	70 voters prefer B to D. 8 voters prefer D to B.	B wins and gets 1 point.
C vs. D	76 voters prefer C to D. 2 voters prefer D to C.	C wins and gets 1 point.

A gets 3 points, B gets 2 points, C gets 1 point, and D gets 0 points. So A wins.

14.

Number of Votes	1500	600	300
First Choice: 4 points	A: $1500 \times 4 = 6000$	B: $600 \times 4 = 2400$	C: $300 \times 4 = 1200$
Second Choice: 3 points	B: $1500 \times 3 = 4500$	D: $600 \times 3 = 1800$	B: $300 \times 3 = 900$
Third Choice: 2 points	C: $1500 \times 2 = 3000$	C: $600 \times 2 = 1200$	D: $300 \times 2 = 600$
Fourth Choice: 1 point	D: $1500 \times 1 = 1500$	A: $600 \times 1 = 600$	A: $300 \times 1 = 300$

A gets $6000 + 600 + 300 = 6900$ points.
B gets $4500 + 2400 + 900 = 7800$ points.
C gets $3000 + 1200 + 1200 = 5400$ points.
D gets $1500 + 1800 + 600 = 3900$ points.

B receives the most points, and wins.

15. A has a majority of first-place votes. In Exercise 14, B wins and so the majority criterion is not satisfied.

16. A is favored above all others using a head-to-head comparison. This is automatically true, since A has a majority of first-place votes. In Exercise 14, B wins and so the head-to-head criterion is not satisfied.

17. There are 2500 voters. 1251 first-place votes are needed for a majority. B has 1500 first-place votes, and is the winner.

18. B is favored above all others using a head-to-head comparison. This is automatically true, since B has a majority of first-place votes. In Exercise 17, B wins and so the head-to-head criterion is satisfied.

19. A receives 180 first-place votes, compared with 100 for B, 30 for C, and 40 for D. Therefore A wins.

20.

Number of Votes	180	100	40	30
First Choice: 4 points	A: $180 \times 4 = 720$	B: $100 \times 4 = 400$	D: $40 \times 4 = 160$	C: $30 \times 4 = 120$
Second Choice: 3 points	B: $180 \times 3 = 540$	D: $100 \times 3 = 300$	B: $40 \times 3 = 120$	B: $30 \times 3 = 90$
Third Choice: 2 points	C: $180 \times 2 = 360$	A: $100 \times 2 = 200$	C: $40 \times 2 = 80$	A: $30 \times 2 = 60$
Fourth Choice: 1 point	D: $180 \times 1 = 180$	C: $100 \times 1 = 100$	A: $40 \times 1 = 40$	D: $30 \times 1 = 30$

A gets $720 + 200 + 40 + 60 = 1020$ points.
B gets $540 + 400 + 120 + 90 = 1150$ points.
C gets $360 + 100 + 80 + 120 = 660$ points.
D gets $180 + 300 + 160 + 30 = 670$ points.

B gets the most points, and wins.

21. There are 350 voters. 176 first-place votes are needed for a majority. A has 180 votes, a majority, and wins.

22. There are 4 candidates, and therefore $\frac{4(4-1)}{2} = 6$ comparisons.

Comparison	Vote Results	Conclusion
A vs. B	180 voters prefer A to B. 170 voters prefer B to A.	A wins and gets 1 point.
A vs. C	280 voters prefer A to C. 70 voters prefer C to A.	A wins and gets 1 point.
A vs. D	210 voters prefer A to D. 140 voters prefer D to A.	A wins and gets 1 point.
B vs. C	320 voters prefer B to C. 30 voters prefer C to B.	B wins and gets 1 point.
B vs. D	310 voters prefer B to D. 40 voters prefer D to B.	B wins and gets 1 point.
C vs. D	210 voters prefer C to D. 140 voters prefer D to C.	C wins and gets 1 point.

A gets 3 points, B gets 2 points, C gets 1 point, and D gets 0 points. Therefore A wins.

23. A has a majority of first-place votes. Based on Exercises 19–22, only the Borda count method violates the majority criterion. B wins by the Borda count method.

24. There are 1450 voters. 726 first-place votes are needed for a majority. No candidate has a majority. A has the fewest first-place votes and is eliminated in the next round.

Number of Votes	900	550
First Choice	B	C
Second Choice	C	B

B has the majority of first-place votes, and wins.

25. There is a new preference table:

Number of Votes	700	400	350
First Choice	B	A	C
Second Choice	C	B	A
Third Choice	A	C	B

No candidate has a majority of first-place votes. C has the fewest first-place votes, and is eliminated in the next round.

Number of Votes	700	750
First Choice	B	A
Second Choice	A	B

A has a majority of first-place votes, and wins. This does not satisfy the monotonicity criterion, since the only change gave B more first-place votes, but after the change B lost the election.

26. A has 400 first-place votes, compared to 200 for B and 250 for C. Therefore A wins.

27.

Number of Votes	400	450
First Choice	A	C
Second Choice	C	A

C has the majority of first-place votes, and wins this election. The irrelevant alternatives criterion is not satisfied, because removing B changes the winner from A to C.

28.

Number of Votes	400	250	200
First Choice: 3 points	A: $400 \times 3 = 1200$	C: $250 \times 3 = 750$	B: $200 \times 3 = 600$
Second Choice: 2 points	B: $400 \times 2 = 800$	B: $250 \times 2 = 500$	C: $200 \times 2 = 400$
Third Choice: 1 point	C: $400 \times 1 = 400$	A: $250 \times 1 = 250$	A: $200 \times 1 = 200$

A gets $1200 + 250 + 200 = 1650$ points.
B gets $800 + 500 + 600 = 1900$ points.
C gets $400 + 750 + 400 = 1550$ points.

B gets the most points, and wins.

29.

Number of Votes	400	450
First Choice: 2 points	A: $400 \times 2 = 800$	B: $450 \times 2 = 900$
Second Choice: 1 point	B: $400 \times 1 = 400$	A: $450 \times 1 = 450$

A gets $800 + 450 = 1250$ points.
B gets $400 + 900 = 1300$ points.

B still gets the most points, and wins. The same thing happens if A drops out instead of C, and so the irrelevant alternatives criterion is satisfied.

30. $\dfrac{275 + 392 + 611 + 724}{40} = \dfrac{2002}{40} = 50.05$

31. With a standard divisor of 50.05:

Clinic	A	B	C	D
Average weekly patient load	275	392	611	724
Standard Quota	5.49	7.83	12.21	14.47

32. Using the results of Exercise 31:

Clinic	Standard Quota	Lower Quota	Upper Quota
A	5.49	5	6
B	7.83	7	8
C	12.21	12	13
D	14.47	14	15

33.

Clinic	Standard Quota	Lower Quota	Fractional Part	Surplus	Final Apportionment
A	5.49	5	0.49	1	6
B	7.83	7	0.83	1	8
C	12.21	12	0.21		12
D	14.47	14	0.47		14
Total	40	38			40

34.

Clinic	Average Weekly Patient Load	Modified Quota $(d = 48)$	Modified Lower Quota	Final Apportionment
A	275	5.73	5	5
B	392	8.17	8	8
C	611	12.73	12	12
D	724	15.08	15	15
Total	2002		40	40

35.

Clinic	Average Weekly Patient Load	Modified Quota $(d = 52)$	Modified Upper Quota	Final Apportionment
A	275	5.29	6	6
B	392	7.54	8	8
C	611	11.75	12	12
D	724	13.92	14	14
Total	2002		40	40

36.

Clinic	Average Weekly Patient Load	Modified Quota $(d = 49.95)$	Modified Rounded Quota	Final Apportionment
A	275	5.51	6	6
B	392	7.85	8	8
C	611	12.23	12	12
D	724	14.49	14	14
Total	2002		40	40

37. Standard divisor: $\dfrac{3320 + 10,060 + 15,020 + 19,600}{200} = \dfrac{48,000}{200} = 240$

State	Population	Standard Quota	Lower Quota	Fractional Part	Surplus	Final Apportionment
A	3320	13.83	13	0.83	1	14
B	10,060	41.92	41	0.92	1	42
C	15,020	62.58	62	0.58		62
D	19,600	81.67	81	0.67	1	82
Total	48,000	200	197			200

38. Try modified divisor $d = 238$.

State	Population	Modified Quota	Modified Lower Quota	Final Apportionment
A	3320	13.95	13	13
B	10,060	42.27	42	42
C	15,020	63.11	63	63
D	19,600	82.35	82	82
Total	48,000		200	200

39. Try modified divisor $d = 242$.

State	Population	Modified Quota	Modified Upper Quota	Final Apportionment
A	3320	13.72	14	14
B	10,060	41.57	42	42
C	15,020	62.07	63	63
D	19,600	80.99	81	81
Total	48,000		200	200

40. Try modified divisor $d = 240.4$.

State	Population	Modified Quota	Modified Rounded Quota	Final Apportionment
A	3320	13.81	14	14
B	10,060	41.85	42	42
C	15,020	62.48	62	62
D	19,600	81.53	82	82
Total	48,000		200	200

41. **a.** Standard divisor: $\dfrac{7500}{150} = 50$

School	Enrollment	Standard Quota	Lower Quota	Fractional Part	Surplus	Final Apportionment
A	370	7.4	7	0.4	1	8
B	3365	67.3	67	0.3		67
C	3765	75.3	75	0.3		75
Total	7500	150	149			150

b. Standard divisor: $\dfrac{7500}{151} = 49.67$

School	Enrollment	Standard Quota	Lower Quota	Fractional Part	Surplus	Final Apportionment
A	370	7.45	7	0.45		7
B	3365	67.75	67	0.75	1	68
C	3765	75.80	75	0.80	1	76
Total	7500	151	149			151

The Alabama paradox occurs. A loses a laptop when the overall number of laptops changes from 150 to 151.

42. **a.** Standard divisor: $\dfrac{200,000}{100} = 2000$

School	Original Population	Standard Quota	Lower Quota	Fractional Part	Surplus	Final Apportionment
A	143,796	71.90	71	0.90	1	72
B	41,090	20.55	20	0.55		20
C	15,114	7.56	7	0.56	1	8
Total	200,000	100.01	98			100

b. Percent increase of B: $\dfrac{41,420 - 41,090}{41,090} \approx 0.0080 = 0.8\%$

Percent increase of C: $\dfrac{15,304 - 15,114}{15,114} \approx 0.0126 \approx 1.3\%$

c. Standard divisor: $\dfrac{200,520}{100} = 2005.2$

School	New Population	Standard Quota	Lower Quota	Fractional Part	Surplus	Final Apportionment
A	143,796	71.71	71	0.71	1	72
B	41,420	20.66	20	0.66	1	21
C	15,304	7.63	7	0.63		7
Total	200,520	100	98			100

The population paradox occurs. C loses a seat to B, even though C is growing faster.

43. a. Standard divisor: $\dfrac{1650}{33} = 50$

Branch	Employees	Standard Quota	Lower Quota	Fractional Part	Surplus	Final Apportionment
A	372	7.44	7	0.44		7
B	1278	25.56	25	0.56	1	26
Total	1650	33	32			33

b. Standard divisor: $\dfrac{2005}{40} = 50.125$

Branch	Employees	Standard Quota	Lower Quota	Fractional Part	Surplus	Final Apportionment
A	372	7.42	7	0.42		7
B	1278	25.50	25	0.50	1	26
C	355	7.08	7	0.08		7
Total	2005	40	39			40

The new-states paradox does not occur. Neither branch A nor branch B loses any promotions.

44. False. Answers will vary.

Chapter 14 Test

1. $1200 + 900 + 900 + 600 = 3600$

2. 600

3. $900 + 600 = 1500$

4. $900 + 600 = 1500$

5. A received 1200 first-place votes, B received 1500, and C received 900. Therefore B wins.

6.

Number of Votes	1200	900	900	600
First Choice: 3 points	A: $1200 \times 3 = 3600$	C: $900 \times 3 = 2700$	B: $900 \times 3 = 2700$	B: $600 \times 3 = 1800$
Second Choice: 2 points	B: $1200 \times 2 = 2400$	A: $900 \times 2 = 1800$	C: $900 \times 2 = 1800$	A: $600 \times 2 = 1200$
Third Choice: 1 point	C: $1200 \times 1 = 1200$	B: $900 \times 1 = 900$	A: $900 \times 1 = 900$	C: $600 \times 1 = 600$

A gets $3600 + 1800 + 900 + 1200 = 7500$ points.
B gets $2400 + 900 + 2700 + 1800 = 7800$ points.
C gets $1200 + 2700 + 1800 + 600 = 6300$ points.

B receives the most points and is the winner.

7. There are 3600 voters. 1801 first-place votes are needed for a majority. No candidate has a majority. C receives the fewest first-place votes and is eliminated in the next round.

Number of Votes	2100	1500
First Choice	A	B
Second Choice	B	A

A receives the majority of first-place votes, and wins.

8. There are 3 candidates. The number of comparisons is $\frac{3(3-1)}{2}$, or 3.

Comparison	Vote Results	Conclusion
A vs. B	2100 voters prefer A to B. 1500 voters prefer B to A.	A wins and gets 1 point.
A vs. C	1800 voters prefer A to C. 1800 voters prefer C to A.	A and C tie. Each gets $\frac{1}{2}$ point.
B vs. C	2700 voters prefer B to C. 900 voters prefer C to B.	B wins and gets 1 point.

A gets $1\frac{1}{2}$ points, B gets 1 point, and C gets $\frac{1}{2}$ point. Therefore A wins.

9.

Number of Votes	240	160	60
First Choice: 4 points	A: $240 \times 4 = 960$	C: $160 \times 4 = 640$	D: $60 \times 4 = 240$
Second Choice: 3 points	B: $240 \times 3 = 720$	B: $160 \times 3 = 480$	A: $60 \times 3 = 180$
Third Choice: 2 points	C: $240 \times 2 = 480$	D: $160 \times 2 = 320$	C: $60 \times 2 = 120$
Fourth Choice: 1 point	D: $240 \times 1 = 240$	A: $160 \times 1 = 160$	B: $60 \times 1 = 60$

A gets $960 + 160 + 180 = 1300$ points.
B gets $720 + 480 + 60 = 1260$ points.
C gets $480 + 640 + 120 = 1240$ points.
D gets $240 + 320 + 240 = 800$ points.

A gets the most points, and wins.

10. A has the majority of first-place votes. Based on Exercise 9, the majority criterion is satisfied.

11. A has 1500 first-place votes, whereas B and C have 1000 each. Therefore A wins.

12. B is favored when compared to A, by 2000 votes to 1500. B is favored when compared to C, by 2500 votes to 1000. So B is favored in each head-to-head comparison. Based on Exercise 11, the head-to-head criterion is not satisfied, because A wins the election.

13. There are 210 voters. 106 votes are needed for a majority. No candidate has a majority. B receives the fewest first-place votes and is eliminated in the next round.

Number of Votes	130	80
First Choice	C	A
Second Choice	A	C

C receives a majority of votes, and wins.

14. New preference table:

Number of Votes	100	60	50
First Choice	C	B	A
Second Choice	A	C	B
Third Choice	B	A	C

No candidate has a majority. A has the fewest first-place votes and is eliminated in the next round.

Number of Votes	100	110
First Choice	C	B
Second Choice	B	C

B has the majority of first-place votes, and wins. The monotonicity criterion is not satisfied, because the only change gave more first-place votes to C, but C lost the second election.

15. B has 90 first-place votes, C has 75, and A has 45. Therefore B wins. If C drops out, there is a new preference table:

Number of Votes	90	120
First Choice	B	A
Second Choice	A	B

A has a majority of first-place votes, and wins. This changed outcome shows that the irrelevant alternatives criterion is not satisfied.

16. $\dfrac{119+165+216}{10}=\dfrac{500}{10}=50$

17. A: $\dfrac{119}{50}=2.38$ B: $\dfrac{165}{50}=3.3$ C: $\dfrac{216}{50}=4.32$

18. A: 2, 3; B: 3, 4; C: 4, 5

19.

Clinic	Average Weekly Patient Load	Standard Quota	Lower Quota	Fractional Part	Surplus	Final Apportionment
A	119	2.38	2	0.38	1	3
B	165	3.3	3	0.3		3
C	216	4.32	4	0.32		4
Total	500	10	9			10

20.

Clinic	Average Weekly Patient Load	Modified Quota ($d = 42$)	Modified Lower Quota	Final Apportionment
A	119	2.83	2	2
B	165	3.93	3	3
C	216	5.14	5	5
Total	500		10	10

21.

Clinic	Average Weekly Patient Load	Modified Quota ($d = 56$)	Modified Upper Quota	Final Apportionment
A	119	2.13	3	3
B	165	2.95	3	3
C	216	3.86	4	4
Total	500		10	10

22.

Clinic	Average Weekly Patient Load	Modified Quota ($d = 47.7$)	Modified Rounded Quota	Final Apportionment
A	119	2.49	2	2
B	165	3.46	3	3
C	216	4.52	5	5
Total	500		10	10

23. New standard divisor: $\dfrac{500}{11} = 45.45$

Clinic	Average Weekly Patient Load	Standard Quota	Lower Quota	Fractional Part	Surplus	Final Apportionment
A	119	2.62	2	0.62		2
B	165	3.63	3	0.63	1	4
C	216	4.75	4	0.75	1	5
Total	500	11	9			11

The Alabama paradox occurs. Clinic A loses one doctor when the total number of doctors is raised from 10 to 11.

24. New standard divisor: $\dfrac{500+110}{12}=\dfrac{610}{12}=50.83$

Clinic	Average Weekly Patient Load	Standard Quota	Lower Quota	Fractional Part	Surplus	Final Apportionment
A	119	2.34	2	0.34	1	3
B	165	3.25	3	0.25		3
C	216	4.25	4	0.25		4
D	110	2.16	2	0.16		2
Total	610	12	11			12

The new-states paradox does not occur. No clinic loses doctors when a new clinic is added.

25. Answers will vary.

Chapter 15
Graph Theory

Check Points 15.1

1. Graphs (a) and (b) both have vertices *A, B, C, D,* and *E*. Also, both graphs have edges *AB, AC, BD, BE, CD, CE,* and *DE*.

   Because the two graphs have the same number of vertices connected to each other in the same way, they are the same. In fact, graph (b) is just graph (a) rotated clockwise and bent out of shape.

2. Draw points for the five land masses and label them *N, S, A, B,* and *C*.

   There is one bridge that connects North Metroville to Island A, so one edge is drawn connecting vertex *N* to vertex *A*. Similarly, one edge connects vertex *A* with vertex *B*, and one edge connects vertex *B* with vertex *C*. Since there are two bridges connecting Island C to South Metroville, two edges connect vertex *C* with vertex *S*.

   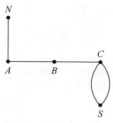

3. We use the abbreviations for the states to label the vertices: ID for Idaho, MT for Montana, WY for Wyoming, UT for Utah, and CO for Colorado. The precise placement of these vertices is not important.

   Whenever two states share a common border, we connect the respective vertices with an edge. For example, Idaho shares a common border with Montana, with Wyoming, and with Utah. Continuing in this manner, we obtain the following graph.

4. We use the letters in Figure 15.13 to label each vertex. Only one door connects the outside, *E*, with room *B*, so we draw one edge from vertex *E* to vertex *B*. Two doors connect the outside, *E*, to room *D*, so we draw two edges from *E* to *D*. Counting doors between the rooms, we complete the following graph.

   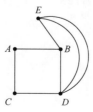

5. We label each of the corners and intersections with an upper-case letter and use points to represent the corners and street intersections. Now we are ready to draw the edges that represent the streets the security guard has to walk. Each street only needs to be walked once, so we draw one edge to represent each street. This results in the following graph.

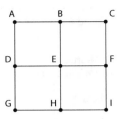

6. We systematically list which pairs of vertices are adjacent, working alphabetically. Thus, the adjacent vertices are *A* and *B*, *A* and *C*, *A* and *D*, *A* and *E*, *B* and *C*, and *E* and *E*.

Exercise Set 15.1

1. There are six edges attached to the Pittsburgh vertex, so Pittsburgh plays six games during the week. One edge connects the Pittsburgh vertex to the St. Louis vertex, so one game is against St. Louis. One edge connects the Pittsburgh vertex to Chicago, so one game is against Chicago. Two edges connect the Pittsburgh vertex to the Philadelphia vertex, so two games are against Philadelphia. Two edges connect the Pittsburgh vertex to the Montreal vertex, so two games are against Montreal.

3. No. Montreal is farther north than New York but is drawn lower on the graph. However, the graph is not drawn incorrectly. Only the games between teams are important, and these are represented by the edges. Geographic position is not relevant.

5. Possible answers:

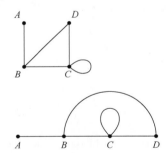

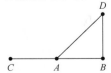

7. Both graphs have vertices *A*, *B*, *C*, and *D* and edges *AB*, *AC*, *AD*, and *BD*. The two graphs have the same number of vertices connected in the same way, so they are the same.

   Possible answer:

   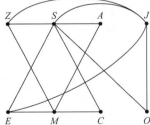

9. We label each student's vertex with the first letter of his or her name. An edge connecting two vertices represents a friendship prior to forming the homework group. The following graph results.

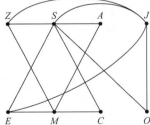

11. Label one vertex *N*, for North Gothamville. Label another *S*, for South Gothamville. Label the islands, from left to right, *A, B,* and *C*. Label three vertices accordingly. Use edges to represent bridges. The following graph results.

13. We use the abbreviations WA, OR, ID, MT, and WY to label the vertices representing Washington, Oregon, Idaho, Montana, and Wyoming. The following graph results.

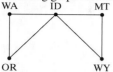

19.

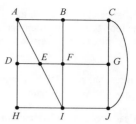

15.

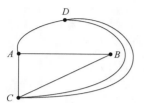

21.

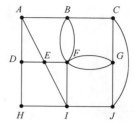

17.

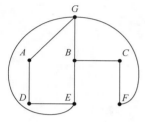

23. The degree of a vertex is the number of edges at that vertex. Thus, vertex A has degree 2, vertex B has degree 2, vertex C has degree 3, vertex D has degree 3, vertex E has degree 3, and vertex F has degree 1. (The loop at E counts for 2.)

25. Vertices B and C each have an edge connecting to A, so B and C are adjacent to A.

27. Starting at vertex A, we proceed to vertex C, then vertex D. This is one path from A to D. For a second path, start at vertex A, then proceed to vertex B, then C, then D.

29. The edges not included are the edge connecting A to C, and the edge connecting D to F.

31. While edge CD is included, the graph is connected. If we remove CD, the graph will be disconnected. Thus, CD is a bridge.

33. Edge DF is also a bridge. With it, the graph is connected. If DF is removed, vertex F stands alone, so the graph is disconnected.

35. Vertices A, B, G, H, and I each have two attached edges, which is an even number of edges. Thus A, B, G, H, and I are even vertices. Vertex C has five attached edges, vertex E has one, and vertices D and F have three. These are odd numbers of edges. Thus C, E, D, and F are odd vertices.

37. Vertex F has edges connecting to vertices D, G, and I. Thus D, G, and I are adjacent to F.

39. Begin at vertex B. Proceed to vertex C, then vertex D, then vertex F. This is one path from B to F. For a second path, begin at B, then proceed to A, then C, then D, then F.

41. Begin at vertex G. proceed to vertex F, then vertex I, then vertex H, then vertex G. This is a circuit. (The counterclockwise order also works.)

43. Begin at vertex *A*. Proceed to vertex *B*, then vertex *C*, then around the loop to *C* again, then vertex *D*, then vertex *F*, then vertex *G*, then vertex *H*, then vertex *I*.

45. *G*, *F*, *D*, *E*, *D* requires that edge *DE* be traversed twice. This is not allowed within a path.

47. *H*, *I*, *F*, *E* is not a path because no edge connects vertices *F* and *E*.

49. Possible answer:

Each vertex has degree 2.

51. Possible answer:

Vertex *A* has degree 1, and the rest have degree 3.

67. Use vertices to represent the six members.

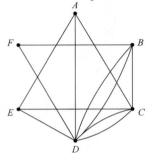

Check Points 15.2

1. We use trial and error to find one such path. The following figure shows a result.

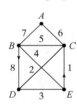

Using vertex letters to name the path, we write *E*, *C*, *D*, *E*, *B*, *C*, *A*, *B*, *D*.

2. We use trial and error to find an Euler circuit that starts at *G*. The following figure shows a result.

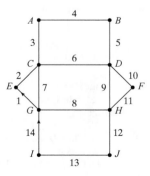

Using vertex letters to name the circuit, we write *G*, *E*, *C*, *A*, *B*, *D*, *C*, *G*, *H*, *D*, *F*, *H*, *J*, *I*, *G*.

3. a.   A walk through every room and the outside, using each door exactly once, means that we are looking for an Euler path or Euler circuit on the graph in Figure 15.34(b). This graph has exactly two odd vertices, namely B and E. By Euler's theorem, the graph has at least one Euler path, but no Euler circuit. It is possible to walk through every room and the outside, using each door exactly once. It is not possible to begin and end the walk in the same place.

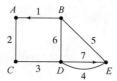

b.   Euler's theorem tells us that a possible Euler path must start at one of the odd vertices and end at the other. We use trial and error to find such a path, starting at vertex B (room B in the floor plan), and ending at vertex E (outside in the floor plan). Possible paths follow.

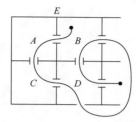

4.   The graph has no odd vertices, so we can begin at any vertex. We choose vertex C as the starting point. From C we can travel to A, B, or D. We choose to travel to D.

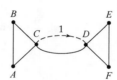

Now the remaining edge CD is a bridge, so we must travel to either E or F. We choose F.

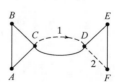

We have no choices for our next three steps, which are bridges. We must travel to E, then D, then C.

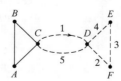

From *C*, we may travel to either *A* or *B*. We choose *B*. Then we must travel to *A*, then back to *C*.

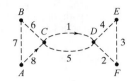

The above figure shows the completed Euler circuit. Written using the letters of the vertices, the path is *C, D, F, E, D, C, B, A, C*.

Exercise Set 15.2

1.

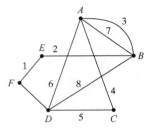

This path does not include edge *FD*, so it is neither an Euler path nor an Euler circuit.

3.

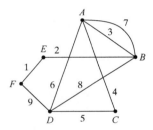

This path travels through each edge of the graph once, and only once. It begins and ends at *F*. Therefore, it is an Euler circuit.

5.

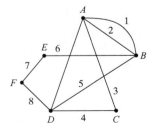

This path does not include edge *AD*, so it is neither an Euler path nor an Euler circuit.

7. a. There are exactly two odd vertices, namely *A* and *B*, so by Euler's theorem there is at least one Euler path.

b.

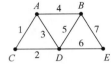

This path begins at *A* and ends at *B*.

9. a. There are no odd vertices, so by Euler's theorem, there is at least one Euler circuit.

b.

This circuit begins and ends at *C*.

11. There are more than two odd vertices, namely *B, D, G,* and *K*. Therefore by Euler's theorem, there are no Euler paths and no Euler circuits.

13. Since the graph has no odd vertices, it must have an Euler circuit, by Euler's theorem.

15. Since the graph has exactly two odd vertices, it has an Euler path, but no Euler circuit, by Euler's theorem.

17. Since the graph has more than two odd vertices, it has neither an Euler path nor an Euler circuit, by Euler's theorem.

19. a. All vertices are even, so there must be an Euler circuit.

b.

A   B
1   8 7   6
C   D   E
2   3 4   5
F   G

21. a. There are exactly two odd vertices, so there must be an Euler path.

b.

A
1
B   4   C
5   3   2
D   6   E
7
F

23. a. There are more than two odd vertices, so there is neither an Euler path nor an Euler circuit.

25. a. There are exactly two odd vertices, so there is an Euler path.

b.

A   2   D
1   3   5   6
B   4   C

27. a. There are no odd vertices, so there is an Euler circuit.

b.

B   1   A
4   2   6
3
C   5   D

29. a. There are more than two odd vertices, so there is neither an Euler path nor an Euler circuit.

31. a. There are exactly two odd vertices, so there is an Euler path.

b.

E   F
4
A   3   5
10
9   C   D   6
7
8   2   G
B   H   1   I

33. The two odd vertices in the graph are A and C. We start with A, so we must progress next to B. From B, we may travel to C, D, or E. We choose C.

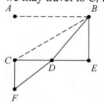

Next we travel to F, then D, then E.

Finally we travel to B, then D, and last, to C. We label each step taken.

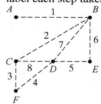

35. The two odd vertices in the graph are *A* and *C*. We start with *A*, then travel to *B*, *C*, and *E*.

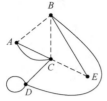

We continue on to *B*, *D*, *D*, *C*, *A*, and *C*. We label each step taken.

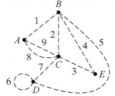

37. We begin with *A*, and travel to *D*, *H*, *G*, *F*, *E*, *B*, and *C*.

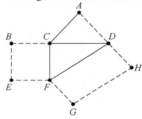

We continue on to *F*, *D*, *C*, and back to *A*. We label each step.

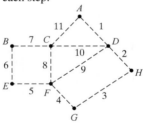

39. We begin with *A*, and travel to *C*, *G*, *K*, *H*, *I*, *L*, *J*, *F*, *B*, *E*, and *D*.

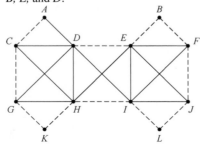

We continue on to *C*, *H*, *G*, *D*, *H*, *E*, *F*, *I*, *J*, *E*, *I*, *D*, and back to *A*. We label each step.

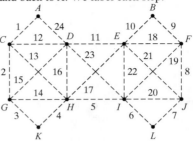

41. a. Remove *FG*.

b. Sample Euler circuit: *EC*, *CB*, *BD*, *DF*, *FA*, *AD*, *DG*, *GH*, *HC*, *CG*, *GB*, *BH*, *HE*

43. a. Remove *BA* and *FJ*.

b. Sample Euler circuit: *CA*, *AD*, *DI*, *IH*, *HG*, *GF*, *FC*, *CD*, *DE*, *EH*, *HJ*, *JG*, *GB*, *BC*

45. The graph that models the neighborhood has no odd vertices, so an Euler circuit exists with any vertex, including *B*, as the starting point.

47.

```
                        Start/Stop
        ┌─────────────────────◄──────────────────┐
        │ A x x x x  B x x x x x x x x x x  C │
        │ x                                  x │
        │ x                                  x │
        │ x                                  x │
        │ x                                  x │
        │ D  x x x x x x  E  x x x x x x x   F │
        │ x  x x x x x x  x  x x x x x x x   x │
        │ x               x x               x │
        │ x               x x               x │
        │ x               x x               x │
        │ G  x x x x x  H  x x x x x x x    I │
        └─────────►───────────────────►────────┘
```

49. a.

A ——— *B* ——————— *C*

D   *E* *F*   *G* *H*

                I *J* *K*

b. There are exactly two odd vertices, namely *E* and *B*. Therefore the guard should begin at one of these vertices and end at the other.

51. a. Label the vertices *N* for North Bank, *S* for South Bank, and *A* and *B* for the two islands. Draw edges to represent bridges.

b. The graph has exactly two odd vertices, *N* and *B*, so residents can walk across all the bridges without crossing the same bridge twice.

c.

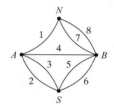

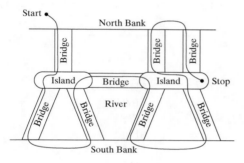

53. Use NJ to label the New Jersey vertex, M for Manhattan, SI for Staten Island, and LI for Long Island. Each edge represents a bridge.

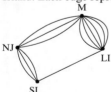

There are exactly two odd vertices, M and LI, so the graph has an Euler path. Therefore it is possible to visit each location, using each bridge or tunnel exactly once.

55. a.

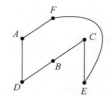

b. There are no odd vertices, so the graph has an Euler circuit. Therefore, it is possible to walk through each room and the outside, using each door exactly once.

c.

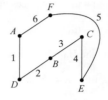

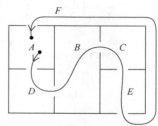

57.

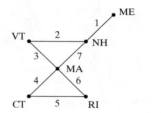

59. a.

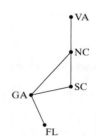

b. There are more than two odd vertices, so no Euler path exists. Therefore it is not possible to travel through these states, crossing each border exactly once.

d. For the same reason as in (b), this is not possible.

Check Points 15.3

1. a. A Hamilton path must pass through each vertex exactly once. The graph has many Hamilton paths. An example of such a path is *E, C, D, G, B, A, F*.

   b. A Hamilton circuit must pass through every vertex exactly once and begin and end at the same vertex. The graph has many Hamilton circuits. An example of such a circuit is *E, C, D, G, B, F, A, E*.

2. In each case, we use the expression $(n-1)!$. For three vertices, substitute 3 for n in the expression. For six and ten vertices, substitute 6 and 10, respectively, for n.

   a. A complete graph with three vertices has $(3-1)! = 2! = 2 \cdot 1 = 2$ Hamilton circuits.

   b. A complete graph with six vertices has $(6-1)! = 5! = 5 \cdot 4 \cdot 3 \cdot 2 \cdot 1 = 120$ Hamilton circuits.

   c. A complete graph with ten vertices has $(10-1)! = 9! = 9 \cdot 8 \cdot 7 \cdot 6 \cdot 5 \cdot 4 \cdot 3 \cdot 2 \cdot 1 = 362{,}880$ Hamilton circuits.

3. The trip described by the Hamilton circuit *A, C, B, D, A* involves the sum of four costs:

$124 + $126 + $155 + $157 = $562.

Here, $124 is the cost of the trip from *A* to *C*; $126 is the cost from *C* to *B*; $155 is the cost from *B* to *D*; and $157 is the cost from *D* to *A*. The total cost of the trip is $562.

4. The graph has four vertices. Thus, using $(n-1)!$, there are $(4-1)! = 3! = 6$ possible Hamilton circuits. The 6 possible Hamilton circuits and their costs are shown.

Hamilton Circuit	Sum of the Weights of the Edges	=	Total Cost
A, B, C, D, A	20 + 15 + 50 + 30	=	$115
A, B, D, C, A	20 + 10 + 50 + 70	=	$150
A, C, B, D, A	70 + 15 + 10 + 30	=	$125
A, C, D, B, A	70 + 50 + 10 + 20	=	$150
A, D, B, C, A	30 + 10 + 15 + 70	=	$125
A, D, C, B, A	30 + 50 + 15 + 20	=	$115

The two Hamilton circuits having the lowest cost of $115 are *A, B, C, D, A* and *A, D, C, B, A*.

5. The Nearest Neighbor method is carried out as follows:

- Start at *A*.

- Choose the edge with the smallest weight: 13. Move along this edge to *B*.

- From *B*, choose the edge with the smallest weight that does not lead to *A*: 5. Move along this edge to *C*.

- From *C*, choose the edge with the smallest weight that does not lead to a city already visited: 12. Move along this edge to *D*.

- From *D*, the only choice is to fly to *E*, the only city not yet visited: 154.

- From *E*, close the circuit and return home to *A*: 14.

An approximate solution is the Hamilton circuit *A, B, C, D, E, A*. The total weight is $13 + 5 + 12 + 154 + 14 = 198$.

Exercise Set 15.3

1. One such path is *A, G, C, F, E, D, B.*

3. One such circuit is *A, B, G, C, F, E, D, A.*

5. One such path is *A, F, G, E, C, B, D.*

7. One such circuit is *A, B, C, E, G, F, D, A.*

9. a. This graph is not complete. For example, no edge connects *A* and *B*. Therefore it may not have Hamilton circuits.

11. a. This graph is complete: there is an edge between each pair of vertices. Therefore it must have Hamilton circuits.

   b. There are 6 vertices, so the number of Hamilton circuits is $(6 - 1)! = 5! = 120$.

13. a. This graph is not complete. For example, no edge connects *G* and *F*. Therefore it may not have Hamilton circuits.

15. $(3 - 1)! = 2! = 2$

17. $(12 - 1)! = 11! = 39,916,800$

19. 11

21. $9 + 8 + 11 + 6 + 2 = 36$

23. $9 + 7 + 6 + 11 + 3 = 36$

25. $40 + 24 + 10 + 14 = 88$

27. $20 + 24 + 12 + 14 = 70$

29. $14 + 12 + 24 + 20 = 70$

31. On a complete graph with four vertices, there are 6 distinct Hamilton circuits. These are listed in Exercises 25–30. We have already computed the weight of each possible Hamilton circuit, as required by the Brute Force Method. The optimal solutions have the smallest weight, 70. They are *A, C, B, D, A,* and *A, D, B, C, A.*

33. Starting from *B*, the edge with smallest weight is *BD*, with weight 12. Therefore, proceed to *D*. From *D*, the edge having smallest weight and not leading back to *B* is *DC*, with weight 10. From *C*, our only choice is *CA*, with weight 20. From *A*, return to *B*. Edge *AB* has weight 40. The total weight of the Hamilton circuit is
$12 + 10 + 20 + 40 = 82$.

35. a. Add *AB*
   Number of Hamilton circuits: $(4-1)! = 3! = 6$

   b. Sample Hamilton circuit: *AD, DB, BC, CA*
   Sample Hamilton circuit: *CA, AB, BD, DC*

   c. Remove *CD*

   d. Sample Euler circuit: *AC, CB, BD, DA*

37. a. Add *AB, AC, BC,* and *DE*
   Number of Hamilton circuits: $(5-1)! = 4! = 24$

   b. Sample Hamilton circuit: *AB, BC, CE, ED, DA*
   Sample Hamilton circuit: *AB, BC, CD, DE, EA*

   c. Remove *BD* and *BE*

   d. Sample Euler circuit: *AE, EC, CD, DA*

39.

Hamilton Circuit	Sum of the Weights of the Edges	=	Total Weight
A, B, C, D, E, A	500 + 305 + 320 + 302 + 205	=	1632
A, B, C, E, D, A	500 + 305 + 165 + 302 + 185	=	1457
A, B, D, C, E, A	500 + 360 + 320 + 165 + 205	=	1550
A, B, D, E, C, A	500 + 360 + 302 + 165 + 200	=	1527
A, B, E, C, D, A	500 + 340 + 165 + 320 + 185	=	1510
A, B, E, D, C, A	500 + 340 + 302 + 320 + 200	=	1662
A, C, B, D, E, A	200 + 305 + 360 + 302 + 205	=	1372
A, C, B, E, D, A	200 + 305 + 340 + 302 + 185	=	1332
A, C, D, B, E, A	200 + 320 + 360 + 340 + 205	=	1425
A, C, E, B, D, A	200 + 165 + 340 + 360 + 185	=	1250
A, D, B, C, E, A	185 + 360 + 305 + 165 + 205	=	1220
A, D, C, B, E, A	185 + 320 + 305 + 340 + 205	=	1355

Using the Brute Force method, we compute the sum of the weights of the edges for each possible Hamilton circuit, as in the table above. The smallest weight sum is 1220, representing a total cost of $1220 for airfare. This results from the Hamilton circuit A, D, B, C, E, A. Thus, the sales director should fly to the cities in this order.

41.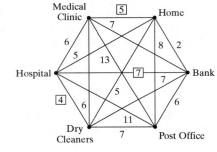

43. $2 + 6 + 7 + 4 + 6 + 5 = 30$

45. Label the vertices H for Home, B for Bank, P for Post Office, and M for Market.

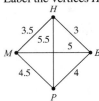

47. From Home, the closest errand is the Bank, 3 miles away. From the Bank, the closest remaining errand is the Post Office, 4 miles away. From the Post Office, the last remaining errand is the Market, 4.5 miles away. From the Market, Home is 3.5 miles away. The total distance for this Hamilton circuit is $3 + 4 + 4.5 + 3.5 = 15$ miles. This is the same route found in Exercise 46.

Check Points 15.4

1. The graph in Figure 15.51(c) is a tree. It is connected and has no circuits. There is only one path joining any two vertices. Every edge is a bridge; if removed, each edge would create a disconnected graph. Finally, the graph has 7 vertices and 7 – 1, or 6, edges.

   The graph in Figure 15.51(a) is not a tree because it is disconnected. There are 7 vertices and only 5 edges, not the 6 edges required for a tree.

   The graph in Figure 15.51(b) is not a tree because it has a circuit, namely A, B, C, D, A. There are 7 vertices and 7 edges, not the 6 edges required for a tree.

2. A spanning tree must contain all six vertices shown in the connected graph in Figure 15.55. The spanning tree must have one edge less than it has vertices, so it must have five edges. The graph in Figure 15.55 has eight edges, so we must remove three edges. We elect to remove the edges of the circuit C, D, E, C. This leaves us the following spanning tree.

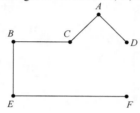

3. Step 1. Find the edge with the smallest weight. This is edge DE; mark it.

   Step 2. Find the next-smallest edge in the graph. This is edge DC; mark it.

   Step 3. Find the next-smallest edge in the graph that does not create a circuit. This is edge DA; mark it.

   Step 4. Find the next-smallest edge in the graph that does not create a circuit. This is AB; mark it.

   The resulting minimum spanning tree is complete. It contains all 5 vertices of the graph, and has 5 – 1, or 4, edges. Its total weight is 12 + 14 + 21 + 22 = 69. It is shown below.

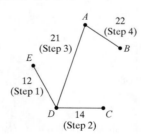

Exercise Set 15.4

1. Yes, this graph is a tree. It has 3 edges on 4 vertices, is connected, and has no circuits. Every edge is a bridge.

3. No, this graph is not a tree. It is disconnected.

5. Yes, this graph is a tree. It has 3 edges on 4 vertices, is connected, and has no circuits. Every edge is a bridge.

7. No, this graph is not a tree. It has a circuit.

9. Yes, this graph is a tree. It has 6 edges on 7 vertices, is connected, and has no circuits. Every edge is a bridge.

11. i; If the graph contained any circuits, some points would have more than one path joining them.

13. ii; A tree with n vertices must have $n - 1$ edges.

15. ii; A tree has no circuits.

17. iii

19.

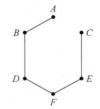

21.

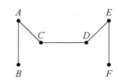

23.

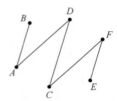

25. Kruskal's algorithm results in the following figure.

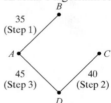

This minimum spanning tree has weight
35 + 40 + 45 = 120.

27. Kruskal's algorithm results in the following figure.

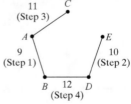

This minimum spanning tree has weight
9 + 10 + 11 + 12 = 42.

29. Kruskal's algorithm results in the following figure.

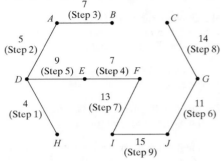

This minimum spanning tree has weight
4 + 5 + 7 + 7 + 9 + 11 + 13 + 14 + 15 = 85.

31. Kruskal's algorithm results in the following figure.

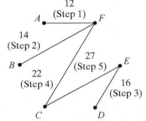

This minimum spanning tree has weight
12 + 14 + 16 + 22 + 27 = 91.

33. Sample Spanning Tree: *AB, AC, CD*
Sample Spanning Tree: *AB, BC, CD*
Sample Spanning Tree: *AB, AD, CD*
Sample Spanning Tree: *AB, AD, BC*

35. Maximum Spanning Tree: *AE, BC, CD, CE*

Total weight is $\overset{AE}{15}+\overset{BC}{14}+\overset{CD}{18}+\overset{CE}{17}=64$

37. Sample Maximum Spanning Tree: *AE, BC, BE, CF, DE, EH, FG, FJ, HI*
Total weight is
$\overset{AE}{10}+\overset{BC}{16}+\overset{BE}{15}+\overset{CF}{17}+\overset{DE}{9}+\overset{EH}{17}+\overset{FG}{16}+\overset{FJ}{19}+\overset{HI}{22}=141$

39.

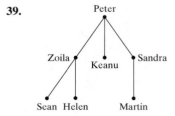

41. a.

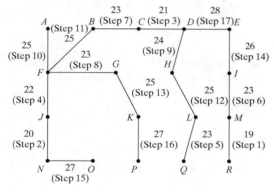

b. Kruskal's algorithm is shown in the figure.

The total length of the sidewalks that need to be sheltered by awnings is
$55 + 80 + 85 + 115 + 135 = 470$ feet.

43. Kruskal's algorithm is shown in the figure.

The smallest number of feet of underground pipes is
$19 + 20 + 21 + 22 + 23 + 23 + 23 + 23 + 24 + 25 +$
$25 + 25 + 25 + 26 + 27 + 27 + 28 = 406$ feet.

Chapter 15 Review Exercises

1. Each graph has 5 vertices, *A, B, C, D,* and *E*. Each has one edge connecting *A* and *B*, one connecting *A* and *C*, one connecting *A* and *D*, one connecting *A* and *E*, and one connecting *B* and *C*. Both graphs have the same number of vertices, and these vertices are connected in the same ways. A third way to draw the same graph is

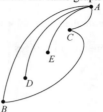

2. *A*: 5 (A loop adds degree 2.); *B*: 4; *C*: 5; *D*: 4; *E*: 2

3. Even: *B, D, E*; odd: *A, C*

4. *B, C,* and *E*

5. Possible answer: *E, D, B, A* and *E, C, A*

6. Possible answer: *E, D, C, E*

7. Yes. A path can be found from any vertex to any other vertex.

8. No. There is no edge which can be removed to leave a disconnected graph.

9. *AD, DE,* and *DF*

10.

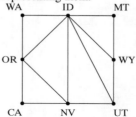

11. Use the states' abbreviations to label the vertices representing them.

12.

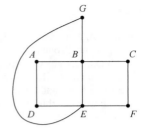

13. a. Neither. There are more than two odd vertices.

14. a. Euler circuit: there are no odd vertices.

   b.

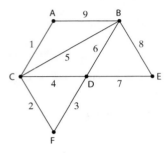

15. a. Euler path: there are exactly two odd vertices.

   b.

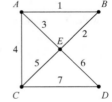

16. There are exactly two odd vertices, *G* and *I*. We start at *G* and continue to *D*, *A*, *B*, *C*, *F*, *I*, *H*, and *E*.

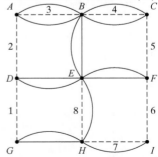

We continue erasing edges as we go, till we have completed an Euler path ending at *I*.

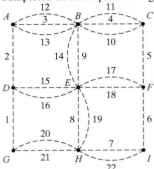

17. We may begin anywhere, since there are no odd vertices. We erase edges as we go, till we have the Euler circuit. We begin at *A*.

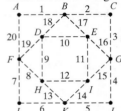

18. a. Yes, they would. The graph has exactly two odd vertices, so there is an Euler path.

   b.

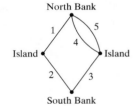

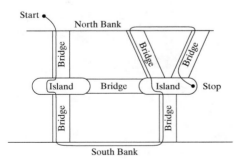

   c. No; there is no such path. Since the graph has odd vertices, it does not have an Euler circuit.

19. Yes, it is possible. There are exactly two odd vertices, and therefore there is an Euler path (but no Euler circuit).

20. **a.** Yes it is possible. There are no odd vertices, so there is an Euler circuit.

   b.

   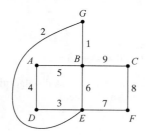

   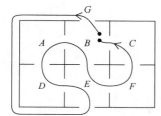

21. **a.**

   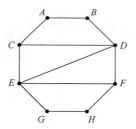

   b. Yes. There are exactly two odd vertices C and F, so there is an Euler path.

   c. The guard should begin at C and end at F, or vice versa.

22. A, E, C, B, D, A

23. D, B, A, E, C, D

24. **a.** No, because this is not a complete graph. It may not have Hamilton circuits.

25. **a.** Yes, because this is a complete graph.

   b. $(4 - 1)! = 3! = 6$

26. **a.** No, because this is not a complete graph. It may not have Hamilton circuits.

27. **a.** Yes, because this is a complete graph.

   b. $(5 - 1)! = 4! = 24$

28.
A, B, C, D, A:	$4 + 6 + 5 + 4$	$= 19$
A, B, D, C, A:	$4 + 7 + 5 + 2$	$= 18$
A, C, B, D, A:	$2 + 6 + 7 + 4$	$= 19$
A, C, D, B, A:	$2 + 5 + 7 + 4$	$= 18$
A, D, B, C, A:	$4 + 7 + 6 + 2$	$= 19$
A, D, C, B, A:	$4 + 5 + 6 + 4$	$= 19$

29. These are the only possible Hamilton circuits on a graph with 4 vertices. The lowest weight, 18, occurs on the circuits A, B, D, C, A and A, C, D, B, A. These are the optimal solutions.

30. Start with A. Then edge AC has the smallest weight, 2, of all edges starting at A. Proceed to C. From C, edge CD has the smallest weight, 5, of edges not returning to A. From D, we must travel DB, with weight 7, to B. We return to A along BA, with weight 4. The total weight of this Hamilton circuit is $2 + 5 + 7 + 4 = 18$.

31. Start with A. Of all paths leading from A, the path with smallest weight is AB, with weight 4. Proceed to B. The path with smallest weight leading from B, but not to A, is BE, with weight 6. The path with smallest weight leading from E, but not to A or B, is ED, with weight 4. From D, we proceed along DC, with weight 3, to C, the only remaining vertex. We then return to A along CA, with weight 7. The total weight is $4 + 6 + 4 + 3 + 7 = 24$.

32.

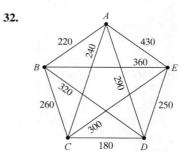

33. Start at A. The lowest cost from A, \$220, is on edge AB. From B, the lowest cost other than returning to A is \$260, on edge BC. From C, the lowest cost to a new city is \$180, on edge CD. From D, the salesman must fly to E for \$250, then return to A for \$430. The total cost of this circuit is $220 + 260 + 180 + 250 + 430 = \$1340$.

34. Yes. It is connected, has no circuits, has 6 edges on 7 vertices, and each edge is a bridge.

35. No. It has a circuit.

36. No. It is disconnected.

37.

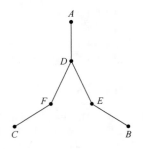

38.

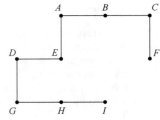

39. Kruskal's algorithm is demonstrated in the figure.

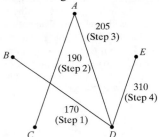

The total weight is
170 + 190 + 205 + 310 = 875.

40. Kruskal's algorithm is demonstrated in the figure.

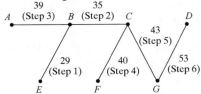

The total weight is
29 + 35 + 39 + 40 + 43 + 53 = 239.

41. The figure demonstrates Kruskal's algorithm and the layout of the cable system.

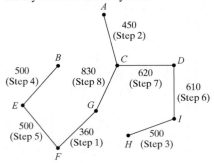

The smallest length of cable needed is
360 + 450 + 500 + 500 + 500 + 610 + 620 + 830 = 4370 miles.

Chapter 15 Test

1. *A*: 2; *B*: 2; *C*: 4; *D*: 3; *E*: 2; *F*: 1

2. *A*, *D*, *E* and *A*, *B*, *C*, *E*

3. *B*, *A*, *D*, *E*, *C*, *B*

4. *CF*

5.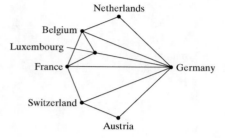

6. **a.** Euler path: there are exactly two odd vertices.

   b.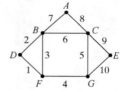

7. **a.** Neither: there are more than two odd vertices.

   b. N/A

8. **a.** Euler circuit: there are no odd vertices.

   b.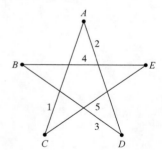

9. We begin at A, then proceed to E, I, H, and so on, erasing edges once they have been crossed. The result is shown in the figure.

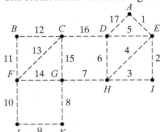

10. a.

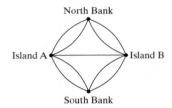

North Bank

Island A   Island B

South Bank

b. Yes: there are exactly two odd vertices.

c. It should begin at one of the islands, and end at the other island.

11. a.

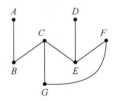

b. No: there are more than two odd vertices.

12. a. Let vertices represent intersections, and let edges represent streets.

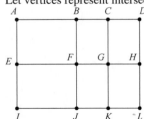

b. No: there are more than two odd vertices.

13. A, B, C, D, G, F, E, A and A, F, G, D, C, B, E, A.

14. $(5-1)! = 4! = 24$

15.

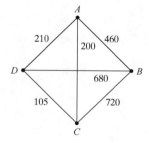

16.

Hamilton Circuit	Sum of the Weights of the Edges	=	Total Cost
A, B, C, D, A	460 + 720 + 105 + 210	=	$1495
A, B, D, C, A	460 + 680 + 105 + 200	=	$1445
A, C, B, D, A	200 + 720 + 680 + 210	=	$1810
A, C, D, B, A	200 + 105 + 680 + 460	=	$1445
A, D, B, C, A	210 + 680 + 720 + 200	=	$1810
A, D, C, B, A	210 + 105 + 720 + 460	=	$1495

The optimal route is *A, B, D, C, A* or *A, C, D, B, A*. The total cost for this route is $1445.

17. Starting from *A*, the edge with smallest weight is *AE*, with weight 5. Proceed to *E*. From *E*, the edge with smallest weight, and not leading back to *A*, is *ED*, with weight 8. From *D*, the edge with smallest weight, and to a new vertex, is *DC*, with weight 4. From *C*, only *B* remains. Edge *CB* has weight 5. Return to *A* by edge *BA*, with weight 11. The total weight of this Hamilton circuit is
$5 + 8 + 4 + 5 + 11 = 33$.

18. No; it has a circuit, namely *C, D, E, C*.

19.

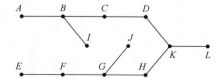

20. Kruskal's algorithm is shown in the figure.

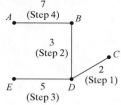

The total weight of the minimum spanning tree is $2 + 3 + 5 + 7 = 17$.